The History of Black Mineworkers in South Africa

Cyril Ramaphosa and James Motlatsi at a press conference to display the range of ammunition used against mineworkers during the 1987 strike.

ORGANISE OR DIE

V.L. ALLEN

assisted by

KATE CAREY

THE HISTORY OF BLACK MINEWORKERS IN SOUTH AFRICA

Volume III

1982–1994

THE MOOR PRESS

and

MERLIN

First edition
South Africa 2003
Hbk ISBN 0 907698 10 7

UK edition 2005
Hbk ISBN 0 85036 567 8

Published by

The Moor Press,
Hainsworth House
Damems Lane,
Keighley,
West Yorkshire BD22 7AR

and

The Merlin Press
PO Box 30705,
London WC2E 8QD

Typeset by Pointset, Randburg, South Africa
Cover design by Bookcraft Ltd, Stroud, Gloucestershire, UK
Printed in Singapore by Fabulous Printers PTE Ltd

This book is dedicated to the small group of black mineworkers who created the National Union of Mineworkers in 1982 and the many thousands who joined it in the years up to the first democratic election in 1994 and who, through their own sacrifices, gave it a vigorous and successful life.

By the same author:

Power in Trade Unions (Longmans, 1954)

Trade Union Leadership (Longmans, 1957)

Trade Unions and the Government (Longmans, 1960)

Militant Trade Unionism (Merlin Press, 1966)

International Bibliography of Trade Unionism (Merlin Press and Maspero, 1968)

The Sociology of Industrial Relations (Longmans, 1971)

Social Analysis: A Marxist Critique and Alternative (Longmans, 1975; The Moor Press, 1982)

The Militancy of British Miners (The Moor Press, 1981)

The Russians are Coming: The Politics of Anti-Sovietism (The Moor Press, 1987)

The History of Black Mineworkers in South Africa, 1871–1994
in three volumes (The Moor Press and Merlin Press, 2005):
Mining in South Africa and the Genesis of Apartheid, 1871–1948
Apartheid Repression and Dissent in the Mines, 1948–1982
Organise or Die, 1982–1994

CONTENTS

LIST OF MAPS

LIST OF TABLES

LIST OF ILLUSTRATIONS

ACKNOWLEDGEMENTS

I would like to thank the following organisations for allowing me to reproduce their photographs: Afrapix of Johannesburg, for the photograph in the frontispiece, five photographs on pages 312, 336, 346, 369 and 379 in chapter 11 and the photograph of Gwede Mantashe on page 614 in chapter 18; *Optima* for the two maps on pages xix and xx; the British NUM for the photograph on page 564 in chapter 16; the NUM for the photographs on page 101 in chapter 4, pages 426, 430, and 437 in chapter 13, pages 522, 524 and 530 in chapter 14, page 546 in chapter 15 and page 605 and the photograph of Senzeni Zokwana on page 614 in chapter 18. The author took the remaining photographs.

MAP ONE[1]

SOUTH AFRICA'S MAJOR TRANSPORT AND MINING INFRASTRUCTURE

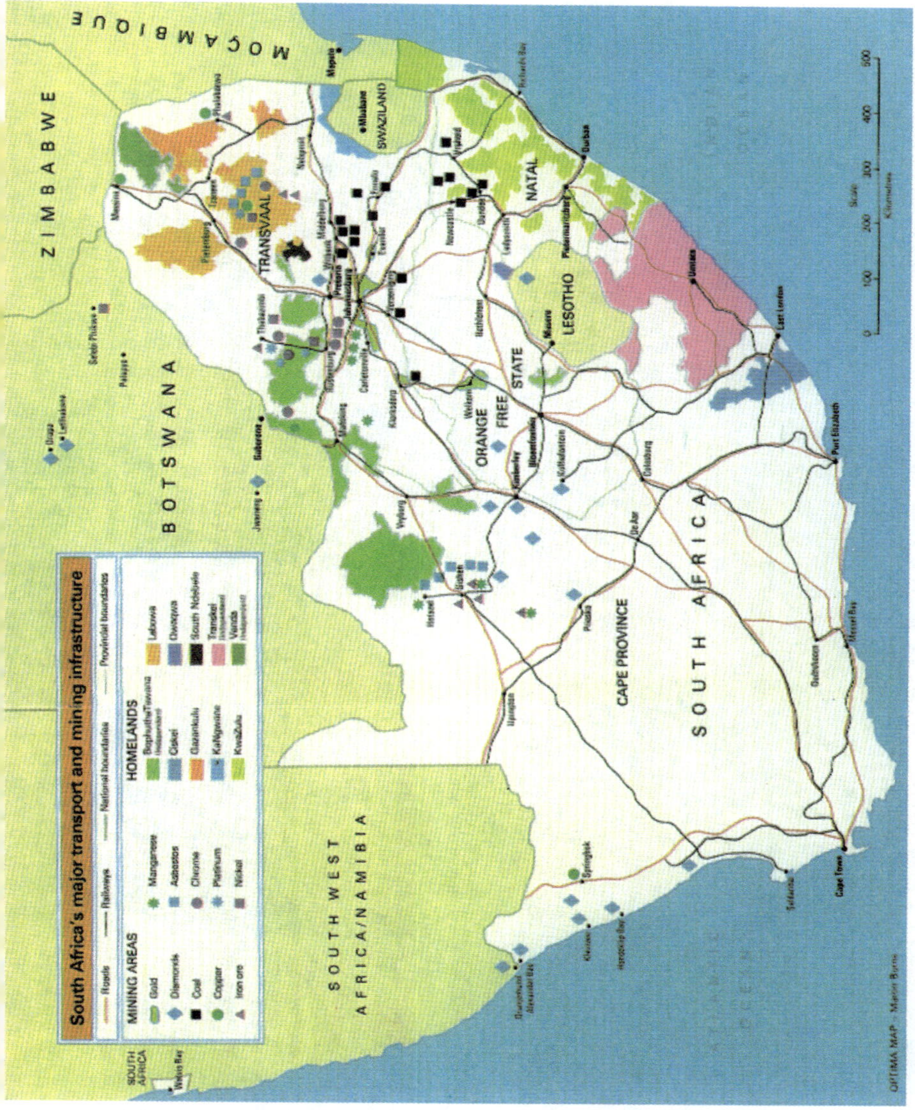

[1]*Optima*, Volume 29, Number Two/Three, November 1980.

MAP TWO[2]

SOUTHERN AFRICA: MIGRANT LABOUR PATTERNS

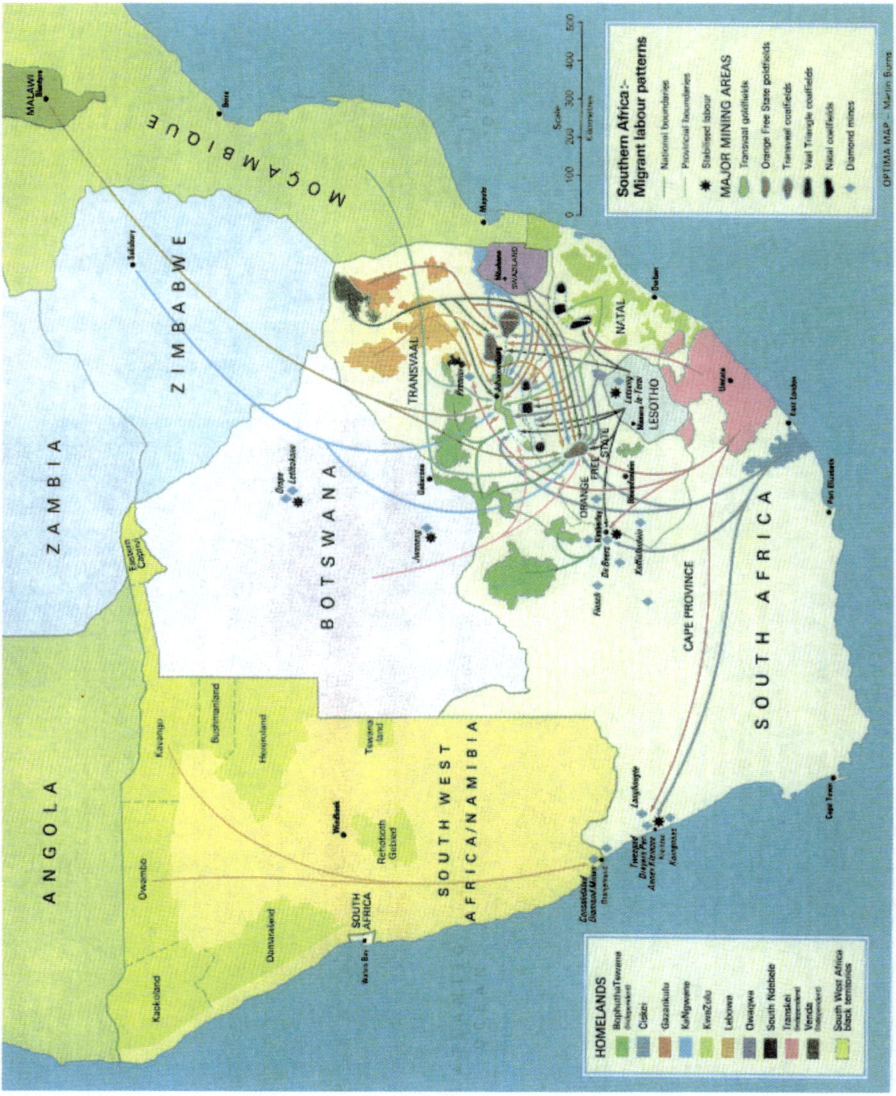

[1]*Optima*, Volume 29,Number Two/Three, November 1980.

TABLE 1

THE AVERAGE NUMBER OF EMPLOYEES IN SOUTH AFRICAN MINES 1982-1994[1]

Year	Gold	Coal	Diamonds	Asbestos	Platinum	Copper	Iron Ore	All Mines[2]
1982	484 888	130 411	20 926	11 536	50 026	14 093	9 289	765 599
1983	495 504	114 382	19 688	10 216	49 742	13 263	7 128	746 408
1984	511 282	117 491	18 848	8 574	62 242	10 978	6 718	774 613
1985	527 947	120 959	18 352	7 567	73 880	10 952	6 826	807 939
1986	551 935	120 214	18 474	6 623	74 682	10 769	6 620	829 235
1987	553 549	114 022	19 346	4 323	82 958	10 288	6 249	829 181
1988	532 559	108 988	19 822	3 614	85 322	9 736	5 921	803 347
1989	516 077	107 170	22 016	4 455	86 833	10 033	6 356	797 231
1990	489 541	103 531	22 982	5 425	97 019	9 440	6 734	778 304
1991	438 822	96 197	21 516	5 282	104 152	8 241	6 714	720 947
1992	411 686	75 858	19 654	3 176	103 871	7 990	6 468	663 303
1993	394 508	61 163	14 880	2 446	102 043	6 426	6 047	619 013
1994	392 327	60 187	15 828	2 230	97 643	5 646	6 085	610 794

[1] Department of Minerals and Energy, *Minerals - South Africa: Statistical Tables 1980-1999, 2000.* After 1980 the government made no distinction between different racial groups in its published statistics. The number of black mineworkers in mining has varied between minerals. For instance, in 1982 they accounted for approximately 92 per cent of the total in the gold mines, 82 per cent in the coal mines, 63 per cent in the diamond mines and 91 per cent in the asbestos mines. The totals contain small numbers of black and white female workers that remained fairly constant. The proportion of all employees in all mines who were females was 2 per cent in 1982 and 1994.

[2] The figures in this column include the numbers employed in all mines and, therefore, are greater than the totals of the numbers in the preceding columns.

TABLE 2

THE TERRITORIAL ORIGIN OF BLACK MINEWORKERS RECRUITED BY THE EMPLOYMENT BUREAU OF AFRICA (TEBA) FOR THE MEMBERS OF THE CHAMBER OF MINES IN SOUTH AFRICA 1982-1994[1]

Year	SA[2]	Mozambique	Lesotho	Malawi	% Foreign[3]
1982	239 215	35 732	65 194	13 876	37,92
1983	230 608	37 152	64 264	13 361	38,35
1984	260 202	42 294	75 787	16 050	38,58
1985	270 872	52 410	83 387	17 613	40,59
1986	272 083	44 250	91 733	19 046	40,96
1987	252 727	39 599	102 133	18 592	43,53
1988	239 969	33 805	99 088	3 490	41,50
1989	217 474	44 075	98 085	17	44,45
1990	189 864	43 515	100 472	13	48,01
1991	172 263	45 545	93 088	0	49,35
1992	153 251	46 809	89 901	0	51,76
1993	137 271	40 847	85 765	0	52,94
1994	132 632	44 044	84 700	0	53,79

COUNTRY OF ORIGIN

[1] The Employment Bureau of Africa Limited, *Reports,* 1982-1994.
[2] This column includes black mineworkers from what were called the independent states of the Transkei, Ciskei, Venda and Bophuthatswana, as well as South Africa.
[3] In this period, foreign mineworkers were those from Lesotho, Botswana, Swaziland, Malawi and Mozambique. Occasionally, a few hundred were recruited in Namibia.

PREFACE

This volume marks the completion of a 3-volume study of black mineworkers in South Africa from 1871 to 1994 and the end of a project that began early in January 1989. The suggestion that I should write it was raised when, whilst sailing to a small island off the coast of Cuba, Cyril Ramaphosa, the general secretary of the South African National Union of Mineworkers (NUM), asked me to drop whatever research commitments I had in order to undertake it. We, along with James Motlatsi, the president of NUM, and John Hendy, the British NUM's legal adviser, were whiling away our time whilst waiting to meet Fidel Castro. We eventually met him on 27 December but before then we had many opportunities for conversation and Ramaphosa used them to the full to make his case. It was not easy for me to give him an answer. I was keenly interested in the project but I had another commitment. I had recently been awarded a Leverhulme Research Fellowship in order to complete an investigation into the determinants of the consciousness of miners in Kentucky and West Virginia in the USA. In addition, it would involve visiting South Africa whilst the apartheid system was still in place. Since the early 1960s I had supported the boycott of South African goods and events and, later, the academic boycott enforced by my union, the Association of University Teachers. I was not eager to visit what I regarded as a pariah state. Moreover, I was not sure that I would be allowed into the country. The South African government had protested to the BBC about a Third Programme talk I had given about a visit I had made to South Africa in 1962 in order to collect data about conditions in the mining industry. Subsequently, I had become friendly with a number of white South African exiles in London, including Ruth First, Joe Slovo and Ruth Mompati, the head of the ANC office in London in the 1970s, who were public opponents of apartheid. Those factors, I thought, would have been sufficient to make me *persona non grata* with the South African government.

Ramaphosa, however, thought otherwise. His request carried with it the authority of the underground ANC executive within South Africa, of which he was a member, and that, in his view, overrode any decisions resulting from the imposition of external sanctions in support of the ANC. Moreover, he did not attach much significance to the possibility of my being banned from entering the country. He faced that risk every time he left the country. He had left South Africa illegally to accompany us to Cuba and did not know whether he would be allowed to return. Defying the government was a feature of everyday life for him, as it was, of course, for most other black South Africans. The only question that concerned him was whether or not I would agree to write the history. He was persuasive. The NUM had existed for only six years but they had been momentous ones. Its executive had asked for the union's brief history to be recorded but he was keen to place it in the context of the treatment of black mineworkers in South Africa since the beginning of commercial mining in 1871, when the systematic recruitment of black migrant labour first began.

By the time we had returned to Havana I had agreed to undertake the project. Events then moved quickly. Within weeks the NUM executive had authorised his proposal and I was in Johannesburg. My apprehension about entering South Africa proved to be groundless. The only question I was asked by the customs officer was whether I had booked my return flight. After that, although the vast majority of my acquaintances were black, the police never accosted me, though I was often aware of their presence. On my first visit, I stayed at the Park Lane hotel in Hillbrow in the centre of Johannesburg, from where I was collected each day by Jerry Majatladi, NUM' press officer, with whom I subsequently established a close relationship. Jerry had a very analytical mind that had been primed by seven years' imprisonment on Robben Island with Govan Mbeki and like-minded people and he used it in our conversations. Then, mostly in the company of either Ramaphosa or Motlatsi, I attended meetings and conferences in Johannesburg and most of the regions of the union, where I met regional and branch officials and rank and file activists until I was acquainted with the structure of the union and the manner of its working. Through successive sojourns in South Africa, this process made it possible for me create a profile of the union as well as gain an understanding of its history. It also enabled me to learn more about the treatment of black mineworkers in general and, thereby, gain an insight into the repressive nature of the apartheid state.

I met many difficulties in obtaining documentary material for the period before the formation of the union in 1982. There were very few written records of the activities of black mineworkers and none that dealt with their collective experiences. In almost every case they had been compiled by white men and reflected their innate racist views. In the records of mining, black mineworkers had no presence, except as 'native labour'. None was ever attributed with a name. Although they had participated in many struggles they had no known leaders except for J B Marks, the leader of the African Mine Workers Union (AMWU) during the black mineworkers' strike in 1946. And he was known, not because of his association with black mineworkers, but through his leadership role in the African National Congress (ANC). Indeed, it was an indication of the paucity of historical knowledge about black mineworkers, that when Ramaphosa and a small group of supporters met at the end of August 1982 to establish a trade union and give it a name, none had been aware that the AMWU had ever existed. If they had known of that union they would have taken its name. One of Ramaphosa's intentions was to remove this degrading omission by countering the white supremacist policy of denying black people any individual and collective sensitivities and aspirations and, thereby, identities.

Despite my initial research problems, I gradually realized that it was going to be impossible to write the history in one volume. A single volume would have constituted a mere precâis of the history of black mineworkers. My terms of reference had been to provide the union members with a comprehensive historical record, written, as far as possible, from the perspective of black mineworkers. It was not necessarily intended for the general public. I discussed the analytical questions raised by asking me, a white, foreign intellectual, to present the perspectives of uneducated, illiterate migrant workers in a unique, repressive society in the Introduction to Volume I. In any event, my assignment could not be fulfilled by a quick, rather superficial tour around the major events in the mining industry. Nor could it be done without placing black mineworkers into the evolving context of the apartheid system. Mining and apartheid, I discovered, were inextricably causally linked. I had to re-examine my plan.

Towards the end of 1991, I divided the material I had collected into three stages, namely 1871-1948, 1948-1982 and 1982 to the present whatever date that happened to be. The first stage covered the period up to general election that put the National Party into power with its apartheid programme. This was dealt with in *The History of Black Mineworkers. The Techniques of Resistance*, 1871-1948, Volume I, which was published in Johannesburg on 4 April 1992. The second stage, 1948-1982, ends at the point where the necessary conditions for the formation of a black mineworkers' union were present and is covered in Volume II, *Dissent and Repression in the Mine Compounds*, 1948-1982. Volume II is being published in Johannesburg, simultaneously with Volume III. This present book deals with the formation of the NUM in 1982 and its subsequent history up to the first democratic elections in April 1994.

There was much material, dealing with the structural character of minerals, the recruitment of migrants, their training and adaptation to mine work, the migrant labour system and its social and health consequences, that could not be dealt with chronologically and did not, therefore, fall into the three time sequences. At first, I envisaged putting this material into a separate, fourth volume but decided not to do that because, though it was important in its own right, its historical significance lay in the extent to which it could explain the intricate environment against which black mineworkers eventually rebelled. It needed, I concluded after much discussion, to be seen as significant for the historical narrative and to be read alongside it. As a consequence, in Volume II there are six chapters at the beginning under the general heading "Preparing for Work with Hostile Rocks" and in this Volume there are three introductory chapters under the heading "The Human Costs of Mine Work" and two concluding ones under the heading "Taking Stock of Unresolved Legacies".

There have been special problems in writing this volume. As there are many people who have lived through the period it covers, there is much oral evidence. I have been able to interview some of those who were involved in the formation of the union and others who were activists during its subsequent history. This added a valuable dimension to documentary data but it needed to be used carefully. Oral evidence by itself does not provide a wholly reliable interpretation of personal experiences. Personal views are usually formed within narrow, highly circumscribed horizons. The mineworkers were often unaware of the comparable experiences of others or of the structural contexts within which they had occurred. They were also, in some instances, unreliable indicators of what actually had taken place due to faulty memories or incomplete evidence for other reasons. There is an illustration in this history where different participants gave three wholly contradictory versions of the same single clearly defined event. It was important, therefore, wherever possible, to have corroborative evidence.

Unfortunately, this was not always available. At different times, the police had confiscated or destroyed union documents during their raids on its head office and some of its regional offices. This had been done so thoroughly that it is amazing that any documents survived. The physical destruction of documents, however, was compounded by an almost systematic neglect in keeping records within the union itself. In so far as documents existed, they had rarely been collected, collated and filed as references for the present or past performances. They simply lay without order in one filing cabinet or another. Or they were missing. The result was that no one union administrator knew precisely what records the union possessed and where they could be found. I had to search the offices in the hope of coming across conference reports, minutes of national executive committee meetings, memoranda from one department or another or confidential correspondence between Ramaphosa and the mine owners. This is not a criticism of the union administrators. The union had necessarily been too preoccupied with surviving the repression of the apartheid state to attach much importance either to minuting the present or recording the past. In addition, due to the Bantu education system and racist job segregation practices, most of its administrators were ignorant and inexperienced in the ways of bureaucracy and had not seen the need or had the skills to record the minutia of its activities. Even in the final years of the period, when apartheid repression against the union had eased, there were significant gaps in the union's records.

The data for this volume has been collected from a variety of sources. It has come from the incomplete union records, a mineworkers' attitude survey conducted in 1989[1], the personal recollections of union officials and activists, the reports of government commissions, statements and reports of the mining houses and the Chamber of Mines, the journals and paper of associations connected with the mining industry, newspapers, academic dissertations and various publications. I continued to use some of the sources I had referred to in the Introduction to Volume II. I was fortunate to find a complete set of papers published by the Association of Mine Managers, from its formation in the early 1920s to the 1980s, in the basement of the library of Leeds University. The Human Resources Laboratory of the Chamber of Mines continued to produce confidential reports for various mining houses that filtered through to academia. The mining houses themselves provided scant information in their official reports and persisted in their attempts to control the flow of information from the mine compounds. Cyril Ramaphosa, however, easily outplayed the mine owners in the field of publicity. He broke their monopoly of news and increasingly publicised details of disputes, accidents and disasters, as well as hostel life. The fascination of the foreign press with the contortions of the apartheid system following the Soweto uprising became more intense in the 1980s and became a conduit for NUM's versions of events. South African historians increasingly wrote in books, monographs, dissertations and journals about different aspects of the liberation struggle within South Africa. Finally, the annual

[1]The survey, undertaken in August and September 1989 covered a sample of 204 mineworkers, drawn from the main mining regions. They were representative of the main mining houses and the principal minerals. Twenty-eight of them were retired. The subject analysis was undertaken by Kate Carey and completed in 1990.

Survey of Race Relations in South Africa, with its synthesis of information from all sources, continued to provide an essential back-up data service. There was not, in the end, either a scarcity of relevant data or a significant lack of balance in its coverage.

The mining industry in the period covered here had characteristics that distinguished it from the earlier periods. The extension of mining beyond gold, diamonds and coal, to encompass asbestos, iron ore, platinum, copper and some other minerals, had come to a close. The expansion of the gold mining industry to new fields and the extension of mining operations to new depths that had dominated the previous period had largely been completed. The technology of mining was reaching a plateau. Industrial relations henceforth provided the principal dynamics of mining life and became a main focus of concern for black mineworkers. From 1982, this concern, combined with that about the continuing physical hazards of mining that cost them their health and lives, dominated their existences for the following 12 years. Mineworkers throughout the industry could not ignore the presence of the NUM and the reactions of the mine employers and the government to its activities, whether or not they were members of it. They experienced greater freedom to move in and out of the compounds and were able, therefore, to get involved in township life and establish links with external organisations, for instance, those within the liberation movement. Through this process they extend the boundaries of their collective freedom to expedite the end of the apartheid system. The NUM became, in consequence, an integral and vital part of the liberation movement. The details of this transition from individual to collective consciousness and the consequences it created are related here in three parts.

Part One describes the human costs of mine work that stemmed from the nature of their work and the manner in which their lives were controlled and manipulated to enhance the profits of mining companies. It portrays black mineworkers as casualties of a migrant labour system which took them away under contract from their rural homes for periods ranging up to eighteen months, incarcerated them in single sex hostels, exposed them to eventually fatal infectious diseases and then returned them to their homes with either no or minimal subsistence support when they were sick, disabled or at the end of their contracts. It describes the conditions of hostel life and the social consequences of incarceration. All of this was done to provide the mines with a constant supply of cheap labour. The details in these three chapters form the basis for the case used by the NUM for the abolition of the migrant labour system.

Part Two is about the formation and subsequent struggles of the NUM within and against the apartheid state. It is the central feature of the book. It describes how the union was formed by Cyril Ramaphosa, the legal officer for the Council of Unions of South Africa (CUSA) with the assistance of Puseletso Salae, a former mineworkers from Lesotho, and a small group of personnel assistants from various Anglo American Corporation gold mines who were attending a training course at the Western Deep Levels gold mine in Carletonville in August 1982. These people formed the nucleus of those who launched the union in December 1982 with fourteen thousand members and then expanded it until it had a membership of 281 000 by 1994. Ramaphosa became its general secretary and remained so until July 1991 when he became the Secretary-General of the ANC. James Motlatsi who had been one of the personnel assistants at Western Deep Levels in August 1982 became its first president and remained so throughout the period covered here. Its first vice president, Elijah Barayi, a personnel assistant from the Blyvooruitzicht gold mine in Carletonville, also remained in office for the duration of those years. Ramaphosa, backed by Motlatsi and Barayi, built the union's organisational structure and headed its confrontations with the Chamber of Mines and the government. The union, however, drew its momentum from the spread of an impatient collective consciousness amongst ordinary mineworkers. Within three years of its foundation, the union's elected secretariat was responding to the initiatives of its rank and file members as they sought to impose solidarity on all black mineworkers, by force if necessary.

Violence was a common feature of life in the hostels. It was the central tenet in the overall maintenance of the apartheid state and in the control of the mine compounds by the mine owners. The years covering the formation and consolidation of the NUM were reputably the most violent in the history of the country since the creation of the Union of South Africa. The growth of solidarity amongst black mineworkers was both a consequence and a catalyst in that situation. From 1985, with the imposition of a state of emergency, South Africa became a country with great political complexity and uncertainty. The contradictions within the system began to transform it with such great pace that the major players were unable to predict how the events would unfold or what the final outcome would be. From 1987, in fairly quick succession, the elderly ANC leaders were released from prison on Robben Island, the bans on the ANC, PAC and the Communist Party were removed and then, as a final gesture, Mandela was released from prison. Negotiations, primarily between the ANC and the governing National Party, over the character of a reconstituted

South Africa, began a chequered course in the early 1990s and ended with agreement to hold the country's first election under universal suffrage on 27 April 1994.

The NUM played a prominent part in these events. In 1985, it was a key player in the formation of the Congress of South African Trade Unions (COSATU) and in its subsequent steps to fill the vacuum in the internal leadership of the liberation movement caused by the destruction of the township organisations by the government security forces. NUM's activists were present at all levels of illegal political activity. Ramaphosa and Kgalema Motlanthe, NUM's education co-ordinator, were members of ANC's small underground internal national executive. Other members took on prominent provincial positions in both the ANC and the SACP. When both of those bodies were allowed to take part in legal political activity, the NUM was virtually denuded of officials as they set about creating grassroots organisations for them. The ultimate sacrifice in this respect was when Ramaphosa was elected as the Secretary-General of the ANC in July 1991. Thereafter, until the democratic election in 1994, a largely new and changing leadership had to cope with a deteriorating industrial situation.

After its defeat in the 1987 national strike, the union had to repair the damage to its organisation caused by mass dismissals and the consequent destruction of branch and regional bodies. Its problems were exacerbated by a decline in the gold mining industry that became more acute each year. The union was faced with the task of providing support for a high and increasing number of retrenched mineworkers. The new leadership after 1991 soon realized that a policy of opposition to retrenchments would have no effect and that it would, therefore, have to consider how to alleviate their impact on both rural communities and individual mineworkers. The situation was that whilst almost unimaginable political progress was being made, the NUM was facing a deteriorating industrial future.

Part Three examines the extent to which the major issues confronted by the union when it was first formed had been resolved by 1994. Those issues were the migrant labour system, the circulation of infectious diseases, the incidence of those diseases, including the impact of HIV/AIDS, and the continuation of injuries and fatal accidents. The final chapter describes the plight of the casualties of asbestos mining.

This Part shows that while, by 1994, the migrant labour system had altered in substance, as mineworkers began to perceive mining in career terms and the mining labour force became stabilised, they still worked on contracts of from twelve to eighteen months, were separated from their families for long periods and, whilst under contract, were incarcerated in single-sex hostels in mine compounds. Thus the circular migration of mineworkers had persisted and, through it, the traditional infectious diseases had continued to ravage the rural areas as well as the mine compounds and the black townships. In important ways the situation had deteriorated through the re-emergence of tuberculosis on epidemic proportions and the addition of HIV/AIDS to the list of destructive infectious diseases. It also shows that whilst the risk of injury in the mines had declined, the possibility of being killed had remained virtually constant. The rate of fatalities in 1994 was only marginally different from what it had been in 1982. Asbestos-related diseases are described separately because they have their own distinctive characteristics and history. Although asbestos mining had virtually ceased by 1994, they had not yet reached their peak, let alone their end. The NUM came in too late to act as a custodian of the interests of asbestos mineworkers but was able to take up their claims for compensation long after they had ceased working in the mines.

Whilst working on this project from 1989 till now, I have relied on the assistance of a fairly constant group of people, most of whom I have named and thanked in both previous volumes. The union, through its office-bearers and its NEC, has unfailingly given me its support from the first to the last day. It made the project possible by enabling me to visit South Africa and travel around the mining areas whenever I considered it to be necessary. It opened doors for me so that I met a range of people with experience and or knowledge of mining in South Africa. It never refused my requests for information or access to its records. Nothing, not even confidential letters and documents were denied me. Importantly, its office-bearers have never asked to see what I had written. They supported me with an enthusiasm that was based on trust. This was the case even though the composition of the leading office-bearers changed after Ramaphosa left to work for the ANC.

I wish to thank Cyril Ramaphosa for initiating the project and supporting me until 1991 and to his successors as general secretary, Kgalema Motlanthe and Gwede Mantashe for continuing that support in principle and in substance. I wish also to thank James Motlatsi, the union president until 2000, who gave unstinting support in difficult as well as good times, and his successor Senzeni Zokwana. Many other officials assisted with varying degrees of involvement and I cannot name them all. But in the early stages, Jerry Majatladi, the press officer, smoothed my way in a constructive and friendly manner while later I depended heavily on Frans Baleni, then a member of the organising staff but now NUM's Education Officer. Before becoming an official, Frans had participated in the making of NUM's history as a mineworker at the Western Holdings gold mine in the mid 1980s and provided me with details about events there. Whenever I had difficulties in connection with any aspect of my work and I was unable to deal directly with them myself, he would step in and help. He has been a genuine, constant friend. Others who assisted me were Archie Luhlabo, the late Jan Selepe, Puseletso Salae, George Nkadimeng and Ikaneng (Director) Matlala, all union officials from the beginning. Edward Ramaila introduced me to Phalaborwa when the apartheid forces were trigger-happy and he had to sleep in the bush for his own safety. I got to know the coal miners in Witbank through the help of Crosby Moni. I am grateful to them all. The arrangements for my visits were handled by Merriam Mashaba, Nomsa Nkwanca, the late Jenny Zintambila and Thule Mofutsanyana, a succession of administrators in NUM's head office. They were always friendly, kind and helpful. Others in the head office who were supportive were departmental specialists, Martin Nicol, May Hermanus, Irene Barandilla, Sazi Jonas and Habofanoe Thomas Ketsise who accompanied me on many trips to the mining areas.

Outside the union, I received much advice and support in my search for relevant data from Kent McNamara, who had written extensively about compound disputes, and from Eddie Webster, Luli Callinicos and Jean Leger in the University of the Witwatersrand. Jean Leger was my prime source for information about health and safety in the South African mines. I also wish to thank Frank Watson, who was responsible for producing the book in South Africa, and Gwen Podbrey in Johannesburg for her expert proof-reading skills. Amongst those in Britain who helped, I would like to thank Sophie Allen who assisted with the administration of the survey of the black mineworkers' attitudes in 1989, Laurie Flynn for allowing me to consult his collection of documents relating to mine safety, the Kinross Disaster and asbestos mining, and Peter Nolan for encouraging me throughout the project, advising me about the organisation of the text and for reading and commenting on the final typescript.

Producing this book has been a long haul, very much longer than I had either planned or even anticipated. The difficulties that accounted for this, however, would, I am sure, have been greater had it not been for the moral support, advice and technical help I received from Kate Carey. Kate was involved in the project from the outset and gave me support that I have acknowledged in the prefaces to the first two volumes. Her role in this book, however, has been pivotal. She has accompanied me on almost all of my visits to South Africa and, using her skills as a social researcher and sociologist, has assisted in the planning of the research, the collection and collation of data and on the form it was to take in the book. She was responsible for the subject analysis in the survey of black mineworkers attitudes in 1990 and advised me when the project seemed to be spilling over into four volumes. She did the bulk of the work for the Bibliography and the Subject Index. Taking the book as a whole, the writing is mine but her imprint is everywhere and for this reason her name appears on the frontispiece of the book. Looking back now on her contribution, this is the least I could do.

Keighley,
West Yorkshire,
England. October 2003.

Part One

THE HUMAN COSTS OF MINE WORK

CHAPTER ONE

THE PERSISTENCE OF THE MIGRANT LABOUR SYSTEM

Migrant Labour

Each year, throughout most of the 20th century, young black men had been recruited from the villages of southern Africa, legally contracted to work from two to 18 months, conveyed by boats, motor vehicles, planes or trains to the central recruitment centre in Johannesburg and then distributed to the mines of South Africa where they were accommodated, as if they were prisoners, in single-sex hostels within closed or partially closed compounds for the duration of their contracts. Over time, changes occurred in this process. Different modes of transport were used to speed up the movement of recruits to the mines. Alterations were made in their manner of living by the construction of new buildings with better facilities for sleeping, eating and leisure and by improvements in the nutritional value of their diets and in medical care. The original essential features of the system, however, had remained unchanged.

After 1904, there had been two substantial hostel-building programmes to revise and improve the living quarters of ordinary mineworkers. The first had been undertaken to accommodate the Chinese migrant workers who had arrived between 1904 and 1910 and the other had occurred when new mines were opened in the Orange Free State, Klerksdorp and the east and west extremities of the Witwatersrand in the 1950s and 1960s. In the hostels of the second phase, sleeping arrangements were less congested and were separated from the provisions for leisure and eating. Full-time mine doctors had been employed from the 1920s and mine health services, which operated through a three-tier mine hospital structure, had been introduced. The mine doctors had been instrumental in improving diets through the provision of a greater variety of foods and insisting on greater care in its preparation. Finally, hostel regulations had been relaxed following the turmoil in the compounds during the middle of the 1970s so that black mineworkers were freer to move in and out of the compounds to visit local townships where they could drink alcohol and meet friends and township women. Regulated provisions had been introduced to enable them to visit their homes before the completion of their contracts. Over time, the lives of the hostel dwellers had become more tolerable.

These changes, however, largely left untouched the fundamental purpose of the migrant labour system that was to provide a cheap black labour force without allowing reason, sentiment or humanity to get in its way. The regulated rotation of mine recruits between their villages and the mines, on two- to 18-month contracts, was to continue until the 1990s. They remained incarcerated in single-sex barracks within compounds controlled and policed by management. They continued to suffer from a rigidly applied system of job segregation. The ability of the mines to pay them poverty-level wages, untouched by wage competition in the rest of South African society, remained largely intact. For this reason, the primary experiences of black mineworkers had remained fairly constant during most of the 20th century.

The controlled flow of labour to the South African mines had been dependent on statutory regulations that were designed to ensure that black people in general would have to seek work as migrants in order to subsist. In other words, there had to be no escape routes which permitted any black persons to cease being migrants by choosing permanent wage work, otherwise the flow to the mines could dry up and the mines would be forced to compete for labour in the open market. Here, the interests of the mine owners and the upholders of white supremacy had coincided for the statutory conditions necessary for maintaining the flow of cheap labour to the mines were also necessary to protect the privileges and power of the white population. The state, therefore, had never had any difficulty in obliging the mine owners by enacting pass laws and influx controls which forbade black people from settling in urban areas or compelled them to make their homes within designated reserves where the land was so limited and infertile that it was incapable of providing adequately for their subsistence. Thus economic circumstances had forced black people to supplement their subsistence through wage work for white employers while the legal constraints had compelled them to do it as migrants. Then, in order to ensure that they were qualified only to undertake the unskilled work allocated to them according to the rules of racial segregation, black people had been denied free and fulfilling primary, secondary and further education, barred from access to any formally skilled work and excluded from all those semi-skilled jobs that had been claimed by white workers as theirs by right.

Within South African society as a whole the co-ordination of this system of racial segregation had been disrupted by the gradual but constantly growing illegal movement of black people into townships and shanty towns which was facilitated by the employment of black workers by white employers irrespective of whether or not they had work or residence permits. Nothing the government did had been able to stop that process and, as a consequence, the contradictions it created began remorselessly to undermine the whole operation. The hostels that served manufacturing industry then became one of various sources of accommodation available to black workers. Without the controls implicit in hostel life, the mechanism to ensure that the migrants returned to their villages failed to work despite the use of penal sanctions by the state. A permanent black wage-earning class living adjacent to the white areas began to be established despite all the efforts of the state to prevent it.

However, even though the total number of black men and women who moved from the rural to the urban areas had kept on increasing, and the controls over the rotation of migrant labour between the white-owned industries and the rural areas had largely broken down, the supply of labour to the mines had not been seriously affected. The mine owners had been able to operate much as before without infringing the statutory laws and management rules that made the system work for their benefit. The hostels, therefore, had remained intact as the fulcrum of the mines' system of migrant labour. For the period up till 1980 almost 99 per cent of the total black labour force in the gold and platinum mines lived in the single-sex hostels. This proportion fell slightly during the 1980s but the three per cent maximum allowed by the government to live in married quarters was not reached by any of the mining houses. The Anglo-American Corporation, for instance, the most liberal of the mining houses in this respect, had almost 98 per cent of its black work force living in hostels by 1987 compared with less than two per cent in married quarters.[1]

Increasingly large concentrations of men were living in mine hostels. The total number of black mineworkers employed in the mines affiliated to the Chamber of Mines had risen from just over 174 000 in 1920 to a peak of 477 000 in 1986.[2] During that time the mineworkers in the gold and platinum mines were employed in larger and larger units. In the 1960s, 22 of the gold mines employed more than 7 000 black

[1]In the succeeding five years the number of residents in Anglo-American hostels fell to 90 per cent of its total black labour force, but that was due largely to an increase in the numbers living-out and residing in their own homes for still only just over two per cent lived in married quarters in the compounds. (Jonathan Crush and Wilmot James, eds, *Crossing Boundaries. Mine Migrancy in a Democratic South Africa*, South Africa, 1995, Table 1, p17).
[2]Jonathan Crush, Alan Jeeves and David Yudelman, *South Africa's Labour Empire*, 1991, Table A4, pp234-235. The total number of black and white mine employees in 1986, according to the Department of Minerals and Energy, was 829 235 *(Statistical Tables 1980-1999, p25).*

mineworkers each while six of them had at least 10 000 in hostel accommodation. The largest gold mine at that time was Randfontein Estates with almost 16 000 living in hostel blocks. As gold was mined at greater depths so the mines continued to expand. By 1980, seven of them were larger than the largest gold mine had been 20 years earlier. Vaal Reefs in Klerksdorp was at the top of the list with a black labour force of nearly 35 000. Harmony in the Orange Free State came next with just over 28 000. Some of the mines continued to grow so that eight years later the Vaal Reefs' labour force had reached 44 000 while Harmony's had grown to 31 000. Western Deep Levels in Carletonville, which was then the deepest gold mine, was the third largest with 26 000 black mineworkers. The platinum mines in the Rustenburg area became as large as some of the biggest gold mines.

The size of the hostels varied with the size of the mines they served and they, in turn, differed between the minerals. There were 140 mine hostels by the early 1990s, ranging in size from less than 100 residents at the De Beers diamond mine at Koffiefontein and at Samancor's Eastern Chrome mine at Delmas in the Eastern Transvaal, to 10 000 at the Impala platinum mine at Wildebeesfontein in Rustenburg. The hostels serving the platinum mines in the Rustenburg region each accommodated between 4 000 and 10 000 residents. The gold mines were divided into shafts and each shaft had its own hostel block with, on the average, from 3 000 to 5 000 residents in each. The Anglo-American Corporation had 35 gold mine hostels, with an average size of 4 400 residents, and eight coal mine hostels, each of which averaged just less than 1 000 residents.[3]

Those who were not in hostels lived in married quarters or with their families in neighbouring townships or squatter camps adjacent to the mines. The black mineworkers who lived independently of the mines were allowed to do so, in the main, because the mines that employed them were relatively small, either because of the nature of their mineral production or because small firms that were independent of the Chamber of Mines owned them. In any event, as those mines employed a minority of the total mine labour force the government was not particularly interested in enforcing its quota of permanent employees on them. They were not threatening the migrant labour system and were allowed to accommodate their workers as they wished. They were not supplied with workers by the large recruiting agencies and had to get their labour from wherever they could in competition with the mining

[3]Crush and James, 1995, pp53-55.

houses and manufacturing industry. The main attraction they could offer was to allow workers to live with their wives and families. This was the situation in many of the small coal mines in Natal where, in some instances, as many as 80 per cent of the black workers lived in family housing on mine property or in nearby townships.[4] Local people were able to volunteer for mine work without leaving their homes and some did so. Other mines, such as in copper and iron ore mining, attracted relatively high numbers of semi-skilled black workers by allowing them to bring their families with them. About 80 per cent of the black mineworkers at the Rio Tinto Zinc's copper-sulphate mine in Phalaborwa were accommodated in the nearby Namakgale township while at the Kimberley diamond mines about 62 per cent of the black workers lived in the local township of Ipopeng. There was a similar accommodation picture at the Sishen iron ore mine in the northern Cape.

The Hostel Model

The model for housing black mine labour in general was that which had been devised by Cecil Rhodes for the diamond mining industry in the 1880s. Rhodes had recruited unskilled, illiterate migrant workers on short-term contracts and had detained them in completely closed, guarded compounds for the duration of their contracts as if they had been prisoners. On the completion of their contracts the mineworkers had been compulsorily repatriated to their home villages. This process of recruitment, confinement and then repatriation was continuous throughout every week of every year. Thus a migrant labour process had been devised which reproduced the controls, servility and repression implicit in slavery but within the framework of a legally free wage labour market.

Because diamonds were small and valuable and were relatively easily secreted out of the mines, extraordinary steps had been taken to prevent them from being stolen. From the outset the white miners had refused to co-operate with any of the mine owners' proposals to stop the illicit diamond trade, with the consequence that only blacks had been constrained and treated as if they were all potential diamond thieves.[5] Black mineworkers were not allowed to leave the compounds for any reason, even to bury a member of a family, until their contracts had

[4]Merle Lipton, "Men of Two Worlds", *Optima*, Volume 29, No 3, 28 November 1980, p125.
[5]See V L Allen, *The History of Black Mineworkers in South Africa*, 1871-1948, Volume 1, 1992, pp119-122 for a discussion of the illicit diamond trade.

been completed. Then, on leaving the compound, they were subjected to humiliating anal searches for stolen diamonds. This involved being dosed with castor oil and confined in closed rooms for about four days, wearing fingerless gloves that were locked at the elbows. The closed compound system with its degrading security measures had been extended over the whole of the dry and alluvial mining industry and remained intact until 1972 when De Beers began to stabilise its labour force in the dry mines around Kimberley and to dispense with migrant labour as a means of meeting its black labour requirements.[6] When an electronic device to detect the secretion of diamonds was invented it enabled De Beers to open its 'closed' compounds in March 1977 and have electronic security checks at the mine exits. The conjunction of these two factors brought an end to the compound system in Kimberley. Thereafter, mineworkers were recruited at the gates of the mines and were allowed to live in local townships. However, in alluvial diamond mining the model adopted from the dry diamond mines persisted.

This model had been copied by the gold mines when they began deep mining in the 1890s but with one difference. The notion of having completely closed compounds was rejected largely because the local traders had wanted access to the compounds in order to exploit the captured consumer markets they contained. Openness, however, was a relative term for the compounds were not intended to be open in so far as the black mineworkers were concerned. This had been made clear when new compounds were built to accommodate Chinese indentured labour from 1904 for they were designed primarily to enable mine managers to keep the residents under constant surveillance and to prevent unauthorised entry or exit. The perimeters of the compounds constituted the outer walls of the buildings that were linked in a rectangular shape as if they were Boer *laagers*.

The new compounds in the 1950s had also been purpose-built for security purposes. They were surrounded by razor wire or electrified fences or walls and were patrolled by armed security guards with dogs.

[6]Duncan Innes, *Anglo*: *Anglo-American and the Rise of Modern South Africa*, Johannesburg, 1984, pp106-107. From 1929, when Ernest Oppenheimer, chairperson of the Anglo-American Corporation, made a successful take-over bid for De Beers, Anglo-American controlled 95 per cent of the diamond production in South Africa. Oppenheimer had already acquired control of the alluvial mines in Namibia and had secured control of the output of diamonds in the Congo, Angola, West Africa and British Guiana. Innes reports that with the take-over of De Beers, "The Anglo-American Group had ... succeeded in bringing under its control all of the known diamondiferous resources in the capitalist world". (*Ibid*, p107.)

Mineworkers in the new hostels experienced more stringent controls than in the old ones because the physical facilities for every activity had been planned so that management could maintain a constant surveillance over them, intercept and separate groups and take prompt action whenever there were disturbances. When mineworkers passed through the controlled checkpoints at the gates of the compounds they entered a world where the employers made the rules, applied them and were the police, judges and jury when they were broken. The restrictions varied between mines in substance and application, often depending on the whims of the mine managers, but nowhere had mineworkers possessed any degree of freedom to influence decisions about their daily lives.

For most of the period, if any mineworker left his compound without permission, for whatever reason, he could have been charged and possibly punished. Each mineworker had to seek special permission, sometimes in writing, before he could visit a nearby township during his free time and usually that was difficult to obtain both because of the high level of illiteracy amongst mineworkers and the intransigence of hostel managers. Even if a black mineworker had wanted to visit another hostel in the same compound he often had to sign out from his own block first. On no account was he allowed to use any mine facilities that were reserved for whites. A black mineworker could not enter a toilet, a change room, a shower, a canteen, a recreation room or a cage that was reserved for the use of whites. He could not drink from a cup that had touched the lips of white miners. Nor was he allowed into any part of the mine compound that had not been designated for the use of blacks. In the white suburbs of the mining towns racial segregation was absolute.

The hostel and compound restrictions, however, were simply a means to an end for they had been part of total control systems designed to ensure that black mineworkers would be available at the right time, in the correct numbers and in an adequate physical condition to meet shift production requirements. This had always entailed regulations but as the mines became deeper and larger and, in consequence, greater numbers of mineworkers were involved, the controls in the compounds had become necessarily more bureaucratic and complex. The layouts at the shaft head had to be redesigned to expedite the movement of larger numbers of people through the crush to the cages and to make surface checks easier and more effective. Before going underground, the mineworkers had to be identified by their numbers, allocated to gangs, given their shift tasks and directed to their places of work. The names of any absentees were noted and searches were made to discover where they were and why they were absent. Managers recorded the times workers went underground, when they returned; why, if at all, some were still underground; whether they had completed their assigned tasks and, if so, whether they were entitled to bonus payments. The use of computers had made it easier to include such tasks within all-inclusive control systems which left individual mineworkers with no discretion or latitude in the ways in which they conducted themselves at work. The controls, however, could never be complete or absolute because the underground geological conditions were unpredictable. Moreover, as the mines became deeper and the stopes moved further away from the shafts, the length of time it took to travel from the hostels to the stopes constantly increased and reduced the amount of time left to spend on production. The white managers and supervisors were, therefore, under constant pressure to streamline the transfer procedures so that the levels of production could be maintained. For black mineworkers that simply meant more impatiently applied regulations.

Conditions in the Hostels

Overall, throughout the mining industry, black mineworkers lived in a variety of hostel conditions. In some of the older mines the hostel provisions remained unchanged whilst others were upgraded in various ways so that there were different structures within a single compound. The new hostels varied in their layouts and provisions. The comfort of the inmates was never a priority. The early hostels had had tin roofs and no ceilings so they were unbearably hot and stuffy in the summer and uncomfortably cold in the winter. The heat had encouraged vermin despite disinfecting measures. Their windows had been positioned well out of reach, as in prisons, so that it was impossible to pass any kind of contraband through them. The consequence for mineworkers was that it was impossible to see through them. In any case, they were usually barred and covered by some kind of netting and offered no protection against the wind and dust. Inside the rooms there were no partitions so no-one had any privacy. The space provided for sleeping also had to serve as a dining room, pantry, recreation room and storage place. The mineworkers had had to keep their private belongings either on an open shelf or suspended from the ceiling for few of them possessed a suitcase to put them in. The poorly sheltered lavatories were usually situated from about 69-92 metres from the hostel rooms though the health regulations under the 1911 Mines and Works Act had stipulated that they should be no more than 16,5 metres away. The occupants were completely exposed for the lavatories had no doors or lights. They rarely had any toilet paper.

Each of the early hostel rooms had been built to hold from 20-100 people but often more than the capacity numbers were crowded in. For instance, in 1943 a room at the Simmer and Jack gold mine measuring just less than eight metres square had accommodated 80 people. A survey of sleeping accommodation on various mine compounds in May of 1931 had found that one-third of them had at least five per cent more occupants than they could properly accommodate.[7] The congestion was worst of all in the gold mining hostels for, rather than getting the existing labour force to work overtime, the mines took on extra labour to meet peak demands for gold and crowded all the additional mineworkers into the existing accommodation. It was different in coal mining where mineworkers were required to work overtime when the demand for coal increased.

At first, the mineworkers had slept on wooden boards without any partitions so that extra bodies could be squeezed in simply by pushing everyone closer together. The Chamber of Mines had agreed after 1916 to put dividers between bunks but 15 years later that practice was still not uniform. In any event, the dividers were only four inches high and had done nothing to stop the spread of infection.[8] The boards were difficult to disinfect and were gradually replaced by concrete bunks with two or three tiers. Each mineworker, then, at least had his own bunk but he had to sleep on cold concrete without a mattress and with only a single blanket to keep him warm. Many mineworkers chose to sleep outside when the weather was warm. Mine doctors had debated whether the concrete could be made warmer by mixing sawdust or some other insulating material in it but they concluded that as the black mineworkers did not complain about the coldness or hardness of ordinary plain concrete there was no reason to alter it. "The point is," the Medical Officer for Rand Mines stated, "that the thin slab of concrete rapidly attains the temperature of the surrounding atmosphere, and although it may feel cold at first, apparently it causes no discomfort to the sleeper once he has settled down on it..."[9]

The Chamber of Mines had rejected a demand from the African Mine Workers' Union for metal bedsteads and mattresses in 1943, asserting that concrete was not appreciably colder than wooden bed

[7]Randall M Packard, *White Plague, Black Labour 1989*, p165.

[8]*Ibid.*

[9]Dr A J Orenstein, *Proceedings*, Transvaal Mine Medical Officers' Association, September 1922, p17.

boards, was warmer than earth or cork and that it was more hygienic because it did not harbour vermin.[10] Concrete bunks, or 'sleeping coffins' as they were called by mineworkers, had been commonly used well into the post-Second World War period and in some cases were still found in hostels in the 1980s.

Each hostel room had been heated by a coal stove in the centre that was also used for cooking the rations of raw meat that the mineworkers were given twice a week. They were not provided with standard rations of fuel for the stoves. In some mines they received coal only on the two days they were given meat. In others they had either daily or weekly allocations of fuel. In any event, they rarely had enough to keep their stoves alight throughout the week and were not given extra in the winter months. They invariably had to supplement their rations with coal bought from the company stores. There were very few changing houses or washing and drying facilities at the shaft head so mineworkers had to hang their clothes, either dirty or wet from washing, from the tin roofs over their beds. That was particularly unpleasant in the case of clothes worn by sanitation workers. The air in the rooms was generally humid, acrid and unhealthy. The mines of the Anglo-American Corporation had installed showers at the shaft heads by the Second World War. Six to eight men had to use each shower so the supply of hot water was quickly exhausted. Mineworkers complained, however, that the water was drawn from underground sources, and, because it was alkaline, was hard a left an unpleasant, itchy white layer on the skin. The general view of mineworkers was that the mines did not provide even the minimum facilities for clean living and did nothing to educate them in the hygiene and health measures that were essential for communal conditions.[11]

The architects of the early hostels had given no thought either to need to prevent draughts or to allow sunlight in. On the contrary, main consideration had been to expose everyone to an abunda of fresh air in the belief that it had a cleansing, fumigating effe Indeed, the government had insisted on the installation of a larg ventilator in each room to ensure that fresh air passed through but whe the air was cold during the winter nights the inmates often had huddle together at one end of the room with blankets over their hea in order to keep warm.[12] Nonetheless, mine doctors insisted on passing even greater volumes of air through the hostel rooms.[13] They claimed it

[10]Allen, 1992, p369.
[11]*Ibid*, pp444-445.
[12]*Ibid*, p241.
[13]Orenstein, 1922.

was a part of African culture for blacks to sleep in huddles, smothered by blankets, and that the mine owners were generous in allowing them to continue doing it.[14]

Experienced mineworkers who transferred to the new mines moved into hostels that, in many ways, were unrecognisable compared with those they had been accustomed to. They found that all the necessary services for living in a compact, centralised and easily accessible way were provided. The mining houses had incorporated some of the suggestions made by the African Mine Workers' Union to the Witwatersrand Gold Mines Native Wages Commission in 1943 and had taken into account some of the proposals made by mine doctors over the years.[15] The new hostels were intended to accommodate more than 2 500 mineworkers though later some were constructed for as many as 6 000 occupants. They were built on two levels with identical provisions on each level. The hostels consisted of four dormitory blocks around a kitchen that was flanked by two large dining rooms, capable of holding up to 1 000 people. On each floor there were 32 dormitories with 10 mineworkers in each one. Every eight dormitories had an ablution block with showers, clothes washing room and toilets. Eighty mineworkers, therefore, shared washing facilities. There were five two-tiered iron bedsteads with wire springs and hard felt mattresses in each room that also had an adjoining storage room where clothes could be dried. All the dormitory blocks were centrally heated from a single boiler so that the temperature was controlled throughout the hostels.[16]

The advantages of the new hostels were obvious and in many ways positive. Black mineworkers lived in relatively small groups and did not have to keep their belongings and dry their clothes in the rooms where they slept. With the provision of dining rooms the practice of eating where they slept was ended. The ration of raw meat was stopped for there were no longer coal stoves in the rooms on which to cook it. They were given metal bedsteads and thin mattresses in place of the 'concrete coffins'. They did not have to wander in the open and the dark when they went to the latrines. Other improvements were introduced, as, for example, on the Welkom mine where subways connected the hostel blocks to the shafts so that mineworkers did not have to walk in the open when they were cold and sweaty after their shifts. In general the new arrangements spatially separated the eating, sleeping, leisure and washing functions so it became

[14]Packard, 1989, pp164-165.

[15]Allen, 1992, *Appendix II*, pp428-470.

[16]Association of Mine Managers of the Transvaal, *Papers and Discussions 1952-53*, pp373-376.

possible for new and more stringent standards of hygiene to be applied.

These relatively neat and functional hostel arrangements, however, were only as effective as the hostel managers made them for they were responsible for applying the standards of hygiene, regulating the quality of cooking and determining the levels of congestion. Mineworkers in the new hostels complained on all three counts. Many said they were persistently plagued by flies in their food and by cockroaches, lice and mice in their sleeping quarters, despite their attempts to clear them away. Their relatively spacious sleeping arrangements were spoiled when hostel managers put boards on the floor in-between the beds to accommodate additional mineworkers. It was still impossible for a person to have privacy. The atmosphere in the rooms was particularly unpleasant when mineworkers argued or fought, as they tended to do when they were drunk. Private possessions were insecure because nothing could be locked away. There was, they said, a lot of theft.[17] The fact that there were only 10 in a room made for less noise and disturbance, but in other ways it made little difference whether 10 or 50 slept in the same room.

They complained, too, about the manner in which some of the new facilities were organised. When mineworkers returned to the surface they were invariably dirty, exhausted, hungry and stressed for it was a daily relief to reach the surface unscathed. It was important, therefore, that they had the time and facilities to wash, eat and relax before the next shift. If they returned on time, at about 14.00 hours, they could manage satisfactorily though there were complaints that some kitchens did not open until late in the afternoon with the result that they sometimes had to wait for two hours in their rooms before being allowed to stand in long, dreary queues to get their meals.[18] Most of them had about two hours of free time after finishing their meals at 17.00 hours for the majority of mineworkers had to be in bed by 19.00 hours. They spent their free time resting in their rooms, washing or repairing their work clothes, visiting friends or drinking in the bar. There was always a significant minority, however, which was delayed in getting back to the hostels from the shaft. Mineworkers who did not arrive back until 16.30 hours usually found that the water in the showers had begun to run cold and that the kitchens were short of food. If they arrived after that time they might find that the

[17]Kate Carey, *Analysis of Survey of Working and Retired Mineworkers*, 1990. The survey was of a sample of 204 retired and working mineworkers drawn from the main mining regions and representative of the mining houses and minerals. It was conducted in August and September 1989.

[18]Transvaal Mine Medical Officers' Association, *Proceedings*, February 1938, p60.

dining rooms were closed and that they were also too late for any kind of leisure activities, such as sport or drinking. Both the compound post offices and pay offices had closing hours that took no account of the needs of those who were late from the shafts. In consequence, some mineworkers, desperate for cash, had to wait for days before they could collect their pay.[19]

Diets and Food Preparation

Mineworkers had been entitled to standard quantities of staple food based on a ration scale established by the government in the 1911 Mines and Works Act and revised in 1922 but what they actually received depended on the whims of the hostel managers. There were considerable variations in feeding arrangements between and even within compounds. A common feature, however, was that the hostel managers had little, if any, scientific understanding of food and the nutritional needs of mineworkers or any knowledge of their dietary habits and tribal customs. They rarely employed anyone in the kitchens with cooking skills. Where there was communal feeding the cooks were usually drawn from the friends of bossboys rather than because of their cooking expertise.

From the late 1930s, cost-cutting studies had been undertaken in the hostels to discover when the mineworkers ate their food, what they customarily ate, how much they wasted and which items of food they tended to throw away. That was the beginning of attempts to standardise dietary control and centralise the provision of food in the belief that the cost of food could be reduced without impairing the health of the mineworkers.[20] The mineworkers themselves were not consulted to discover their preferences. In the words of one mine medical officer, "as regards tastes and customs, it behoves us, of course, to pay great attention to them while endeavouring at the same time tactfully to overcome those which cause objections to the partaking of foods which we know the Natives ought to eat for the sake of health".[21]

Nonetheless, mineworkers were frequently given food that was badly cooked, comprising ingredients they did not choose or like and

[19]P de Vries and J Radise, "The Effects of Mine Procedures and Practices on the Quality of Life of Black Mineworkers on Four Mines", Human Resources Laboratory, Chamber of Mines, *Research Report No 5/80*, February 1980, pp20-27.

[20]L S Williams, "The Food and Feeding of Mine Native Labourers", Transvaal Mine Medical Officers' Association, *Proceedings*, July and August 1941. This study was based on five mines belonging to New Consolidated Gold Fields, Ltd.

[21]*Ibid*, p5.

doled out in quantities irrespective of their tastes or appetites. There was, as a result, much waste. In one compound, with 4 000 mineworkers it had been calculated, in 1941, that one-and-a-half tons of food, enough to feed another 400 people, were wasted each day.[22] Most of the food thrown away was edible mealie meal and beans. Another survey in 1955 had produced a similar result. At the Vlakfontein gold mine on the Witwatersrand, where 2 400 mineworkers were employed, every month 50 drums, each with a capacity of 44 gallons, were filled with edible food and disposed of as swill. The drums were placed outside the hostel rooms so that the mineworkers could throw away the porridge, beans, samp, vegetables and bread they were unable or unwilling to eat. This level of wastage occurred even though in many mines the mineworkers had their meal tickets marked before each meal to prevent them from getting second helpings.

However, some positive innovations were introduced. The balance of protein, vitamins and carbohydrates in the diets was altered in order to make up for deficiencies caused by the lack of fresh meat, fruit and vegetables. Fish was added to some diets though many mineworkers refused to eat it. Some compound managers experimented with pilot studies amongst selected workers and launched propaganda campaigns to try to reduce the amount of wasted food. In some instances the tribal police reprimanded those who wasted their food. There were some mines where it was a disciplinary offence for mineworkers to leave food on their plates.[23]

But none of these actions significantly improved the provision of acceptable, edible and nourishing food. The government had intervened in 1964 when, in the Bantu Labour Act, it revised the quantity and quality of food in the compounds by specifying the number of grams of mealie meal, bread, beans, meat and fish, dripping, sugar and vegetables each mineworker had to receive. Six years later the Human Sciences Laboratory of the Chamber of Mines had reported that the feeding arrangements in the mines needed to be investigated so the Gold Producers' Committee appointed a medical consultant to conduct an examination of the diets of mineworkers. The mineworkers themselves were generally critical of management proposals and provisions. When they were told to waste less, they said that as the food was part of their wages they could do what they liked with it. In any case, few of them were satisfied with the food they were given. In a survey in 1976 nearly all of those questioned had

[22]*Ibid.*
[23]Carey, 1990.

complained about it. They said they were given samp that caused stomach pains, mixed vegetable soup that was like fatty water, bones but no meat and, if they arrived late, water was added in order to increase the quantity.[24] Many of them described their meals as 'pig swill' consisting of uncut, unpeeled vegetables, offal and meat they could not identify. They frequently had to satisfy their appetites by buying food from the company stores.[25]

Single-Sex Hostels or Mine Villages

When the new gold mines were opened in the 1950s the mine owners were given an exceptional opportunity to start dismantling the single-sex hostel system by building mining townships where mineworkers could live under stable conditions with their families. There were sound economic and social reasons for doing so. They would have broken the circle of disease infection. And they would have done themselves a good turn by creating a stable black labour force with accumulating experiences and skills in mining. The costs of operating a migrant labour system were high. The labour turnover on the mines was over 100 per cent each year for mineworkers stayed at work, on average, for less than 12 months on each contract. In order to maintain an adequate labour force, the mine owners, therefore, had had to depend on an elaborate and expensive recruiting organisation in which, with a few exceptions, every person who signed a contract was treated as a new employee, irrespective of his previous experience. This meant that every recruit had to be tested, trained, inducted and acclimatised for deep mining as if entering wage-employment for the first time. Once the mineworkers had grown accustomed to their new jobs, often in unfamiliar mines with new managers, strange work mates and different geological conditions, there was relatively little time for them to acquire skills before their contracts expired. Against the costs of operating such a system were the benefits it gave mine owners. It enabled them to keep wages consistently down to a bare subsistence level, impose a total control system over the whole black labour force and, thereby, repress dissent.

The question before the mine owners was whether or not the mining industry should have a stabilised black labour force. That would have involved either extending the existing mining towns to accommodate black families or building new townships for black mineworkers and

[24]De Vries and Radise, February 1980, p23.
[25]Carey, 1990.

their families in the same way as towns were constructed throughout the mining areas for white miners and their families. The recruiting organisations would have been defunct because a stable labour force could only have been created through a voluntary movement to the mines by black workers with their families. There were many coercive features in the process of urbanisation but its context had to be a free labour market. This, of course, would have contradicted the principle of enforced separation between blacks and whites that underpinned the apartheid system.

The idea of establishing permanent mining villages where mineworkers could live with their families had been suggested as early as 1914 by Colonel William C Gorgas as a means of eliminating the scourge of pneumonia in the mining compounds where the mortality rate was about 14 for every 1 000 workers. Gorgas, who was the Chief Sanitary Officer of the United States Army Medical Corps in the Panama, had been invited by the Chamber of Mines to report on the morbidity problem in the Rand mines because of his role in significantly reducing the mortality rate for pneumonia amongst black workers building the Panama Canal between 1906 and 1912.[26] In his report, amongst other proposals, he suggested that mineworkers should be allowed to build small huts in which they lived with their wives and families. The Chamber of Mines did not give the suggestion serious consideration.

The matter had not been raised again until the African Mine Workers' Union campaigned in the 1930s and 1940s for the end of the migrant labour system. In its evidence to the Witwatersrand Mine Natives' Wages Commission in 1943 it called for "the total abolition of the compound system and its replacement by some system similar to that prevailing in Northern Rhodesia, i.e. the establishment of townships on mining properties on a plan based on conditions of the Rand. About 60 per cent of the workers recruited to the Northern Rhodesian mines are married men who have their families with them. The mining companies provide houses of one to three rooms depending on the size of the family."[27] The Commission had rejected this demand, claiming that the gold mining industry needed cheap labour that could only be supplied through a migrant labour system.

The question of constructing mine villages was next raised when the

[26]A P Cartwright, *Doctors of the Mines*, Cape Town, 1971, p28.

[27]Statement submitted to the Witwatersrand Gold Mines Native Wages Commission (Allen, 1992, Appendix II).

new mines were being planned in the Orange Free State and Sir Ernest Oppenheimer, chairman of the Anglo-American Corporation, which had built the mine villages in the Northern Rhodesian Copperbelt, began to question whether the advantages of the migrant labour system were worthwhile. Oppenheimer's intervention in support of mine villages instead of migrants' hostels had not been received well by the National Party, then advocating its policy of apartheid in preparation for the 1948 election. One of its members in the House of Assembly, Mr H S Erasmus, raised the issue when he stated that Sir Ernest Oppenheimer had claimed that it was the policy of the Chamber of Mines "to create Native villages in the Free State to house Native mineworkers, instead of the customary compounds". He added that, "if the Natives were allowed to bring their families the Native population in the Free State would eventually run into millions ... If the mines were ever worked out, the Free State would be responsible for the unemployed mineworkers."[28] The matter did not end there. The ANC had taken it up when its President-General, Dr A B Xuma, advocated the construction of mine villages in March 1947.[29] Sir Ernest Oppenheimer had repeated his views at the sod-cutting ceremony for No 2 shaft at the Welkom gold mine on 16 May 1947, when he stated: "Personally, I am convinced it is in the interests of the Natives and of the mines to establish villages for married Native people ..."[30] The Anglo-American Corporation was the major company in the new gold field and needed to attract many thousands of new recruits. Young black men, however, avoided mine work if they could after the war and Anglo-American saw the construction of mine villages as one way of overcoming their aversion to it.

Sir Ernest Oppenheimer, however, had not, at any time, envisaged ending the migrant labour system. The Anglo-American Corporation intended to continue recruiting the majority of its workers from outside the boundaries of South Africa and to house them in single-sex hostels. The villages were only for the accommodation of married black mineworkers from South Africa. The proposal was eventually modified to refer only to married workers with special responsibilities, such as clerks and policemen. These numbered about 30 000 or 10 per cent of the total labour force but many of them were foreigners, particularly Shangaans from Mozambique, who were not legally entitled to live in

[28]*Rand Daily Mail*, 8 February 1947.
[29]*Ibid*, 13 March 1947.
[30]Theodore Gregory, *Sir Ernest Oppenheimer*, 1962, p574.

permanent mine villages while some of the South Africans did not qualify because they were not married. In reality, then, the Anglo-American Corporation proposal had been both modest and cautious and in no way threatened the survival of the migrant labour system. Indeed, the company eventually decided to proceed with its experiment by building one 'Native village' of only 100 houses which it subsequently expanded until it had 500 houses.[31]

Before the Anglo-American Corporation could make any headway it had had to obtain the permission of the government that by now was a Nationalist one pursuing the policy of apartheid. Mr Harry Oppenheimer, then the Member of Parliament for Kimberley, had spoken for Anglo-American when he asked the Minister of Native Affairs on 12 June 1952 if he would "define his attitude in regard to the experiment which is being made on certain of the new gold mines in the Orange Free State by the establishment of villages on the mines for married Native employees". The Minister gave his unequivocal reply the next day. He said:

> "In regard to married quarters at the mines on the Free State gold-fields, I want to state quite unequivocally that I am opposed to that development ... and my department has been instructed to investigate the position very thoroughly and to stop the development of such villages as far as possible. Let me give my reasons for this. Within that Free State gold mining area every mine can then establish its own Native town with married quarters. That will then mean a series of towns. While we are already establishing Native towns in the vicinity of the big cities to provide housing for the Natives, it will mean that in addition a large number of black spots will be spread out throughout the whole Free State mining area. Now we must bear in mind that when the mines stop working one day that large numbers of towns will remain there spread out over that area. They may amount to 20 or 30 ... They do not fall under our control either. The Department of Native Affairs has no jurisdiction over those Native towns on mining land ... In addition it is our view that it must be borne in mind that the Natives who work in the mines are not the sophisticated Natives. They are usually Natives who have been recruited by agents in the reserves, but usually their children who grow up in those mining towns do not want to work on the mines. So when the married quarters have been created ... the children will be pushed on to the neighbouring general community. They will disappear from those mining towns in the course of time and they will become a burden to the neighbouring towns or cities where they will have to be accommodated in the locations ... It is an unhealthy development if it occurs without control for the sake of the non-European labour of one industry..."[32]

Subsequently, there was a long correspondence between the

[31] *Ibid*, p578.
[32] *Ibid*, pp578-579.

Anglo-American Corporation and the Department of Native Affairs but without producing any change in government policy. Eventually, the Department of Native Affairs had laid down the government's policy by reiterating the essence of Verwoerd's speech to Parliament. It favoured the continuation of the traditional policy for the gold mines of using migrant labour accommodated in single-sex hostels and wholly disapproved of providing family houses for married workers. It was prepared, however, to make a special concession by allowing mines to build married quarters for "Native personnel who are employed on a permanent basis because of specialised training and who are required to be continuously on the mine". It added that no "foreign Natives may be permitted this privilege and they may under no circumstances be allowed to import their families". The government ruled that no more than three per cent of the total labour force in any mine could be accommodated in married quarters and added that "it must be clearly understood that the department will only agree to additional married quarters up to the said three per cent when it is satisfied that the extension of the concession is fully justified by the circumstances of each case".[33] At that time, much less than one per cent of the mining labour force lived in married quarters.

The Chamber of Mines, contrary to Oppenheimer's alleged claim, had also been opposed to the Anglo-American Corporation's request for mine villages. This was made clear by J A Gemmill, representing the Chamber of Mines before the Native Laws Commission in 1947, when he said it was "not desirable or practical to increase the number of 'detribalised' Natives".[34] The mine owners had no intention of disturbing the system that had served their interests so well during the previous half-century. They all, including eventually the Anglo-American Corporation, fell into line with the government's three per cent limit but none of them got anywhere near it. Their applications for additional housing were generally rejected by the Department of Native Affairs so that by 1970 still less than one per cent of the black labour force lived in married mine quarters.

Then, in 1974, as the Chamber of Mines had begun to experience severe problems in recruiting foreign workers and was forced to try to recruit more from South Africa, it had decided, like Anglo-American in the late 1940s, that the provision of married quarters might induce them to work in the industry. It had, therefore, requested the government to

[33]*Ibid*, p580.
[34]*Rand Daily Mail*, 27 February 1947.

raise the three per cent limit. The next year, in its evidence to the Inter-Departmental Committee of Inquiry Into Riots on the Mines, it had pressed for the right to stabilise 10 per cent of the mining labour force. The Anglo-American Corporation set itself a 10 per cent target. The government responded by stating that it would only allow the mines to accommodate the full three per cent in family quarters but by 1977 the actual level was still below one per cent.[35] There were some black mineworkers, estimated to be about one per cent of the labour force in the gold mines in 1979, who had been allowed to live with their families through the regulations of Section 10 (1) of the Bantu Urban Areas Act of 1945.[36] But the Section 10 regulations, in effect, had ensured that only senior long-serving mineworkers were able to qualify for residence near the mines. The only exception was that where mines were situated either near to or in the Bantustans, mineworkers from them could live legally with their families and commute daily to work. In that way, a small number of black mineworkers in the Rustenburg platinum mines had been able to live in nearby Bophuthatswana and some who worked in the Natal coal mines had managed to commute to work from their homes in KwaZulu.

However, during this period, the Chamber of Mines introduced changes in the conditions of employment of black mineworkers that, in effect, helped to fortify the migrant labour system. It encouraged the stabilisation of its black workforce without altering its migrant character, in the first instance, by transforming the gaps between contracts into annual leave provisions. From 1973 black mineworkers were given cash in lieu of rail fares for their annual leave and were allowed to accumulate up to 21 days' leave over a two-year period. It ceased to be a criminal offence

[35]Lipton, 1980, p114.

[36]*Ibid*, p83. Under Section 10 no black males could remain in prescribed, white urban areas for more than 72 hours unless they had resided in an area since birth, or had worked continuously for the same employer for 10 years or for different employers for 15 years. Those with one or other of these qualifications were entitled to have the right of permanent residence extended to their wives, unmarried daughters and sons under the age of 18 years. However, the difficulty all blacks with Section 10 rights experienced was in getting housing for their families near to where they worked. Many who possessed the right of permanent residence remained migrants in practice because their wives still lived in their rural homes.(Lipton gives details of Section 10 rights on pp85-86.) Until the mid-1970s the government did its utmost to prevent blacks from qualifying for Section 10 rights by compelling them to return annually to their homes at the end of contracts, thus breaking their continuity of service. It also tightened up the Pass Laws by adding the system of Labour Bureaux to control the allocation of jobs to blacks and increasing police controls. The number of prosecutions for pass offences increased from 232 420 in 1951 to 670 000 in 1968.

to break a contract. They were given cash inducements to return to their respective mines on the expiry of their contracts and, in 1977, were allowed periodic weekend leave to visit their homes, where that was possible. They began to have much greater freedom in the compounds and were able to move in and out of them and between them and the townships with greater ease. In effect, black mineworkers were in the process of becoming permanent workers who were still legally obliged to retain links with their rural homes. They were migrant workers who spent most of their adult working lives in the urban environment of the mine compounds and nearby townships where their way of life was not simply an alternative to that in the villages, but a contradiction of it. This situation continued into the 1990s.

CHAPTER TWO

THE CIRCLE OF INFECTION

The Spread of Infectious Diseases

The rotating migrant labour system made the mining industry the primary vehicle for initiating and sustaining the spread of urban infectious diseases in South Africa. The large numbers of early mine recruits who left their rural homes in various parts of southern Africa had not had any previous contacts with Europeans and did not possess any form of immunity from the principal urban killing diseases, namely pneumonia and tuberculosis, that were commonly associated with the rapid and unregulated urbanisation of western European countries in the mid-19th and early 20th centuries. European settlers, migrants and travellers had carried these diseases into the southern African region. They were taken into the mines by the white immigrant miners from various western countries but particularly from Britain. British Cornish miners constituted the majority of white miners on the Rand at the beginning of the 20th century. Many of them were already consumptive or suffered from active phthisis.[1] The black recruits who had worked alongside them were in such a vulnerable condition through malnourishment and indigenous debilitating diseases, such as scurvy, hookworm and bilharzia, that they had little resistance when they came into contact with the new diseases which spread rapidly because of dirty, overcrowded living conditions and quickly became endemic in the mining compounds.[2] The infections had been transmitted to the rural areas by returning mineworkers who had been repatriated because they were sick or injured or had simply completed their

[1]See Gillian Burke and Peter Richardson, "The Migration of Miners' Phthisis Between Cornwall and the Transvaal, 1876-1918" in Belinda Bozzoli, (ed) *Labour, Townships and Protest*, Johannesburg, 1979. The term 'phthisis' was used in South Africa to describe silicosis complicated by tuberculosis. In Britain, however, it only described pulmonary tuberculosis. For the purpose of international clarification, the term 'phthisis' was rejected and replaced in 1950 by 'silicosis'.

[2]As far as can be ascertained without written records, it appears that a number of common infectious diseases were brought to southern Africa by European settlers and travellers. Tuberculosis, for instance, was unknown there until the era of colonisation though it had been endemic in Europe and Asia for millennia. (H M Coovadia and S R Benatar, (eds), *A Century of Tuberculosis. South African Perspectives*, 1991, p20). This view, described as the "virgin soil theory", was questioned in relation to tuberculosis, but not refuted, by Packard, 1989, Chapter I. Packard wrote: "The evidence supporting the virgin soil theory is

contracts. In their home villages equally vulnerable unsuspecting members of their families and close acquaintances were infected and they, in turn, passed the diseases on to others so that eventually new migrants carried the infections back to the mine compounds and thence to the urban centres. Thus through the migrant labour system the mineworkers and their families had become trapped in a circle of infection which was kept in rotation, on the one hand, by the poverty and the awful neglect of health provisions in the rural areas and, on the other hand, by dietary deficiencies and insanitary conditions in the mine hostels.

The diseases quickly took their toll. It was estimated that in 1883, as many as one in every 15 blacks was sick at the Kimberley Central Diamond Mine. After the closed compounds were introduced, the overall death rate for the diamond mines rose from 80 per 1 000 mineworkers in 1878 to 100 per 1 000 in 1888.[3] They were, therefore, death traps from their inception. The situation was remarkably similar in gold mining from the 1890s. There was little, if any, migration of black or white workers between diamond and gold mining yet their mortality rates immediately after the Anglo-Boer War were almost identical. In 1903 the mortality rate for black mineworkers was 57,7 per 1 000 in the Rand, while in Kimberley it was 57,2 per 1 000.[4] After that date the incidence of deaths in the Rand compounds increased. The gold mines from their outset recruited many workers from Mozambique and adjacent tropical areas to the north who had no immunity from the infectious diseases and who were also susceptible to colds, influenza and pneumonia associated with the colder climate of the Rand. Their mortality rates were very high, as was illustrated by the case of the Simmer and Jack mine in Johannesburg in February and March 1904. Forty-two out of a contingent of 206 Shangaans from Mozambique died within a month of their arrival. By June, 116 of them were dead. Throughout the mining industry the mortality rate for all mineworkers from north of latitude 22 degrees south was 130,1 per 1 000 for the year ending June 1905.[5]

hardly conclusive. There is, in fact, reason to believe that TB has existed in South Africa, albeit at a low level of endemicity, for a very long time ... There can, at the same time, be little doubt that in most areas cases of the disease among Africans were relatively rare prior to the late 19th century ... It is equally clear ... that the susceptibility that Africans exhibited as a result of their relative inexperience with the disease, was exacerbated by the extremely adverse conditions under which they were made to live and by their lack of access to medical attention" (Packard, 1989, p31).

[3]Packard, 1989, p75.

[4]Cartwright, 1971, pp17-18.

[5]*Ibid*, p19.

The incidence of deaths amongst tropical mineworkers had developed into one of epidemic proportions during the years leading up to the First World War. The principal casualties came from mosquito-infested areas in Mozambique, Nyasaland, Tanganyika and Southern Rhodesia where the balance between hunted wild life and arable production had already been destroyed and where pestilence and droughts had made subsistence for rural black people precarious.[6] Many mine recruits arrived, therefore, seriously malnourished and already sick from a variety of tropical diseases. Many died of dysentery, meningitis, measles and enteric fever but their cause of death was usually described as pneumonia. The fact that almost one-third of all the 'tropical Natives' on the mines died in this way created alarm amongst the mine owners and the government.

The mortality rate amongst black mineworkers in South Africa was exceeded only by that amongst black labourers employed to build the Panama Canal in Central America in the late 19th and early 20th centuries. When the French canal builder, Ferdinand de Lesseps, had attempted to build the canal between 1881 and 1889 as many as 22 200 black workers died. Another 1 000 black workers transported from West Africa died within six months of arriving. The French were forced to withdraw from the project because of the mortality rate, estimated to be 240 per 1 000, and the USA government took over in 1903. Three years afterwards, the death rate from all diseases was still relatively high at about 39 per 1 000 workers. By 1912, however, it had fallen to little more than 6 per 1 000.[7] The death rate from pneumonia amongst black workers fell from 18,45 per 1 000 workers to 2,3 during the same period. This dramatic improvement in the health of the workers was attributed to the work of Colonel William C Gorgas of the United States Army Medical Corps who had been installed as the Chief Sanitary Officer for the project and was responsible for the health of the 50 000 black workers from the Caribbean islands and West Africa who worked on it.

Towards the end of 1912, Samuel Evans, chairperson of the recently floated Crown Mines in Johannesburg, spent a month in the Canal Zone to discover what Gorgas was doing. He was so impressed by his achievement that he urged the Chamber of Mines to invite him to South Africa to report on the appalling health problems in the Rand mines.

[6]Nyasaland became Malawi on achieving independence in 1966. Tanganyika became the United Republic of Tanzania in 1965 on getting its independence. Southern Rhodesia first became Rhodesia under white rule in 1965 and then Zimbabwe when it achieved complete independence in 1980.
[7]Cartwright, 1971, p28.

Evan's memorandum making out the case for the invitation stated: "It is perfectly clear that the things we are doing now, such as attention to clothing, diet, change houses, covered ways, diminution of draughts in the compound rooms, etc, have not helped in the least ... As far as I can see there is no ground for believing that multiplying doctors, or appointing whole-time doctors, will help so long as we continue on the present lines. It must be remembered that the heaviest death rate of all is in the WNLA compound where the medical attention is ample. The mines that already have whole-time doctors are, if anything, worse off judging by the death-rate, than many mines where comparatively little attention is paid to the health of workers."[8] The Chamber of Mines cabled Gorgas: "Invite you to come here advise sanitation mining district and prevent high mortality coloured miners. Fee £2 000 and all expenses."[9] Gorgas accepted the invitation, visited South Africa, and reported back to the Chamber of Mines in February 1914.

Gorgas recommended that as infectious diseases spread through overcrowding, mineworkers should be housed in small huts with their families and where that was not possible they should have more room space. In that way, he added, they would not be forced into close contact with infected persons. He stated that much had to be done to eliminate the sources of infection by ensuring that the living quarters of workers were cleaner and that water-borne sewerage was introduced in place of the 'bucket' system that was common in most mines. He insisted that the diets of the workers had to be improved and raised to at least the equivalent of that of soldiers in the British army. Finally, Gorgas believed that the mine-by-mine patchwork development of medical services was wasteful and that in order to eliminate duplication and improve efficiency the hospital services should be centralised and expanded. The mining houses did not welcome these recommendations and none was applied as Gorgas intended though small incremental improvements, more pronounced in some mines than others, were eventually made.[10]

The responsibility for making the compound improvements largely lay with the medical doctors who worked in the mines. Until after the First World War they were mostly part-time, working with facilities that were described as being by far the least efficient part of the mine operations.[11] At that time, however, they carried a difficult burden. The

[8]*Ibid*, p30.
[9]*Ibid*, p31.
[10]*Ibid*, pp29-36. See also Packard, 1989, pp161-164.
[11]Cartwright, 1971, p42.

54 mines on the Rand in 1913 employed 38 part-time doctors who ran 62 small compound hospitals catering for 2 150 patients whose ailments ranged from pneumonia, enteric fever, tuberculosis and severe surgical conditions down to eye complaints, cuts and bruises and other minor complaints. Each doctor did all the surgery, medical work, pathology and x-rays. Those who were part-time also ran busy private practices from which they derived their main incomes. The burden of their private work often made it necessary for them to delegate their responsibilities in the mines to young, newly qualified and, therefore, inexperienced medical assistants. The hospitals were usually converted hostels under the control of untrained, wholly inexperienced white 'hospital superintendents' who were assisted by equally untrained black nursing orderlies.[12] They had no proper operating facilities. Injured limbs were usually amputated rather than healed. Medication consisted of purgatives, stock cough mixtures and a few miscellaneous pills. Black mineworkers rightly had a deep suspicion of 'white' medicine. They ignored prescriptions and refused to enter hospitals with the result that each of the early hospitals, with secure perimeter fences and padlocked gates, were designed more to keep patients in than to keep strangers out.[13]

However, the improvements that had been made in the living conditions in the compounds were offset by the continuing deterioration in the villages where diseases persisted, were consolidated and became as endemic as in the townships. The rural areas had barely any health care provisions to curb the spread of diseases. On average during the 1920s, there was one doctor for every 5 250 black people throughout the country as a whole.[14] The doctors for the black population, however, were concentrated in the townships. Very few were prepared to work in the villages and, when they did, they were dispersed over such wide areas that many black people had virtually no access to them. These provisions worsened after 1948 when the rudiments of a national health

[12]In 1917, Dr A J Orenstein, chief medical officer for Rand Mines, insisted, for the first time in South Africa, on training black women as hospital nurses under the supervision of white hospital sisters against heavy opposition from the nursing profession *(ibid*, pp44-49).

[13]Mine doctors enhanced their status by forming the Transvaal Mine Medical Officers' Association in 1921. The Association held regular meetings at which reports and papers were read that dealt with various facets of compound life that directly affected health, such as nutrition and diet and feeding provisions. It published those proceedings. Following the establishment of the Association, the mining houses began to appoint full-time doctors and to centralise, rather tentatively, their own respective mine-based health services.

[14]The figures for black and white people were 900 doctors serving 4,7 million black people compared with 1 500 doctors serving 1,5 million white people. (John Farley, *Bilharzia. A History of Imperial Tropical Medicine,* Cambridge, 1991, p131.)

service were dismantled by the government and replaced by a two-tier system of health care that provided for an extensive fee-paying system for whites and a meagre state provision of health centres for blacks. The South African government, however, claimed it had no responsibility for health care provisions in the nominally independent black 'homelands'. As the 'homeland' governments had virtually no financial resources of their own with which to establish health centres, they were almost denuded of them. Those black mineworkers, therefore, who came from the 'homelands' and who were systematically repatriated to them by the mines when they became sick or injured, by and large had no further treatment or care after leaving the mines. Neglected, they were left to re-infect.

In addition to reducing the incidence of disease by making marginal improvements in compound environments, the government and the mines had tried to take out those novice recruits and mineworkers who were most susceptible to infections. The first step had been the government ban on recruits from all areas north of latitude 22 degrees south in 1913. The ban continued until 1934 when, under pressure from the mine owners, the government allowed 2 000 tropical workers to be employed in the mines on an experimental basis. The ban had been followed by a gradual fall in the incidence of compound diseases and the experiment was aimed at discovering whether or not it would rise if tropical recruitment were restarted. The disease rates did not suddenly rise. This, plus the fact that compound conditions had been improved and antibiotic medicines had become available, persuaded the government to amend the Immigration Act of 1913 to allow recruits from the tropical areas to work again in the mines. From 1938 they entered the mines under agreed quota arrangements between the various governments and the Chamber of Mines. By the following year their number had reached the 1913 level of just over 19 000, after which it increased steadily, but unevenly, until 1972 when it was 84 000 and amounted to almost a quarter of the total mine labour force. The return of the tropical mineworkers, however, was accompanied by a steady rise in the incidence of tuberculosis though whether or not they were causally related was difficult to establish because so many other variables were present.[15] The next step to check the spread of disease had been to subject all recruits, whether novices or not, to a series of medical checks, starting at the recruitment offices where they were initially recruited and ending at the mines to which they were allocated.

[15] *Ibid.*

The primary purpose of the checks was to ensure that those who were taken on could survive the rigours of underground mining. The checks at the first stage were cursory and mainly visual and resulted in the rejection of between two to three per cent because of obvious abnormalities or easily diagnosed ailments. The rest were sent on to the central recruitment compound in Johannesburg where they were given a further medical examination. Until after the Second World War the diagnosis of chest ailments had been largely dependent upon the use of stethoscopes but it was a laborious, labour-intensive and not very accurate way of examining chests, especially when as many as 1 000 recruits might be waiting to be examined at any one time. The mine doctors, therefore, had used their stethoscopes selectively for people who showed physical signs of being susceptible to respiratory diseases. In practice that meant examining those who were small and underweight, or sickly-looking with sunken eyes, and who might possibly end up by claiming compensation from the mines. When labour was scarce, however, some mines enrolled underweight recruits and tried to feed them up for about two weeks before putting them to work. However, a black worker who was rejected on medical grounds after going through the recruitment procedures at the WNLA centre could go straight to another mine as a volunteer and enter without going through such a rigorous selection process. Or a mine contractor could give him a job without subjecting him to a medical check at all. Thus some workers who might be particularly susceptible to infectious diseases managed to filter through the system.

In general, because of differences in recruitment practices and in the quality of health service provisions at mine level, there were considerable variations in the incidence of respiratory diseases between the mines. For instance, in the 31 gold mines on the Witwatersrand, the number of black mineworkers in each mine who were ill from respiratory diseases in 1933 varied between 23,5 and 176,2 per 1 000 workers.[16] The range in other mines was even wider. The variations in the incidence of respiratory diseases in the 16 coal mines in the Witbank region ranged from 12 to 194,4 for every 1 000 mineworkers every month over an eight-year period.[17] From the point of view of the health of black mineworkers, their allocation to mines was like a lottery. Quite clearly, by standardising provisions in their own groups of mines and eliminating

[16]Dr Daubenton, Transvaal Mine Medical Officers' Association, *Proceedings*, June, July, August 1934, p12.

[17]F J Allen, "Morbidity and Mortality on Coal and Gold Mines for 1941", *ibid*, October 1943, p66.

duplication between their own respective mine hospitals, the mining houses had much scope to make general overall improvements in health care. Apart from the employment of black female nurses, black people played virtually no part in the process of standardisation and development until the 1940s when a supply of black male "medical aides" began to enter the mines from the five-year course that had been established at the South African Native College at Fort Hare by a grant of £75 000 from the Chamber of Mines in 1936.

No black mineworkers had been x-rayed in mine hospitals until the 1930s because x-ray machines were expensive, cautiously introduced and used almost exclusively for white miners. Then gradually they were used selectively for black mineworkers. The Randfontein Estates, for instance, which was one of the first gold mines to obtain an x-ray machine, used it only on three categories of black mineworkers, namely those with 60 months' cumulative underground work, those with obviously defective lungs and those who showed signs of significant loss of weight. All the rest, including the vast majority of new recruits, were medically tested in other ways. But even such a limited use of x-rays could save a mine a substantial amount of money by reducing the number of mineworkers who might subsequently contract TB and to whom it might be obliged to pay compensation.[18]

The widespread use of x-rays was dependent upon the invention of mass miniature radiological techniques. Miniature photography had been possible for many years but early in the Second World War it was developed as a radiographic technique for use with large numbers of people in the armed forces. The improvements were such that the details on miniature radiography closely resembled those produced by full-scale x-rays though they required careful reading, preferably by two doctors, to prevent diagnostic mistakes being made. The possibility of making errors in identifying small lesions on the lungs from miniature x-rays, however, was notoriously high.[19] With the use of mass radiology the main WNLA depot in Johannesburg was able by 1962 to x-ray as many as 850 recruits an hour.[20] The Pneumoconiosis Act of 1956 had stipulated that every mineworker had to have his chest x-rayed on

[18]Dr Peall at Randfontein Estates stated that between May 1930 and December 1936, 15 237 recruits were x-rayed at a cost of £2 444,5s. As a result of the x-rays 201 recruits were rejected, thus saving the mine £10 050 in possible compensation payments (Transvaal Mine Medical Officers' Association, *Proceedings*, November 1937).

[19]Dr G K Sluis-Cremer, *South African Medical Journal*, 12 July 1980, p59.

[20]Discussion between the author and the WNLA Compound Manager in Johannesburg, August 1962.

recruitment and every six months thereafter. This meant that more than one million radiographs were taken on the mines each year.

The initial x-rays were taken as part of the process of registration and induction at the main WNLA compound. All recruits went through the process naked and at a trot. They ran into the radiographic department where they hurriedly pressed their chests on to x-ray screens before trotting off on to the next stage. The resulting miniature films were examined within an hour and anyone showing a lung abnormality was then given a normal x-ray examination. If the abnormality was confirmed then the recruit was sent to the compound hospital for a further examination before a decision was taken as to whether he should be repatriated or offered surface work. The practice of placing some of the workers with defective lungs in surface jobs resulted in the morbidity rate on the surface being twice as high as amongst underground workers. The repatriation of infected workers reinforced the transfer of tuberculosis to the villages because 95 per cent of the repatriates did not receive any further treatment at home and were neither cured nor stabilised.

By and large, the youngest and obviously fittest of the recruits were allocated to do underground jobs because they were the least likely to contract diseases and the most likely to survive the rigours of underground mine work. The mines, however, continued to have to choose from populations that did not meet those standards because they were generally under-nourished and debilitated by indigenous diseases. Some recruits had acquired immunity to the main compound diseases but there were others who had still not come into contact with either the major urban diseases or even the common childhood ones such as mumps, measles and chickenpox.[21] Moreover, because of their newness to mining and their first acquaintance with the frightening hazards of underground mine work, many mineworkers suffered from acute stress and that increased their vulnerability to diseases. In any event, the mineworkers were crowded together so tightly, not only in the hostels but also during all the stages to work and back to the hostels, that they had the maximum exposure to infection.

The journey to work was dominated by frustrating queues and long waits. It started at about three o'clock in the morning with mass awakenings in all of the hostels through the use by the security police of loudspeakers or the compound tannoy systems. This was done irrespective of the time the mineworkers had to be underground with the result that some of them had to wait from two to three hours before

[21]Mine Medical Officers' Association, *Proceedings*, 22 April 1965, p103.

they started work. It often took that amount of time, however, for them to get to work. When they left the hostels for the beginning of a shift they went to the mine buildings at the head of the main shaft where they were provided with lamps and given their identity discs before being crammed into an enclosed area known as the 'crush' in order to wait for a cage to descend down the mine. When one arrived that was not reserved for white miners they were packed tightly into it and travelled to the shaft bottom, coughing and spitting in each other's faces and sharing germs. They also had to queue for the locos that became increasingly necessary to take them part of the way to their working places as the distances between the shafts and the stopes lengthened. They could not, by law, start work until a white miner had turned up to inspect the working places so they frequently had to stand at their designated 'waiting places' because white miners were often late in turning up for work. A survey of the gold mines along the Witwatersrand in 1938 showed that half of the black mineworkers were actually underground by 5 am and that all but about six per cent of the remainder were at work by 6 am. This had not changed 35 years later when the Human Resources Laboratory of the Chamber of Mines conducted an investigation into the quality of life of black mineworkers.[22] Then, the average amount of time they spent in getting to their workplaces was one hour and 49 minutes but some took as long as three-and-a-half hours. If that meant that they were late for work, then they were likely to be accused of having slept in and risked being assaulted by the white miners. They could also be sent back to the surface and lose a day's pay.[23]

The nature of the work made them vulnerable to infection for it invariably resulted in weight loss, a prime indicator of vulnerability. Mine work was exceedingly physically exhausting, particularly for those who were extracting ore in the stopes and were under constant management pressure to increase production. Every one of them sweated when working underground but the drillers and lashers sweated most. It was estimated that a driller, handling a heavy metal drill and working to set targets, could lose up to two-and-a-quarter kilograms in weight during a single shift.[24] A lasher in a coal mine in the mid-1930s loaded

[22]For a synthesis of the studies, see Mine Medical Officers' Association, *Proceedings*, January-December 1982, pp6-15 and P de Vries and J Radise, "The Effects of Mine Procedures and Practices on the Quality of Life of Black Mineworkers on Four Mines", Human Resources Laboratory, *Research Report*, No. 5/80, February 1980.
[23]De Vries and Radise, 1980, p6 and p9.
[24]Orenstein, 1922, p7.

about 20 tons of coal a shift and had to raise each full shovel almost a metre to get over the edge of the tub. On average he lost just over three-and-a-half kilograms in weight during every shift but sometimes he could lose as much as five kilograms.[25] Over the period of a contract, however, mineworkers generally put on a little extra weight, though this could have been because most of them were in the age range when they would have gained weight anyway.

If the daily weight loss was not made up, the mineworkers were not only likely to have insufficient energy to work their next shifts effectively but also were less able to resist infections. However, not all who became ill were diagnosed and repatriated to their villages. Some managed to keep on working, shift after shift, though often this was due more to the strictness of the supervision than anything else. It was impossible for anyone to hide away in a hostel room unless he had a compelling medical reason for doing so. It was not surprising, therefore, that barely two per cent of black mineworkers were absent from work on an average working day in any year until the 1970s, compared with the 14 per cent absence rate for white miners.[26]

After at least eight hours' work the return journey was worse than the inward one for they were tired, wet with sweat, exhausted and frustrated by further long queues and waits for the cage. Before the Second World War, although the majority of black mineworkers had finished their shifts by between 1 pm and 2 pm, more than half of them had to wait until 4 pm at the bottom of the shaft, which was the draughtiest and coldest part of the mine, before being hoisted to the surface.[27] It was much the same in the post-war period. In 1976, it still took the average mineworker one hour and 30 minutes to get back to the hostel. Some took almost three hours to return. These waits were always in damp, chilled clothes. The stress was increased by the discriminatory practices of the mines for white miners, who always seemed to arrive at the shaft bottom last, were invariably given priority in leaving. A white miner could walk to the head of the queue, no matter how long the black mineworkers had been waiting, and be taken to the surface all by himself. When that happened black mineworkers might have had to wait up to an hour for the cage to return. Blacks and whites were not allowed to mix in the cages in any circumstances.

[25]Dr F J Allen, October 1943, p62.

[26]Francis Wilson, *Labour in the South African Gold Mines*, 1972, p168.

[27]Transvaal Mine Medical Officers' Association, *Proceedings*, January-February 1938, pp59-61.

All in all, by 1976, the average black mineworker spent just over 10 hours travelling to his working place, working and returning to his hostel. During that time he had no food because mineworkers were not allowed meal breaks underground. Some of them, however, went without food for 24 hours because they set off without breakfast. Before the Second World War some mines had refused to provide breakfasts because they believed it was "contrary to Native customs". For that, and other reasons, only about one-third of all mineworkers ate any kind of food before they went underground.[28] A general complaint was that, in order to collect food for their breakfasts to eat in their rooms, they had to walk from about 180 metres to 360 metres in the open from their rooms to a central kitchen, when the weather was sometimes wet and cold. An attempt was made to overcome this problem by equipping special rooms with breakfast facilities at the shaft head so that mineworkers could spend 15 minutes eating just before they descended the shaft. But neither this, nor the provision of central dining rooms adjacent to the hostel blocks in the new compounds in the 1950s, persuaded everyone to eat a breakfast. Mine doctors persisted in trying to get every underground worker to eat something before going to work for if they were without food for 24 hours they were likely to suffer from hypoglycaemia, resulting in headaches, faintness, a lack of concentration and mental confusion in general, during the second half of the shift. Hypoglycaemia, however, was rarely recognised as a possible cause of accidents.[29]

The Changing Pattern of Compound Diseases

A broad picture of the character of diseases amongst black mineworkers showed that they suffered primarily from bilharzia, scurvy, pneumonia and tuberculosis, often in combination with one of the dust-related diseases, namely silicosis, pneumoconiosis or asbestosis. It was difficult, however, to present a detailed or even a broadly accurate picture of the incidence of each disease and the precise relationships between them. In the first place, there were serious defects in the nature of the data that had been collected in that it did not reflect the extent of the diseases. All of the diseases had been systematically under-diagnosed. Incidence rates concerning dust diseases, for instance, were based on

[28]Transvaal Mine Medical Officers' Association, *Proceedings*, July-August 1941, p6.
[29]Mine Medical Officers' Association, *Proceedings*, January-August 1978, p33.

the daily average, not the actual number of mineworkers in dusty occupations; no service profiles of the work forces were provided so it was impossible to differentiate between occupational or age groups. The periodic medical checks they underwent, moreover, did not always detect the early signs of a disease. Systematic under-diagnosis was caused by the failure to detect symptoms which emerged after the mineworkers had returned to their rural homes, by the fact that ill workers often excluded themselves from the statistics by failing to return to the mines after the end of their contracts and by the use of faulty diagnostic techniques, such as the use of mini rather than full x-rays, incompetent readings and inadequate lung function tests.[30]

Mine recruits from most of the regions suffered from bilharzia or 'human redwater disease', as it was described in the villages. The disease, though it was at first expressed through minor symptoms, seriously undermined their ability to cope with the demands of underground work. It resulted in forgetfulness, indifference to punishment, laziness, mental fatigue and nervous irritability.[31] Later, if the eggs of the parasitic worms that caused it, and which lived in the gut, liver and bladder, increased then there were other symptoms such as blood in the urine, painful urination and impotency. The mental symptoms made it difficult for new recruits to adapt to the rigours of training and the acclimatisation process. Once they were underground, bilharzia reduced the acute mental awareness necessary to avoid accidents.

The disease was transmitted by fresh water snails which discharged parasitic worms that swam freely in the water and entered the skin of humans. The snails were indigenous to most of the inland waters of Africa and were particularly prevalent in South Africa. Wherever, therefore, people were without piped water supplies and had to drink and wash in the untreated waters of pools, lakes, dams and rivers they were exposed to it. Relatively few black people in South Africa had ever had piped water. Those who lived in the rural areas were dependent on the contaminated supplies of inland waters. Not surprisingly, bilharzia had been recorded amongst mine recruits from the earliest days of the century. In 1910 it was noted that as many as 20 per cent of the recruits from Mozambique, Malawi and Zululand suffered from it.

[30]Jean Leger, "Key Issues in Safety and Health in South African Mines", *South African Sociological Review*, Vol 2, No 2.

[31]Farley, 1991, p245. It was possible for rural people to have these symptoms without them seriously affecting their lives; indeed, without them even recognising that they were suffering from an infection. Women were the most likely sufferers because they were the water-carriers and clothes-washers. After the initial stage, bilharzia made women sterile.

In the inter-war years it was said to be endemic in the Northern Transvaal, Natal and Cape Province. After 1948 it ran a completely unchecked course in the Transvaal and assumed "gigantic proportions because for all practical purposes virtually the whole Native population (was) suffering from bilharziasis".[32]

It was possible to treat the disease by giving intravenous injections of tartar emetic, but that alone was futile unless it was accompanied by socio-economic measures. Treatment campaigns, however, were considered to be too expensive for the black population while the socio-economic measures would have involved major investments and a political commitment to remove poverty and deprivation in the rural areas that was inconsistent with the government's intention to maintain white supremacy and apartheid.[33] Bilharzia, in consequence, was allowed to continue its debilitating course amongst black mineworkers.

The deprived economic and social conditions, which facilitated the spread of bilharzia, were also the source of scurvy. Recruits who were infected by one disease were likely, therefore, to be afflicted by the other. Scurvy was not an infectious disease but was caused by the absence from diets of ascorbic acid, called vitamin C, generally as a result of a lack of fresh fruit and vegetables but also as a consequence of the over-cooking of vegetables, which was a common feature of hostel life. Thus cabbage, a common hostel food, which was rich in vitamin C, lost 80 per cent of it when it was boiled for 20 minutes or more. Where, as in the Kimberley diamond mines, mineworkers provided and cooked their own food, they developed scurvy because they could not afford an adequate and balanced diet. But where they lived in hostels they suffered because they were given prepared meals containing insufficient fresh fruit, an excess of mealie meal, and vegetables that were usually spoiled by overcooking. The balance between vitamin sufficiency and scurvy was so tenuous that a neglectful cook or a disruption in the supply of vegetables because of a transport breakdown or simply the onset of winter could result in a scurvy outbreak. Scurvy had been one of the earliest manifestations of the consequences of compound life. It usually only entered the compounds in a latent form because once it was manifest it displayed visible

[32]*Ibid*, p137 and p243.

[33]Bilharzia had been found to be common amongst the children of poor white Boers in the 1920s. The government then jumped into action by providing treatment, building bilharzia-free swimming pools and introducing measures to alleviate the poverty of poor whites. The government succeeded in removing bilharzia from the white population but refused to apply the same measures for black people.

physical symptoms and those who were suffering from them, therefore, were usually rejected during the medical checks. The latent infections were activated in the compounds by the combination of persistent dietary deficiencies and exhaustion from constant hard work. The symptoms usually appeared after the recruits had been eating vitamin C-deficient food for some months and, as large numbers were given the same food, many of them simultaneously began to show signs of it such as anaemia, bleeding under the skin, bleeding gums, the loss of teeth and difficulty in healing infected sores and cuts. This gave rise to the mistaken belief that scurvy was an infectious disease. By the time it was discovered that it was not infectious, it was already pervasive in the compounds. It often resulted in death. Indeed, it was estimated that in the early years of mining twice as many black mineworkers died of scurvy as were killed in accidents.[34]

Scurvy continued to be widespread for the first four decades of the 20th century, though in the later years not always in such a lethal form. Indeed, it was so prevalent that one mine medical officer, after reviewing 1 000 records of scurvy between 1929 and 1939, described it as a "disease of mine life".[35] This view was supported by the fact that in some instances the symptoms disappeared when mineworkers returned home in between contracts because they were able to eat vitamin C-rich foods. They re-appeared, however, after their return to compound diets. The mortality rate from scurvy fell almost to zero in the post-Second World War period as more attention was given to the provision of adequate diets and ascorbic acid was added to marewu, the acidulated cold beverage made by the fermentation of maize flour, which the mineworkers liked to drink. But the disease was not entirely eradicated for, despite the improved diets, mineworkers continued to suffer from a vitamin C deficiency as a result of the stress of underground work.[36] This in itself did not disable them but made them more susceptible to other diseases such as pneumonia and tuberculosis.

Pneumonia caused special diagnostic problems in the mines for, unlike most other diseases, it did not progress from a simple infection until it eventually destroyed the organs of the body. It was the term used

[34]F W Fox, "The Food and Feeding of Mine Native Labourers", Transvaal Mine Medical Officers' Association, *Proceedings*, September 1941, p25.
[35]Packard, 1989, p169.
[36]M E Visagie, J P du Plessis, G Groothoe, A Alberts and N F Laubscher, "Changes in Vitamin A and C Levels in Black Mine-Workers", *South African Medical Journal*, 11 December 1974, pp2 502-2 506.

to describe the condition where the lungs were so damaged by inflammation that they gradually filled with air. It was possible for this to result from a variety of respiratory ailments, such as colds or influenza, or it could take hold when the patient was weak from malnutrition or from one dust disease or another. It was, therefore, a mode of death, the end stage, rather than the precipitator of death. So it was possible for mineworkers weakened by scurvy to die of pneumonia. It could also be lethal when combined with a dust-related disease. This raised a question about the accuracy and usefulness of morbidity statistics for when pneumonia was present it created confusion about the prime causes of death.

Mine doctors chose to pursue medical solutions firstly through the use of vaccines and then antibiotic drugs. The principal vaccines were those that had been developed at the South African Institute for Medical Research, established by the Chamber of Mines in 1911. Almroth Wright, a prominent British bacteriologist and immunologist, developed the first pneumonia vaccine for the immunisation of black mineworkers in 1911 but it had a short life and a disputed effect. Wright was succeeded by Spencer Lister, a mine doctor at the Premier Diamond Mine in Pretoria who joined the Institute and improved the quality of the vaccine.[37] However, because of the mobile and disparate character of the labour force it was always difficult for any vaccines to be properly tested so that their effectiveness was always in dispute.[38]As a result, during the 1930s and 1940s, some mine doctors immunised recruits while others did not. Others did it intermittently. By and large, however, it gradually came to be recognised that though vaccines might help in some cases they were not an effective means of combating the disease in general.[39] The mines eventually adopted a pluralistic approach by combining the use of vaccines and antibiotics with a policy of excluding from the mines everyone whom they thought might fall prey to pneumonia and of returning to their homes those who became sick during their contracts of employment. The result was that in the decade leading up to the Second World War the incidence of pneumonia

[37]See Peter C English, "Therapeutic Strategies to Combat Pneumococcal Disease: Repeated Failure of Physicians to Adopt Pneumococcal Vaccine, 1900-1945", *Perspectives in Biology and Medicine*, 30, 2, Winter 1987, p176; also Cartwright, 1971, p26 *et seq*.
[38]English, 1987, pp181-182, maintained that vaccines had positive effects but were rejected by mine doctors for a variety of reasons such as prejudice, deficient testing and the advent of antibiotic drugs.
[39]See Transvaal Mine Medical Officers' Association, *Proceedings*, June, July and August 1934 and May and June 1937.

was almost halved. It rose again during the war, in part due to the reintroduction into the mines of 'tropical workers' who were in a poor state of health and were unable to cope with the relatively wet and cold conditions on the Rand. They also suffered from the deteriorating conditions in the hostels and lax wartime medical checks.[40]

The overall situation, however, began to improve after 1945. There was a better understanding of the disease after it was discovered that mineworkers suffered from different strains of pneumonia, some of which were resistant to treatment. The development of sulpha drugs and antibiotics improved the chances of cure in general. In addition, the introduction of mass x-rays by WNLA meant that a number of potential victims were rejected before they could be exposed to infection in the compounds. There were also some environmental improvements. Government regulations concerning the housing and feeding of black mineworkers were promulgated, amended and redrafted. These factors made it easier for the mine owners to stem the spread of the disease and, thereby, to reduce the mortality rate, but they had a limited effect that did not extend very much beyond their initial impact. The problem was more complicated than had been thought.

When the incidence of acute respiratory diseases, including pneumonia, was analysed in the mid-1970s it was shown that there was a direct correlation between attacks of pneumonia and changes in temperature, giving rise to sharp seasonal variations in the numbers affected. Each year, with little variation, mineworkers had suffered when the temperature fell during the winter months, especially during June and July. No account had ever been taken of this. In particular, nothing special had been done to protect them either from the cold when they walked from the shafts to the hostels and were wet with sweat, or in the hostel at night during the winter months. It had been proposed as early as 1915 by the newly appointed medical officer of Rand Mines that black mineworkers should be issued with better clothing, especially in the deep mines, to keep them warm when moving from the hot underground environment to the cooler temperatures on the surface but the proposal had been rejected because it was considered to be too costly to implement. Other mines had reacted similarly.

Gradually, however, some mines had begun to issue tunics to black mineworkers although they deducted the cost of them from their wages. Then in 1935 the Gold Producers' Committee decided to issue free

[40]It was an anomaly that amongst mineworkers the tropical workers were the most susceptible to pneumonia but the least likely to contract tuberculosis.

tunics as part of their campaign to convince the government that the mines were safe for tropical labour.[41] The mineworkers, however, were issued with only one tunic at a time so that often they had to go to work wearing one which was wet either from the previous day's sweat or from washing. They were also able to buy extra blankets from mineworkers who acted as agents for outside retailers, but few could afford more than one. Blankets for wearing rather than for use on beds, however, was largely confined to the Sotho mineworkers. But tunics and blankets gave only marginal protection to the novice recruits in the deep gold mines who contracted pneumonia because they were least able to withstand the transition from the hot, humid conditions underground to the cold atmosphere outside. They usually developed colds, then influenza and, after about two months, fell ill with pneumonia. It was difficult for them to avoid infections from the batches of additional novice recruits who, just before the onset of winter, were crowded into the hostels, making them even more congested. At night, the inmates were compelled, more than usual, to huddle together for warmth at one end of the room and thereby share each other's ailments.[42]

Although by the 1980s mine doctors had the medical means to contain the spread of pneumonia and to prevent epidemics breaking out, the tardy responses of the mining houses to proposals to tackle the causes of respiratory ailments in general, combined with persistent malnutrition in the reserves, ensured that pneumonia would continue to trouble mineworkers. There was a perpetual reluctance by the mining houses and the government to accept that there was a social basis to any of the infectious diseases because to have done so would have presented them with the need to alter policies that were pillars of the apartheid state, namely the migrant labour system and poverty in the reserves.

Tuberculosis, the scourge of black mineworkers, was another case in point. This, the most social of all diseases, was contracted mainly through the inhalation of airborne germs from persons who already suffered from it. It was also possible for people, especially young children, to be infected through the milk of cattle suffering from bovine tuberculosis, though in South Africa that was not a common source. [43] It spread remorselessly, though not quickly, in crowded conditions. It was possible for a person carrying the germ to pass it on to between 10

[41]See Packard, 1989, pp346-347.

[42]C H Wyndham, R Gonin and R D W Reid, "Seasonal Variations in Acute Respiratory Diseases and Meningitis in Black Miners Living in Hostels", *South African Medical Journal*, 26 August 1978.

[43]Packard, 1989, pxviii.

and 20 other persons and about 70 per cent of his or her close contacts in the bedroom and home by the time of diagnosis.[44] It was claimed that in crowded conditions there was a 30 per cent risk of becoming infeced after long or frequent exposure.[45] An infection, however, did not necessarily lead to an active disease for it was estimated that only between 10 and 15 per cent of those carrying the germ showed symptoms within two years. It was also possible for a person to have a latent infection that became activated much later in life.

If active tuberculosis was left untreated then within six to 12 months it spread throughout the lungs and possibly to other parts of the body through the blood or lymph systems. Sometimes it acted much more quickly than that. It would start with a slight cough and fatigue, then weight loss and coughing up blood. Once it started progressing then other symptoms emerged, such as periods of fever, joint pains, hearing loss, diarrhoea and chest pains, which caused extensive damage to the body's organs and eventually led to death. The speed of progression depended on the physical state of the infected person. If a person had no immunity to tuberculosis because he or she had never been exposed to it before or had an immune system that was impaired by other diseases such as measles or malaria, or was debilitated by malnutrition or stress or exhaustion, then the progression to death was hastened.

Tuberculosis had flourished in the hastily built insanitary towns during the industrialisation of Europe from the late 18th to the early 20th century. As the poor and deprived workers and peasants emigrated from Europe they carried their infectious diseases, tuberculosis included, with them. In that way, tuberculosis had travelled with white immigrant workers to the mines of South Africa, was passed on to the black mineworkers, took root in the mine hostels and was transmitted to their friends and families when they returned to their villages at the end of their contracts. It was a disease that had flourished through the circle of infection. Consumptive tourists, who sought relief and cure within the country's idyllic climate, also brought the disease in with them though it was unlikely that their particular germs entered the mine compounds.[46] Although there was some dispute as to whether or not South Africa in pre-colonial times was entirely free of tuberculosis,

[44]Coovadia and Benatar, 1991, p243.

[45]Packard, 1989, pxviii.

[46]The government became concerned about the importation of TB by white consumptives in search of a sanatorium in the sun through the advertisement of South Africa as a health resort and it set up the Tuberculosis Commission in 1912 to examine the issue. It responded

there is clear evidence that it was, at the least, an uncommon disease.[47]

The mine doctors endorsed this view for when they began to show serious concern about the incidence of tuberculosis in the 1920s they discovered that it was rare to find new recruits with the disease. The vast majority of mineworkers appeared to have developed symptoms only after working and living alongside infected persons for some months after entering the mines. Moreover, when the mine doctors looked back to the early years of the century they recollected knowing of very few cases either in the reserves or the mines. One doctor, for instance, had stated in 1921 that 20 years previously he hardly ever saw a case of tuberculosis. Another declared that when he was investigating its prevalence throughout southern Mozambique in 1905 he had found only 14 cases, 10 of which involved men who had worked on the mines.[48] This led to the belief that tuberculosis was acquired on the mines and that the compounds were the focus of infection.[49]

This particular circle of infection was well established during the inter-war years for by then tuberculosis was rampant in all areas of black life though it was most prevalent in the townships where it reached epidemic proportions.[50] The rapid increase in illegal urbanisation had brought with it all the factors, such as intensified poverty, homelessness and the construction of shantytowns, that facilitated its spread. In the reserves the disease was endemic. The worst affected rural area was the Transkei, followed by Swaziland and Botswana, each of which was a major source for mine recruits. The problem in the Transkei had been caused primarily by malnutrition amongst children. In the mid-1970s, almost 11 per cent of all children there up to the age of 10 were infected with it though the highest incidence was amongst the children of the poorest families.[51] Many of the children had a latent

to the Commission's report by passing the 1913 Immigrants Regulation Act that prohibited the entry of people with clinically recognisable tuberculosis, except by permit and subject to certain conditions. By the Second World War TB barely bothered the whites. Their rate of infection was generally lower than in England and was amongst the lowest in the world. A different slant is provided by the statistic that the notification of TB cases for whites by the late 1960s was one-20th of those for blacks. (Gill Westcott and Francis Wilson (eds), *Economics of Health in South Africa*, 1979, Volume I, p141.)

[47]See Packard, 1989, Chapter I, "Pre-industrial South Africa: A Virgin Soil for Tuberculosis?"

[48]Transvaal Mine Medical Officers' Association, *Proceedings*, August 1921, p5 and p10.

[49]*Ibid.*

[50]Even as early as 1912 the Tuberculosis Commission had reported that tuberculosis was three times more prevalent in the towns than in the rural areas.

[51]Peter Burney and Suheil Shahyar, "Tuberculosis in the Transkei", in Westcott and Wilson

form of the disease that the boys amongst them carried to the mines. The rate of black sufferers in general soared and continued to do so after 1948.[52] There was, however, a drop in the death rate from it due to the introduction of effective anti-tuberculosis drugs. The drop was less for black sufferers than for whites because it was considered too expensive to give drugs to them and then provide them with supervised treatment afterwards.

It was difficult, however, to get a true picture of the extent of the disease in the black communities. Their deaths were not registered, post-mortems were rare and many people refused to report to the white medical authorities when they were ill. The incidence of tuberculosis could only be assessed on the basis of notified cases and there were a number of reasons why blacks avoided going through the process of notification. For instance, many people in the townships had no legal right to be there and would have been sent to prison or to farms as forced labour if they had reported their illness. Then from the late 1970s they became increasingly unwilling to report to the clinics because of pressure from black liberation groups. Whenever there was political unrest, black people tended to avoid places that were serviced by whites, such as clinics, in case they were labelled as collaborators. In the rural areas, the method of notification itself had deterred people from reporting their illness. In the Transkei and the Ciskei, for instance, people had had to wait at stopping points, sometimes for hours, for a mobile clinic to turn up with its medical team. Because of an acute shortage of doctors, one such clinic might have had to serve up to 10 000 people.[53] Moreover, after 1948, the government policy of forcibly removing blacks from their homes in the shantytowns, adjoining the black townships, and relocating them in the rural areas increased the burden on the mobile clinics by transferring the sick with everyone else. The dislocation this caused also worsened the conditions under which tuberculosis spread

The incidence of diseases in the mines usually mirrored that in the society outside but in this case there was a difference. The incidence of tuberculosis was lower in the mines than in the rest of society, not only because the mines had become more proficient at diagnosis and in preventing deaths through the administration of drugs, but also because

(eds), 1979, Volume II, p345.

[52]For details of the incidence of TB in the reserves and townships, see Coovadia and Benatar, 1991; Packard, 1989 and Westcott and Wilson, 1979, Volume I.

[53]Westcott and Wilson, 1979, Volume I, p20.

they became more stringent in their selection of new recruits. They focused their attention on the physique of recruits. Mine doctors had always been convinced that the weight of a recruit was an effective means of detecting tuberculosis in its early stages and they were encouraged in their diagnosis by the Miners' Phthisis Bureau, which, from 1921, had established a standard of physical types representing "the weight of healthy men of different heights and at different ages". Mine doctors officially used the standard as a basic guide but in their daily practices they amended it to satisfy their own prejudices and rarely agreed as to what the appropriate weight scales should be. In any event, all recruits were weighed at the WNLA compound and then again every four or five weeks at the mine. According to the Miners' Phthisis Act of 1925 any mineworker who lost 2,3 kg or more between two consecutive weighings had to have a stethoscopic examination, though again, in practice, there was no agreement amongst mine doctors as to what the maximum permissible weight loss should be.[54] In general the absence of any agreement about the best weight for mine working and about weight loss figures was convenient for mine managers for it allowed them to adjust their medical standards to accommodate the varying labour needs of their respective mines. They invariably lowered the standards when they needed more labour.

The Randfontein Estates Gold Mine had had its own scheme during the 1930s that its mine doctor, Dr P A Peall, had described as "the scientific management of Native labourers".[55] The recruits were put into three weight grades that were related to different categories of work. Every black recruit then had to appear naked before a medical officer, carrying a card on which were written his tribe, previous mining history and weight. He was then medically examined to see whether he was capable of doing the work his weight suggested he ought to be capable of doing. He was subsequently weighed every month and if his weight rose or fell sufficiently for him to be regraded then a 'misfit' form was filled in so that he could be moved into another grade of work, at the discretion of management. Studies of the significance of weight as a guide to the health of mineworkers concluded, however, that weight was not very helpful. The medical officer for a group of Witbank collieries found in 1943 that illness amongst new recruits appeared "to have no relation to the size or weight of the individual".[56] This was confirmed by a

[54]Packard, 1989, p180.
[55]Transvaal Mine Medical Officers' Association, *Proceedings*, November 1937, p51.
[56]*Ibid*, October 1943, p62.

Chamber of Mines' survey of tuberculosis in the mining industry in 1963 which concluded that "physique on engagement was not found to be useful in predicting tuberculosis".[57] Nonetheless, some emphasis on weight continued because weight loss was the only common objective symptom of tuberculosis that seemed to have any immediate relevance. However, as advances in radiology enabled the mine doctors to make more accurate assessments of the condition of lungs, so weight became a less used criterion.

The improved facility to detect lung defects was one of the factors responsible for increasing the reported incidence of tuberculosis from the relatively low level of 2 per 1 000 mineworkers in 1952 to almost 5 per 1 000 in 1961.[58] It was this upsurge which had provoked the Chamber of Mines in 1963 to initiate an investigation into tuberculosis in the mines to discover what other factors were involved.[59] The investigation tested the effectiveness of screening methods that isolated the carriers in the compounds and highlighted the high-risk age and ethnic groups.[60] It recommended that the high-risk groups should be identified and that their specific needs should be met through medical surveillance and the use of the BCG (Bacilus Calmette Guerin) vaccine or chemoprophylaxis. But, by the time the report was published, these methods had already been tried and were being phased out. One mining house inserted a drug called calcium methaniazide into the drink of its black employees in 1962 and was, by 1968, treating 42 000 black mineworkers in that way. By 1970, however, the drug was being restricted to high-risk groups and administered only for the first three months of a contract. Another mining house had used the BCG vaccination in one mine for a trial period between 1965 and 1968.[61] It was claimed that both methods had positive effects but neither was continued in use for long or applied extensively. BCG was being used extensively on children in Western countries at that time but was of little use in populations with high levels of tuberculosis. It was also relatively expensive and did not attack the basic causes of the disease.

[57]Mine Medical Officers' Association, *Proceedings*, May-August 1968, p15. A survey in 1925 had found that only in 26,2 per cent of the cases did weight loss assist in the diagnosis of TB. (Packard, 1989, pp180-181.)

[58]*Ibid*, p8.

[59]See *ibid*, p19 for a summary of the report.

[60]The investigation found that the person at least risk was a man from Malawi with normal lungs, under 35 years of age, working underground but with less than four years' experience, while the one at greatest risk came from the Transkei, was older than 35 years, had defective lungs and was, in consequence, employed on the surface.

[61]Sluis-Cremer, 1980, p60.

The concern about the spread of tuberculosis persuaded some mines to treat infected mineworkers rather than repatriate them immediately after diagnosis. They encouraged ailing mineworkers to remain for treatment for from six to nine months though, except for those in the diamond mines, without pay. Those who worked at the De Beers diamond mines who were diagnosed as having tuberculosis were transferred to the Public Health Hospital and put on half pay. When their disease reached the quiescent stage they were given sheltered jobs.[62] Mineworkers in the gold mines had to balance the possibility of a lengthy treatment with no pay against the need to feed their families. They usually opted to return to their families in the hope that they could obtain some income from either farming or other waged work in the rural areas. The mine authorities sometimes saw this as a justification for not providing more treatment facilities and regarded it as an illustration of the primitive mentality of blacks.[63] The result was that workers in the gold mines were only treated for the time it took to arrange for the payment of compensation and that was usually less than three months. After that they were repatriated to the debilitating conditions of the reserves.

It was not until the late 1970s that the Chamber of Mines altered its policy for mineworkers in gold mines to accord more closely to that in the diamond mines. In the meantime, the morbidity rate had risen well above the figure that had disturbed it so much in 1963. The rate had reached almost seven for every 1 000 mineworkers by 1976 and then, according to one calculation, touched 13 per 1 000 by 1980, making it the highest known rate in the 20th century.[64] There were various reasons for the rise. The Chamber of Mines' investigation had shown that tropical workers, though susceptible to pneumonia, were much less likely to contract tuberculosis than those from the non-tropical territories in southern Africa and that the highest incidence by far was amongst mineworkers from the Transkei, followed by Swaziland and Botswana.[65] In fact, Xhosa mineworkers were more than three times as likely to suffer from tuberculosis than were recruits from Malawi, Zambia and northern Mozambique. From 1974, however, the number of tropical workers declined dramatically due to a foreign labour crisis and the Chamber of Mines sought to replace them with workers from

[62]Packard, 1989, p289.

[63]*Ibid*, pp288-289.

[64]Leger, *South African Sociological Review,* Volume 2, No 2, p25.

[65]Mine Medical Officers' Association, *Proceedings*, May-August 1968, p14.

the Transkei and Ciskei where, incidentally, tuberculosis was widespread. Also, partly as a result of this recruitment switch, the mining labour force, from the late 1970s, though still migrant, had begun to become stabilised and, in consequence, its average age increased. The older mineworkers, meaning anyone over the age of 35 years, were the most vulnerable to infection.[66] Then the economic conditions outside worsened as the South African economy began to move into recession. Poverty deepened and the extent of tuberculosis became more acute.

Mine Doctors

All in all, when viewing the incidence of all infectious diseases, it was clear that black mineworkers had benefited from the extended role of mine doctors who had encouraged the introduction of policies to alter hostel structures, improve the provision of meals and the substance of diets and to install more effective health care infrastructures in the compounds. Through their insistence that the level of production was directly dependent on the health of black mineworkers, the mining houses had invested in health care provisions. The result was that by the 1980s, black mineworkers were served by a mine health service which was incomparably superior to the very best available to black people in the townships and the rural areas.

The service was pyramidal with a broad base consisting of thousands of mineworkers holding valid Red Cross First Aid certificates who worked from underground and surface dressing stations. Within the hostels there were medical stations, run by registered nurses, that had spaces for examination, dressing and hearing and vision testing and a sick bay. At the top of the pyramid was Rand Mutual Hospital, established by the Chamber of Mines in 1979, which provided specialist medical and surgical facilities. In between were 41 mine hospitals with a total of 7 000 beds, representing seven per cent of all the general hospital beds in South Africa. Two regional hospitals, the Ernest Oppenheimer Hospital in Welkom in the Orange Free State and the Leslie Williams Memorial Hospital in Carletonville, provided specialist medical and surgical facilities. The Rand Mutual and the Ernest Oppenheimer hospitals had spinal injury units where cases were

[66]In the Chamber of Mines' investigation it was discovered that the morbidity rate for the 35-55 age group was 34,7 per 1 000 mineworkers while the rate for the 18-35 age group was only 7,2 per 1 000. The mineworkers least likely to contract tuberculosis were aged between 30 and 35. Their morbidity rate was 4,3 per 1 000. (See *ibid.*)

treated from injury to rehabilitation. The service overall was run by a health care staff of 3 500 plus about 300 full-time medical doctors. The doctors staffed the mine hospitals, undertook the duties of hygiene officers and were responsible for the general health of the hostel residents. They were pivots around which the mine health service revolved.

Mine doctors, however, were influenced in the judgements both by their white medical heritage and their place in the hierarchy of mine management. Though they were stigmatised by being called 'kaffir' doctors within the medical profession, they observed its values by regarding black indigenous culture as primitive and by deriding tribal medical practices. Their understanding of tribal life was often based on myths that sometimes led them to make serious mistakes, such as prescribing diets that were so deficient of protein and vitamins that they actually encouraged diseases. For instance, at one stage, milk and milk dishes were not provided to patients because the mine doctors believed that "Bantu tribal laws and customs said that men would not drink milk or eat milky food because milk is baby food and the adult males believed they would become unmanned if it formed part of their diet".[67] Moreover, in common with the South African medical profession, they espoused the germ theory of causation. Medical research in South Africa was long preoccupied with a "parasite and disease-oriented approach in which specific diseases rather than malnutrition and socio-economic conditions took precedence, and where fundamental biological research came to occupy a central role".[68] In the mine compounds, this 'germ theory' approach fostered the belief that the tuberculosis problem could be resolved by removing from the labour force those who were likely to be vulnerable to the tubercular bacilli, by vaccinating all or selected sections of the black labour force as a protection against it, by repatriating those they identified as germ carriers and by treating the sick ones who remained with drugs.

The place of mine doctors in the management structure, however, was their greatest handicap when serving black mineworkers for it confined their formal legitimate attention to factors, which though important in controlling the spread of diseases, were peripheral to tackling their causes. They had to comply with the official policy of the Chamber of Mines that preferred the repatriation of the sick to the rural areas rather than treat them in the mines. Though they invariably talked of the need for 'happy, healthy Natives', this meant providing a supply

[67] Cartwright, 1971, p41.
[68] Farley, 1991, p233.

of workers who were fit for the rigours of underground mining. Their primary concern had to be to improve the efficiency of the mines by reducing the cost of sickness and raising the stamina, and therefore physical efforts, of the mineworkers.

They were compelled, as part of the management structure, to be persistently cost-conscious and to present detailed budgets for any proposals they made. They knew that their suggestions would be regarded as worthless unless they satisfied the financial criteria set by management. In consequence, they too frequently used cost as the criterion to determine whether or not a new health measure should be introduced. It was, for instance, the fact that concrete bunks were reckoned to be cheaper than wooden ones that convinced management to use them.[69] Although mineworkers were critically in need of vitamins A and D in the inter-war years, it was argued that whether or not they could be supplied was a matter of finance as even a slight addition in the cost of an individual ration would aggregate to a very large sum.[70] It was even felt necessary to calculate the expense of issuing soap to wash blankets as a prelude to recommending that it should be done. In 1943 the medical officer for a number of Witbank collieries recommended that other mines should follow the example of his and give mineworkers a free ration of soap to wash their blankets three times a year. The cost, he said, was only £1 a year for every 100 mineworkers.[71]

The management constraints on mine doctors prevented them from raising structural issues, such as the migrant labour system, even if they had thought they were relevant. As it was, they continued to give their attention to diets, meal times, medical checks and aspects of preventative medicine such as the use of vaccines and drugs. Their almost exclusive preoccupation with medical solutions led them to see advances in the quality of health achieved through improving the effectiveness of drugs, increasing the number of doctors employed in the mines, raising the standard of their own qualifications, improving and increasing their research achievements and extending the provision of medical facilities such as hospitals and equipment. In this, within certain cost limits, they were encouraged by the mining houses that liked to display their respective health care facilities and their showpiece hospitals in the public arena. Individual mineworkers who needed treatment for injuries or disease, benefited, of course, from the improved medical

[69]Orenstein, 1922, p6.
[70]Packard, 1989, p171.
[71]Transvaal Mine Medical Officers' Association, *Proceedings*, October 1943, pp61-62.

facilities but medical care by itself had never significantly improved the health of groups of people whether in compounds, communities or societies in general.[72] Measures directed at helping individuals usually had little effect on populations. The consequence of this emphasis by the mining houses was that the mine doctors were able to contain but never eradicate compound diseases and the circle of infection continued its uninterrupted course, eventually bringing within its orbit even more devastating diseases than those it already contained.

[72]See Michael Savage, "The Political Economy of Health in South Africa" in Westcott and Wilson, 1979, Volume 1.

CHAPTER THREE

DAMAGED LIVES

Contact With Home

The lives of the vast majority of mineworkers comprised two distinct spheres of activity and interest, namely their rural tribal homes and the mine compounds. In the mine compounds, the focus of their existence also switched between two distinct spheres, namely work and the hostels. When at work their relationships were with white overseers and members of tribally mixed work gangs while in the hostels they related to their narrow circles of friends and kin and wider social groups of common tribesmen. Until the late 1970s they had rarely been able to establish relationships outside the compounds. Indeed, it was not until then that they had been permitted to break the continuity of their contracts by going home except in cases of emergency and then only for short periods at their own expense.[1] Even compassionate leave was denied them in most instances. Even if they wanted to leave their respective compounds for the whole or a part of their free day, on a Sunday, they needed the permission of their hostel manager. In general, their contact with the outside world was governed almost entirely by the regulations and whims of the hostel managers. Most of them only received outside news when new recruits arrived.

In most instances, black mineworkers did not even have any contact with or news about their families for the duration of their contracts and that, for most foreign mineworkers at least, meant for periods of up to 18 months. About 80 per cent of them were functionally illiterate and could not, therefore, write their own letters home. Sometimes an illiterate person would be able to persuade a literate friend in his hostel to write for him or he would pay a hostel scribe to do it. But this mode of communication inhibited expressions of emotion and intimacy between mineworkers and their families. One wife, for instance, said: "Even when I did receive letters, each letter had a different handwriting and I then realised what a setback illiteracy is. I could no longer be open and frank in my letters to him because they were not private. They have to be read to my husband by someone else."[2] But communications were difficult even for literate mineworkers for the postal services were slow and cumbersome. It was possible for messages to be sent from

[1]Association of Mine Managers of the Transvaal, *Papers and Discussions*, 1942-45, p795.
[2]*Work for Justice*, Transformation Resource Centre, Maseru, Lesotho, December 1985.

homes through local recruitment centres to individual mines but it could take up to two weeks even for a telegram to arrive informing a mineworker of a death in his family. It had been estimated in the mid-1970s that, on average, it took 10 days for a letter to arrive from home and 12 days to get a reply back.[3] Letter-writing was a wholly inadequate way of resolving domestic crises and thus reducing the tensions of hostel life.

The isolation of the hostels was relaxed after the introduction of an 11-shift fortnight in April 1977 for then, for the first time, mineworkers within commuting distance of their homes were able to visit their families on their free weekends. Black mineworkers, on average, had from 60 to 64 free hours, including a Saturday, from the end of their last shift at the close of an 11-day stint to the beginning of the next one, which they could spend either in the local townships or travelling to visit their homes. In this way, some black mineworkers were able to interact on a relatively continuous basis with their families and village events or make acquaintances and become involved in activities in the nearby townships.

The new working arrangement did not apply to foreign mineworkers from the periphery of South Africa who, in any event, lived too far away to go home for weekends. Nor were they satisfactory for the many Xhosa-speakers within South Africa whose homes were inaccessible for weekend commuting. Some of those who could reach their homes were prevented from doing so by the cost as the lowest possible return fare came to about 16 per cent of their monthly wage. It was estimated in 1980 that, overall, about 23 per cent of mineworkers visited their homes for weekend leave.[4] The proportion was higher on festive occasions. For instance, during December 1977 as many as 40 per cent of all the black mineworkers employed by the Anglo-American Corporation in the Orange Free State went home for a weekend.[5]

Those who were best able to take advantage of the leave facility were the mineworkers from the western areas of Lesotho who worked in the Orange Free State mines. It took them about four hours to get to Maseru though, on average, they spent another four or five hours travelling before they reached their homes for many of them lived in isolated homesteads that were poorly served by transport and could only be reached in the final stages by walking or riding a horse. Some

[3]Mine Medical Officers' Association, January-December 1982, p8.

[4]Chamber of Mines, "Black Migrant Mineworkers' Contact with Home", *Human Resources Monitor*, Volume 4, No 5, Johannesburg, 1980.

[5]Lipton, 1980, p163.

wives of mineworkers tried to make it easier for their husbands by moving their homes from the rural mountains in Lesotho to the border areas nearest to the Orange Free State but they often found it difficult to make the transition from their traditional village lives to living in concrete settlements along the main roads or on the outskirts of towns.[6] The amount of time the mineworkers actually spent at home on each occasion varied from 12 to 42 hours. This did not give them much time to adapt back into family routines, resolve family difficulties or undertake homestead chores but it none the less made a qualitative difference to compound life for some of the hostel inmates.[7]

Social Life

The lives of all black mineworkers, except the few who lived in married quarters, were dominated by the lack of any meaningful opportunities for self-expression, the absence of social activities compatible with their cultures, long periods of enforced celibacy and by sheer boredom. Those who survived illness and avoided accidents had to devise some means of coping with the highly regulated existence that managements imposed on them, like animals in intensive farms, in their attempts to make them into more efficient sources of labour. They had little scope for deviating openly from the commands of white supervisors for that resulted in penalties and dismissal so they generally expressed their independence by reacting negatively, even when the orders might have been of some benefit to them. For instance, they engaged in subtle sabotage, refused medical treatment, ignored medical advice, destroyed medicines and, where the penalty for deviation was immediate and painful, practised malingering by responding dolefully as if they had no intelligent understanding of what was required of them. This reinforcement of the stereotype that white miners and managers held of them, was, in fact, frequently a protest against it.

Black mineworkers, in order to retain their sanity, struggled to create a tolerable social existence within severely constraining conditions. Their social activities were structured by the edicts of white compound managers whose task was to "organise and encourage the optimum use of spare time (in order) to keep the men occupied and to prevent

[6]*Work for Justice*, June 1987, pp4-5.

[7]For a description of weekend visits, see C M Laburn and J K McNamara, "Weekend Visits Home by Migrant Workers", *Research Report No 7/81*, Chamber of Mines, 1981. Also see Mine Medical Officers' Association, January-December 1982.

them straying off to nearby Bantu townships, where... they became intoxicated and (were) assaulted or injured in brawls".[8] Sports such as football, rugby, boxing or skittles were organised for them. Competitive war dancing was arranged for Sunday mornings when white miners could take their families and watch. Sometimes, specially selected diversionary films such as American Westerns were shown in the evenings. In some compounds, literacy classes were arranged. A tiny minority, usually Shangaans, spent their spare time earning extra money by repairing clothes with treadle sewing machines or repairing boots. The majority, however, did not participate in any activities and were either spectators, or quietly hiding from view in their hostel rooms or drinking in the liquor outlets.

There were, by and large, two social groups in the hostels. There were the largely unschooled traditionalists and those who had been mission-educated. This distinction was most markedly seen amongst the Xhosa mineworkers. The 'Red' traditionalist Xhosas from the rural areas perpetuated cultural facets of their indigenous way of life, including their Xhosa religion, when they moved away from home. They were not interested in the organisations of the white people and they ignored the associational activities of compound life created by the compound managers.[9] The traditionalists comprised the majority amongst all of the tribal groups. Most black mineworkers, therefore, did not regard literacy in English or Afrikaans as an essential qualification for living their mine lives satisfactorily. They did not need it in their work or social activities and, therefore, had no incentive to acquire it. For instance, in the mid-1960s only between one and two per cent of them had attended the compound literacy classes in one or other of these two languages.[10] They preferred, whenever they could, to live according to their own cultures but were only allowed to do so within the constraints imposed by the hostel managers. The other group comprised those who had attended mission schools. They were literate and were influenced by Western Christian values and aspirations.Amongst the Xhosas these were called the 'school people'. They usually got the most prestigious black mining jobs as policemen, clerks and personnel assistants and rarely worked for long in the unskilled underground jobs. They, too, were the ones who most frequently participated in the compound sports and social activities.

[8]Mine Medical Officers' Association, 22 April 1965, p107.

[9]See Philip Mayer, *Townsmen or Tribesmen*, 1961, for a detailed analysis of the behaviour of the Xhosas away from their rural homes.

[10]Mine Medical Officers' Association, *Proceedings*, 22 April 1965, p107.

The Use and Abuse of Liquor

Beer-drinking played an important part in the traditional ceremonial and social lives of black men in their rural homes. They mostly drank home-brewed beers made from sorghum or maize, or khali made from syrup and a local root. The womenfolk in the homes made these drinks easily and quickly because they required little fermentation. They usually had a significant nutritional value, were low in alcoholic content and were drunk in large quantities. Gradually, from the late 19th century, European-type liquor had become available to black people and formed an increasingly important part of their drinking behaviour. But this was much more expensive and potent than what was called 'kaffir' beer. When the Shangaans began migrating to the mines in large numbers they were already substantial consumers of cheap Portuguese wine and spirits. Many of them were addicted drinkers. Cheap rum distilled in Mozambique and noxious German potato spirits were imported into South Africa and directed at the black population. Various concoctions were made and sold under false labels. It was possible to buy 'Kaffir Whisky', 'Dutch Gin for Kaffirs' and 'Kaffir Brandy' under the reputable Hennessy and Martell labels but which bore no resemblance to the original products in quality. The *Transvaal Leader* had reported in 1899: "Do not be aghast at the prospect of drinking sulphate of copper and green tea, acetic acid and oil of Neroli as Martell's Five Star Liqueur Brandy; you have done it often enough."[11] Some of the concoctions were actually poisonous. A Transvaal chemist described the imported German potato spirit in 1895 as being unfit for internal consumption.[12] The advent of these spirits encouraged a taste for stronger drinks than they had been accustomed to. The women began to ferment sorghum until it became stronger and concocted many potent but quickly and easily made mixtures in the townships.[13] As the drinks became more intoxicating, the drinking became more important for its own sake. Many more black people experienced drunkenness while some of them became alcoholics.

In the early years, the Chamber of Mines attempted to use alcoholic

[11]For the early history of liquor in South Africa, see Charles van Onselen, "Randlords and Rotgut, 1886-1903", *Studies in the Social and Economic History of the Witwatersrand*, 1886-1914, 1982, Volume 1. Van Onselen describes the composition of some of these concoctions on pp56-57.

[12]*Ibid*, p55.

[13]The distillery was located just 10 miles east of Pretoria. See La Hausse, *Brewers, Beerhalls and Boycotts. A History of Liquor in South Africa*, Johannesburg, 1988, p47.

drinks as a means of stabilising the labour force. It encouraged mineworkers to drink in the belief that if they spent their wages in that way they would be induced to extend their contracts. It also regarded giving black migrants the chance to obtain liquor through working in the mines to be as big a recruitment attraction as making guns available had been in the 19th century diamond mines in Kimberley. These factors, plus the knowledge that liquor production was a profitable enterprise, persuaded the local mining capitalists to invest in the Hatherley distillery, which had the sole right to distil liquor in the Transvaal, when it became a public company in 1892. Some mine managers had already put their money into the canteens that sold the liquor.[14] In a few cases they even administered them along with their other tasks.

But this interest in increasing liquor consumption amongst mineworkers soon developed into a contradiction as the mining companies and mine managers recognised that the costs of drunkenness in the mines by far outweighed any benefits liquor might bring as an aid to recruitment. The mine owners estimated that between 15 and 25 per cent of their workers were disabled by drink each day so that in order to maintain a full complement of workers they had to employ many more than they needed. Recruitment, already a problem, became more difficult. As they began to experience the costs of drunkenness they became ambivalent about selling it to black mineworkers. They still wanted them to be able to obtain liquor but in a controlled way, with restrictions on the issue of liquor licences, police action against illicit sales outlets and with the consumption of liquor confined to those with permits which the mining companies themselves issued. However, in July 1896, the Chamber of Mines reversed its policy entirely by advocating the total prohibition of the sale of alcohol to black people in mining districts and surrounding areas.[15] As a sign of their influence over government, the following year a new law came into force on the Witwatersrand that made it illegal for any black person to buy, sell or drink any kind of liquor.

During the subsequent period of prohibition, the mining houses banned the consumption of alcohol in their hostels and encouraged police action against mineworkers who sought to get it illicitly in the

[14]Van Onselen, 1982, p60.

[15]This did not mean that the mining houses severed their interests in the liquor industry. In 1984 Anglo-American had a 32% interest in South African Breweries, Ltd, the largest group of breweries in South Africa, and a 49% interest in Northern Breweries, Ltd, which monopolised the Zambian liquor industry (Innes, 1984, Appendix 2).

townships. Compound managers, however, did not follow the letter of the law and used their discretion as to whether mineworkers should be allowed to brew liquor. They knew that they could attract volunteer migrants by allowing drink in the hostels because black workers sought out those mines with the most lax rules in this respect.[16]

There was a special liquor problem in the coal mines of northern Natal because of the token credit system which operated there. When mineworkers had used up all of their cash they were able to apply to their managers for advances on their future earnings and were given token coins with different values. The tokens were used in the mine stores to buy clothes and food and to purchase liquor in the beer halls. They enabled mineworkers to buy liquor constantly in between paydays even though they had no money. However, they had to settle their accounts with management each payday. Sometimes they were so much in debt that they received no actual pay and had to borrow from one payday to another. Gradually they became so indebted that they had to extend their contracts to clear themselves. Thus by encouraging beer-drinking the coal mine managers succeeded in tying mineworkers to the mines.[17] Some workers in the coal mines, however, were not constrained by compound regulations because they lived in nearby locations where it was relatively easy for them to obtain illicit liquor, provided, of course, that they had cash.

In the gold mines it was difficult for mineworkers to leave the compounds and get liquor in the townships. But it was not impossible. The compound managers could not entirely prevent them from searching for illicit liquor without closing the compounds completely and they were not allowed to do that. Some mines, moreover, had married quarters just outside of the compounds where the wives or partners of mineworkers brewed beer and sold it from their homes, even though the police frequently raided them.[18] But neither the approbation of the mine owners nor the vigilance of the police was enough to prevent mineworkers buying alcohol when they wanted it. Prohibition in South Africa, as in the USA, simply gave rise to a flourishing illicit liquor market. It did, however, have the effect in the mines of reducing

[16]The Native Grievances Inquiry stated in 1914 that some compound managers allowed khali, a fermented liquor in which golden syrup and a local root were the active agents, to be made. It was made much more quickly than 'Kaffir beer' and, though it had a low alcoholic content, it caused much drunkenness. (*Report of the Native Grievances Inquiry*, 13-14, Cape Town, 1914, UG 37- '14, paras 514-524.)

[17]La Hausse, 1988, p52. In the mid-1930s a campaign was launched by church leaders in the coal fields to expose the token system as exploitation and to have it abolished.

[18]*Report of the Native Grievances Inquiry*, 1914, paras 134 and 139.

the extent of drunkenness and, in consequence, of lowering the rate of absenteeism.

The government altered the prohibition law in 1937 when it inserted a clause in the Native Laws Amendment Act that made sorghum beer legal and allowed municipalities to control beer sales either by establishing municipal monopolies, licensing selected black traders to brew and sell it or by permitting home brewing. Most municipalities chose to establish their own monopolies. They built beer halls or 'drinking cages' throughout the country through which they made massive profits.[19] This did not resolve the problem of illicit drinking. The shebeens, where highly intoxicating liquor was provided during the days of complete prohibition, had become an essential part of township life. They grew and spread after the Second World War despite persistent police raids and many arrests for distilling, selling and drinking hard liquor.

The mines had altered their practices after the change in the law in 1937. Some of them had already allowed beer to be brewed in the compound kitchens but, following the establishment of beer halls, they created formal liquor outlets in the compounds. Each afternoon at the end of a shift the bars would be packed with mineworkers buying liquor to take back to their rooms. When the new hostels were built with their centralised dining rooms, beer was issued with the evening meal in an attempt to cut out private and uncontrolled drinking. But private drinking continued. In general, black mineworkers experienced more freedom to obtain and consume 'kaffir' beer but were still constrained in their search for liquor. During the decade of repression that had followed Sharpeville in 1961 it became more difficult for them to leave the compounds and when they did, if they were looking for drink, they were in increased danger of being arrested.[20] Moreover, as their wages had barely changed from about 1948 till the early 1970s, while the cost of living had risen, they had very little spare cash for drink.

This situation changed after the wages explosion in the 1970s. A survey of the drinking habits of black mineworkers in the Free State

[19]Some municipalities used the profits to finance local health and social services for black people. Dr A B Xuma, the President of the ANC, attacked that policy. He said it meant that, "if Natives want milk for their children, they must drink more beer. If they want adequate health and social welfare services from the municipalities with beer halls, they must drink beer." (La Hausse, 1988, pp49-50.)

[20]When visiting a hostel in the East Geduld gold mine in 1962 the author was told that no black mineworkers in that mine were allowed to leave the compound in between contracts.

Geduld mine in 1977 revealed that 72 per cent of the black labour force drank alcohol. That was not unusual amongst manual workers in heavy industries and the rate was, moreover, lower than it had been in 1930.[21] What was unusual was that almost a half of those who drank, comprising roughly one-third of the black labour force, did so until they were inebriated. Each drinker consumed, on average, about 50 litres of beer and spirits each month.[22] Most of this was sorghum beer but the intoxicating Western-type beers, spirits and wines, drunk by the whites, were becoming increasingly popular, particularly amongst the monthly-paid black staff. The problem was that mineworkers often switched to the more intoxicating drinks without altering the drinking habits they had acquired at home where heavy drinking was a social custom that did not carry any stigmas. Though they drank during the week, they usually stayed sober. On their free days or weekends, however, their tendency was to drink without a break, as they would have done at home. Their drinking excesses were often marred by violent fights and followed by absenteeism.

The increase in drunkenness resulted from a combination of factors. Mineworkers had more money to spend on drink, which, in turn, was more freely available. They used drink to relieve the stresses and strains of mining life caused by the increasingly hard and exhausting work as the mines got deeper, hotter and generally more dangerous. It was an escape from the onerous restrictions in the hostels, the undiluted boredom of being confined in the same rooms with the same people for months on end, the long separation from their families and worries about their welfare. There were added strains in the 1970s with new, dramatic escalating violence in the hostels, political and violent turmoil in the townships and political repression in their homelands. The partially self-governing bantustans, created in the former reserves after 1959, had become centres of financial corruption and political oppression as the collaborating chiefs and rich merchants fought each other for the pickings.[23] This situation created additional anxieties for the absent mineworkers and led to the emergence of new alliances amongst them. Thus from the confines of their hostels, they saw a highly complicated scenario outside which led them not simply to try and obtain some solace from drink but to use the right to drink as and when they wanted

[21]Native Economic Commission, 1930-1932, *Report*, UG22, 1932, para 770. It was estimated that 80 per cent of the mineworkers in a Johannesburg mine had drunk alcohol in that year.

[22]Mine Medical Officers' Association, *Proceedings*, January-August 1978, pp29-33.

[23]John Pampallis, *Foundations of the New South Africa*, 1991, pp186-188 and pp233-234.

as an expression of political dissent. Going to the illicit shebeens and drinking illicit hard liquor were acts of political defiance in the environment of the late 1970s.

Some mine managers tried to reduce the incidence of drunkenness by changing the atmosphere in the liquor outlets from that of drinking dens to places where mineworkers could also eat, talk and play. They encouraged mineworkers to eat whilst they drank by offering snacks for sale in the bars. They installed game tables, such as draughts and marabaraba, and provided television sets as distractions. Educational programmes about the evils of drink were introduced during the induction processes and at various times afterwards. Video-cassette educational programmes were supplied with the televisions. Anti-drink propaganda films were shown at the mine cinemas and messages about the evils of drinking alcohol were relayed over the public address systems in the hostels.[24] Black social workers were appointed in some hostels to counsel those with drink problems. Managers set in place procedures to identify such people by their standard of work, their response to discipline, their state of health and their general social demeanour. It was then decided whether they should be medically treated or disciplined.

Excessive drinking, however, persisted. It was estimated that one in every 10 heavy drinkers became an alcoholic. The admission rate for men under the influence of drink at one of the mine hospitals in the Orange Free State rose 11-fold between 1973 and 1977.[25] But this represented only a part of the problem. The money they spent on drink was money that could not be sent to their families at home. Indeed, worries about the plight of their families at home that tended to encourage mineworkers to drink in the first place, often resulted in their families being worse off.[26] Some of those who drank excessively, and did not send remittances to their families, deserted them altogether whilst they were in the mines. But it was difficult for them to cut themselves off from their homes entirely because at the end of their mineworking lives their only source of social security lay with their families at home. Without work, debilitated by disease and disabled by drink, they often had no other alternative but to return in the hope that their behaviour would be forgiven and that their wives would take them in again. The

[24]Mine Medical Officers' Association, *Proceedings,* January-December 1982, pp1-2.
[25]*Ibid*, January-August 1978, p32.
[26]Colin Murray examines the economic effects of the migration of men from Lesotho on the families left behind in detail in his book *Families Divided. The Impact of Migrant Labour in Lesotho*, Cambridge, 1981. Murray does not deal specifically with mineworkers though the majority of Lesotho migrant males went to the mines.

heavy drinkers who returned in between contracts took their problems with them and disrupted their family and village activities with their unruly behaviour. All in all, this neglect by some mineworkers of their families exacerbated an already precarious situation by creating a new sub-standard of suffering. One mineworker's wife, whose case contained many common elements, described her experiences in the following terms:

> As a migrant's wife, life is very fearful, lonesome, unpredictable and difficult. As a Mosotho woman, I battle daily getting water, gathering firewood from the eroded hills, and weeding and harvesting from the infertile fields which have been badly ploughed and stricken by drought. I get very little from the fields ... My dear husband left promising to write to us, to send us money and to visit us as regularly as he could afford ...In the past I often waited six or more weeks, going down daily to our local shop where we collect our mail, for a letter and money from my husband. Sometimes when a letter came ... there was a yellow piece of paper enclosed (a TEBA Remittance voucher). The voucher was a promise that money from my husband would eventually come. In Sesotho the voucher promised that if my husband worked in the mines, I could not be poor. But in English it stated that I should wait for 10 days before I could go to the remitting office to cash it. But I always needed the money immediately for my family. That is why I often went early to the remitting office – (a bus drive away) – only to be turned back and told to return on the correct date. Once my husband came home. We could no longer understand each other. While I talked about our present drought problems and the forthcoming Balimo ceremony in my family, he talked about buying a black and white television set. He continually complained about ... how hard it was sleeping on the floor. What is it that the mines have taught him? ... I can no longer complain to him because he had been gone now for two-and-a-half years, and during that time he has never written or sent us money. I am now too busy trying to make ends meet. My life is centred around my five children whom I am trying to bring up alone ... I am now one of the poorest women in this village, which is already poor ... My heaviest burden now is having to support a permanently disabled husband, returned to me, discarded by the mines. He no longer has any skills suitable for village life ... How shall we survive right here in this village? Is there a worse fate than this?[27]

Relationships Within the Hostels

As in all large single-sex institutions where people are confined for long periods, the mineworkers tried to cope by making meaningful relations with each other. They did so by forming networks of relationships within the same tribal group, often from the same village and sometimes with their own kin. The congestion in the hostel rooms, the complete absence of privacy in all their activities and unbroken time in each other's company, tended to create strong ties between some of

[27] *Work for Justice*, No 9, December 1985.

them. The constant physical intimacy that was forced on them by their circumstances sometimes resulted in them engaging in sexual activity. According to some reports, it was commonplace and was practised without stigma or embarrassment in the rooms in the presence of others. If a mineworker visited a room where homosexual activity was taking place, it is likely that he would ignore it and act as if it were normal.[28] The African Mine Workers' Union stated in its evidence to the Witwatersrand Mine Natives' Wages Commission in 1943 that sodomy was common in the hostels and that the returning mineworkers were introducing it into the reserves.[29]

Homosexuality in the hostels, however, could be an exploitative activity, reflecting power relations amongst the mineworkers, in much the same way as heterosexual activity was in the townships. Some young mineworkers made money out of it by prostituting themselves. They were described as the 'women of the compound' and charged money, blankets or other goods for their services. There was also a form of homosexuality that was confined to the senior blacks, the boss-boys, *indunas*, police and clerks who had some powers of patronage and were able to give rewards and impose punishments. They used their powers to obtain 'surrogate wives' from amongst the youngest mineworkers. These young men acted as sexual partners at night and provided domestic services, such as washing and ironing clothes, cleaning, cooking, running errands and a variety of odd jobs, during the day.[30] The sexual activity they practised was copied from the sexual pre-marital play of Nguni adolescent boys and girls.[31] It did not involve penetration and was confined to thigh contact. The senior mineworker who had a 'surrogate wife' would lie on him and move and ejaculate between his thighs. A 'surrogate wife' was required to be passive and not to respond or reciprocate in any way during sexual activity.

The senior mineworkers who engaged in this practice sought out novice recruits who were young, fat and generally looked like women,

[28]*Another Blanket*, An AIM Report on the Migrant Labour Situation, 1976, p22.

[29]Allen, 1992, p467.

[30]The evidence about 'mine marriages' is largely anecdotal. Mineworkers knew that they occurred though the participants usually acted discreetly and the matter was not discussed either in the mines or at home in the villages. The most detailed examination of the phenomenon is "Migrancy and Male Sexuality on the South African Gold Mines" by T Dunbar Moodie in the *Journal of Southern African Studies*, Volume 14, No 2, January 1988. The practice of having young boy servants playing the surrogate female role in Rhodesian mines is described by Charles van Onselen in *CHIBARO, African Mine Labour in Southern Rhodesia* 1900-1933, 1976, p176.

[31]See T Dunbar Moodie with Vivienne Ndatshe, *Going for Gold*, 1994, p121.

and then engaged in ritual 'mine marriages' with them. The liaisons were not necessarily temporary and the partners were not casual. They usually lasted for the period of a contract and were sometimes extended from one contract to another. In some cases the partners extended their mine contracts simply in order to maintain their relationships. They were usually terminated when the 'surrogate wife' felt he had outgrown the role, wanted a partner for himself, desired to go with women in the townships or had accumulated enough money to enable him to leave mining. But this did not mean that the liaisons were strictly voluntary. When young boys first entered mining they were vulnerable, susceptible to pressure and in awe of authority. Some young recruits, particularly Shangaans amongst whom the practice of 'mine marriages' was relatively common, anticipated being propositioned when they arrived for work in the mines. In any case, they were usually given little opportunity to refuse the advances by the senior blacks who bribed them with money and the offer of easy jobs. They would invariably be well looked after with food, presents, money and protection. It was known for an *induna* or boss-boy to hand over from one-half to the whole of his wage to a 'mine wife'. It was a quick way to earn money for those who wished to buy cattle or improve their homesteads and some accepted the surrogate role for that purpose. If they rejected advances, however, they could be harassed and subjected to a complex and subtle form of coercion.

The Early Relationship Between Mineworkers and Township Women

In the early years of mining the migrants, both black and white, had been overwhelmingly male, resulting in a marked imbalance between the sexes in all of the mining areas, including Johannesburg itself. In 1896 there were almost two white men for every white woman in the most developed part of Johannesburg but for blacks the imbalance was very much greater. It was estimated that in that year there were 40 855 black men living within three miles of the Johannesburg mine compounds compared with only 1 678 black women. Thus for every woman there were 24 men.[32] The disparity between the sexes narrowed as increasing numbers of women moved illegally into the townships but

[32]An authoritative account of prostitution on the Witwatersrand up till 1914 is given by Charles van Onselen, 1982, Volume 1, Chapter 3.

it remained significant. In 1948 there were more than three men for every woman throughout the urban areas of South Africa in the age group 20-49 years. The disparity was slightly greater for mineworkers in Johannesburg where there were at least four men for every adult woman. Along the Rand the imbalance was greatest in the mining town of Springs.[33]

But this was only part of the story for whereas most of the men were unattached, a majority of the women were not. More than half of all the women in the mining townships in 1948 were married and were living with their husbands though this did not necessarily mean that they did not associate with migrant workers. There was a great fear amongst men married to relatively young women that their wives would be seduced by visiting mineworkers. The townships were generally regarded as sexually promiscuous places though in reality there was little difference in this respect between them and the rural villages.[34] Still, many of the married women did live monogamous lives, thus reducing the opportunities for mineworkers to contract casual liaisons. Only a quarter of the total number of men in the townships were married with the result that there were six times as many unattached men as there were unattached women. These proportions narrowed in the years following 1948 but with the persistence of the migrant labour system, the imbalance between the sexes and the consequent practice of casual sex persisted.

The intense competition for the company of women had inevitably attracted prostitutes into the mining areas. In the early years of mining some of them were black but mostly they were white immigrants who came from Western European countries, either voluntarily as working prostitutes or involuntarily through an organised white slave trade. Johannesburg had almost as many white prostitutes as there were black women living in the whole of the mining districts around it. Indeed, it was estimated that in Johannesburg itself as many as one in every 10 white women over the age of 15 was a prostitute.[35] Many of the white prostitutes went with black mineworkers and this caused sharp white establishment reactions against sexual intercourse between black men and white women in the Cape, the Transvaal and Natal. Prostitution had been made illegal in the Transvaal in 1897. The Morality Act in the Cape in 1902 made sexual intercourse between black men and white

[33]S L Kark, "The Social Pathology of Syphilis in Africans", *South African Medical Journal*, 23 (5), Jan 1949, p80. The figures were based on the 1936 census returns.
[34]Mayer, 1961, p253.
[35]*Ibid*, p104.

prostitutes illegal while in the Transvaal in the following year a similar Act made it illegal for a black man to have sexual intercourse with any white woman and not just prostitutes.

Prostitution, however, had continued and flourished. Until the expansion of black townships with their own female populations, it was the main source of sex for black men away from their homes. Recruits to the mines met prostitutes around their local recruitment offices, on the trains to and from the main recruitment centre in Johannesburg, in the streets of the city and in farms near to the mines. Prostitutes were particularly numerous on the return train journeys for they knew that the mineworkers would have been without female company for some months and had cash with them. Prostitutes also plied their trade around the mine compounds. White prostitutes from overseas had been displaced after the Anglo-Boer war by young Afrikaner women who sought to alleviate the desperate poverty of their families but they were soon outnumbered by black women as poverty spread in the reserves. By 1910 there were from 200-300 full-time black prostitutes operating around the mine compounds on the outskirts of Johannesburg.[36] Prostitution around the mining compounds became a permanent by-product of the migrant labour system.

The Plight of Unattached Women

The women whom black mineworkers met in the townships were usually those who had migrated to escape from the poverty caused by being unattached. They were either not yet married, widowed, divorced, abandoned or had left their migrant husbands because they no longer received remittances from them. Many had small children to maintain. They were all marginalised in their village societies and, as a result, were amongst the poorest members of their communities. For women, marriage meant a lifelong means of economic support for through it they derived rights to subsistence from the farms, to their husbands' remittances and to the labour of their children. Marital failure for them was inseparably related to economic distress for without marriage they were dependent upon their fathers or brothers for charity.[37] But even with marriage they faced hardships if their husbands were migrant workers for they had to manage rural households that

[36] *Ibid*, p146.

[37] See Judy Gay, "Wage Employment of Rural Basuto Women: A Case Study", *South African Labour Bulletin*, Vol 6, No 4, November 1980, p43.

were generally incapable of providing subsistence for their families. In the late 1970s as many as 70 per cent of the households in Lesotho were managed by women in the absence of their migrant worker husbands.[38] Cash remittances from South Africa, then, were critical for those women. It was an irony that when the cash did not arrive or was inadequate because their husbands had squandered the money on drink or township women they were sometimes forced to engage in the same kind of money-making extra-marital relationships as the township women in order to survive. Some simply became prostitutes in their villages even though their tribal elders disapproved of it.

There was little reason, then, for divorced and abandoned women to remain in their rural homes. But if they left to earn cash in the townships they were disadvantaged by the fact that they were classified as illegal immigrants and had to live illicit lives. Following legislation in 1952 all women, as well as men, had to have permits to be in a town for longer than 72 hours. The law was applied unevenly but in any event it created great uncertainty amongst them for at any time they could be arrested, screened and deported back to the rural areas.[39] From January 1959 the full range of influx controls was applied to women. Thus, leaving their rural homes, entering urban areas and getting legitimate employment became almost a legal impossibility for women which meant, in effect, that when they migrated they moved illegally and had to earn money illicitly.

Relatively few Xhosa-speaking women were prepared to risk the long illicit journey from their homes in the Transkei to the mining areas but, in any event, they were discouraged from migrating by their tribal authorities. The few who did migrate went mainly to the nearby Eastern Cape townships. The majority of the women living near the mines were from the Sotho-speaking areas, mainly in Lesotho, which were adjacent to the mines in the Orange Free State. They suffered from the general restrictions imposed on all black South African women until 1963 when migrants from Lesotho were defined as 'foreign' and they were subjected to exceptionally restrictive entry conditions at the newly established border posts on the Lesotho and South African frontier. Though women from Lesotho were not allowed to enter South Africa legally, many continued to make the illegal trip over the border. The new restrictions, however, had a dramatic effect upon the proportion of

[38]Colin Murray, "The Effects of Migrant Labour: A Review of the Evidence From Lesotho", *South African Labour Bulletin*, Vol 6, No 4, November 1980, p26.
[39]See Ellen Hellman, "The Application of the Concept of Separate Development to Urban Areas" in K Kirkwood (ed), *St Anthony's Papers*, Volume X, *African Affairs*, 1959.

township women who came from Lesotho. In 1956, only 27 per cent of the 155 000 Lesotho migrants in South Africa were women. This percentage had fallen to 17 by 1966 and to 15 by 1976. It continued to fall thereafter, though not so dramatically.[40] At the same time, an increasing number of Xhosa-speaking women migrated to the townships and began to take the places of those from Lesotho.

Those women who entered the urban locations illegally lived clandestine lives. They had to hide away from the police. They were afraid of visiting the town centres in case they were asked for their passes or reference books. Once in the townships the women from Lesotho in particular rarely left to visit their homes because they risked arrest on the way back and could be prevented from re-entering South Africa.[41] They were unable to get legal employment because employers faced prosecution for employing migrants without valid documents. New regulations were introduced in 1979 that increased the fines for those who employed 'illegal blacks' and provided for the repatriation of the women concerned. This placed those women who managed to get wage work in a highly exploitative situation. Many were forced to work long hours for the lowest of wages under the threat of being reported to the authorities unless they complied.

Many women in the townships, therefore, existed in the twilight zones where everything was illegal. They set up as petty traders, brewed beer, operated illicit stills and engaged in prostitution. Relatively few of them, however, became full-time prostitutes. Some women were encouraged to engage in prostitution by white farmers who provided facilities for squatter settlements called *mekhukhu*. There were a number of such settlements around Welkom. The women lived in tin-roofed pig pens or one-roomed brick houses or wood and corrugated iron structures or four-roomed houses where each room was occupied by two women and their male companions. They paid rent to the farmers and were allowed to entertain mineworkers with alcohol, food, music and sex. The police often turned a blind eye to these activities and did not harass the women.

Most town women, however, subsisted through a system of extra-marital relationships that was called *bonyatsi* in Sotho. These women became the township partners of migrant workers for whom they cooked and cleaned and with whom they had sexual relations in return for a degree of financial security and companionship. Some women had

[40]Gay, 1980, p40.

[41]*Ibid*, p48. Gay estimated that in 1977 male Basuto migrants returned home two or three times a year whereas the women got home, on average, once every two years.

a main partner and liaisons with two or more others who gave them presents and money, though they were not regarded as payments for sex.[42] There was a strict order of priority between the main and subsidiary partners which the men themselves respected. A subsidiary partner, for instance, would be discreet and not challenge the rights of the main one. In fact the township women developed a complicated set of fairly routine sexual relationships with migrant workers that involved changing partners with varying degrees of frequency.

Not surprisingly, in the rural areas there was a strong antipathy towards town women, whom mine recruits were advised to avoid. But there were additional constraints. According to traditional Xhosa philosophy, both men and women were entitled to sexual satisfaction whether or not they were married. The Red Xhosa men believed that husbands could take lovers openly without endangering their marital relations and without feeling any sense of guilt or shame. This meant that they freely engaged in extra-marital relationships though they would not go with anyone.[43] When they went into the mining townships they preferred to meet Red Xhosa women but there were very few of them there until after the Lesotho women were declared prohibited immigrants in South Africa. This constraint was reflected in the relatively low incidence of syphilis amongst Xhosa mineworkers. At the WNLA centre in Johannesburg in 1930 only two per cent of them were tested positive for syphilis whereas the Sothos, who were preferred by the Sotho township women, had an incidence of almost 30 per cent.[44]

Apart from these tribal differences, however, mineworkers who wanted to meet women for sex could do so with very little trouble provided they had money. They met prostitutes *en route* to and from the mines and in the bars and shebeens in the townships where they went for liquor. There was a strong connection between liquor-drinking and prostitution. Women were employed in some of the shebeens as a means of attracting men to them so that even if mineworkers went there to socialise and drink there was the likelihood that they would end the

[42]See Karen Jochelson, Monyaola Mothibeli and Jean Leger, "Heading for Crisis? The Implications of the Migrant Labour System for the HIV Epidemic", Sociology of Work Programme, University of the Witwatersrand, 1988, for an examination of the pressures that caused women to enter into different kinds of sexual liaisons with black mineworkers.
[43]See Mayer, 1961, Chapter 16.
[44]Kark, 1949, p78. See also S V Humphries, "A Consideration of the Advisability of Treating Syphilis in Mine Natives", Transvaal Mine Medical Officers' Association, *Proceedings*, February 1937. Humphries reported that tests taken in 1927 showed that 29,45 per cent of Sotho mine recruits had syphilis.

A waiting place for wives and girlfriends at Western Holdings,
No 4 shaft.

evening in the company of women. After the compound restrictions were relaxed from the mid-1970s, 'waiting places' were designated by the mines just outside the compound gates where wives could stand around and wait for their husbands. Other women also collected there in the hope of meeting mineworkers to arrange to engage in casual sex for money. Some of them were regular prostitutes. The 'waiting places' were especially busy on paydays. Not all of the prostitute/client relationships, however, were one-off affairs for some mineworkers developed regular relationships with prostitutes and paid them monthly amounts.[45] On the other hand, a mineworker might give a woman a sum of money which entitled him to visit her many times.[46]

All the other relationships that mineworkers established with women involved varying degrees of commitment and obligation. Sometimes, it happened that they met women in bars and shebeens and established short-term friendships with them that were in addition to any long-term relationships they might have with other women. But, in general, it was difficult for those who made only fleeting visits to the townships to form relationships with women except through the medium of kin or friends. Irrespective of how they met, however, there was usually a financial nexus between them. Mineworkers normally gave their casual female friends gifts or relatively small monthly sums of money. Where they became the 'main lovers' of town women and lived as their husbands for the duration of their contracts, they paid sufficient money each month to maintain the women and their children and received domestic services and sex in return. In some instances, they also managed to have two or more casual affairs at the same time for which they paid out small sums. Those who did that, of course, spent almost all of their earnings on sex and had none left to remit home.

The Spread of Sexually Transmitted Diseases

The spread of venereal diseases had been the inevitable consequence of the advent of prostitution in the early years of mining in Johannesburg. Syphilis was firmly established there by 1895 yet, like tuberculosis, it had been unknown in South Africa before it had been colonised by the Europeans.[47] There was no word for syphilis, for instance, in the Nguni

[45] *AIM*, 1976, p22. The monthly payments provided sex at a discount.

[46] Jochelson, Mothibeli and Leger, 1988, p24.

[47] In other parts of Africa, such as East Africa, which had been exposed to Western contacts through Arab traders and white slavers, syphilis was reportedly widespread before

language other than *isisfo sabelungu*, meaning 'the disease of the white men', or *isifo sedolopi*, meaning 'the disease of the town'. Syphilis arrived with white immigrant men and women and was largely confined to their communities until white men transmitted it to black women and white prostitutes infected black men. Then black women and black men infected and re-infected each other.

It was, then, through contacts with prostitutes, that black mineworkers had first become infected with syphilis and gonorrhoea. They had carried the diseases back to the reserves and passed them on to their wives and girlfriends. It was rarely the other way round. Married men were hardly ever infected by their wives but by extra-marital affairs. Women with the diseases then passed them on to others in their communities, including their babies. A study of expectant mothers in the Polela Health Unit in Natal between 1941 and 1946 had revealed that more than 35 per cent of them were infected. In the same district of Natal, 21 per cent of their babies had congenital syphilis and were born with physical defects, such as blindness and a susceptibility to other infections.[48] In 1938 and 1939, it was estimated that 23 per cent of black children in both rural and urban areas throughout the country showed signs of suffering from venereal diseases. A random sample of school children in Bloemfontein revealed an incidence of 43 per cent. Some children were carriers of a latent form of syphilis which, later in life, they passed on to others. As a result, it was not long before sexually transmitted diseases became endemic in both the rural areas and the black townships. The overall consequence was that, by 1948, 25 per cent of the black population was said to have chronic syphilis.[49] This led to the comment the following year that "the incidence of syphilis in the Union is enormous ... Not only are we dealing with a large mass of latent syphilis in the African populations, but also with a very high incidence of new infections each year. This process is taking place in highly urbanised areas, as well as in the more remote rural districts ... Few countries can have a higher incidence of the disease than has South Africa."[50] The reason was not simply because the migrant labour system acted as a transmission belt but also because the government took no preventative action to stop the diseases spreading

any European settlement. See Lord Hailey, *An African Survey,* 1938, ppl 143-1 146.

[48]See Kark, 1949, for details of the Polela studies.

[49]Jochelson, Mothibeli and Leger, 1988.

[50]Kark, 1949, p77. Other southern African countries were severely affected. In Botswana in 1932 it was calculated that between 60 and 80 per cent of the total population was infected with a venereal disease.

amongst black people. As early as 1937, a mine medical officer had stated that he knew of no other civilised community that neglected venereal disease, as did South Africa amongst its black population.[51] Government neglect persisted and the scourge continued its momentum through to the 1980s, as a number of small independent studies indicated.[52] This transmission belt of disease, of course, involved all migrant workers but black mineworkers figured prominently because they, more than any other group, were regularly rotating between their work places and their homes.

Before the invention of penicillin, the treatment of venereal diseases took up to two years, was relatively expensive and, in South Africa, was not available to black people. Moreover, it was not always successful. Penicillin, which became available in the post-Second World War period, was a revolutionary remedy that was effective against both syphilis and gonorrhoea. It began its life as an antibiotic and, after an experimental stage when its effectiveness was questioned, was used extensively from the early 1950s. Penicillin could make syphilis non-contagious with a single injection though it needed a course of treatment lasting up to three weeks to effect a cure.[53] Gonorrhoea, however, could be cured by a single injection. With prompt attention and consistent application, therefore, it had become possible to eradicate both syphilis and gonorrhoea as major contagious diseases.

Indeed, that was what happened in the developed countries where it was possible to finance the extensive use of penicillin and which possessed all the medical facilities and infrastructure necessary to effect cures. In order to administer the treatment successfully the patients had to be confined in hospitals or clinics.[54] None of these conditions could be met in black Africa. The position in South Africa was confused by its segregation policies. White people received treatment that was as effective as in the Western countries but penicillin was considered to be much too costly to be used on black patients. In any event, in the rural areas and the black townships there were virtually no basic health care facilities, let alone special clinics, for the administration of the treatment of sexually transmitted diseases.

In the years following 1948 black mineworkers had received privileged treatment for venereal diseases compared with that given to

[51]Humphries, 1937.

[52]Jochelson, Mothibeli and Leger, 1988, p16.

[53]See James Marshall, "The Present-Day Treatment of Syphilis", *South African Medical Journal*, 29 January 1949, pp84-85 for a discussion of different methods of treatment.

[54]By the end of the 1970s the incidence of secondary syphilis in the United Kingdom was

black people in the country at large for they had access to a medical infrastructure, including hospitals, which was denied to those outside the compounds. The mine facilities, however, were still inadequate for their needs because the mine hospitals were able to treat only about half of those who reported the disease, either as in-patients or out-patients.[55] These cases were usually in advanced stages of the diseases because black mineworkers were generally reluctant to seek treatment from mine hospital clinics associated with management until it was obvious that they had no alternative. Their reluctance was spurred on by the belief that if managers knew they were infected they were likely to lose their jobs. Some of those who were unwilling to report to the clinics chose to preserve their anonymity by getting private treatment from general practitioners. However, neither cure nor avoidance seemed particularly important to them compared with other dangers in their lives.

The statistics relating to sexually transmitted diseases, though underestimates, were the only guide to trends in infection and they showed a significant overall downward movement after the introduction of penicillin. By the end of the 1970s the average morbidity rate for syphilis and gonorrhoea was about half of what it had been before the introduction of penicillin. There were, however, sharp variations between individual gold mines and between the mines for different minerals. In the gold mines, the incidence rate for gonorrhoea in the late 1970s varied between 8 per 1 000 employees at Gold Fields' mines to 41 per 1 000 at the Anglo-American Welkom mine and 39 per 1 000 at Western Deep Levels.[56] The rates for syphilis were generally lower in all of the gold mines. The highest incidence by far was in the coal mines of Natal followed by the Kimberley diamond mines. In the Anglo-American coal mines in Natal, for instance, the morbidity rate for sexually transmitted diseases varied between 109 per 1 000 and 134 per 1 000 in the years 1978-1982. This showed little improvement over the pre-penicillin position. On the other hand, the Transvaal coal mines had rates as low as those prevalent in the gold mines.[57]

down to 0,03 for every 1 000 people while in the USA it was 0,11 for every 1 000 people. The rates for gonorrhoea were slightly higher at 1,1 and 4,2 per 1 000 people in the UK and the USA respectively. The comparable statistic for the incidence of gonorrhoea in the large cities of Africa, into which category South African townships fell, was 100 to 300 cases for every 1 000 people. (Jochelson, Mothibeli and Leger, 1988, pp18-19.)

[55]Mine Medical Officers' Association, *Proceedings*, May-December 1980, p11.

[56]*Ibid.*

[57]Jochelson, Mothibeli and Leger, 1988. The figures in this monograph were conservative estimates because they did not include those mineworkers who went for private treatment and whose cases were not, therefore, recorded. The rates given by the Anglo-American

The Hidden Damage

The extent to which the lives of black mineworkers and their families had been blighted by injuries, diseases and alcoholism caused and facilitated by the perpetual rotation of mineworkers between their economically deprived rural homes and the congested mine hostels was hidden from the public view and from the workers themselves. It was concealed, in the first instance, by the sheer size of the labour force and the conditions under which it was employed, involving vast movements of workers every month.

In 1982, approximately 700 000 black mineworkers were employed in the mines. Vital statistics, however, had been collected showing the incidence and distribution of damage to the lives of only the 480 000 of those who were in actual employment in mines affiliated to the Chamber of Mines. In those mines the statistical picture was highly confusing. Between 31 000 and 48 000 had left every month from January to December and had been replaced to varying extents by monthly batches of recruits made up from mineworkers who were either renewing their contracts or were entirely new entrants, described as novice recruits.[58] The novice recruits, totalling more than 65 000 in 1982, took the places of those who had retired through age or who had preferred not to return to mine work; the 21 000 who had deserted; the 2 000 who had died or had been killed; most of the 13 000 who had been injured in accidents and an unrecorded number who were sick. These rough figures do not give the full picture of the damage done because they do not take into account the sick and injured who had struggled to continue working because they feared losing their jobs if their ailments became known to the mine managers. Nor do they include any information about the damage done to the lives of the approximate 300 000 migrants who had worked each year in the relatively small, remote mines not affiliated to the Chamber of Mines.

Those who left the mines, never to return, were dispersed unobtrusively to their remote villages throughout southern Africa. Only their families and neighbours knew their state of health and even they often had limited, sometimes misleading, knowledge. The casualties did not return with case notes showing their medical histories. Nor did they have further diagnoses and treatment. In many instances they knew

Corporation for its coal mines in Natal were 128,04, 143,01 and 123,55 per 1 000 employees for the years 1978, 1979 and 1980 respectively.

[58]The Employment Bureau of Africa (TEBA), *Report and Financial Statements*, 31 December 1982.

nothing because they suffered from diseases with gestation periods of from 10-20 years and had returned with no immediately visible symptoms. When all of these people left the mines, as individuals they moved into obscurity. When they died that obscurity became absolute because the causes of their deaths were not recorded. Their deaths were not even registered. There were no death certificates for them. None was given a post-mortem. Indeed, obscurity was the logic behind the mine owners' policy of repatriating ailing mineworkers back to their homes because it meant that no questions were ever asked about the responsibility of the mine owners for the damage done to people's lives. But even in aggregate they carried little significance to the mine owners because unknowing sons, brothers and friends of those who had been forced to return replaced them with ease. Only if the impact of the damage had caused labour replacement problems would the mining houses have been compelled to examine the causes. As it was, apart from deaths and injuries from accidents, they published no figures from which the scale of the damage to the health of black mineworkers could have been calculated. Nor did the government ever undertake a survey of the sick and disabled in the rural areas from whence black mineworkers were recruited and to which they returned. There were, of course, no vital health statistics on which such surveys could have been based. As for the mining industry, partly in deference to the mine owners and partly because it did not consider the matter of importance, the government let the damage continue from year to year in obscurity.

The result was that there was no public knowledge of the damage done to the health of black mineworkers in pursuit of mineral production and, therefore, no public consciousness about the issue. Given the attitude of the white establishment towards black people at that time, this lack of public awareness was relatively unimportant. The important factor in this respect was that the awareness of black mineworkers themselves was highly fragmented. In general, they were as ignorant of the scale of the sacrifices they were making for the mining industry and the national economy, as was the general public outside. It was not until they were able to organise collectively that they became aware of the enormity of the burdens they shared. Their collective awareness stimulated investigations into the extent and development of those burdens and gave them an historical perspective. Through this process, a collective consciousness began to emerge about mine safety, the ravages of contagious diseases and the social consequences of the migrant labour system, which overcame the structural causes of fragmentation and which was not dependent on the whims and self-protecting policies of either the mine owners or the government.

Part Two

THE NATIONAL UNION OF MINEWORKERS
AUGUST 1982 - APRIL 1994

CHAPTER FOUR

THE FOUNDATION OF THE NATIONAL UNION OF MINEWORKERS 1982-1985

Black Mineworkers and the Spread of Trade Unionism

On 30 June 1982, the Chamber of Mines announced that the starting wages for black mine surface workers were to be raised by 11 per cent and those for black underground workers by 12 per cent. This was a continuation of the ritual that had taken place for many decades. Each year, at about the same time, the mine owners decided in consultation with each other what wages they were going to pay black mineworkers. They had never asked for the opinions of black mineworkers. Nor had they ever revealed the deliberations of the Gold Producers' Committee that led to the announcement of one level of increase rather than another. The figures, however, usually had some logic, at least for the mine owners. The eight white mining unions, represented by the Council of Mining Unions, had recently demanded a 16 per cent increase in the wages of their members but had been offered only three per cent. The unions had reacted by threatening to strike and had forced the Chamber of Mines to compromise. An offer of 12 per cent was made and accepted. That figure had guided the Gold Producers' Committee in its deliberations over black mineworkers' wages.

Although the black wages ritual had continued without interruption, its form had changed. Cracks in the unanimity of the mine owners about black wages had begun to appear since the mid-1970s and were pronounced by 1982. They already interpreted the wage rates announced by the Chamber of Mines to match their own respective policies regarding payments for skill and seniority. Some preferred to build up the base rate while others chose to increase payment for skills and supervision. On this occasion, there was serious disagreement about the size of the increases pronounced by the Chamber of Mines. Three of the seven affiliates of the Chamber of Mines, namely Gold Fields, Gencor and Anglo Vaal, gave their black employees the precise increases pronounced by the Chamber of Mines while Anglo-American, Johannesburg Consolidated Investment (JCI) and Rand Mines increased their base rates by 19 per cent for surface workers and just over 15 per cent for underground workers.[1] Those black mineworkers who were paid the lower rates promptly expressed their dissatisfaction. On 2 July, two days after the wages announcement, strikes began at the Gold Fields mines at East and West Driefontein in Carletonville and at Gencor's Stilfontein and Buffelsfontein mines in Klerksdorp and Grootvlei mine near Springs. Altogether about 40 000 mineworkers in eight mines were involved, making it the biggest wages strike since 1946. Approximately 13 500 mineworkers refused to go underground at West Driefontein. Then an estimated 4 000 demonstrated in the compound and burned down an administration block. Similar events occurred at the other mines. White miners' cars were stoned and offices were burned down. At Buffelsfontein, the police fired teargas into the crowds and arrested about 300 strikers. The public relations officer for the mine described the strikers as like "savages dancing around a fire".[2] When the police were called to help the local security guards at Grootvlei and West Driefontein they shot and killed six mineworkers and injured many others.[3] Altogether about 2 000 mineworkers were bussed back to their homes. The managers of neither the Gold Fields nor the Gencor mines made any concessions to the strikers who then returned to work for the morning shift on Monday, 5 July.

That, however, was but the beginning of the strikes. As the strikers returned to work in the Carletonville mines, the mineworkers at the rich Gold Fields mine at Kloof in Westonaria went on strike. This was followed by a strike at the Venterspost gold mine, also in Westonaria, when about 120 mineworkers barricaded themselves in a mine tunnel about one kilometre from the surface and refused to return until a police colonel went down to mediate and asked them to elect a committee who would be allowed to negotiate on their behalf. Then 2 000 mineworkers went on strike for two days at the Impala platinum mine at Rustenburg. Several people were arrested, scores were injured and about one-third of the strikers returned to their homes. There were also wages strikes at the Durban Navigation Colliery in Natal and at the three Donhauser Collieries, also at Durban, where the entire work force refused to work and attacked the compound beer halls.

[1]The seventh affiliate was Lonrho which operated Western Platinum and two coal mines, but no gold mines.

[2]*Rand Daily Mail,* 2 and 3 July, 1982. A burning mattress was, allegedly, the site of the fire around which the dancing took place. See also South African Institute of Race Relations, *Survey of Race Relations in South Africa*, 1982, p114.

[3]*Rand Daily Mail*, 6 July 1982.

The strike at Kloof had begun when 9 000 mineworkers, dissatisfied with their wage increase, refused to go underground on the Sunday evening shift on 4 July and on the following Monday morning shift. On Monday evening, the mine manager gave them an ultimatum that they either return to work or face dismissal. Their wages, he said, were not negotiable. Nonetheless, the strike continued for another two days, during which time there was much violence, mainly against those who refused to support the strike. Three mineworkers were killed, 25 were injured and many fled the compound. The police used teargas to round up hundreds of those who had escaped to a hill near the compound. They were herded on to the mine football pitch where the West Rand Divisional Commissioner of Police asked them to tell him their grievances. No-one replied, nor would they disperse and return to the compound. Fifty-eight were subsequently arrested and 2 000 were dismissed, packed on to police trucks and taken to Westonaria railway station where they were put on trains under heavy police guard and sent back to their homes. All of the arrested mineworkers were refused bail and kept in prison for more than two months until they appeared before a court in October, when the charges against them were dismissed. A further 29 black mineworkers, who had been arrested during an illegal strike at the Winkelhaak gold mine near Evander in April, were also remanded in prison until November. Their bail had been set at the exorbitant level of R1 000 each which ensured that they stayed in prison until at least their trial. Overall, at least 10 people were killed.

The strikes had followed the pattern of earlier protests by black mineworkers. They were spontaneous, countered by armed members of the South African Police (SAP) and the mine security forces and, in consequence, were marked by sporadic violence involving people and property. There was no evidence of even the semblance of a strike organisation and no sign of any leaders. For instance, the police claimed to have arrested the alleged ringleaders of the Buffelsfontein strike, though, according to the chairperson of Gold Fields, no-one in management knew who they were because "no natural leaders" had come forward to explain the causes of the unrest.[4] The strikes, however, revealed a high degree of solidarity in pursuit of a clear industrial objective and had all the hallmarks of trade union action, though without the backing of an organisation. The Kloof strikers, in particular, recognised this as a weakness. They had been forced back to work without any concessions and had no means of pursuing the redress of

[4]*Rand Daily Mail*, 3 July, 1982.

their grievance. A group of them, therefore, approached the Council of Unions of South Africa (CUSA) during its National Congress on 14 July and asked for help in forming a trade union for black mineworkers. [5]

The request was neither exceptional nor unexpected. Indeed, there had been speculation in some white quarters that a cause of the Kloof strike was the absence of a trade union through which the mineworkers could channel their grievances.[6] Johann Liebenberg, the Industrial Relations Adviser to the Chamber of Mines, in response, had said that it was not the mine owners' fault that there was no union to represent the interests of black mineworkers. The mine owners, he asserted, were pragmatic about the emergence of black trade unionism. "Mine employers," he stated, "do not encourage or seek to retard the growth of black unions and the one such union which has asked for access to mine workers has been granted it."[7] That union had been the Black Mine Workers' Union, sponsored by the white South African Boilermakers' Society to enable it to recruit in the mine compounds. It had been formed by Chillian Motha, a former official of the Black Allied Workers' Union, in April 1980.[8] He had submitted a draft constitution to the mainly white Trade Union Council for its endorsement and had appealed to it and to the mine owners for financial donations, but without success, even though it posed no threat to them. Its aim was to "educate and conscientise black mineworkers and make them aware of their significance and power in a job situation; and to obtain just wages and settle disputes with employers by peaceful means". The union had no intention, Motha declared, of indoctrinating

[5]CUSA and the Federation of South African Trade Unions (FOSATU) represented two distinct ideological wings of the black trade union movement. CUSA was oriented towards black consciousness while FOSATU supported non-racial trade union and political action. For a description of the formation of these two central federations, see V L Allen, *The History of Black Mineworkers*, 2003, volume II, chapter 13.

[6]South African Institute of Race Relations, 1982, p114.

[7]*Rand Daily Mail*, 8 July, 1982.

[8]The Black Allied Workers' Union was established following the 1973 strikes in Durban as part of the Black Consciousness Movement. It was opposed to mobilising workers in class terms and was against illegal strikes. It aimed to "raise the productivity of black workers by sponsoring training courses and training centres for black youth". It received substantial financial support from the American Labour Movement and had strong links with social democratic organisations in West Europe. The union refused to affiliate to FOSATU and had split in March 1979 when a dissident group, dissatisfied with its racially exclusive policies, formed the South African Allied Workers' Union. There was a further split in BAWU in 1981. (Rob Davies, Dan O'Meara and Sipho Dlamini, *The Struggle for South Africa. A Reference Guide*, 1985, p305 and p337.) Motha was involved in this splitting process.

mineworkers in the political sense.[9] The Chamber of Mines signed an Access Agreement with his union in May 1982 to allow it to recruit in the hostels.

The mine owners had found it difficult to resist pressures to adopt a more relaxed attitude towards black trade unions following the establishment of the Commission of Inquiry Into Labour Legislation established by the State President on 21 June 1977 under the chairmanship of Professor Nicolaas Wiehahn.[10] The Commission's first Report, published on 1 May 1979, had recommended the legal registration and, therefore, the recognition, of black trade unions. It had, thereby, altered the climate of industrial relations for black workers. These changes, however, did not begin to affect the mining industry until after the Commission's Sixth Report, dealing exclusively with the mining industry, was published in September 1981, although the Wiehahn Commission had refused to exempt the mining industry from its proposals for labour legislation in its First Report two years earlier.[11] After the publication of the Sixth Report and given that the government had actually legislated to enable black unions to be recognised, the position of the mine owners had become untenable. Some of them, moreover, could be accused of applying double standards because they owned non-mining companies where black trade unions had been recognised.

An early sign of the reversal of the hostility of the mine owners to the organisation of black mineworkers was when De Beers, the

[9] South African Institute of Race Relations, 1980, p171.

[10] The Commission's terms of reference were to "inquire into, report upon and make recommendations in connection with the existing labour legislation" with specific reference to making it more effective "for the needs of our changing times", adjusting the machinery for the prevention and settlement of disputes, eliminating bottlenecks within the entire sphere of labour and creating a foundation for the expansion of "sound labour relations". These terms of reference excluded any mention of the mines, but on 16 August 1977 they were altered "to include the Mines and Works Act, 1956 or any other Act administered by the Department of Mines". The Commission dealt with the question of industrial relations in the mines in its Sixth Report. In establishing the Commission the government was responding to the turmoil in the townships, manufacturing industry and the mines and was prepared to alter its policies on job reservation and black trade unions in its effort to ameliorate black discontent. See Allen, 2003, volume II, chapter 13, for a full description of the Wiehahn Commission and its impact.

[11] As late as 1978 the Chamber of Mines had asked the Minister of Manpower, Mr Fanie Botha, not to extend the legislation establishing works and liaison committees to the diamond and platinum mines because it believed they were too developed and not because, as was the case, they were discredited amongst black workers. (Steven Friedman, "Chamber of Mines' Policy and the Emerging Miners' Unions", *South African Labour Bulletin*, Volume 8, No 5, April 1983, p27.)

diamond monopolist, granted negotiating rights to a union representing 'non-white' mineworkers in February 1982. De Beers was a part of the Anglo-American Corporation but it was not a member of the Chamber of Mines and this gave it the freedom to apply its own, more liberal policy with regard to the organisation of black mineworkers. It signed an agreement with both the South African Boilermakers' Society and the Federated Mining, Explosives and Chemical Workers' Union to negotiate on behalf of about 6 000 mineworkers in all except artisan grades at four Kimberley diamond mines and the Finsch and Koffiefontein diamond mines. The Boilermakers' Society claimed to have recruited about half of the black mineworkers in those mines while the Federated Union insisted that it had recruited 95 per cent of the coloured workers there.

The coloured mineworkers had been the first 'non-white' group in the mines to take advantage of the new dispensation to organise after 1979, but there were relatively few of them. In 1979 they numbered only about 14 000, representing less than four per cent of the total labour force. They had never been subjected to the legal restrictions in the mines that had applied to black mineworkers and some of them worked in strategic posts alongside white miners. In the north-western Cape they worked as artisans while in the gold mines of the Orange Free State and Klerksdorp they were usually employed as computer processors, engineering draughtsmen and process operators. Since 1956, a small number had been organised by the white artisan mining unions but they were forced to leave them when the government made racially mixed unions illegal. In August 1979, the coloured mineworkers, some of whom had been members of the South African Boilermakers' Society, formed their own union, the Federated Mining, Explosives and Chemical Workers' Union, with about 600 members, but they did so with the help of the Boilermakers' Society, with which they shared a general secretary. Another white union, the Mine Surface Officials' Association, also formed a parallel union, the Coloured Mine Employees' Association, in July 1979, but it had only about 250 members, mainly in the Stilfontein area of Klerksdorp. The agreements reached in 1982 covered wage negotiating rights, health and safety clauses, redundancy procedures and allowed for the appointment of shop stewards to implement and monitor the agreements. By the end of June the Federated Union had also been given recognition rights by the Anglo-American Corporation to negotiate for particular categories of coloured workers who had been recruited at two of its gold mines and for black mineworkers it was attempting to organise there. With government registration, a white sponsor and a membership core of skilled workers, the parallel unions appeared as a safe option for management recognition.[12]

Matters, however, were by no means as straightforward as the public picture seemed to indicate. In the first place, there were sharp divisions amongst the mine owners about the wisdom of recognising black unions. Both Gold Fields and Gencor were opposed to the very idea of it. Moreover, whatever the mining houses decided, the mine managers had, and applied, their own views and used their own discretion about the use of force to suppress unionising activities. Management attitudes at mine level were likely to be the last to be liberalised. It was also clear that despite the claims of those unions with access or recognition agreements, they were barely touching the fringe of the black labour force. There were large areas where the organisation of black mineworkers had not even been contemplated even though the conditions for unionisation were present, in that they had relatively stabilised employment with career aspirations in mining and had concerns, therefore, about longer run wage levels and conditions of work that could only be assured by negotiated agreements. In some instances, this was accounted for by the fact that many mineworkers had strong reservations about responding to overtures from white-sponsored trade unions. The Kloof strikers, for instance, chose to approach the black consciousness-oriented CUSA without a record of organising in the mines rather than any parallel unions that did have that experience.

Underlying these factors were hostile economic conditions that favoured employers in the class struggle. Although the economy had been in recession to varying degrees since the mid-1970s, the coal and gold mining industries had been largely insulated from its effects. The coal industry had thrived through the oil crisis while the dramatic rise in the price of gold had cushioned the gold mining industry. From 1981, however, there had been a down-turn in the economy with inflation running high at 16 per cent. The price of gold fell from its highest-ever point at $800 an ounce in 1980 to $300 in 1982. This created a sharp fall in South Africa's foreign reserves and led to a balance of payments crisis that rebounded on the living standards of black workers and put many of their jobs in jeopardy. Black mineworkers suffered directly from the fall in the price of gold. The price of $300 an ounce gave sufficient profit in about two-thirds of the mines to encourage further expansion but in 12 marginal mines, employing about 100 000 workers, the prospects were bleak.[13] Production, for instance, was cut

[12]Friedman, 1983, p31.

[13]See John Lang, *Bullion Johannesburg: Men, Mines and the Challenge of Conflict*, 1986, p459.

back at the West Rand Consolidated gold mine in the Gencor group with the loss of 3 340 jobs. Diamond mining was hit by the general worldwide recession which resulted in both a fall in the demand for diamonds and the need to stockpile them. This caused 498 mineworkers to be retrenched at the Premier diamond mine near Pretoria and the closure of the diamond mine at Koffiefontein near Kimberley in June. Due to the recession in the steel industry, Anglo-American closed a coal mine in Northern Natal with the loss of 800 jobs.[14] Such bleak job prospects had generally inhibited industrial action because workers in fear of losing their jobs had usually been hesitant about offending management by supporting it. But that was not necessarily the case in the mining industry. Black mineworkers were accustomed to losing their jobs in the ritual at the end of contracts. And although they showed that they valued their jobs by consistently renewing their contracts and treating the periods between as holiday breaks, this appeared to be insufficient to deter them from seeking to redress their grievances, as the strikes at the Gold Fields and Gencor mines had clearly shown.

The Creation of a New Union

The delegates at CUSA's National Congress in July 1982 unanimously resolved to support the request from the Kloof strikers to form a union for black mineworkers. CUSA then asked the Scandinavian unions for financial assistance but they were unwilling to help because, they declared, they had already lost money in a similar venture. The year before, FOSATU had asked its affiliate, the Metal and Allied Workers' Union (MAWU), to form a black mineworkers' union and had succeeded in obtaining funds for the project from the trade union movement in Scandinavia. MAWU, however, had approached the task methodically by proceeding on a mine-by-mine basis and had failed to make any progress in any one. It had been unable to obtain access agreements from any of the mining houses to enable its organisers to enter the hostels and was unable, therefore, to agitate and ignite an interest in the formation of a union. CUSA's appeal for financial assistance, however, did not entirely fall on deaf ears. After some time, the Industrial Workers' Union (FNV), in the Netherlands, heard about it and decided to support the CUSA project itself by providing the equivalent of $250 000 for CUSA to use on organising black mineworkers.

[14]South African Institute of Race Relations, 1982, pp111-112.

The news that a black consciousness union federation was attempting to form a black mineworkers' union disturbed unions in FOSATU that were broadly supportive of the ANC for there was much hostility between it and CUSA. When Piroshaw Camay, the general secretary of CUSA, visited London, an ANC delegation met him to discuss the issue but he made no concessions to it. He insisted that the decision by CUSA to organise mineworkers would be implemented. Not long afterwards, the ANC received a message from some FOSATU affiliates stating that they could not trust CUSA and were, therefore, considering trying again to organise a union in the mines. The South African Congress of Trade Unions (SACTU), in exile, was equally distrustful of CUSA and supported this initiative. The ANC, however, was unhappy with the idea of forming a competitive union and provoked a debate in which it argued that it would be wrong to duplicate the efforts of CUSA because a mineworkers' union, irrespective of who formed it, would inevitably, eventually, be a progressive force in South Africa. Their primary aim, therefore, should be to support CUSA's efforts. The ANC won the argument and CUSA was allowed to proceed without any destructive competition from SACTU- or FOSATU-sponsored unions, though it did not have their co-operation.[15]

All previous attempts to organise black mineworkers since 1946 had illustrated the enormous gap between passing and implementing resolutions. Both SACTU and FOSATU, which at different times had passed such resolutions, had to operate in defiance of the mine owners and without any recognisable support from within the compounds. At no time did they seriously succeed in penetrating the barriers which prevented them from establishing communication links with the hostels. CUSA faced a similar, though not quite so intractable, situation. All of the mining houses, except for Anglo-American, were privately hostile to the presence of a union in their compounds though their public hostility had been moderated by the effects of the Wiehahn recommendations. The problem of implementation, it seemed, was almost as great for Phiroshaw Camay as it had been for those who had preceded him in similar circumstances. But it did not remain directly his problem for long for he handed the resolution to Cyril Ramaphosa, who, six months before, had been appointed as the head of CUSA's Legal Unit, and asked him, with limited financial backing, to tackle the seemingly impossible task of implementing it.

[15]See Keynote Address by President Thabo Mbeki to the 10th National Congress of the NUM, Pretoria, 24 April 2000.

When Ramaphosa started his task in August 1982, he had had no experience of trade unionism except that witnessed during the previous six months from CUSA's legal office. Moreover, he knew absolutely nothing about the mining industry. His only political experience had been as a student activist at the University of the North (Turfloop) where he had identified with the Black Consciousness Movement. His family came from the small northern province of Venda but he had been born in Soweto where his father was a policeman. When he began his studies at the university in 1972 he joined the Students' Christian Movement and the South African Students' Organisation (SASO) and was elected as the chairperson of the university branches of both organisations in 1974. In that year, he organised a pro-Frelimo rally for which he was arrested under Section Six of the Terrorism Act and detained for 11 months. On his release he joined the Black People's Convention (BPC) through which he was active in support of the Soweto uprising in 1976. He was again arrested under the Terrorism Act, detained for six months at the John Vorster Square police station in Johannesburg and expelled from the university. After his expulsion, he played no part in black politics for about three years. He became an articled clerk in a solicitor's office and completed his legal training through a correspondence course with the University of South Africa. He also studied for a Diploma in Industrial Relations. When he went to work for CUSA he was 29 years of age. At that time he was beginning to reappraise his understanding of black politics by perceiving black consciousness as a limited, parochial approach to the problem of apartheid. Ramaphosa, however, was unfamiliar with the activities of the ANC and had never actually met anyone who belonged to it. It was in that frame of mind that he started his job with CUSA, which, though it was allied with the Black Consciousness Movement, was primarily involved in the day-to-day task of sustaining and extending the boundaries of black trade unionism.

Phiroshaw Camay gave Ramaphosa his new organising assignment on a Saturday morning and he immediately went to have his first glimpse of a mine shaft and a mine hostel in the Johannesburg area. On the following Monday morning he went to the public library in Johannesburg to learn about the organisation of the Chamber of Mines and the ownership, number, size and location of the mines. When he began the practical work of organising he was given one assistant, Puseletso Salae, a dismissed mineworker from Lesotho who was an active member of the Basutoland Congress Party (BCP). Salae had been in dispute over his wages with the manager of the head office of TEBA in Johannesburg where he worked as a clerk in the fingerprint department. In consequence, when his contract expired he was refused a Valid Re-Engagement Certificate and, therefore, was denied any kind of work in the mines. In his dilemma he consulted a Catholic priest about what he should do and was advised to offer his services to CUSA where Phiroshaw Camay asked him to assist Ramaphosa. With the money from the Netherlands, Ramaphosa was able to buy a combie van which Salae drove around, accompanied by four other dismissed mineworkers, in order to throw bundles of leaflets into the hostels at night, usually pursued by the police. Salae had previously worked for a short time in the Vaal Reefs gold mine, but neither he nor Ramaphosa had any friendly contacts in the mines whom they could use in order to gain surreptitious access to the hostels and they were frustrated whenever they attempted to make any. No progress at all was made during the first month. Then, through a fortuitous meeting on the road home after a fruitless effort to gain access to the mines in Klerksdorp, Ramaphosa's fortunes changed.

Alfred Mphahlele, who worked as a personnel assistant at Western Deep Levels, came from the Northern Transvaal but his family lived in Soweto and, whenever he could, he went to see them. He invariably walked from his hostel in Carletonville to the main Klerksdorp-to-Johannesburg road where he would flag down a car for a lift to Soweto. He did that one Friday evening towards the end of August and as he climbed into the car he recognised the driver as Cyril Ramaphosa, whose picture he had seen in a Northern Transvaal newspaper during the student protests in the mid-1970s. A conversation began and Ramaphosa asked who he was and where he worked. When Ramaphosa learned that he was a mineworker he told him that he was trying to organise a union for black mineworkers. Mphahlele said he was interested and told him about the course he was attending, where they had recently discussed that very same topic.

Mphahlele was attending a training course for black personnel assistants organised by the Anglo-American Corporation in August 1982 at its Western Deep Levels mine in Carletonville. It was the first course of its kind and catered for black personnel assistants from its mines at Welkom, Klerksdorp and Carletonville. During the course, the 30 students were given a sympathetic lecture on industrial relations during which the issue of black trade unionism was discussed and the question of a union for black mineworkers was raised. A few of the students from the eastern Cape had heard of trade unions but none of them had experience or even knowledge of them. They had not met anyone from any of the unions allegedly interested in organising black mineworkers. Nonetheless, the idea of establishing a union in the mines appealed to them. Mphahlele told Ramaphosa about that interest, whereupon Ramaphosa said he would meet him in Soweto and give him a batch of union application forms to distribute to the other students.

The first person Mphahlele met when he arrived back at the mine on Monday morning was his friend, James Motlatsi, who was also a personnel assistant in his department and was attending the same course. Motlatsi had read in the press about CUSA's assignment and was interested in learning more about it. When, therefore, Mphahlele told him of the encounter with Ramaphosa, he agreed to sign a membership form and help to distribute the rest of the forms amongst the students. Thus a ripple of interest in forming a union began. A number of the students from various mines, but notably Patrick Hamnca and Ikaneng Matlala from the Western Holdings gold mine and Abel Ntoi from the Saaiplaas gold mine, also signed up and decided to assist in the distribution of the leaflets. Within a couple of days, Motlatsi advertised their intention by telling a local newspaper reporter that they were trying to form a union at Western Deep Levels. When Ramaphosa read this he arranged to meet Motlatsi at the mine on the last day of the course. Ntoi saw them together and called Hamnca over to "meet these union guys". After a discussion they took the forms and receipt books back to Western Holdings and Saaiplaas where they handed them out to their friends. In Western Holdings, Hamnca and Matlala slept in the same room as Morebuli Selepe and Korie Maloka, who were working underground on a course to qualify for their blasting certificates. They all became enthusiastic supporters of a black mineworkers' union.[16] The four of them were well known in the mine and were able to move freely from room to room in the hostel, which housed between 2 000-3 000 mineworkers, and thereby spread the news about the union. They avoided the rooms where the *indunas* stayed so it was some time before the hostel manager learned about their activities.

James Motlatsi and Alfred Mphahlele undertook a similar task at Western Deep Levels but they immediately ran into problems. When they openly approached other mineworkers with their application forms neither of them realised that they needed to have the consent of management before doing so. But they quickly learned this after

[16]They were members of the first batch of black mineworkers who were allowed to train for blasting certificates. They completed their course successfully but were not allowed to receive their certificates because of the colour bar. Selepe returned to a surface job as a clerk while Maloka left the mine to work as a union organiser with Ramaphosa.

another personnel assistant reported them to the mine manager, who then told Mphahlele that he was acting illegally. Though neither he nor Motlatsi was disciplined for their alleged misdemeanour, he was discouraged from taking any further part in organising his mine. He remained a supporter of the union but left the task of organising to Motlatsi. James Motlatsi, however, had another quality that was valuable to Ramaphosa. He fulfilled the role of a roving emissary for the union.

During his spare time and leave periods Motlatsi travelled around the mines in the Orange Free State, Westonaria, Klerksdorp, Carletonville and Witbank regions where he sold soft goods, such as trousers and blankets, mainly to Sotho mineworkers, through agents. In this way he earned enough extra cash to buy a new Volkswagen Passat car for his travels. After he met Ramaphosa he combined selling blankets with spreading the idea of a union and he utilised his car and agents for both purposes. He became as well-known as Ramaphosa throughout the mining industry. Justice Hlalele, for instance, who later became a member of the union's national education committee, from the Free State Geduld gold mine in the Orange Free State, remembered having secret meetings with him in order to avoid the police.[17]

After a frustrating first month, Ramaphosa had gathered around him a small group of people from a few Anglo-American mines, mainly Western Holdings, Saaiplaas and Western Deep Levels, who were strategically placed within their respective hostels to spread the news about the formation of a new union. He recognised the limited scope of this activity. He had no assistance in any of the mines belonging to the other mining houses and, as most of the organisers he had collected around him were Sothos, he was appealing only to Sothos in the first instance and then to Xhosas, but not to the large slice of the labour force from Mozambique. But with Salae in his combie and Motlatsi in his Volkswagen spreading the news further afield, there was enough activity to convince Ramaphosa that it was time to launch the union with a name and a constitution. He decided not to approach the Chamber of Mines in August for any kind of agreement because he felt it would compromise the independence of the union. By September, however, he realised that unless he had permission from management to enter the hostels he would make little progress. He then applied to the Chamber of Mines for an Access Agreement.

The union was launched at the end of August 1982 at a meeting at

[17]Hlalele had joined the union in 1982. Interview, Johannesburg, 24 January 1992.

Hammanskraal, just north of Pretoria, attended by Phiroshaw Camay, Cyril Ramaphosa and a few others, mainly those who had attended the Anglo-American training course at Western Deep Levels. Much of the proposed constitution was copied from that of the British National Union of Mineworkers and was not controversial. Choosing the name, however, was a highly political issue. No-one in the meeting had ever heard of the African Mine Workers' Union which had led the 1946 strike so its name was not considered. Many, however, admired the British National Union of Mineworkers and preferred its name. CUSA's intention, however, was to form a union for black mineworkers and it insisted that the word 'black' should be in the title. Motlatsi, on the other hand, wanted a name that would be acceptable in Britain and internationally and preferred the NUM. Others supported him because they wanted to emphasise the national as against the ethnic character of the union. The matter was discussed for five hours before it was agreed to call the union the National Union of Mineworkers on the understanding that it could be changed at the union's inaugural congress. By the time that was held, however, the matter was sealed, for the union's note paper carried the NUM logo.

Not long after the meeting at Hammanskraal, tensions began to develop within CUSA about Ramaphosa's activities. He was still a staff member of CUSA which paid his salary and he remained under the control of its National Executive Committee. There was in CUSA, as in trade union federations elsewhere, a recognition of the need to employ intellectuals for their specialist services despite having a deep-seated distrust of their personal and, sometimes, political motives. The worker leaders of the affiliated unions wanted constraints to be imposed on staff members so that they functioned strictly to serve them and did not behave as if they too were leaders.

Ramaphosa did not readily fit into this pattern. He organised workers as they did, but with much greater success. In doing so, he both attracted the attention of the press and revealed a capacity to exploit the publicity it provided. The CUSA leaders, in consequence, saw him as a staff member who had got out of hand and they tried to curb him. Early in October, Camay, the general secretary, was directed by the CUSA executive to stop Ramaphosa making press statements but Camay, protective of Ramaphosa, reacted by issuing a directive to stop all staff members from speaking to the press. Then, on 11 October, a member of the executive told him directly what the problem was. "The trouble with you," he was told, "is that you are an intellectual and you intellectuals always want to have positions so that you can have power. We uneducated people don't want power. We serve the masses in their struggle."[18]

This hurt Ramaphosa. He recognised that in building the union he had been constructing a power base for himself. But he had sacrificed a career as a private legal practitioner to work for CUSA, was still one of its employees and had never aspired to be the general secretary of the new union. So, in the evening of 11 October, he sent a detailed report of his activities in the NUM, the plans of its Interim Planning Committee and a letter of resignation to Camay to take effect immediately. His intention was to resign from his post in CUSA and to withdraw from the NUM simultaneously.[19] But, even at that early stage, it was too late for an easy resignation. Both Camay, whom Ramaphosa claimed had "talked him into taking the job", and the small group of activists in the mines insisted that he should continue to organise mineworkers. Two days after he had sent his resignation letter, but still employed by CUSA, Ramaphosa signed the first Access Agreement with the Chamber of Mines. He continued to operate as before but the incident had lingering consequences for it accelerated Ramaphosa's disenchantment with CUSA and its brand of black consciousness and led to a complete break two years later.

This internal bickering, plus the difficulty of gaining access to the compounds, slowed down the progress in building the union. However, once his own position was secured and the Access Agreement with the Chamber of Mines had been signed on 13 October 1982, there was a renewed surge of activity. The Agreement laid down the general principle of access but left the precise details to the individual mining houses with the result that there were vastly different responses. The Anglo-American Corporation and Rand Mines gave the union official access to the hostels but the others, particularly Gencor and Gold Fields, made the lives of union activists very difficult. But even within a single mining house there were different responses for the real authority as to what happened in the mines on a day-to-day basis lay with the mine managers. This was to be made explicit by the Memorandum of Agreement between the NUM and the Chamber of Mines eight months later when it was stated "that official representatives of the Union may have access to any mine listed in the attached schedule for the purpose of conducting their business, subject to such conditions as individual mine managers may determine".[20] The early union activists believed

[18]Letter from Cyril Ramaphosa to Phiroshaw Camay, 11 October 1982.

[19]*Ibid.*

[20]Memorandum of Agreement between the Gold Producers' Committee of the Chamber of Mines of South Africa and the National Union of Mineworkers, Johannesburg, 9 June 1983, para 4.4.

that the Access Agreement signed by the mining houses not only did not reflect the views of the mine managers but was not intended to tie them to any particular policy.[21] This was so even in the Anglo-American Corporation where the relatively liberal views of its executives about black trade unionism simply set guidelines for the managers without necessarily impinging on day-to-day practices. The manager of Western Deep Levels, for example, was hostile to the union while the one at the President Brand gold mine was less so. At Western Holdings the manager seemed to be indifferent and allowed the union to operate much as it wished.

In general, even with an Access Agreement, entry into the hostels was restricted in a variety of ways. Mine managers tended to prohibit mass meetings or refuse to allow even small meetings to be held in the compounds. They restricted the distribution of union literature or the use of public address systems. Some mine managers continued in their old ways and intimidated and arrested those trying to establish the union. They used informers so they could learn about the intentions of the union activists and attempted to damage the union's image by declaring that it was the political wing of the ANC. Such was the climate of suspicion that when Cyril Ramaphosa visited the Western Holdings gold mine for the first time some of the mineworkers there believed that he was an undercover agent for the Special Branch of the Police.[22] There were mine managers who simply refused to allow any union activity in the compounds even where they were covered by the Agreement. In so far as there was a general managerial view it was that the attempt to form a union would fail in the long run. In any event, they believed that even if one were formed they could tame it and use it to control the work force. Few managers believed that a fully independent black trade union would be allowed on the mines. There had been no overnight conversions amongst them to any form of workplace democracy and they had no intention of losing control of the compounds to elected workers' committees.

Managers, however, were not the only ones who presented problems. The first converts to the union were all personnel assistants or clerks who had access to ordinary mineworkers at work and in the hostels. More than any others they had the freedom to move around and

[21]Interviews with Emmanuel Masita, Justice Hlalele, Vitalis Refuoe Ngoae and Justice Sello Tsukulu, Johannesburg, 24 January 1992. They had each joined the union during this period and had subsequently become national and regional officials.

[22]Interview with Ikaneng Matlala who was working at Western Holdings at that time, 5 October 1990.

communicate underground and on the surface and were thus strategically placed to recruit members. Their task, nonetheless, was not easy. In the first place there were very few of them. The majority of the senior blacks acted for white management and obstructed moves to establish a union. But they also faced difficulties in persuading the ordinary mineworkers that a trade union would work in their interests. The mineworkers were initially hostile to any requests to join anything which involved handing over money for, in most of the mines, they had been cheated by bogus insurance representatives who had taken their money and then disappeared. They said: "These youngsters just want to rob us of our money."[23] Some mineworkers had to be persuaded for other reasons. Although managerial authority had been damaged by black revolts in and out of the mine compounds since 1973, there were still some mineworkers who believed it was impossible to confront the mine owners head on. Others were hesitant because they knew that dismissal was often the price of protest. The foreign mineworkers were most fearful of being dismissed for they automatically lost their right to be in South Africa when they lost their jobs and they had little possibility of obtaining work at home. It was, however, mainly the Mozambicans who adopted this attitude for most of the early union organisers were Sothos who also risked deportation if they were dismissed.[24] In general, it was a risky business for any of the advocates of a union to speak openly about it in the hostels even after the Access Agreement had been signed.

The embryonic union organisers, therefore, needed to use persuasion and guile in their recruitment activities. They handed out application forms but did not ask for subscriptions until much later when information about the new union was available. Because of the need for secrecy they made contacts through friendship networks and met for discussions with mineworkers from other mines in the concessionary stores and at football matches. Their tactics varied according to the different attitudes of the mine managers and their use of the mine security though, in the main, the mine security did not interfere because at that time it was quite lax in Anglo-American mines. At Western Holdings, the four activists moved from room to room to gain recruits. Patrick Hamnca also used other means. He was responsible for a work section and possessed a list of the names and numbers of all the mineworkers in it. He posted a notice requesting some of them to call in his office and

[23]Interview with Morebuli Selepe, Johannesburg, 10 May 1990.
[24]Three of the four activists from Western Holdings were Sothos.

when they arrived he asked each one to sign a form. When they asked why, he said he would explain it to them that night in the hostel. They all signed. Later he and Selepe went around the rooms and told them that they had joined the union. No-one, however, took offence at being recruited in that way. Quite soon enough mineworkers in the hostel had joined to enable them to elect a committee of 10 who became union advocates amongst their friends in the other hostels and mines. The situation at Western Holdings received a stimulus when Ramaphosa made Maloka a full-time organiser after he was diagnosed as an epileptic and was refused permission to work underground. The first branch of the union was formed there. The manager of the mine did not obstruct the union but he was not wholly at ease with its methods. When, for instance, he found union application forms in the mine office where Patrick Hamnca worked, he demoted him, whereupon Hamnca left and was given an organiser's job alongside Maloka.

Once the branch at Western Holdings was properly established Hamnca and Maloka both went to adjoining mines. At the President Brand mine they established a branch which grew rapidly after they promised mineworkers a long list of services including financial loans and investigations into their complaints about hostel accommodation and food. The organisers, for instance, told mineworkers not to throw bad food away but to take it to them so it could be taken back to management. This branch was the first in the union to gain recognition from the Chamber of Mines. Hamnca and Maloka then went on to form a branch at St Helena, the first Gencor mine to be organised. Recruitment at Western Deep Levels was difficult at first because mineworkers there were more sceptical of the union due to the widespread activities of bogus insurance representatives. Moreover, because of Motlatsi's roving duties he was left with little time to help build the union organisation in his own mine. Nonetheless, Western Deep Levels No 3 Shaft, where he worked, became one of the main centres of union organisation with a reputation for agitation and defiance against management.

In general, once the Access Agreement was signed, it became easier to gain recruits for the union because it had an important symbolic significance for ordinary uncommitted black mineworkers. It was seen as a legitimation of the union and they became less suspicious about union organisers. They were more willing to attend meetings, listen to advice, sign application forms and pay union subscriptions. The spread of the union developed its own momentum which was actually assisted by the negative reactions of managers. The determining factor in those circumstances was the presence within the mines of people who were willing to campaign for the union despite the risks to themselves.

There were further developments at Western Holdings when the mine manager, in the hope that he could learn more about its activities, allowed the union to hold a general meeting in the compound hall at the mine's training centre. Cyril Ramaphosa used the occasion to show graphically how the union would help those injured at work by bringing to the meeting a combie full of paraplegics who were being treated at the Ernest Oppenheimer Hospital in Welkom. After the meeting, recruiting tables were placed outside each of the hostels. Each new member was given a target of 40 others to recruit within three days. It was not long before the NUM at Western Holdings had organised between 60 and 70 per cent of its entire black labour force. No 6 shaft was the most highly organised part of the mine. An official of the branch was posted to sit outside the gate of No 6 shaft through which all of the mineworkers had to pass after being paid their wages. As each one passed through he was asked to show his pay slip and if it did not show a deduction for the NUM he was fined R50 on the spot. The Western Holdings branch became richer than the OFS regional office.

The union received a publicity boost in October 1982 when the 58 mineworkers who had been arrested and remanded in prison in July after the wage strike at the Kloof gold mine were freed after the intervention of the NUM. Ramaphosa had taken up their case and had arranged for them to be represented in court by S Monyatsi, the CUSA attorney, with the result that when they were put on trial the charges against them were dismissed. Ramaphosa then hired combies to take the freed mineworkers back to Kloof where he demanded that, as they were innocent, they should be given their jobs back. The mine manager agreed and they were immediately re-employed. This was the first act of substance to support the union's claim to protect the rights of mineworkers. The publicity this legal victory received persuaded many mineworkers to join the union. Some of the Kloof strikers became prominent members of it. One of them, Justice Tsukulu, went on to become the chairperson of the Westonaria region of the union. Gold Fields, however, obstructed the union and refused to allow NUM officials into the Kloof hostels even after the Access Agreement had been signed. The climate there was so hostile that some of the reinstated mineworkers refused to join the union and all attempts to establish a strong branch at Kloof failed.

The union spread to the Klerksdorp gold mines when Puseletso Salae returned there in October to establish a branch at the Anglo-American mine at Vaal Reefs. After he and Ramaphosa met the mine manager, Salae was allowed access to the compound but confined to one place and not allowed to move away from it. After two months he was provided with an office but it was so near to the mine manager's office that mineworkers were reluctant to visit him. Then the mine manager called the security police in to question him about his union activities. The police wanted to know why he, from Lesotho, was organising mineworkers in Klerksdorp. The questioning turned into harassment when he was arrested and fined R15 for not having a lodger's permit to live in the Jouberton township which was near Klerksdorp.[25] However, he was encouraged in his union work by the local Catholic priest who allowed him to hold meetings at the Jouberton Catholic Church hall. He managed to organise a union branch at Vaal Reefs around a core of fellow members of the Basutoland Congress Party (BCP).

Gradually, between October and December, despite the piecemeal fashion in which branches were being formed, a union took shape. It extended beyond Western Holdings, Saaiplaas, Vaal Reefs and Western Deep Levels to a small number of other mines but remained within the gold mining industry. There was some agitation in the Kriel, Matla and New Denmark coal mines but no branches had been formed there. The shaping process was hazardous because the coterie of union protagonists lacked organisational and administrative experience. Indeed, every day was a learning experience for them. They had little understanding of conventional organisational prerequisites. A few of them from the Eastern Cape had heard of trade unions but none of them knew anything about the pyramidal structures of unions stretching from shaft stewards' committees to branches through regional offices to a national executive committee as the union constitution prescribed. They were unaware of the normal duties of secretaries or chairmen involving the conduct of meetings, making reports and keeping minutes. None of them had ever previously held positions of influence. They were moving in a world which, in their experience, had hitherto been occupied exclusively by white people.

The Inaugural Congress

The congress to inaugurate the fledgling union was held in the Roman Catholic Church Hall at Jouberton township near Klerksdorp on 4 and 5 December 1982. It was an important event for it transformed the

[25] *CUSA News*, February 1983.

CUSA resolution into a reality and converted a loose collection of branches, held together by a self-selected Interim Planning Committee under the direction of Ramaphosa, into an integrated organisation with a democratic constitution. The union, at the time, had 14 000 members in nine branches, seven of which were at Anglo-American gold mines in the Orange Free State and the West Rand.[26] The union had four regions but no regional officials or regional offices. However, there were 25 organisers who, between them, covered all of the regions. Each of the four regions was allowed 50 voting delegates though many more than 200 people attended the Congress.[27] Indeed, Ramaphosa claimed that as many as 1 800 delegates were present. The general secretary of CUSA, Phiroshaw Camay, was the guest speaker but none of CUSA's affiliated unions was represented to celebrate the event.[28]

It was a new experience for every mineworker present to participate openly in a Congress and they had to be guided carefully through the procedures in a manner which enabled them to get through the business without violating their own cultural traditions. Many of the delegates were illiterate and spoke only their indigenous languages. Every issue, therefore, had to be explained orally in a language they understood. This was Ramaphosa's task, which he did with the help of Sotho and Xhosa interpreters. He controlled the business of the Congress and explained in detail its objectives and procedures. He read the proposed constitution and explained how it was intended to extend the base of the structure by electing shaft stewards and forming shaft stewards' councils and committees. The Congress then debated resolutions submitted by the regions until such time as they reached a consensus without voting. This led to lengthy discussions which lasted through the night but it became union practice to debate in that way. They decided not to register under the law, even though the Chamber of Mines had declared it would not negotiate with an unregistered union. It also opposed the virtually unobtainable terms set for recognition by the Chamber of Mines which were that the union had to organise at least 30 per cent in

[26]The founding branches were: Saaiplaas, President Brand, Welkom, Vaal Reefs, Elandsrand, Western Holdings, Blyvooruitzicht, Western Deep Levels and Kloof. All except Blyvooruitzicht (Rand Mines) and Kloof (Gold Fields) were Anglo-American mines. Some at the Congress came from Stilfontein (Gencor), Hartebeesfontein (Anglovaal) and West Driefontein (Gold Fields). All of them were gold mines in either the Orange Free State, Klerksdorp, Carletonville or Westonaria.

[27]Ramaphosa, in an interview with *CUSA News*, February 1983.

[28]None of the unions affiliated to CUSA gave the NUM any assistance at any time. Their officials continued, allegedly, to be envious of the NUM's success and to regard Ramaphosa as an upstart.

each job category throughout the industry.

Finally the Congress elected a small National Executive Committee (NEC) which appointed Cyril Ramaphosa to be the general secretary, the only national full-time post. It then went on to elect the three national lay officials. James Motlatsi was elected as president. He was a mineworker who came from a family of mineworkers. He retained his job as a personnel assistant at Western Deep Levels and had an arrangement with the mine manager that enabled him to undertake union work. Motlatsi, who came from Mohale's Hoek in Lesotho, had been a mineworker since 1970. On his first contract he had worked at Welkom gold mine, first as an underground drain cleaner, then as a box attendant and a store attendant. For his second contract he was employed as a winch driver until he was promoted to be a team leader. He left Welkom, he said, because the white supervisors were demanding bribes from the team leaders. Those who refused to pay were penalised by being given the poorest-paid jobs. He then went to Western Deep Levels and was present during the outbreak of violence there in September 1973 when 11 mineworkers were shot and killed by the police.[29] He worked at Western Deep Levels as a winch driver, then as a stope team leader and a senior team leader. Late in 1975 he was asked to work in the Personnel Department as a personnel assistant. It was whilst working in that capacity that he met Ramaphosa and, along with him, became the most well-known personality in the union.

The result of the election of the second national lay official, the vice-president, was unexpected. The successful candidate was Elijah Barayi, a 52-year-old personnel assistant from the Blyvooruitzicht gold mine in Carletonville, where he had established a union branch after reading in the press of the attempts to form a mineworkers' union. He was much older than the other activists and had had experiences in the ANC which none of them possessed. Indeed, he was the first ANC member Ramaphosa had met and that was most probably the same for the other delegates. He came from Cradock in the eastern Cape which was an area well-known for its support of the ANC and which had produced a number of prominent ANC activists. He was entirely unknown before the Congress but achieved instant popularity through his impressive oratory, his political commitment and his unrestrained attacks on the apartheid governement. When it was a criminal offence to say such things he described himself as an old ANC stalwart. His election gave a sign that there was an underlying sympathy for the

[29]For details of the dispute see Allen, 2003, volume II, chapter 12.

Elijah Barayi, vice-president of NUM, from its beginning in 1982 until his death in January 1994. He had also been the president of COSATU from its formation in 1985 until 1991.

ANC amongst the delegates. The post of national treasurer was, at that time, largely symbolic, and went to Ishmael Thulo. With its organisation in place, the union was officially established on 1 January 1983.

The Recognition Agreement

Whilst these activities were going on, the Chamber of Mines had begun to soften its position. In September 1982, it gave limited recognition rights to the Federated Mining, Explosives and Chemical Workers' Union which organised coloured mineworkers. The Chamber of Mines described this as an historic move for it was the first time in its 95-year history that it had formally recognised a union representing workers other than whites. Under the agreement the union was allowed to negotiate on behalf of vehicle drivers, handymen and painters at the Hartebeesfontein gold mine belonging to Anglovaal, and Anglo-American's Vaal Reefs gold mine. The union was informed that it would be allowed to negotiate on behalf of black mineworkers when it could show that it adequately represented them.

On 9 December, within days of the Congress, the Chamber of Mines responded with new recognition guidelines that largely satisfied the NUM's demands.[30] It dropped the stipulation that a union had to be registered before being granted recognition and announced that it would recognise any black union in any mine where it could claim at least 30 per cent membership in any job category. Thus it no longer insisted on terms for industry-wide bargaining. Any branch would be able, by its own efforts, to obtain recognition. This removed a crucial obstacle to the expansion of the NUM though the union, at its Congress, still thought that 30 per cent was too high for a qualifying rate. The branches, however, would not have the right to negotiate with mine management. That was to be done centrally on their behalf by the national officials of the NUM in negotiations with the Chamber of Mines which remained responsible for the wages and conditions of all mineworkers in each mine affiliated to it.

From the outset the union concentrated on building an organisation

[30]The Chamber of Mines' explanation as to why it amended its guidelines for the fourth time was the proposed amendment to the Labour Relations Act which would permit an unregistered, but representative, trade union to apply for the appointment of a conciliation board. It added: "The amended criteria will serve as the basis for recognition by the Chamber of newly emergent unions or of recognised unions which become significantly representative of new categories of employees." (Chamber of Mines of South Africa, *Annual Report*, 1982, p39.)

and establishing itself as an integral part of the black trade union movement. For that purpose it deliberately avoided confrontational issues with the mine owners that might lead to strike action and confined itself to the legal representation of victimised workers, raising health and safety issues and engaging in legitimate negotiations with the Chamber of Mines. As soon as the Chamber of Mines had published its new guidelines the NUM applied for recognition for its branches at Western Holdings, Elandsrand and President Brand. Five other branches were added to the list later. This process was long-winded and continued until 9 June, 1983, when a Memorandum of Agreement was signed between the union and the Gold Producers' Committee of the Chamber of Mines.

The Agreement, which was legally binding, gave the union the right to negotiate for its members in the occupations and mines where it had achieved recognition. At that time they numbered less than 6 000 out of a labour force of just over 418 000, on nine of the 41 gold mines controlled by the Chamber of Mines. It allowed for the coverage of the Agreement to be extended as the union gained recognition in more mines or for additional occupations and for it to be reduced where the union could not prove that it was "significantly representative" of the occupations listed in the schedule to the Agreement. It allowed for the union to negotiate over (i) conditions of employment; (ii) the recognition of shaft stewards and their role; (iii) the security of employment of members, including retrenchment and dismissal procedures; (iv) procedures for the resolution of individual and collective grievances; (v) procedures for the resolution of disputes over the Agreement; (vi) procedures for monitoring the Agreement and (vii) deciding which issues should be negotiated at mine level and which at the level of the Chamber of Mines. It stipulated that the mines should not prevent any mineworker from joining the union but that, at the same time, the union should not victimise any mineworker who refused to join. The Agreement also made provision in the mines where the union had recognition agreements for the check-off to be introduced whereby the management deducted union subscriptions from the wages of mineworkers, provided each member had signed a 'stop order' form authorising it to do so.[31] It gave black mineworkers the right to strike provided all the procedures for the resolution of disputes had been exhausted but, after a prolonged argument between the parties, the mine

[31]See Appendix A, "Stop Order for Deduction of Trade Union Fees From my Remuneration".

hospital and catering staff were denied that right. Another issue which the union lost concerned the status of black security personnel for the union was not allowed to negotiate on their behalf. The signing of the Agreement was an historical occasion for it was the first time in the history of South African mining that black mineworkers had been allowed by the Chamber of Mines to play any part in influencing their conditions of employment.

The union had began to prepare itself for the subsequent negotiations in April, well before the Agreement was signed, by summoning shaft stewards from the various mines to form a Wage Policy Committee to decide which issues should be raised with the employers. They demanded "a 30 per cent wage increase and an end to racial discrimination on the Chamber's mines; annual 'paternity leave' for married migrants to enable them to see their families; equivalent pay for African mineworkers doing work legally reserved for whites; extension of the white unions' pension fund to Africans; protection for workers if new technology was introduced; and a change in the leave system and hours of work".[32] Negotiations opened on 14 June when T-shirted black mineworkers entered the head office of the Chamber of Mines as members of the NUM team. They ended three days later. The team reached an agreement over wages and accepted a Chamber of Mines assurance that it would eliminate racial discrimination at work. There was no agreement over the other issues raised by the union. The Chamber of Mines insisted that it required more time to examine them but Ramaphosa thought this was just a delaying tactic. He did not, however, report the partial rejection of the union's demands back to the members in case they rejected the whole agreement and put the fledgling union in a state of confrontation with the mine owners at a time when it could not have survived a strike or lockout. He was criticised for this within the negotiating team.

The wages agreement allowed for increases that ranged between 14 and 16 per cent on minimum rates but they were not the result of real negotiations. The process was really a charade for the mine owners conceded only that which they had been prepared to give to all the 465 000 black, Asian and coloured mineworkers on all of their gold, coal and platinum mines without negotiations. Nothing had changed. Even the traditional date on which the Chamber of Mines made its annual wage awards remained the same.[33] At the end of June,

[32]South African Institute of Race Relations, 1983, p155.

[33]"The Chamber stated that "these negotiated increases were in line with what the Chamber was prepared to grant to the 465 000 ... employees ... with effect from 1 July 1983". (Chamber of Mines, *Annual Report*, 1983, p18.)

Ramaphosa, believing that he had been duped, went back to the Chamber of Mines to demand that the wage talks should be re-started because it had been bargaining in bad faith. The Chamber denied this and the matter was dropped. Then, in mid-July, it agreed to discuss issues other than wages, including the question of ending racial discrimination, but no progress was made on any of them.

In August the union declared a dispute with the Chamber when its demand for a 40 per cent wage increase for 250 workers at the Rand Refinery in Germiston was rejected and it was offered eight per cent instead. The government appointed a conciliation board to investigate and it raised the offer to nine per cent. But though this was also rejected at first, the union eventually accepted it. The union went through the formal motion of protesting but, in reality, it had no other option but to accept the terms set by the mine owners. It was still very much a small-time player and a confrontation would have been disastrous for its organisation at that stage of its development. The union's NEC recognised the weakness of its position because each time Ramaphosa reported back that the negotiations had reached an impasse, irrespective of the issue, he was told to try again but to give way if necessary.

The West Driefontein Incident

The area of mining activity where it was possible for the union to make an impact in 1983 was that which concerned health and safety matters. It was handicapped, however, by the fact that it possessed neither the organisational facilities nor the expertise and financial resources to pursue them effectively. The union did not have a health and safety department. In fact, there were simply Ramaphosa, Motlatsi and their small coterie of supporters. Ramaphosa had little knowledge of mining in general and no experience at all of mining accidents or disasters. Motlatsi had personal experience of underground hazards and of one accident to himself but he did not possess the technological or geological knowledge necessary for evaluating underground conditions in the manner of the Mines Inspectorate. The union's resources were restricted to a limited amount of overseas aid, the services of a small legal civil rights firm in Johannesburg and Ramaphosa's flair for publicity.

In April 1983 the union had applied to the Government Mining Engineer for representation on the Mine Safety Committee, set up under the terms of the Mines and Works Act. Its application was rejected, allegedly because the union was both unregistered and unrepresentative although the Committee's constitution did not require unions to be either of those things. Only the white unions had members on it. The

NUM, however, was drawn into health and safety matters in a much more direct and horrific way when on 12 September 1983 a methane gas explosion at the Hlobane Colliery in northern Natal killed 68 mineworkers and seriously injured another eight. The union was immediately and publicly involved in the events that followed.[34] Ramaphosa skilfully used the press and exploited the transient public sympathy for the victims in order to press the case for the involvement of mineworkers in devising and implementing safety measures.

A few days after the disaster the NUM became involved in a controversy with Gold Fields about whether black mineworkers had the right to refuse to work in dangerous places. It raised a principle of fundamental importance that was to preoccupy the union for many years afterwards. Ramaphosa had given it second place in the Miners' Bill of Rights to improve mine safety he had proposed after the Hlobane disaster.[35] But it was also a matter of immediate practical relevance for the safety of underground mineworkers. The annual death toll in the gold mines was high and overwhelmingly the casualties were black. In the previous year, 579 men had been killed whilst at work in the mines. Given the impact of the Hlobane tragedy, the carnage was likely to be worse in 1983. It was not known how many had been killed in work situations which the black mineworkers knew were dangerous but it was common for them to be ordered to work after complaining that it was dangerous to do so. The following year, the fledgling NUM commissioned the Department of Sociology at the University of the Witwatersrand to undertake research into the causes of the high accident rate.[36] The research revealed that, within its sample, 89 per cent of the team leaders and 79 per cent of the machine drillers had refused to work in dangerous conditions and that in 55 per cent of the cases their objections had been ignored. "Accidents happened to almost a half of the stope informants who felt they had been coerced to return to work after their initial refusal."[37] The situation was epitomised by what the white miners described as the 'rule of the mine', namely 'work first, report later' whenever black mineworkers challenged them about their

[34]See chapter 5 for a description of the disaster and the involvement of the union in its aftermath.

[35]*City Press*, 17 September 1983.

[36]The research was undertaken by Jean Leger in the Sociology of Work Programme. Ninety underground black mineworkers with at least one year's underground experience from four gold mines were interviewed between August 1984 and January 1985. See J P Leger, "Safety and the Organisation of Work in South African Gold Mines. A Crisis of Control", *International Labour Review*, Volume 125, No 5, September-October 1986.

[37]*Ibid*, p598.

approach to dangerous conditions.[38]

The official procedures were clear but unsatisfactory in that they gave black mineworkers only a limited statutory right to withdraw from dangerous places. Under the Mines and Works Act of 1956, mine managers had the responsibility for making every working place safe at the beginning of every shift before allowing any black mineworkers to start work. If any black mineworker believed that his allotted working place was dangerous he was protected by the Act which stated that "if any non-scheduled or other person working under the supervision of a ganger or miner complains that the part of the working place where he is required to work is dangerous, the ganger or miner shall withdraw all workmen therefrom until he has personally, with such assistance as he may require, made it safe, or until he has had it examined by another person who shall be the holder of a Blasting Certificate and has obtained the concurrence of such person as to the safety of such part. Until he has complied with this requirement, the ganger or miner shall take all reasonable precautions to prevent the entry of any person to the part complained of, except such persons as are necessary to carry out or assist in carrying out the examination of the making safe."[39] This meant that in order to comply with the regulation one white miner had to consult another.

The procedure, then, was apparently simple. A mineworker just had to complain and action would be taken immediately. The reality, however, was different. All the complainants were black and those who decided whether or not a working place was safe were white miners whose decisions were legally binding commands. The white miners always and invariably had the first say, in that they decided whether or not complaints were *bona fide*. Their general practice was to ignore complaints that they considered not justified. If the mineworkers continued to complain, and, therefore, persisted in ignoring white instructions, they were disciplined and that usually meant dismissal. Ramaphosa emphasised this point when he spoke at the Commemorative Service for the victims of the Hlobane disaster. He said that three of the injured mineworkers, whom he had visited in hospital, had told him that as their working place had not been tested

[38]J P Leger and M Mothibeli, "South African Gold Miners' Perceptions of Safety, 1984-1987", *Mine Safety and Health Congress*, Johannesburg, 19-20 November 1987, p7.

[39]Mines and Works Act, 1956, Regulation 8.3.2. J P Leger and M Mothibeli discussed the issue in *Participation in South African Mining Safety: A Postal Survey of Black Migrant Workers' Organisations and Management*, International Migration for Employment, Working Paper, ILO, Geneva, June 1989, pp9-11.

for gas they knew it could be a fatal move to work there, "but because a refusal or objection to work would result in victimisation or instant dismissal they went" nonetheless.[40] That was also the position at West Driefontein in September 1983.

The incident began when three members of a gang working a stope at the West Driefontein gold mine in Carletonville were injured by rock falls following an incident in which a white miner was injured. The gang, in consequence, became uneasy about the state of the rocks and was reluctant to continue working there. The manager responded by decided to switch the gang with one from another working place and which knew nothing about the problem. As the new gang of about 40 black mineworkers was making its way to the waiting place, where they were due to hang around until the white shift boss arrived, they met the senior team leader of the previous gang who informed them about the accidents and said that the area was dangerous. After some delays the new team leaders went to examine the condition of the working place and decided that the roof of the work area was hanging dangerously, that the roof supports were tilted at an angle, that rocks kept falling when slightly touched by their helmets and that there were unusually wide cracks in the roof. They also heard rumblings in the rock, indicating the possibility of a rock burst. The new gang then refused to work.

In the events that followed, the gang's views were put by the team leaders who were called in to explain the situation to the white supervisors.[41] The management gratuitously dismissed the assessment of the team leaders by stating that "novices are not in a position to 'recognise portents of danger from rockfall' in any sense. The science of rock mechanics is an elaborate one and a great deal of experience has been built up by the mining industry in South Africa."[42] When the team leaders continued to insist to the shift boss that the area was too dangerous to work he verbally abused them and threatened to cut their pay. They had a similar reception from the underground manager who said they would be demoted if they continued to refuse to work. On each occasion these allegations were denied by the white supervisors but they fitted the pattern of treatment accorded to intransigent blacks. The winch drivers said they would be killed if they worked under the prevailing conditions whereupon the shift boss lost his temper, was

[40]Commemorative Service Speech, 16 September 1983, NUM Head Office.
[41]A detailed description of the events that followed are contained in the *Industrial Law Journal*, South Africa, 1984, Volume 5, Part 2, pp101-150.
[42]*Ibid*, p125.

abusive and insulting and said he did not care if they were killed.[43]

The first crack in the solidarity of the gang came when the team leaders agreed to work after the mine captain had told them they would be stripped of their rank and possibly dismissed if they continued with their opposition. But, though he threatened the other members of the gang and allegedly assaulted one of them, they all maintained their refusal to work. He met the complete gang on the fourth day of their strike and asked those who wanted to work to step to one side, but no-one did so. Then each member of the gang was interviewed individually in the manager's office but they all gave the same negative reply. A group of white supervisors told the gang that they had inspected the disputed working area and had found it safe to work in but by then the black mineworkers were refusing to believe anything the white miners told them. On that day, 20 September, Ramaphosa heard about the dispute for the first time.

The NUM had some members at the mine, though not enough to gain recognition, which meant that Ramaphosa did not have automatic access to the mine manager. He wrote, therefore, to the personnel manager of Gold Fields asking to be able to discuss the problem. Despite the intervention of the personnel manager, the acting mine manager refused to see him. Ramaphosa, therefore, rang Colin Fenton, the Executive Director of Gold Fields who, out of courtesy he said, arranged a meeting with the acting mine manager for the following day.[44] Ramaphosa went along with Lucas Sigela, the union organiser for the Mid-Western Transvaal Region. No mineworkers were allowed to attend the meeting.

For the second time that year, Ramaphosa showed his skills as a publicist and protagonist. When he went to the meeting he was fully briefed and acted as one who had the backing of all the striking mineworkers. He explained their case and said they would go underground only if the union advised them that it was safe to do so. Even

[43]James Motlatsi described an incident from his own experience in 1970 which illustrated the manner in which white miners treated black mineworkers. He said: "In 1970 I was assaulted by the shift overseer at the Welkom mine. He assaulted me because I was the new worker in the gang - for no other reason. The other workers told me that the shift overseer would assault me until another new worker came. I still remember trying to find a way to make a complaint. I went to the mine overseer and he called me in. I told him about my complaint. *Do you have any cattle at home?* he asked me. I said yes. *Do you ever plough?* he asked. I said yes. *What do you do when one of the cattle doesn't do what you want?* I said I would then kick it. *Yes*, he said, *that is why the shift overseer kicked you.*" (NUM, *A Thousand Ways to Die*, 1986, p5.)

[44]Fenton was also the Chairman of the Chamber of Mines for that year.

though he had rarely been underground before he suggested that he and another union official should be allowed to inspect the working area. This was turned down. He then asked for the Inspector for Mines to be allowed to give his opinion about the conditions, but that too was rejected on the grounds that qualified people had already said they were safe. Ramaphosa was not informed that the Chief Inspector for Mines in Carletonville had already arranged to visit the mine that evening to conduct an inspection.

He was told, however, that unless the gang members returned to work by six o'clock the next morning they would be dismissed. This ultimatum was put to the mineworkers and rejected. Early the next morning the acting mine manager met the strikers and told them that the area had been declared safe by the Inspector for Mines the previous evening but, such was their suspicion of management intentions, they refused to believe him. Their suspicions were reinforced when the mine manager refused to allow Ramaphosa to see a copy of the Inspector's report. After a week, the strikers were again asked individually whether they were prepared to work. Seventeen still refused and were immediately dismissed and ordered out of the compound because they had 'unlawfully' persisted in refusing to work.

Two days after the dismissals, a mineworker who had agreed to work rather than be dismissed was injured in a rock fall. Two other mineworkers were also injured. The union then applied for an inquiry under the Mines and Works Act which allowed a trade union to request the Government Mining Engineer to hold an investigation when it thought there was a danger to health and safety in a mine. It also declared a dispute with Gold Fields, under the Labour Relations Act, claiming that the dismissals were an unfair labour practice. This was the first time the union had taken legal action against a mining house.

The inquiry under the Mines and Works Act was held before the Assistant Chief Mining Inspector on 9 and 10 November. His report, 206 pages long, confirmed that the working place was safe for mining. Four days later the matter, including the evidence from the inquiry, was presented before the Industrial Court whose task was to decide whether or not the mine was guilty of an unfair labour practice. The NUM did not dispute the Inspector's report declaring the area safe but argued that the management had failed to take proper steps beforehand to reassure the mineworkers that the area was indeed safe. Management, on the other hand, denied the allegation and contended that it had correctly dismissed the 17 mineworkers because they had disobeyed reasonable instructions. The Industrial Court dismissed the claim that the mine was guilty of an unfair labour practice but acknowledged that as the dispute was important for the welfare of all mineworkers the dismissed mineworkers should be temporarily reinstated as from the day they were dismissed, pending a full investigation of the case. At that investigation it was decided that, as it was not possible to say many weeks after the event that the area was technically unsafe at the time of the complaint, the dismissal of the mineworkers was valid. The NUM accepted that decision and withdrew its case, so the mineworkers remained dismissed. This left unresolved the primary question as to whether mineworkers had the right to refuse to work in what they regarded as dangerous conditions.

Before the West Driefontein case was settled, the NUM had to return to pursue the case for the dependants of the Hlobane disaster. But thereafter it neither forgot nor neglected the safety issue. In between West Driefontein and the Second Congress of the union on 10 December 1983 there were three disasters. One was in September in a chrome mine at Potgietersrus in the Northern Transvaal when six were killed. The other two were in Anglo-American mines. In October, 10 died in a fire at Vaal Reefs near Orkney while the following month six were killed at Western Deep Levels. The union was involved in investigating an allegation that the disaster at Western Deep Levels was caused by blasting while six mineworkers were still working underground.

All of the events following Hlobane were in the minds of the union delegates when they met at their Second Congress and resolved to pledge solidarity with all mineworkers who were killed at work. They called on the Chamber of Mines to involve the union on all committees dealing with safety matters and requested the Minister of Mineral and Energy Affairs to appoint a commission of inquiry to investigate safety in the mines. They stated that if the mines refused to negotiate safety agreements with the union it would instruct its members to refuse to work in unsafe places. The union, however, was not strong enough to translate its words into deeds. It had to tackle each disaster as it came though there were elements of continuity in the issues they raised. Just over two weeks after the Congress there was a spate of accidents in the Carletonville area which claimed the lives of 11 mineworkers. In one of them, on 27 December 1983, again at West Driefontein, a rockburst killed five mineworkers and injured 92.

The question concerning the ability of black mineworkers to understand the relationship between the sounds and movements of rocks, that had been raised and derided by management at West Driefontein just over three months before, was raised again at the same mine in December. Jean Leger, who had just started working on the

NUM-sponsored safety and health in mining project at the University of Witwatersrand, visited the scene of the accident and spoke with the survivors. "They recounted," he said, "how the rocks underground had 'talked', and claimed that it had been obvious that a disaster was about to take place." They blamed the accident on the mine's management for not 'listening' to the rocks. When the inquiry was held, however, this folklore of the mineworkers was ignored. No-one was blamed and the rock burst was explained, in the words of Leger, through "a paradigm which defined itself with abstract theories buttressed by a mystifying terminology of seismic events, responsible persons, hypocentres, stabilising pillars, energy release rates, Richter scales, destressed areas, active and passive supports".[45] In order to put the mineworkers' case, therefore, the union had to take on a many-faceted role which entailed demystifying official, pseudo-scientific jargon so that black mineworkers could follow safety proceedings.

Problems of Expansion

By the end of 1983, the NUM had spread its activities to 48 mines in eight regions and its membership had increased to 55 000. However, it was being faced with increasing competition from newly emerging unions. During 1983 the number of unions claiming to organise black, coloured and Asian mineworkers rose to 10 though only three of them, including the NUM, had been given access to the mines. The Azanian People's Organisation (AZAPO), which had been formed in 1978 to fill the leadership gap in the Black Consciousness Movement caused by the wholesale banning of its organisations the year before, decided at its Congress in February 1983 to form links with local trade unions and, where they did not exist, to establish new ones.[46] As a result it, formed the Black Allied Mining and Construction Workers' Union with Letsatsi Mosala, the Labour Secretary of AZAPO, as its President.[47] The union was unregistered but was allowed to recruit in the gold and coal mines under the aegis of the Chamber of Mines and claimed to have 7 500 paid-up members in 1983. Its main success was in organising

[45]J P Leger, "*Talking Rocks": An Investigation of the Pit Sense of Rock Fall Accidents Amongst Underground Gold Miners*, unpublished thesis for the degree of Doctor of Philosophy submitted to the Faculty of Arts, University of the Witwatersrand, Johannesburg, 1992, pvii.

[46]AZAPO at that time consisted mainly of black intellectuals and members of the black urban petty bourgeoisie. See Davies, O'Meara and Dlamini, 1985, pp308-310.

[47]Mosala insisted that the union was independent of AZAPO. (South African Institute of Race Relations, 1983, p156.)

about 90 per cent of the mineworkers at the Lavino chrome mine in the north-eastern Transvaal. It applied for 'stop order' facilities at the mine during 1984 but was turned down by the Minister of Manpower because it was not registered. The Black Allied Workers' Union had begun to organise mineworkers in the coal mines of Northern Natal but most of the mines there were small, and few were members of the Chamber of Mines. The Chemical Workers' Industrial Union had penetrated the mining industry through SASOL, the state-owned oil-from-coal and chemical-producing corporation which possessed its own coal mines. By 1984 it had organised most of the mineworkers at the four SASOL collieries in the Secunda area but it made no attempt to spread its influence to the other sectors of mining. That situation was accepted by the NUM.

The NUM's initial recruiting successes were with the higher grades of clerks, team leaders and personnel assistants but the reason for this was historical and not a consequence of strategy. The rules of the union stipulated that at least 100 members were required before a branch could be formed and, after the fourth revision of the Chamber's guidelines, between one-third and one-half of all the employees in each job category had to be organised in order to obtain recognition. To reach these levels the activists could not easily pick and choose which categories of mineworkers to organise. Much depended on the respective group attitudes of the workers themselves. The machine operators who had been more collectively conscious than others since wage differentials were introduced in the early 1970s often invited the union to organise them with the result that they soon formed the dominant occupational sector in the union.[48] Indeed, by the end of 1983 the union had stopped being exclusively concerned with gold mines and had began to organise all types of workers employed by Chamber-affiliated companies. It had also altered its method of recruitment and had begun to respond to requests for membership from groups of mineworkers rather than to try and recruit anyone without any understanding of their particular group needs. For instance, it organised diamond mineworkers from Namaqualand and phosphate mineworkers from Phalaborwa after receiving requests from them to join the union.[49]

During 1984 the occupational character of the union's membership

[48]The experience of trade unions in Britain has been that skilled workers with most to protect were always the first to recognise the value of collective action. The low-paid, unskilled workers were invariably difficult to organise and the last to be recruited into trade unions.

[49]National Executive Committee Minutes, 12 February 1984.

became more varied. It gained recognition from the Chamber to represent workers at TEBA's central recruitment complex in Johannesburg, the Rand Refinery, the Rand Mutual Hospital and the Chamber of Mines' Sports Club. Following an agreement to represent mineworkers at Douglas Collieries, owned by Rand Mines, in October 1983, the union established itself at a number of other collieries, including Matla, New Denmark, Coalbrook and Kriel. The union also sent organisers to the platinum mines in Rustenburg and Bophuthatswana, but in Bophuthatswana the government had passed a law making it illegal for externally based unions to operate in its territory. However, the white miners' union (MWU) was exempted from the terms of the Act and allowed to organise as before. There were two brief strikes against the banning at Union Carbide's UCAR Minerals Mine on 23 and 24 January 1984. Union Carbide had been willing to recognise and deal with the NUM and negotiations were at an advanced stage when the Bophuthatswana government stepped in and stopped them.[50] Ramaphosa went north to Phalaborwa in November 1984 to organise mineworkers at the Rio Tinto Zinc's Phalaborwa Mining Company where 30 workers had been dismissed for refusing to work overtime.[51] By 1985, the NUM had begun to have a presence in all of the main mining areas.

The rapid growth and spread of the union, however, created organisational problems. First, the rate of recruitment outpaced the growth of the union's administrative infrastructure so that it was incapable of servicing them effectively. This, in turn, was handicapped by the lack of education and administrative experience amongst the young black men and women who operated it. The most persistently active union organisers were those who worked from a commitment to trade unionism and they mostly came from the original group of personnel assistants in the Anglo-American mines. Other organisers came from CUSA but they, as well as the additional ones recruited as the union expanded, were usually young blacks whose sole qualification was that they had reached standard eight to 10 at school and were literate in English. The union was so persistently short of staff that Ramaphosa had to accept almost anyone who offered to work for the union. There was a high turnover amongst the organisers so the union was continually

[50] *Star*, 31 August 1984.

[51] When he went to Phalaborwa, Ramaphosa was arrested and imprisoned for a night for addressing an illegal meeting. He had, however, been arrested under the Riotous Assemblies Act which was defunct and he was, in consequence, released without being charged.

having to search for replacements. The shortage was illustrated by the fact that whereas the union grew substantially during 1983 the number of full-time organisers fell from 25 to 13.

Many of these young organisers had no knowledge or understanding of trade unionism. Nor were they able to distinguish between their own needs and those of the union as an organisation. Union equipment was both wasted and neglected. They were perpetually reckless, for instance, in their use of cars which accounted for almost a quarter of the total expenses of the union in 1983. They failed to keep accurate accounts of their working time and expenses or of union subscriptions which, at that time, were collected by hand. There were instances where individual members, who were not officials, collected subscriptions without the authorisation of the union and did not pass them on to it. Similar difficulties were faced in the recruitment of head office staff members. No-one amongst the four people employed in the head office was qualified to act as a bookkeeper to maintain records and co-ordinate regional financial activities. Although Ramaphosa had announced the establishment of an Education Department and a Health and Safety Department he had no-one to run either of them. The Swedish trade union movement had agreed to finance a project to train 600 shaft stewards, but, though it had been initiated under the direction of an Education Co-ordinator, it had still not started after a year because no material had been prepared for the initial training courses. The problem of building an infrastructure that would effectively deliver services to the members was, of course, related to the deprived condition of black people in general in South African society and, therefore, at that time, did not have a solution.

Nonetheless, the union continued to expand and so did its standing in the black trade union movement. It had affiliated to CUSA and obtained a seat on its National Executive Committee. It also became internationally known and recognised. Twice during 1983 Ramaphosa went on foreign tours, established relations with foreign trade union centres and mining unions and returned with promises of material and financial support. At the end of March, Camay, the General Secretary of CUSA, arranged through the American Embassy for him to visit the United States where he met Richard Trumpka, the President of the United Mine Workers of America, and called at the ANC office in New York. From there he went to Britain and Belgium. The outcome of that visit was that the union agreed to affiliate to the Miners' International Federation in May. Then in October he visited Britain, West Germany and Sweden. He was not, however, welcomed wherever he went. The leaders of the exiled South African liberation organisations, SACTU, the ANC and the

SACP were suspicious of him because he was not one of them and they knew nothing about him.[52] Both he and Motlatsi faced the added difficulty when they were abroad of being ignorant of foreign politics. They knew very little about the credentials of the various political groups that wanted to associate with the union and were not aware, therefore, that by accepting one invitation they were possibly jeopardising others. For instance, when Motlatsi went to Holland early in 1984 he was invited to England by the Militant Tendency of the Labour Party and that severely restricted the scope of his contacts there.[53] Ramaphosa, however, was officially and enthusiastically received in Scandinavia. The Swedish trade unions decided to finance a continuing education project for the union. By the end of 1983, the NUM was receiving grants on a recurring basis from various Scandinavian union centres to such an extent that 74 per cent of its annual income came from foreign sources.

The First Legal Strike [54]

Although the leadership of the NUM tried to avoid strikes in the early years of its existence, they occurred nonetheless and in one way or another implicated the union. Its very existence aroused collectivist reactions amongst the mineworkers and, in some circumstances, encouraged them to strike. There were struggles to establish the union and attempts to get recognition, as in Bophuthatswana in January 1984. There were also spontaneous protests led by NUM members as, for example, the strike against racism by 1 400 workers at Gencor's Impala Platinum Refineries in the middle of January 1984. The management responded by dismissing the entire striking work force, including all the NUM members. The union was also involved in disputes because of deliberate management provocation designed to destroy it. The mine managers at Gold Fields, after a brief experience of life alongside a

[52]Interview with Ramaphosa, 6 June 1990.

[53]The Militant Tendency was a small but highly vocal Trotskyist group that operated from within the British Labour Party but which was under attack from its leadership.

[54]See Phillip van Niekerk, "NUM First Legal Mine Strike", *Labour Bulletin*, Vol 10, No 2, Oct/Nov 1984. During the period of the strike there were numerous reports in the *Rand Daily Mail, Star, Citizen, The Sunday Times, The Sowetan, Financial Mail* and *City Press*. The NUM also distributed its own press statements. Faith McDonald wrote a detailed editorial feature called "The Mining Dispute: Issues Within the Collective Bargaining Context". The NUM's copy is typewritten with no date. Although many statements were made about the unrepresentative character of the union, the press publicity it received was unprecedented.

black union, closed all of the union offices on their mines early in 1984. This was a prelude to many years of hostility towards the NUM.

This fragmented approach to the problems in the mines was overshadowed by the second series of national wage negotiations between the union and the Chamber of Mines that began in May 1984. A majority of the delegates at the union's Second National Congress had been dissatisfied with the 1983 wages award and complained that they had not been given the opportunity to express their views about it before it was implemented. The union leadership took steps, therefore, to involve the members at the critical stages of the negotiating process. Firstly they altered the timing of the negotiations so that the members could participate before the implementation date of 1 July. The union had to submit its wage claim in the first week of April and inform the Chamber of Mines so that the negotiations could start at the beginning of May and be concluded by 1 June. Secondly, the Congress instructed the National Wage Policy Committee to arrange report-back meetings throughout the negotiations. Thirdly, if the Chamber of Mines failed to make an acceptable offer the Committee had to convene a National Delegate Conference to make a final decision. Lastly, if that Conference rejected the offer then the members had to be consulted about taking industrial action. Given that the Chamber of Mines was completely unaccustomed to negotiating and compromising with black mineworkers and that the union was seeking to make substantial improvements in wages, this procedure was a recipe for a confrontation.

The 1983 wages demand was not finally settled until 16 April 1984. On the same day the union submitted a new demand on behalf of its members in eight recognised gold mines for a 60 per cent wage increase for the unskilled workers, 40 per cent for semi-skilled workers and 30 per cent for all monthly-paid black workers.[55] The first meeting to discuss the demand was held early in May 1984 when the Chamber of Mines rejected it and made its own proposal of eight per cent for everyone across all of the grades. Both sides changed their positions as the negotiations proceeded. Ramaphosa brought the union's demand down to 25 per cent, which was the lowest his mandate would allow. The Chamber of Mines raised its offer to just over 10 per cent but, after the 600 delegates at the special NUM National Conference on 2 June had rejected that also, it raised it further to just under 14 per cent. But the members, exhilarated by the experience of exercising some control

[55]They were Elandsrand, Western Holdings, Welkom, Saaiplaas, President Brand, Vaal Reefs West, Vaal Reefs East and Kloof.

over their wages for the first time, turned that offer down too. In separate negotiations for the five coal mines where the NUM had recognition, the final offer from the Chamber of Mines was also rejected. Ramaphosa had no alternative but to declare that the union was in dispute with the Chamber of Mines in both cases.

The Chamber of Mines chose to break the deadlock by repeating its practice of previous years, namely implementing its wages offer for all mineworkers on 1 July irrespective of what they thought about it. Its explanation was that as "the vast majority - certainly over 90 per cent - of black workers on the gold and coal mines are not members of (the NUM and the Federated Mining Union) and are expecting their annual increases on 1 July 1984, it (has been) decided to implement increases and improvements in other conditions of employment from that date... These ... constitute the offers which were rejected ..."[56] The Chamber was confident that it could get away with this because it had measured the support for the union's wages demand by the size of its paid-up membership. The union claimed a signed-up membership of 90 000 but only 13 300 had actually paid their subscriptions. The Chamber of Mines reckoned that the number of paid-up members was much lower even than that. It stated that in the recognised mines, with a labour force of 75 000, the NUM had only 8 500 paid-up union members.[57] The same point was emphasised by the manager at the Kloof gold mine when he said that only 350 out of its labour force of 13 500 were members of the NUM. The Chamber of Mines added that even these low figures exaggerated the union's influence because the majority of union members were either clerks or surface workers so that the union would find it difficult to get support from the underground workers.

No-one took Ramaphosa seriously when he warned that if the Chamber of Mines acted unilaterally there would be widespread illegal and legal strikes. But his forecast was correct. The strikes began in the coal mines on 25 June 1984 when 1 700 mineworkers at Anglo-American's Vryheid Coronation Colliery in Natal refused to work in anticipation of the imposed wage increases. The mineworkers gathered near the mine offices early in the morning and asked to speak to the manager, but he refused to discuss anything with them. Instead, he

[56]Chamber of Mines, *Annual Report*, 1984, p22.

[57]*Citizen*, 8 September 1984. Ramaphosa admitted the discrepancy but said it had been caused by the fact that the Department of Manpower had taken seven months to give the union permission to have stop-order facilities and that during that time there had been an accumulation of forms which the management had not processed.

called in the police who tried to disperse them with dogs and tear gas. The mineworkers retaliated by stoning the police and damaging buildings and vehicles. While this was happening white miners in a nearby white residential area armed themselves and, when the black mineworkers were returning to their hostels later in the night, fired on them. Two mineworkers were shot dead and four were wounded. No white miners were arrested. On the contrary, Colonel S J van der Merwe, who was in charge of the police, stated that the white miners had "all the necessary reason" to do what they did.[58] On the other hand, 77 black mineworkers were arrested and charged with public violence.

That strike was followed by a number of others. Altogether about 5 000 mineworkers, mostly from the Douglas, Kriel and Goedehoop collieries, were on strike over a 10-day period from 2 July. Then on 4 July the strikes spread to the asbestos mineworkers at Penge in the north-eastern Transvaal, even though their wages were not directly set by the Chamber of Mines and they were organised by the Black Allied Mining and Construction Workers' Union in competition with the NUM. The Griqualand Exploration and Finance Company which owned the mine dismissed the whole labour force but the mineworkers refused to vacate their hostels until, after three weeks, the company won a court case in Pretoria to have them evicted.

The NUM declared disputes with the Minister of Manpower in respect of the deadlocked negotiations with the Chamber of Mines in the gold and coal mines, the non-chamber collieries at Duvha and Rietspruit, the Rand Refinery, TEBA's central office in Johannesburg and the Rand Mutual Hospital. The Minister then set up separate Conciliation Boards, each of which comprised a Department of Manpower official and representatives of the union and employers. If a Board failed to reach an agreement within 30 days then the union was legally entitled to hold a strike ballot. At one stage, Ramaphosa was involved simultaneously in four sets of negotiations. Deadlock was reached after the first meeting of the Board dealing with gold mine wages. The NUM lowered its wage demand to 18 per cent while the Chamber of Mines improved its offer of fringe benefits but refused to give way over wages. The union, however, was only interested in obtaining a wage increase. There was a similar deadlock in the coal mine negotiations.

The union then held report-back meetings over a period of three

[58]Van Niekerk, 1984, p17. The police alleged that only one black mineworker was killed. See South African Institute of Race Relations, 1984, p333.

weeks so that its members could decide whether or not they should strike. Whilst they were being held, the union gave no indication in public of their mood. On 6 September, however, it issued a statement that its members in the eight recognised gold mines were still prepared to accept a reasonable offer but, failing that, they would take strike action from 17 September. It added that though, as an unregistered union, it was not legally required to hold strike ballots, they would be held on all the recognised mines to determine the extent of the support for the union's demand amongst all the mineworkers and to indicate "to the bosses that a large majority of workers intend to strike unless their reasonable demand is met".[59] The balloting started on 10 September and was spread over about four days. It took place in only seven of the mines as the union refused to hold a ballot at Kloof because the manager there insisted it should be completed in 24 hours. Ramaphosa described how queues of mineworkers waited to cast their votes but there was no verification of this from other sources because the mine managers refused to allow press reporters into the compounds.

Once the two parties had reached deadlock there was much discussion in the media about the impact of a legal strike in the mines on the South African economy. Some commentators speculated that a prolonged gold strike could bring catastrophe to the economy by causing a dramatic fall in the value of the rand. Ramaphosa's response was that that was the responsibility of those in charge of the economy, not the NUM. He said that considerations about the economy did not enter his mind when making demands on behalf of his members. His concern was to get higher wages. And when the Chamber of Mines complained that although wages had increased there had been virtually no movement in productivity, he countered by saying that so long as the mine owners paid starvation wages, then productivity would remain low for low wages and poor organisation acted as constraints on productivity.

Whilst this conflict was developing there were dramatic events outside of the compounds. More than 625 000 black students boycotted their classes, burned their schools and attacked the homes and cars of black education officials. There was a series of stay-aways and boycotts throughout the Vaal Triangle which stemmed initially from local protests against rent increases but which rapidly grew into a national one. In the background, organising and acting as a catalyst for protests, was the United Democratic Front (UDF), the largest mass resistance movement since the 1950s, that had been formed in August 1983 as an

[59] *Sowetan*, 7 September 1984.

alliance of civic associations, trade unions, religious organisations, women's and students' groups and other democratic organisations. There were demonstrations in Soweto where the police were stoned and buildings were burned. There was a real fear that a mining strike would generate violence which would link up with that in the townships. The government's response was to ban all anti-government meetings.

Ramaphosa did not share that fear. The union was not associated with the UDF and had made no statements about it, though the UDF, the Release Mandela Committee and various black political organisations had expressed their support for the mineworkers. He believed that as the union had strictly followed the legal guidelines set by the Labour Relations Act neither the mine security forces nor the police would intervene and that, therefore, there should be no violence. Mine managers, on the other hand, prepared to counter the legal strike as if it were an illegal one. In some instances they harassed union officials and tried to disrupt the ballot. They stockpiled ore and took what they regarded as appropriate security precautions both to protect mine property and their supplies of strike-breaking labour. The mines retained the common law right to dismiss legal strikers provided they warned them beforehand that by striking they failed to fulfil their contractual obligations. The Chamber of Mines stated that mass dismissals should not be ruled out. Gold Fields went further and distributed leaflets in all of its mines stating that "strikers will be dismissed whether the strike is legal or illegal".[60] They were not worried about getting replacements even if all the strikers were dismissed. Mr Colin Fenton, the head of the Gold Division of Gold Fields, explained that they "could replace striking mineworkers readily from the thousands of unemployed former mineworkers in the homelands".[61] The mine managers liaised with TEBA which, endorsing Fenton's claim, stated that it would have no difficulty in obtaining as many as 300 000 extra recruits if that were necessary.[62]

Shortly before declaring the ballot results on 14 September, the NUM stated that its members at the five collieries in dispute had accepted an offer from the Chamber of Mines. This left Ramaphosa free to concentrate on the gold mines where the results showed that about 70 per cent of the total labour force in seven Anglo-American mines were in favour of strike action. Only 207 of those who voted opposed the

[60]*Rand Daily Mail*, 17 September 1984.
[61]*Ibid*, 15 August 1984.
[62]*Ibid*.

strike, which, Ramaphosa declared, would start on the night shift on Sunday 16 September. However, when he informed the Chamber of the ballot result he added that he was still willing to negotiate in a last bid to avert the strike. His offer was taken up and, after seven hours of negotiations over the weekend, the mine owners agreed to raise holiday leave allowances. Ramaphosa, in turn, agreed to recommend this 11th-hour proposal to his members. However, it came too late on Sunday evening to be put to the mineworkers in all of the affected mines before the strike deadline. Only those at Elandsrand in Carletonville learned of it early enough to stop the strike from starting. The rest, numbering 75 000 in 22 different hostels, hundreds of kilometres apart, went on strike as planned.

Union organisers tried desperately to arrange report-back meetings at the Vaal Reefs, President Brand and Western Holdings mines where about 45 000 mineworkers were on strike but they were hampered by managers who had called in the police to disperse demonstrators and remove pickets. The police used batons, rubber bullets, tear gas and dogs, causing almost 200 mineworkers to be admitted to hospital with ruptured spleens and livers, fractured skulls and, in two cases, the loss of an eye.[63] At the Welkom Division of Western Holdings three Casspir armoured troop carriers and 10 police vehicles were moved into the compound. The managers at the Vaal Reefs mines near Orkney provoked violence when they prevented the union from holding meetings and sent security officials to the hostels in the early morning to order the strikers to go to work. When they refused, tear gas canisters were fired and dogs were set on them.[64]

The violence became more intense on 17 and 18 September after the strike had spilled over to three mines where the NUM was not recognised. Mineworkers at the Durban Roodepoort Deep gold mine in the West Rand and the Western Areas gold mine in Westonaria struck in sympathy with the legal strikers but suffered heavily for doing so. About 111 mineworkers were injured at Durban Roodepoort Deep while at Western Areas seven were killed and 250 were injured after striking mineworkers refused to return to their hostels. There was violence, too, at the Hartebeesfontein mine, owned by Anglovaal, when almost 90 per cent of the total labour force of about 18 000 went on strike on 18 September to demand the instant recognition of the NUM. Anglovaal refused to talk with Ramaphosa or to acknowledge his claim

[63]South African Institute of Race Relations, 1984, p334.
[64]*Star*, 18 September 1984.

that the union had about 7 000 members in the mine. When the strikers rejected an ultimatum to return to work, 10 000 were dismissed and returned to their rural homes.

The legal strike was settled on that same day. The revised offer from the Chamber of Mines was restricted to seven Anglo-American mines and accepted by the mineworkers at report-back meetings. Despite the strike, the police action, the violence and 16 deaths, Anglo-American hailed the settlement as a "victory for collective bargaining". Ramaphosa saw it differently. For the first and not the last time, he questioned whether there was any point in sticking to the rules and conducting a legal strike. "We are starting to ask ourselves," he said, "whether the agreement is worth the paper it is written on ... it's an agreement stained with blood."[65] He believed that the mine owners had not been serious about reaching a settlement because their underlying intention was to destroy the union before it became an even bigger threat to them. His view was supported by the subsequent behaviour of the Chamber of Mines before the end of the year. Disputes broke out over wages in at least 11 mines and in each case the Chamber of Mines forced the workers to accept their terms by threats of dismissals and police violence.[66]

However, by the end of 1984, the NUM seemed to be on an unstoppable course of expansion. Though it had gained a wage increase in only seven mines, it had forced the Chamber of Mines to more than double its original offer. Where it was defeated it advertised its presence by its willingness to stand up to the mine owners. Ramaphosa had become a household name in the townships and villages throughout southern Africa.[67] In the space of two years the NUM had come to symbolise opposition to white management. It had shown that it was able to confront the mine owners and not simply survive, but grow in authority. This was reflected in the confident attitude of Ramaphosa who said that "in future negotiations our union will not settle for compromises in the last hour ... Once we have taken the decision to strike, we will go ahead."[68]

[65]*City Press*, 21 September 1984.

[66]See South African Institute of Race Relations, 1984, pp334-335 for details.

[67]Phillip van Niekerk told the story of a black Sunday newspaper which ran its own crossword and had, as one of its clues, "General Secretary of the National Union of Mineworkers". He said: "Thousands of entries poured in from places as far afield as Mdantsane, Katature and Soweto. 'Not one,' said the man whose job was to sort out a large cardboard carton filled with postcards to find a winning entry, 'got Ramaphosa wrong.'" (*Rand Daily Mail*, 24 September 1984.)

[68]*Rand Daily Mail*, 19 September 1984.

The union had no problem with recruitment for those black mineworkers who came from within South Africa rushed to join it. When its Third Congress met in January 1985 it had a signed-up membership of 120 000 which gave it representation in 85 mines, compared with only 25 mines the year before. It had recognition agreements in 28 of them. But it faced a twofold problem. First the government was reluctant to grant it permission to have stop-order facilities so that management could deduct union dues from wages. The NUM, as an unregistered union, had to apply to the Minister of Manpower for stop-over facilities each time it signed a recognition agreement. Registered unions, however, received permission automatically. This issue persuaded the union to reconsider its attitude towards registration. It reversed its policy in January 1985 and decided to register. "We feel," Ramaphosa said in justification, "that registration is no longer a political issue."[69]

The second part of the problem was that even where stop-order facilities had been granted, managers sabotaged the processing of them with the result that only 54 000 of its signed-up members had paid their dues. Where this happened it was usually accompanied by other obstructive acts such as refusing to allow the union facilities to communicate with mineworkers in the compounds. In this respect there were already marked differences between the mining houses, ranging from a broad acceptance of the union by Anglo-American to outright hostility towards it by Gold Fields and Gencor. This was reflected in the distribution of the union's membership for more than 80 per cent of it worked in Anglo-American gold and coal mines.[70]

[69] *Rand Daily Mail*, 10 January 1985.

[70] By mid-1985 "about 40,5 per cent of the 163 000 black workers on Anglo's gold mines and 18 per cent of the 20 000 on its collieries were NUM members. Paid-up memberships at the other houses were: Rand Mines – 3,5 per cent of 68 200 workers on its gold mines and 25 per cent of 7 000 on its collieries; Anglovaal – 8,2 per cent of 26 800 on its gold mines; Gold Fields – 11,9 per cent of 69 500 workers on its gold mines; and Gencor – 0,5 per cent of 97 000 on its gold mines and 7,1 per cent of 18 400 on its collieries." (South African Institute of Race Relations, 1985, p199.)

APPENDIX A

STOP ORDER FOR DEDUCTION OF TRADE UNION FEES FROM MY REMUNERATION

THE MINE MANAGER
Libanon Gold Mining Company
WESTONARIA

Dear Sir

In terms of Section 78 (1B) (a) of the Labour Relations Act, 1956
I, ...
COMPANY NO: ....
being a member of the **NATIONAL UNION OF MINEWORKERS, hereby authorise you to** deduct an amount of R1,00 commencing.......................................
from my remuneration in respect of membership fees payable to the **NATIONAL UNION OF MINEWORKERS** and forward the said amount to the union on my behalf.

Yours Faithfully

WITNESSES

____________________ 1. ____________________

SIGNATURE

DATE: ____________________ 2. ____________________

I certify that the above mentioned person is a member of the **NATIONAL UNION OF MINEWORKERS** and that his monthly membership fee is R1,00

DATE: ____________________

SECRETARY OF THE UNION

MEMBERSHIP NUMBER: ____________________

CHAPTER FIVE

A BAPTISM OF HORROR THE HLOBANE DISASTER 12 SEPTEMBER 1983

The Disaster

Early in 1983, Cyril Ramaphosa had visited Britain for a discussion with the British National Union of Mineworkers about health and safety issues. The British union suggested that it should send a mining engineer to South Africa to assess the situation and provide the union with advice. This was agreed but before he could arrive the nine-month-old union was faced with a major disaster at the Hlobane colliery, about 27 kilometres east of Vryheid in northern Natal.

Three seams outcropped from under the Hlobane Mountain. The top seam, called Alfred, had been worked out. Underneath that was the Gus seam that was still being worked but was nearing exhaustion and was losing some of its pillars through extraction. The main coal-getting activities, and the future of the colliery, lay in the Dundas seam, 16 metres below the Gus seam and 185 metres from the surface, just below the base of the mountain.[1] There were about 1 500 mineworkers in the mine, 85 of whom worked in sections 5 and 10 of the Dundas seam, about three kilometres from the shaft at the north-west extremity of the mine.

The black mineworkers, who were directed to work in sections 5 and 10, stood in readiness at their waiting places at about 6.30 am on Monday, 12 September 1983. They had to stay there until their two white supervisors had tested for methane gas and had made sure that the faces were safe to work. Once they were assured that the faces were safe, the gangs went to their working places and started mining. About an hour later there was a methane gas explosion in the north-western extremity of section 5 that was so powerful that it killed or injured

[1]The data for this chapter has been mainly obtained from the *Report of the Joint Inquest and Inquiry Into the Hlobane Disaster, 7 to 17 February 1984*, Department of Mineral and Energy Affairs, 1984 and Paul Benjamin, *Annexure to Application for Additional Compensation Under Section 43 of the Workmen's Compensation Act, June 1984*. The application had been submitted by Paul Benjamin for the National Union of Mineworkers on behalf of a dependent of a diseased mineworker. Data has also been obtained from newspaper reports and subsequent analyses of the disaster that appeared in various journals.

everyone in its wake for 400 metres. Sixty-four people were killed instantly. Most were killed by multiple injuries, though 11 died from carbon monoxide poisoning. Twelve mineworkers were seriously injured but four of them, with burns covering at least 80 per cent of their bodies, subsequently died of their injuries. Only nine men who were working just beyond the limits of the impact of the explosion survived without injury. The lives of 300 mineworkers who were working on the Gus seam above were threatened when smoke and poisonous fumes seeped up from the seam below but, after waiting for three hours, they were led to safety without any serious casualties.[2]

The access to the scene of the explosion, through an incline in the mountainside, was difficult because of the presence of thick smoke and poisonous gases. Indeed, it was 10 hours before the four with the most serious injuries reached the Rand Mutual Hospital in Johannesburg, despite having the use of a military helicopter to convey them there. They arrived swathed in bandages and attached to intravenous drips. Three of them were immediately put into intensive care units. The rest of the injured, with burns covering from 20-80 per cent of their bodies, were later taken to the hospital by a Dakota aircraft. Once the injured were receiving treatment, the dead were removed to a makeshift mortuary at the mine hospital. Their bodies lay on the floor, still dressed in their boots and overalls. Some were burned beyond recognition.[3]

It took days to identify the dead bodies. It should have been possible to identify them through the lamp discs which each of the mineworkers had handed in on receipt of a lamp before proceeding underground but the records in the lamp room were so chaotic that no-one knew how many lamps had been issued or who had received them. The Joint Inquest and Inquiry into the disaster commented on the lamp room chaos and stated that if "the explosion had been more violent, this omission could have had very serious consequences".[4] It was discovered that 21 of the victims came from Lesotho. Most of the remainder were from the Transkei, Ciskei and KwaZulu.

Few survivors could remember anything about the explosion. It was over in seconds. One of them heard a huge blast and then woke up in hospital with extensive burns. Three of them recalled that they were

[2]See *The Star, The Citizen, Sowetan* and the *Rand Daily Mail* from 13 to 15 September 1983 for details about the casualties. Also see the *Joint Inquest and Inquiry*, February 1984, para 1.

[3]*Star*, 13 September 1983.

[4]*Joint Inquest and Inquiry*, February 1984, para 6(a).

ordered to work in section 10 without any gas tests being made. One said that he saw one of the white miners, Thinus Bezuidenhout, standing at the waiting place without a flame safety lamp, indicating that he had not tested for gas in section 5. There was no other oral evidence because everyone who had seen anything was killed. As soon as the alarm had been sounded, steps were taken to evacuate the whole mine. The following day some mineworkers were reluctant to go underground but they were persuaded to do so and, except for the sections 5 and 10, the mine began to operate normally, as if nothing had happened.

The Causes of the Explosion

Hlobane was classified as a 'fiery' mine, meaning that methane was expected to be found there but that there would be occasions when it could not be detected.[5] For that reason, it was not regarded as being a very dangerous mine due to the presence of gas. Nonetheless gas was an ever-present danger and had exploded on a number of occasions beforehand. There had been a major disaster in the mine in 1944 when 57 men were killed in a methane explosion. On that occasion, gas had been released as the coal seam was broken up and, as it was lighter than air, it had risen and lodged among the crevices in the roof. It had exploded when a miner forced open his safety lamp to check for a defect and had caused a spark. It was management's responsibility to prevent such occurrences by taking three precautionary steps. These were to ventilate the mine so that both gas and coal dust were dispersed; to check for their presence before and during every shift and to make sure that there were no ways in which the gas or the coal dust could be ignited. For an explosion to occur, there had to be a failure on all three counts.

A Ventilation Failure

In 1983, the bord-and-pillar method of mining was used in the mine so that its layout consisted of a series of parallel tunnels 30 metres apart that intersected at right angles with similarly spaced tunnels. The effect was a squared network of rooms separated by pillars that supported the

[5]In international mining circles, the gas was sometimes called 'firedamp' which is a synonym for methane.

roof. The height of the tunnels was determined by the thickness of the seams that varied between 1,1 and 1,3 metres. As some of the working faces were as much as three kilometres from the shaft the ventilation system had to supply a movement of air over a lot of ground in a multiplicity of confined spaces.

The purpose of the ventilation system was to disperse and dilute both respirable dust and harmful gases throughout the mine to the last of the through roads and working faces. It worked by extracting foul or used air with four main exhaust fans, thus reducing the pressure in the mine and causing fresh air to flow into it. As the air came in through the down-shaft it was directed along tunnels called intake airways to the working faces. It was management's task to control the flow of air and to get it to where it was required with as high a velocity as possible. It was important, therefore, to prevent it from being dissipated through indiscriminate circulation. All possible leakages and diversions had to be stopped. This was done in the first instance by building brick stoppings across the tunnels that were no longer in use. But where the tunnels were being used to enable men and machines to gain access to the faces, brattice-cloth made from polypropylene was used to close them off and direct the air along the quickest route to the working faces.

At Hlobane the brattice-cloth stoppings were made of a lightweight material that was easily damaged by the heavy machines that constantly brushed through them. It was the responsibility of the underground managers to repair the brattices in the vicinity of the faces but they did not treat the matter of repair seriously.[6] Sometimes the brattices were rolled up to allow vehicles, and, therefore, the air, to pass through unimpeded. Section 5 had four brattice-cloth stoppings that were not properly maintained, with the result that air passed through them and was sucked out of the mine without ever having been used to dilute and disperse methane gas where the mineworkers were working. The ventilation officer for the mine had submitted numerous reports to the production managers about these sub-standard ventilation conditions but nothing had been done about them.[7] In consequence, the volume of air reaching the faces was frequently considerably below that required by the regulations.

The ventilation problem was compounded by an incident in section 10 on the morning of Saturday, 10 September when a hole was inadvertently cut from one tunnel into another causing a gap of about

[6] *Joint Inquest and Inquiry*, February 1984, para 4.
[7] *Ibid*, para 6 (d) (iii)

1,1 metres high and 5,5 metres wide through which air was diverted with such pressure that it reversed the direction of the flow of air throughout the whole section. Ideally, each section should have had its own intake and return airways. But although it was regarded as bad practice, section 5 was served by the same air supply as section 10 so that the complete ventilation system in both sections was disrupted.

Moreover, in another act of bad practice, the ventilation systems of the Dundas and Gus seams were linked in that used air from sections 5 and 10 was directed to ventilate the seams above. It was through this conduit that the smoke and poisonous gases had passed to threaten the lives of the 300 men working on the Gus seam immediately after the explosion. In such circumstances the fault with the ventilation in section 10 had immediate repercussions over a wide area. The total effect in section 5 was to reduce the supply of air by between 60 per cent and 80 per cent. At that level it was impossible to dilute and disperse any gas that collected in the crevices or near the roof of the face workings.

The holing that was so critical had not come as a surprise. Indeed, it was anticipated at a planning meeting the week before when it was noted that if a hole were cut it would short-circuit the ventilation and should, therefore, be closed off immediately. Instructions to that effect were then given to the white miner, R J Morgan, who was in charge of section 10, and to P J Klassens, Morgan's shift boss. As soon as the holing occurred Morgan informed Klassens, who told him to close it with a brattice-cloth. On that Saturday only one shift worked from 4.30 am to 11.30 am, which meant that the mineworkers would leave section 10 at about 11 am. This time restriction created a difficulty for Morgan because by the time the hole had been blasted and cleaned it was 10 am and he was left with only an hour to prepare and install the brattice-cloth. He complained to the shift boss that, because he was short of time and of labour and could not easily get a roof bolter whom he thought would be needed to install the brattice, he would not be able to close off the hole that day. Klassens informed P E Swanepoel, the mine overseer, that there was a hole in section 10 and that he had ordered Morgan to close it off. He did not say anything about the fact that there would be a delay in closing it off.

Klassens made a note of Morgan's explanation about not having enough time to complete the job. Then, without inspecting section 10 himself, he went off duty. Morgan returned to his section under the impression that the shift boss had taken over the responsibility of informing his senior manager that the hole would not be closed before the end of the shift. He, too, went off duty. No managers, therefore, knew anything about the hole though both Morgan and Klassens were aware that it would disrupt the mine's ventilation system if it were left open and could have serious consequences. All the senior managers questioned about the matter later asserted that a permanent brick stopping should have been put in place immediately. As it was, sections 5 and 10 were left virtually devoid of air for the weekend. It was a scenario for a catastrophe in the making.

Two other factors relating to the ventilation issue contributed to the intensity of the disaster. The first was that the brattice-cloth stoppings were made of polypropylene that was inflammable and gave off poisonous carbon monoxide and carbon dioxide gases when it burned. There was already evidence from underground fires in Britain and the USA that polypropylene was a dangerous substance.[8] In this case it undoubtedly helped to spread the flame over a much wider area than would have been the case if methane alone had been involved. Its poisonous fumes, moreover, killed at least 11 of the mineworkers and then filtered upwards to the Gus seam.

The second factor concerned the daily measures to prevent the spread of coal dust. If dust was left to lie around it contributed significantly to the extent and intensity of an explosion. Ventilation, stone-dusting and water were used to stop dust becoming a hazard. But ventilation played virtually no part in dispersing dust at Hlobane because so little air got through to the faces. The mine was normally stone-dusted effectively but during the week before the explosion the water supply to the two sections had been disrupted and the stone-dusting was suspended. This, however, was not allowed to interfere with the production of coal so there was more airborne dust in the two sections than there should have been. It was evident that coal dust caused the fire to spread from section 5 along the tunnels to section 10. If the faces had been properly stone-dusted the mineworkers in section 10 might have survived.

Testing for Gas

The possibility that gas might accumulate, even in a properly ventilated mine, led to the Mines and Works Act stipulating that strict inspection

[8]Polypropylene and polyurethane were two related substances that were used in underground mines with equally disastrous consequences. They differed only marginally in their chemical composition. The polyurethane fire that occurred at Kinross gold mine in 1986 is described in chapter 9.

procedures had to be carried out at the beginning of every shift before work could start. Originally the inspection had to be done by what were described as 'scheduled persons', that is, Europeans, Cape coloureds, Malays or Mauritius Creoles born in South Africa, for only they were allowed to obtain the blasting certificates that were the necessary qualification for making inspections. No black mineworker in the whole of the coal mining industry held a blasting certificate at that time and none, therefore, was entitled to test for gas. However, as the result of an agreement between the Chamber of Mines and the Mine Workers' Union, which was exclusively concerned with white miners, a mine could obtain an exemption from the Chief Inspector of Mines to enable a black team leader to inspect a face and allow work to start, provided a white miner re-examined the area within 90 minutes. The Hlobane mine had been given such an exemption but it had expired just over three months before the explosion. The black team leaders, therefore, waited with the gangs until the two white miners who were responsible for sections 5 and 10 turned up to do the necessary inspections.

The procedure was clear and simple. Each white miner had to fill in a 'fireman's report' stating whether he had discovered any methane gas when making safe his section prior to starting the shift. The reports were then passed on to the shift boss, usually two hours later at about 8 am. In addition, the white miners were required to chalk their initials on the roof of each working place they had inspected and found free of methane gas. In British and American mines the tests had to be done within three hours of the beginning of every shift and that, in the South African mines, would have meant by 3 am. In the South African mines, however, the white miners were usually the last to arrive underground and the first to leave. The testing, therefore, was always done in a hurry and at the last minute before the shift was due to start.

On that morning, the two white miners responsible for sections 5 and 10, Thinus Bezuidenhout and Edison Venter, had signed their reports indicating that they had tested for gas and had found none but only Venter from section 10 had initialled the roofs of his working places to confirm that he had tested there. He had, in fact, done only about 60 per cent of them. In Bezuidenhout's section 5, however, there were no initials. When Bezuidenhout, moreover, was seen standing at the waiting place for section 5 prior to the start of the shift, he was not holding a flame safety lamp as would be expected if he had tested for gas. The belief was that he had not done any tests though there was no real evidence to confirm that he had not tried. There was so much gas about that Bezuidenhout could not have failed to detect it if he had made a test. But it was difficult to imagine him - or, indeed, any other miner - being so negligent as to fail to carry out formal tests for gas for it was such an essential routine matter. Nor would the black mineworkers standing at their waiting places have dreamed of starting work until they had been told that the faces were safe to enter. There was nothing in the situation, therefore, that would have encouraged a miner to forget to carry out the tests. It seemed that Bezuidenhout must have done what he was supposed to have done but had failed to detect gas. If that was the case, then the only possible explanation was that his flame safety lamp was defective.

The flame safety lamp, invented in 1891, was the standard means of testing for gas at the Hlobane colliery. It indicated the presence of gas through variations in the character of the flame as it was slowly moved from the floor to the roof but it needed a trained eye to make sense of it. The warning sign was when the flame was capped with a transparent but faintly luminous aura for that indicated that there was about two-and-a-half per cent gas in the air. At that point either work had to stop and the gas cleared or the face had to be evacuated. The problem with the lamps was that they could not make accurate readings of less than two per cent of gas, nor could they detect relatively thin layers of it against the roof. This was potentially a major defect because it was calculated that one-and-a-half per cent of methane gas was highly explosive in air containing coal dust or other explosive gases.[9] For this reason, mining operations had to stop in all British and USA mines if as little as one per cent of gas was present at a face.[10] If one-and-a-half per cent of gas was detected, then all mineworkers had to be withdrawn from the area as well. In order to recognise that amount of gas in the air, the British and US mines used battery-driven methanometers for they gave much more accurate readings of methane in the face as a whole and in the roof layers in particular. The flame lamp had largely been displaced by the methanometer in the mines of the Western industrialised countries but it had not been rejected altogether. In Britain, for instance, mine deputies, who were roughly equivalent in their duties to the white miners, usually carried both a flame lamp and a methanometer. The flame lamp was used to measure the level of oxygen so that miners did not enter areas where there was not enough of it to enable them to breathe properly while the methanometer

[9]See J Davitt McAteer, *Miners' Manual*, Washington, USA, 1985, pp152-159 for safety precautions in relation to methane gas.

[10]If one per cent of gas is found in mines in the USA, the law states that ventilation must be increased, no electrical equipment must be operated, no intentional roof falls are permitted and no blasting is allowed. (*Ibid*, p153.)

measured the presence of methane gas.

At the time of the explosion the Hlobane colliery possessed 138 flame safety lamps but only five methanometers. The performance of a flame lamp was improved substantially by fitting it with a 'probe attachment' that enabled it to detect layers of gas lying near the roof. Indeed, without it the lamp was regarded in Britain and the USA as being of little use. The attachment, however, was clumsy to carry, particularly where the seams were narrow and the working spaces were confined as at Hlobane, and miners there preferred not to carry them. Moreover, not all of the lamps at Hlobane were fitted with the attachments, although the manager of the mine claimed in his evidence to the Joint Inquest and Inquiry that they were. During an adjournment, however, he was asked to make a check and found that three of the four lamps issued to those working in sections 5 and 10, including that used by Bezuidenhout, were without attachments. So even if Bezuidenhout had assiduously searched for gas he could have left enough undetected to have caused an explosion.

There had been negligence in testing for gas at Hlobane long before the explosion. According to the Regulations of the Mines and Works Act any finding of methane gas where it had not been detected over the previous three months had to be reported to the manager, who then had to inform the Inspector of Mines. A week before the explosion Bezuidenhout had reported in his logbook that there was a methane gas level of four per cent in section 5. This was only fractionally below the level at which the gas became extremely explosive. The last recording of gas in excess of the upper limit of two per cent in section 5 had been 29 months previously in March 1981. Yet Bezuidenhout's report was not passed on to the mine manager. Moreover, the shift boss and the mine overseer had erased his entry in the logbook. When they were questioned about it, at first they prevaricated. One of them denied any knowledge of it. Then they admitted that they did it because they wanted "to avoid hassles with the loss control department" which was responsible for the safety rating system at the mine. For public relations purposes, they wanted the mine to be labelled "the safest in the world" with a five-star rating.[11]

The erasure was a major factor in the events that led to the disaster for it meant the elimination of a critical warning sign. But it was

[11]Laurie Flynn, referring to a report in the *Rand Daily Mail*, gave another explanation, which was that the "mine management believed the white miner was lying and only mentioned methane in order to excuse poor coal production in his gang". (Laurie Flynn, *Studded With Diamonds and Paved With Gold*, 1992, p269.)

consistent with the general attitude in the colliery towards methane gas readings. No-one had kept a record of the presence of gas in sections 5 and 10 for at least a year. Before then the ventilation manager had kept a logbook showing gas levels but when he heard that the mine captain was doing the same thing, he stopped it to avoid duplication. However, the Inquiry was told that the mine captain had not kept a gas logbook. Neither of them had consulted with each other. As there was no record of gas in sections 5 and 10 before the disaster, those with statutory obligations to make the mine safe were not able to examine trends and identify a build-up of gas before it reached dangerous proportions.[12] The seriousness of the omission was revealed later when 39 reports of methane gas levels in excess of two per cent in the colliery were made in the five months immediately following the explosion.

The Source of the Ignition

The presence of gas alone was dangerous because it could crowd out the oxygen and, thereby, cause death by suffocation but in order for it to explode, it had to be ignited. There was much concern about the source of the spark that caused the fire at Hlobane. A used matchstick was found at the scene of the disaster. There was no evidence to link it to the explosion though it did give rise to concern about the management's failure to prevent anyone taking 'contraband' underground.[13] The most likely source was a spark from a machine. There were many ways to cause sparks to fly at Hlobane. Undercutting the seam with a coal cutter started the extraction process. After that electric coal drills were used to drill the coal under the stone band that separated the seam from the roof to prepare for blasting. The coal was blasted and loaded on to battery-driven scoop-trams. The next stage was to remove the stone band so that the coal above it was accessible. It was drilled with jackhammers and blasted with the remainder of the coal. That, too, was removed by scoop-trams. It was possible, therefore, for sparks to be caused by the friction between shovels and rocks, faulty electricity contacts, badly maintained cutters, drills or vehicles and defective flame-proofing mechanisms.

The primary safety precaution was to make all the electrical

[12]Benjamin, June 1984, paras 82 and 83.

[13]It was illegal to take matches underground in a fiery mine. The manager was required by regulation to arrange for everyone going underground to be questioned and searched for contraband articles.

equipment flame-proof. Regulation 21.16 of the Mines and Works Act stated: "Where there may be risks of igniting gas, coal dust or other explosive material, only flame-proof, explosion-proof or intrinsically safe apparatus of a design and construction approved by the Government Mining Engineer shall be used." The job of making the equipment flame-proof rested with the mine's engineering department which was also responsible for "the visual examination of machinery, electric cables and safety devices and the protection of machinery, including electric cables, from damage by blasting or other mining operations or strata movement..." But the mine engineers were not allowed to interfere "with any fuses, control mechanisms, cables, conductors or any portion of electrical apparatus or other apparatus which is normally enclosed or kept locked ..."[14] Yet, despite this regulation, when the Inspector of Machinery entered the mine after the disaster he found that 14 of the 28 items of machinery in the two sections were not flame-proofed and that three of them that had been flame-proofed were faulty. All of those pieces of equipment had constituted a danger underground and any one of them could have ignited the gas.

The Inspector of Machinery had entered the mine as a part of the investigation team. He decided that the spark had come from the No 56 scoop-tram that was found after the disaster near the source of the explosion with its engine still running. Although it had been twice taken in for repairs during the previous three days it had not been flame-proofed. On the first occasion it had failed to work during the night shift. Its flame-proof panel was removed for inspection but it was put back in a faulty position without one of its fastening bolts so that a gap was left through which a flame could shoot. In addition, a cable was badly replaced. It was pinched and could have caused a spark. But the most likely igniting source were the contacts inside the panel that produced sparks whenever the machine was accelerated or decelerated. It should have been impossible for these sparks to escape because the scoop was fitted with a safety switch which cut off the power whenever the machine was not properly flame-proofed. The switch, however, had been taped over during the repairs and was left in that state when the scoop was sent back underground so that the safety mechanism could not operate. It seemed, therefore, that sparks had passed through the gap where the flame-proof panel was improperly fitted.

[14]Benjamin, June 1984, para 42.

A Catalogue of Errors

The failure to ventilate, to test for gas and to prevent sources of ignition took place within a context of neglect that involved the repeated infringement of 21 regulations of the Mines and Works Act.[15] The Deputy Government Mining Engineer, who conducted the official inquiry into the disaster, listed what he considered to be the most important of these infringements. They amounted to a catalogue of errors. There was such chaotic record-keeping in the lamp room that the records were of no value in identifying the bodies. There had been a number of malpractices concerning the use of explosives. Half of the items of equipment in sections 5 and 10 had not been flame-proofed. Then there were the defective maintenance of equipment, the use of easily damaged brattices, the presence of inflammable polypropylene as brattice material, serious ventilation defects in the coursing of air from one section and one seam to another, the dangerous use of a booster fan, inadequate precautions to control the level of dust, a refusal to utilise modern gas sensors, an unreliable intercommunications system, a failure to check the competence of the two white miners who were killed and deficient supervision in all areas of mining activity.[16] The actual list was longer but the indictment was complete. There was negligence, it appeared, at every level of management. It had been hazardous to go underground at the Hlobane colliery at any time.

A Four-Star Mine

Yet the managers of the Hlobane colliery were proud of its safety record. When the disaster happened it was preparing a function to celebrate in style the "millionaire's award" given to it three weeks earlier by the Chamber of Mines for having one million fatality-free shifts underground.[17] Hlobane, moreover, had a four-star rating under the International Safety Rating System (ISRS) that had been adopted by the Chamber of Mines in 1978. According to the ISRS, Hlobane was only marginally less safe than the very safest mine in the world. As the acts of negligence at the mine unfolded it became clear, therefore, that there was an enormous discrepancy between the proclaimed public image of Hlobane and its reality underground.

[15]Leger, *SA Sociological Review*, Vol 2, No 2, p31.
[16]*Joint Inquest and Inquiry*, February 1984, pp4-7.
[17]*Citizen,* 14 September 1983.

Mr Frank Bird, an American with no experience of mining who ran the International Loss Control Institute in Loganville, Georgia in the USA, had devised the ISRS. The idea of using such a rating in South Africa had been raised by a group of mine managers after attending a course in Georgia run by Bird. He was subsequently invited to South Africa to assist the Chamber of Mines in adapting the ISRS to meet South African mining conditions. At first it was applied only by the gold mines. The coal mines took it up three years later, in 1981, so that when the disaster occurred it had been in use at Hlobane for about two years. In South Africa the system was run entirely by mine managers without any external supervision. According to the Chamber of Mines it had two objectives, namely to identify and set "standards for the management work that has to be done to control injuries to persons" and to enable "performance of this work to be measured".[18] The standards were identified and set out in a manual. It contained a detailed list of accident prevention measures for the guidance of the mine managers. Their success in implementing them was measured by a safety audit for which points were awarded. The points were graded as stars ranging from one to five in ascending order of merit. Thus a five-star award was the highest rating and was indicative of a near-perfect safety record. These star ratings were given much publicity in South Africa through billboards and banners at mine entrances, the ceremonious presentation of certificates and speeches of self-adulation by mine managers and representatives of the Chamber of Mines. The media of South Africa enthusiastically joined in this charade.[19]

Mines with a five-star rating claimed to be the safest in the world, yet the system was in no sense an international one. In 1983 no mining authorities anywhere outside of the USA and South Africa had ever heard of the system or of its founder, Frank Bird.[20] South Africa's mining industry was, in fact, the only one that practised it at that time. Moreover, because of the very wide differences in the collation of accident statistics between countries and the uniqueness, not simply of mineral deposits, but of individual mines in terms of geology, work forces, cultures and markets, it was not valid to make comparisons of

[18]Quoted in H S Eisner and J P Leger, "The International Safety Rating System in South African Mining", *Journal of Occupational Accidents*, 10, (1988a) p142. This article gives a detailed and highly critical account of the working of the system. When read in conjunction with Flynn, 1992, Chapter 14, which underscores the criticisms with perceptive observations of the system and Bird's role in formulating it, the system appears as a cruelly deceptive public relations exercise.

[19]Flynn, 1992, pp276-277, comments acidly on this phenomenon.

[20]*Ibid*, p279 and Eisner and Leger, 1988a, pp142-143.

any kind between one mine and another in a different country.[21] But the problem with the star ratings system went much deeper than that for the ratings were distortions of reality even in relation to the mines to which they directly referred. The billboards and banner proclamations simply reflected an uncritical enthusiasm for unfounded claims and generated a managerial complacency about safety that had dangerous consequences for black mineworkers.

This was made clear in the case of Hlobane. The safety audit for the scheme was a joint exercise conducted by a representative of the Chamber of Mines with the manager of the mine. It started with the mine's internal audit where the manager answered the numerous questions listed in the handbook in preparation for the scoring. Dr Herbert Eisner, who had travelled from Britain to South Africa to give expert advice at Hlobane, stated that the audit "was a very major paper exercise asking question after question in theory and involving precious few real checks on the ground. Indeed, it was so voluminous that in order for a mine manager or section head to work through it, he'd have almost to fudge things in order to get to the end."[22] In fact, the manager had to write detailed replies to nearly 1 000 questions, many of which required multiple answers. The inquiry into the Hlobane disaster revealed that the manager was unable to fill in the questionnaire because he knew so little about the actual conditions underground and, in any event, had sometimes been supplied with false information about them by his subordinates.

When the external auditor from the Chamber of Mines visited the colliery he had to verify the answers, inspect the relevant documents, interview mine personnel and make a physical inspection of the mine.[23] In the case of mines with four- or-five star ratings this process took about eight days to complete. In order to finish the audit in that time the external auditor was not able to conduct his own empirical investigations but had to rely heavily on the word of the manager. The effect was to enable the manager to award his own star rating to his own mine. It was not surprising, therefore, that of the nine coal mines operating the scheme, no mine had a one-star rating while seven had star ratings of

[21]For a critical examination of such comparisons, see H S Eisner and J P Leger, "Safety in South African Mines: An Analysis of Accident Statistics", *Journal of the South African Institute of Mining and Metallurgy*, Vol 88, No 1, January 1988b, pp1-7. Opposing views by staff employees of the Chamber of Mines were published in the same journal in the July 1988 issue.

[22]Flynn, 1992, p280. Dr Eisner had recently retired as Director of the Explosion and Flame Laboratory of the British Safety in Mines Research Establishment.

[23]Eisner and Leger, 1988a, pp150-151.

between three and five.[24] This process of assessment explained why the numerous safety defects in the Hlobane mine were not identified. "Hlobane," Dr Eisner stated, "was way below standard. The failure to test for methane, the lack of flame-proofing of machinery, the defective brattice-cloth in the ventilation system ... should have been picked up by any serious safety audit."[25] The competitive nature of the scheme amongst the South African mines had led "to anything from, at best, wishful thinking to, at worst, falsification of records".[26] At Hlobane the worst had happened.

The Intervention of the NUM

There was much press comment about the lives of the two white miners, T J Bezuidenhout and E Venter and the electrician, A Cronje, who were among the dead. Stories were published describing their backgrounds, their wives and children and their fears about mining.[27] Their families received much public sympathy. The black mineworkers at Hlobane, however, lived and died anonymously. To the society beyond their compound they were simply men from the rural homelands who worked in the mine and lived in hostels.[28] They were not known in the nearby white town of Vryheid, except, commented a white female resident, as blacks who occasionally "worked in our gardens and visited our maids".[29] It was only after death that they were given names but this was simply to link them with families who would assume the responsibility for their burial. Their tragic circumstances attracted no publicity. Nothing was reported about the poverty of their families. No tear-jerking tales were written about the destitution amongst their children as a result of the loss of the only wage-earning members of their large extended families.

The black mineworkers, however, though unsung, had representation

[24]Dr Eisner calculated in 1985 that there "were 21 five-star mines, 10 four-star, 10 three-star, 11 two-star and only one-star mine"(Flynn, 1992, p281). A mine had to achieve a minimum score of 90 per cent in each element to obtain a five-star rating so that 21 of them were said to be near-perfect. Given the fatality rates in the South African mines, this appeared to be an exaggerated claim.

[25]Flynn, 1992, p280.

[26]Eisner and Leger, 1988a, p157.

[27]*The Citizen* had an article on 14 September 1983 entitled "Wives waited in fear for news of miners" while the *Sunday Tribune* reported on 18 September about the three whites as "Victims [who] didn't want to work on fateful day".

[28]See *Sowetan*, 13 and 15 September 1983.

[29]*Rand Daily Mail*, 14 September 1983.

as never before in a mining disaster. At all disasters prior to Hlobane, except Coalbrook, the interests of those killed or injured and their families were represented at inquests and inquiries by one of the recruiting agencies of the Chamber of Mines. It was an extension of the managerial control of black mineworkers till after death. In 1983, however, the families of the black mineworkers were represented by the NUM which, though it had existed for only nine months, was able to assemble a legal team and foreign advisers to protect their interests.

The union had only a rudimentary organisational structure at the time of the disaster. Nonetheless, its involvement in its investigation became a catalyst for the recruitment of members and the development of its organisation. Cyril Ramaphosa argued the union case in public from the day following the disaster until it was displaced from the press headlines by other graphic events. He blamed the management from the beginning and demanded a commission of inquiry, on which mineworkers' representatives should sit, to investigate the disaster and publish its findings. He gained first-hand information from the injured members of his union whom he visited in hospital and which he then passed on to the press. They told him, for instance, that no gas tests had been made before the shift started and that, in any event, the white miners carried outdated flame safety lamps. The mine manager, the Chamber of Mines and the government condemned his criticisms of Hlobane's safety measures but they were nonetheless reported in the press. For the first time since 1962 the mine owners failed to control the flow of information to the public about a disaster.[30] They lost their absolute control over access to the mines, interviews with mine personnel and opportunities to talk with the injured victims. Through the NUM, the tragedy ceased to be the exclusively private concern of the mine owners as most previous disasters had been.

With each public statement Ramaphosa repeated his demand for a public inquiry. He also asked for the union to be allowed to take part in the official investigations that the Government Mining Engineer had started, but his request was refused. He declared that the union would present its evidence to the subsequent official inquiry. Within three days of the disaster he had drawn up a 10-point Bill of Rights which, he said, would provide a framework for mine safety in the future by enabling black mineworkers to participate at every stage of formulating and applying mine safety measures. His Bill of Rights demanded that:

[30]The mining house, JCI, owned the *Rand Daily Mail*, one of the most influential newspapers on the Witwatersrand.

1. Mineworkers had the right to have their own safety representatives. 2. That they had the right to refuse to work under conditions or practices believed to be unsafe, unhealthy or illegal. 3. That they could demand a special government inspection of suspected violations of safety regulations or imminent dangers. 4.That they should be able to review the failure of Government Inspectors to conduct inspections into safety. 5. That they had the right to exercise rights on behalf of the safety and health of others. 6. They should be able to institute and testify in proceedings without interference or victimisation. 7. That they should be allowed to accompany Government Inspectors during inspections without loss of pay. 8. That they should participate in the development of mining procedure plans. 9. That there should be adequate health and safety training for mineworkers. 10. That mineworkers should be protected from discrimination for exercising their statutory rights.[31]

On Friday, 16 September, the NUM organised a lunchtime memorial service in Johannesburg in collaboration with the South African Council of Churches. This gave Ramaphosa an opportunity to speak at length about mine safety. He related the events at Hlobane as told by the injured mineworkers and then attacked the record of the Chamber of Mines, blaming it for putting production before safety:

"In the past 10 years," he said, "more than 8 000 miners have died in our mines: 98 per cent of those were black miners. Last year alone about 800 miners died. When 165 people die in US mines it is a national tragedy. When 800 die in South Africa the Chamber of Mines and the government say it is a good record ... The Chamber of Mines constantly deceives the public that all is well in our mines. They give figures which they claim compare well with other countries. The contrary is true ... our safety record in spite of the experience and profits the mine bosses generate is the worst in the world." He added: "All the safety committees are over-populated by white miners who are not interested in the welfare of the black miner. There is a clear and distinct conspiracy against the black miner."

Ramaphosa ended with a warning that transformed the safety question into an industrial relations issue. He said: "We must warn the Government Mining Engineer, the government, the Chamber of Mines and the mining bosses that unless a Commission of Inquiry is instituted to look into safety on the mines, the industry will know no industrial peace. We will mourn our dead in deeds."

Union members handed out leaflets during the service, declaring: "Greedy mine bosses were putting production figures before safety" and listing a series of union demands, including the Bill of Rights. A disaster fund was set up for the victims with a R1 000 donation from the South African Council of Churches. The Chamber of Mines was quite unused to this kind of public reprimand. It could not retaliate, as it had done in the past with others, by penalising the critic for Ramaphosa was neither employed by the mines nor acting illegally. All it could do was to issue rebukes in a tone that implied that it had learned no lessons from the experiences of the past nine months.

[31]Press statement by the NUM reported in the *City Press*, 17 September 1983.

The shaft stewards in the mines organised by the NUM were summoned to Johannesburg during the first weekend after the disaster to discuss a strategy for Hlobane. They agreed to call for all union members to observe Monday, 26 September as a day of mourning and to take part in a work stoppage for half an hour from 9 am to 9.30 am on the same day. Their call was endorsed by CUSA. The union sent letters to the Chamber of Mines and all the mining houses asking them to follow the union's advice.

The mine owners were in a dilemma. They could not publicly refuse to participate in mourning the dead while, at the same time, they did not want either to be seen to be following the lead of the NUM. or encouraging a work stoppage. The dilemma resulted in contradictory responses. The general manager of the Chamber of Mines stated that the disruption of mining operations was "quite inappropriate" as an expression of mourning and could be construed as an action against management. He warned that the stoppage would constitute an illegal strike and that management might take action to prevent it.[32] At the same time, the president of the Chamber of Mines, Mr Colin Fenton, said that managers would not object to mineworkers' observing a "private moment of silence" between 9 am and 9.30 am.[33] In the mines, however, the managers told workers that they would be allowed to observe only 'one minute's' silence. Indeed, it was alleged that some managers distorted the contents of the union's letter by reading it to mineworkers as a request for a moment's silence only.[34] Ramaphosa reacted by saying: "We do not need the permission of mine managements to mourn our dead."[35]

The work stoppage went ahead as planned. The NUM claimed that 30 000 black mineworkers took part in it from the nine mines, including two collieries, Rietspruit and Matla, where it had members. The Chamber of Mines tried to play down the demonstration by saying that it was its "impression that there was no stoppage at all". But as very few white miners were actually present where blacks were working underground, it is unlikely that they would have noticed anything unusual anyway.[36]

Blame for the Disaster

The Joint Inquest and Inquiry into the disaster started on 7 February

[32] *City Press*, 25 September 1983.
[33] *Rand Daily Mail*, 24 September 1983.
[34] *Ibid*, 27 September 1983.
[35] *Ibid*, 24 September 1983.
[36] *Star*, 27 September 1983.

1984 and lasted for seven days. The NUM was supported by two foreign advisers who gave expert advice on various mining topics.[37] They revealed that the claim that the South African mines were amongst the safest in the world was fraudulent. A legal team, led by Paul Benjamin from the Johannesburg civil rights firm, Cheadle Thompson and Haysom, however, provided the impetus for the NUM case. Benjamin conducted the cross-examination of the 13 witnesses whose testimony provided both the evidence of the catalogue of errors which led to the disaster and the names of the people who had committed them.

The oral evidence exposed a chain of irresponsibility that stretched from Gencor, which had owned the colliery until two months previously, to the present owner, the Vryheid Railway Coal and Iron Company, and 15 of its managers and supervisors. The final report of the inquiry, published by the Department of Mineral and Energy Affairs, however, named only the three people involved in the holing incident and the two white miners who had been killed in the explosion. And although it described the events surrounding the explosion and listed the mistakes that had contributed to it, the report made no direct accusations of negligence against anyone except Thinus Bezuidenhout, who had been killed. It concluded that the final act of negligence was that "miner Bezuidenhout, who was in charge of section 5, failed to carry out the statutory tests and that he was thus primarily responsible for the accident".[38] His area of responsibility, however, was confined to testing for gas and that was only one of the three main contributory acts of negligence. Paul Benjamin, who wrote his own report on behalf of the NUM in order to make a claim for additional compensation for a dependant of one of the victims, named nine people whose negligence was alleged to have caused the disaster.

Questions were raised during the inquiry about the morale and general standard of competence of the managers and supervisors in the colliery in an attempt to try and understand why they had neglected even elementary mine safety precautions. One factor, it was thought, might have been the fears and apprehension caused by the speculation throughout 1982 that the colliery might have to close due to the worldwide recession in the steel industry. It was asserted that the uncertainty about their future continued when the ownership of the

[37]The advisers were: Dr Herbert Eisner, the former Director of the Explosion and Flame Laboratory of the British Safety in Mines Research Establishment, and Dr Robert van Dolah, who had been the Research Director of the United States Bureau of Mining.
[38]*Joint Inquest and Inquiry*, February 1984, para 5 (b).

colliery changed hands early in 1983 and jobs were lost in a management restructuring exercise. Low morale, however, could not satisfactorily explain why acts of criminal negligence had been committed.

The competence of the white supervisors was an entirely different matter. The principal factor here was the colour bar. The two white miners who were killed, Thinus Bezuidenhout and Edison Venter, were each responsible for about 25 black mineworkers. Both were young and without any real experience of mining or of exercising managerial authority. They owed their positions entirely to the fact that they were white. Bezuidenhout, who had allegedly neglected to test for gas, was 23 years of age. He had obtained his blasting certificate only three months before the disaster. Prior to that he had worked in a bank. Venter was 26 years of age and had received his blasting certificate on the Friday before the disaster. He, like Bezuidenhout, had no previous experience of mining.[39] He had arrived in South Africa in 1982 after serving in the army in Zimbabwe for eight years. Neither was familiar with the fearful underground mining conditions that made men perpetually vigilant. During the short time they had been in the mines they had acquired an instrumental approach to their jobs which led them to make mistakes, neglect some things and forget others and, therefore, to endanger the lives of those around them.

Following the completion of the Joint Inquest and Inquiry, the Attorney General instituted criminal proceedings against the Vryheid (Natal) Railway, Coal and Iron Company and five of its employees, namely the mine manager, the shift boss and miner who had been responsible for the holing incident, a mine overseer and the lampman who had been responsible for the chaotic state of the records in the lamp room. The accused comprised a random selection. Only three of them had a direct connection with the explosion. The two who were accused of neglecting the lamp room records had not been called to give evidence in the inquiry, did not appear on Paul Benjamin's list and were not, in any event, ultimately responsible for the state of the lamp room. No-one with the specific responsibility for ventilation, flame-proofing and the maintenance of the machinery was charged. All

[39]The pit deputies in the British mines were invariably recruited from the ranks of underground miners with many years' experience. In any event, the basic qualification for a deputy was to be at least 25 years of age with five years' experience underground, two of which had to be spent at the face. For a critical comparison of recruitment policies for supervisory posts in British and South African mines, see H S Eisner and J P Leger, "Safety in South African Mines II: Aspects of Supervision and Inspection", *Mine Safety and Health Congress Proceedings,* 19 November 1987, Chamber of Mines, pp11-19.

the offences related to contraventions of regulations issued under the Mines and Works Act.

The trial was held on 25 July 1984. The company was found guilty on three counts of contravening the regulations in that it failed to flame-proof the machinery, installed an auxiliary fan illegally and did not equip a conveyor belt with an automatic brake. The last two counts had little to do with the explosion. It was fined a total of R400. The mine manager who was charged with the same offences was cautioned and discharged. The shift boss and the miner were found guilty of "failing to obey an order given in the interests of safety or health", which meant that they had not stopped the hole in section 10 as instructed. They were each fined R200. The company, the manager, mine overseer and lampman were found guilty of failing "to keep proper records in the lamp room of persons to whom portable lamps had been issued", and were fined R100 each. According to the Mines and Works Act an individual act of negligence which endangered the safety of persons carried a maximum fine of R1 000. In the Hlobane case, the nine fines for contributing to the deaths of 68 persons amounted, in total, to R1 200. The NUM had succeeded in getting publicity for its cause, but not justice for the dead mineworkers and their families.

Discriminatory Compensation Payments

The matter did not end with the trial, for the NUM took up the question of the compensation paid to the dependants of the victims. The system of giving trivial lump sum payments as compensation for blacks had been altered from 1 June 1977 when a lump sum of approximately R400 was paid towards the funeral costs, a "widow's lump sum" of about R300 was given to the dependants and a pension, amounting to 30 per cent of a mineworker's earnings at the time of death, was paid.[40] The minimum wage set by the Chamber of Mines for

[40]There were many complaints about the lump sum payments. For instance, when a mineworker died on a mine, the mine owners did not always return his body to his home and it often cost more to transport the body than was granted for the whole of the funeral costs. In June 1981, the Chamber of Mines introduced a new, contributory death benefit scheme designed to provide a payment of 24 times the monthly wage plus an estimated sum for the value of food and quarters. R720 was to be paid immediately to assist with funeral expenses and the balance was to be paid later. For those killed at work this scheme operated in addition to the benefits provided by the Workmen's Compensation Act. This new scheme which required black mineworkers to contribute from their already low earnings was imposed on them without any consultation and provoked much unrest

underground workers in 1983 was R147 per month. A widow, then, received a pension of just over R44 each month to maintain herself and her children, irrespective of how many there were.[41] Provision had been made in the Workmen's Compensation Act of 1941 for periodical adjustments in pensions in the light of the rate of inflation, but nothing was stipulated about when they should be adjusted or by what proportions they should be increased. The increases granted by the Commissioner were arbitrary ones and did not keep pace with the rate of inflation which, in the 1980s, ranged between 12 and 18 per cent each year. The consequence was that the real purchasing power of pensions fell by 40 per cent between 1977 and 1983.[42]

The Workmen's Compensation Act contained a provision that enabled the recipients of pensions to claim increased compensation if they could establish that managerial negligence had contributed to the accident or disaster. Paul Benjamin, one of NUM's legal advisers, submitted such a claim for the dependants of one Hlobane victim, in June 1984, alleging negligence by the nine managers and supervisors who had given evidence at the inquiry. The Compensation Commissioner agreed to the claim in July with the consequence that the families of 40 of the deceased mineworkers were reimbursed the full financial loss the deaths had caused them. In effect, this meant not only that they received the full equivalent of the mineworkers' monthly earnings but also had them adjusted to take inflation into account. The Commissioner also agreed that the injured mineworkers should benefit from this ruling. Rand Mutual, the company formed by the Chamber of Mines to administer its insurance policies, was given the task of providing up to R2 million to finance the award.

The intention of the Workmen's Compensation Act was to eliminate the need for litigation in compensation claims. An employer, in consequence, could not be sued for damages. It remained possible, however, to sue negligent 'co-employees' for damages under common law. The NUM, therefore, initiated a civil court action on behalf of four survivors of the disaster. It sued the manager, the resident engineer and the mine overseer for damages totalling R65 000 for "pain, shock, suffering, loss of the amenities of life, disfigurement and general

involving strikes and riots. (See Margaret Knight, "The Mines Death Benefit Scheme", *South African Labour Bulletin*, Volume 9, No 4, February, 1984.)

[41]The Supplementary Living Level, which was the higher of two poverty datum lines, calculated by the University of South Africa, for August 1983 was R81,72 per week compared with R11 a week received by the widow of a mineworker.

[42]Leger, *SA Sociological Review*, Vol 2, No 2, p14.

disability".[43] The defendants did not go to court. The owners of the mine intervened and offered to meet between 50 per cent and 60 per cent of the claim in a quick out-of-court settlement. This was accepted by the NUM. Overall, because of the presence of the NUM, the Hlobane disaster was the beginning of a new, though long and incredibly slow, learning process for the owners and managers of mines where safety matters were concerned.

[43]Jean Leger, "From Hlobane to Kinross: Disasters and the Struggle for Health and Safety on the Mines", *South African Review*, 4, Southern African Research Service, 1987, pp292-304.

CHAPTER SIX

GREATER UNITY AND THE PROBLEMS OF GROWTH 1985-1986

Political Awareness

Although, by 1985, the NUM was visibly influencing the lives of mineworkers, it was operating within circumstances over which it had little control. It was part of a wider trade union movement that had its own agenda and which was, in turn, an implicit member of an even wider emerging network of organisations waging struggles for the liberation of black people. In various ways this context made demands on the NUM and imposed physical and moral constraints on its activities in such a way as to prevent it from determining its own courses of action. Its most immediate context, arising from its relationship with the Chamber of Mines and the mining houses, was the most restrictive and demanding. The union's affairs, even after such a short time, were dominated by what had come to be a yearly ritual in which it presented its demands to the Chamber of Mines and then, depending upon the Chamber's response, followed the grievance procedures stipulated by the Labour Relations Act. This process could and did run over from one wage demand to another and, therefore, from one year to another. In all of its activities, wage negotiations were the only ones where the union found it could make an impression. On all the other problematic issues, such as job segregation and the hostel system, the combined forces of the government, the mining houses and the white mining trade unions were generally arraigned against it.

The experience of this unequal and complex struggle both enhanced the solidarity of mineworkers in their respective union branches and clarified and moulded their overall political perspectives. Although they were faced by white mine owners it did not take them long to realise that the mine owners' source of power was their ownership of the mines and not their colour. As this happened, they moved perceptively closer to the ANC's non-racial policy of opposition to apartheid that was epitomised in the union by Elijah Barayi who had advocated it from the outset. This transition was illustrated by the changes in Cyril Ramaphosa's own perception of South African politics. He modified his political understanding by acknowledging that the pursuit of black advancement alone, or even as a priority, would not radically alter the standard of life of his members. This did not weaken his condemnation of apartheid or alter his belief that racial discrimination had a vital bearing on the everyday lives of black people. It simply resulted in him putting those elements into a broad class context so that he saw racism just as one repressive factor within the capitalist system. In his unfolding perception, Ramaphosa saw the NUM struggling against capitalist mine owners who happened to be white. In both ideological terms and practical everyday political struggles, this meant moving away from the Black Consciousness Movement in general, and CUSA in particular, and identifying more closely with the ANC and the South African Communist Party (SACP).

James Motlatsi did not have to make that same political journey. He had come to trade unionism via a different route. He had been unaffected by the Black Consciousness Movement because his country, Lesotho, was already politically independent and controlled by the Sothos. His politics stemmed from the class struggle within Lesotho, on the side of the exiled socialist Basotholand Congress Party, and from his experiences as an underground mineworker. He had admired Elijah Barayi for his politics from the moment he heard him at the Inaugural Congress and never seriously disagreed with his forthright, vociferous and uncompromising support for the ANC, even though his political allegiances, at that stage, lay outside the country.

These changes in political perception were reflected in a gradual movement by the NUM towards political engagement. It had been the initial intention of Ramaphosa and Motlatsi to try to steer the NUM clear of direct political involvement until it possessed a sufficiently strong institutional base to survive government repression. They were helped at that time by the fact that all of the political action was taking place in the townships where the union had no presence. From late 1984, however, this situation had begun to change as powerful pressures from the townships drew the union, despite its unpreparedness, into direct involvement in the anti-apartheid struggle. The majority of the members of the union came from South Africa and few of them were untouched by the wave of rent, consumer and school boycotts in all of the main urban areas of the country that had escalated during the second half of 1984.[1] Sometimes their own kin and friends were involved, even as casualties. But in any event, they learned about the

[1]See Kirk Helliker, Andre Roux and Roland White, "*Asithengi*: Recent Consumer Boycotts" and Matthew Chaskalson, Karen Jochelson and Jeremy Seekings, "Rent Boycotts and the Urban Political Economy" in the *South African Review* 4, Southern Africa Research Service, 1987.

struggles against repression from the flow of new recruits, during their periodic weekend visits to their homes and when they went to the nearby townships during their free time.

The NUM, however, was not immediately and directly affected by the escalation in the township violence, as were those black trade unions that drew the bulk of their memberships from the townships. The members and officials of those unions were invariably drawn into the township protests though the core participants and organisers were mainly the unemployed young men and women acting through their makeshift street and community organisations. In addition, they were directly involved in the organisation of the consumer boycotts in Soweto, Alexandra, Tembisa and the townships in the eastern Transvaal. They also initiated their own boycott activities, such as that launched against the products of Simba-Quix by the Sweet, Food and Allied Workers' Union in September 1984 after 422 of its members had been dismissed for striking in support of three sacked workers. This boycott had important collective consequences because all the main trade union bodies, anti-apartheid organisations and other black interest groups supported it. It quickly took hold in the townships and had particularly strong support in the Witwatersrand region.

The boycott overlapped with extensive school protests organised by COSAS the following month though after COSAS appealed for workers to join in with it the character of the protests began to change. The first response to the appeal was in the black township of KwaThema, near Springs, when virtually the whole community stayed away from work on 22 October. Less than a week later, a Transvaal regional stay-away committee was formed by a coalition of black protest groups, including the Federation of South African Trade Unions (FOSATU), CUSA and some independent unions, to organise a two-day strike on 5 and 6 November 1984 in support of the student demands for democratically elected students' representative councils, the withdrawal of security forces from the townships, the release of all detainees and the reinstatement of the Simba-Quix workers. It was estimated that between 300 000 and 800 000 workers took part in the stay-away.

Some of the union leaders were detained by the police during the stay-away but the employers, in general, did not take retaliatory action against the strikers. The principal exception was SASOL, the part state-owned plant for producing oil from coal, which dismissed all of the 6 000 striking members of the Chemical Workers' Industrial Union (CWIU). The NUM played no part in the stay-away but it expressed strong feelings about SASOL's decision to dismiss the strikers even though they belonged to a competitive union that was recruiting mineworkers in SASOL's coal mines in the Secunda region. A special meeting of the NUM's Central Committee was called in December 1984 to discuss what action should be taken in support of the CWIU but no decision was reached. The issue was raised again at the Third Congress of the NUM on 19 January 1985 when the delegates decided that the union would call a national strike against the company unless it reinstated all the dismissed workers. By that time, however, the dispute had crumbled. The SASOL management had encouraged the dismissed strikers to reapply for their jobs and at least 4 000 had done so before the congress was held. The NUM strike threat, however, formed a part of a national and international campaign to get SASOL to reinstate all of the dismissed workers. In response, the company agreed to take back 70 per cent of them, though that was little more than the number who had already voluntarily reapplied to go back. The SASOL management, however, instituted a selective re-engagement process to ensure that the strike activists would be excluded. Nonetheless, in a contradictory move, it granted recognition rights to the CWIU. Though the union was unhappy with the settlement package, and only reluctantly accepted it, the dispute itself persuaded a majority of the mineworkers employed by SASOL to join the CWIU.[2]

The Diversification of Strike Issues

The union membership was growing rapidly without a corresponding expansion of its regional and branch structures so that there was no effective administrative infrastructure through which there could be an interchange of communications between the branches and the national officials. There were no mechanisms to ensure either that nationally determined decisions were carried out or that branches could co-ordinate their actions through regional and national committees. Mineworkers were flooding into the union without any experience of trade unionism and, therefore, without any real awareness of the need for co-ordination and discipline to give collective action its strength. They perceived union statements as straightforward exhortations to take action by any means possible with the result that they responded by engaging in their own fragmented, unofficial and illegal protests. As the union was beginning to take a much more pronounced political posture in its resolutions and statements, there was greater scope for branch initiatives.

[2]Chemical Workers' Industrial Union, *Struggle for Workers' Rights: A History of the Chemical Workers' Industrial Union*, 1994, pp32-36.

For instance, union branches responded to a resolution at the Third Annual Congress in January 1985 that condemned the colour bar and threatened strikes against it if necessary, by calling strikes without any discussion with the union head office. There was, in consequence, a vast increase in the number of illegal strikes that, by and large, failed to achieve any of the union's objectives. In the first quarter of 1985 there were two-and-a-half times more days lost through unlawful strikes in mining than in all the other industries put together.[3]

The expansion of the union at the Vaal Reefs gold mine in the Orange Free State in 1985 provides an illustration of the localised, unco-ordinated manner in which the union was growing. When Puseletso Salae went to Vaal Reefs he collected around him a group of young mineworkers who were eager to establish the union. Between them they formed branches for separate shafts or combinations of shafts. The branches, however, were organisational skeletons with officials and relatively few members. For instance, the branch that was based at shafts No 9 and No 10, with a combined labour force of 24 000, had only about 2 000 members by late 1984. Its meetings were poorly attended and never attracted more than 100 members.

The chairperson of that branch was Lira Setona, a winch driver at the Vaal Reefs No 9 shaft, who had been recruited by Salae. Setona was an extrovert character. Though he was illiterate he disguised the fact by always carrying a ballpoint pen in his hand and a file under his arm. When he talked he tended to make extravagant gestures. Setona believed that the management's resistance to the union was responsible for the low membership. The managers never had direct contact with the mineworkers because they worked through the *indunas*, the team leaders and the personnel assistants. Setona reasoned that if, in some way, he could publicly humiliate those people, then their status amongst the ordinary mineworkers would be undermined and management's resistance to the union would be damaged.

He recruited a group of 50 young mineworkers to assist him with his union tasks. He gave them special assignments during which they were responsible to him personally and not to the branch. Then, towards the end of 1984 he summoned a meeting of the union branch in No 9 shaft. Very few people attended so he sent his special group to request the chief *induna* to attend the meeting and discuss issues with him. The *induna* was indignant that he should be asked to go to meet a mere winch driver, particularly at a union meeting, and he refused. Setona

[3]Friedman, 1987, p375.

then instructed his group to return to the hostel and bring the *induna* to the meeting "without his feet touching the floor". They did that to the great annoyance and humiliation of the *induna* and were followed by hundreds of mineworkers who witnessed the spectacle. He then sent the group to ask the senior personnel assistant to attend the meeting. That was a different matter. The senior personnel assistant was the most influential black employee in the mine with considerable powers of patronage for he distributed jobs and determined promotions. He sat in the bar surrounded by other personnel assistants, team leaders and clerks who were trying to curry his favour by plying him with a constant supply of drinks. He acted in an imperious manner and would drink half a can of beer, then push it to one side for someone else to finish it. No-one questioned his authority. So when the group told him of Setona's request he reacted even more abrasively than the *induna* had done. As before, the group informed Setona and then returned to collect the senior personnel assistant "without his feet touching the floor". Everyone in the bar followed him. The news about the humiliation of these two men spread throughout the hostel and the union meeting grew until about 5 000 mineworkers were present. The two officials were so frightened by Setona's defiance and his influence over the meeting that they cringed and apologised for refusing to attend. He told them that they should be present at a meeting at the personnel office at 2 pm the following day and that they should bring all the complaint sheets containing charges against individual mineworkers with them so they could be destroyed.

No-one believed that Setona would turn up for the meeting, but he did. This time the white personnel officer was present to use his authority in support of the two black officials but he, too, was intimidated by Setona's uncompromising approach and the presence of a large number of his enthusiastic supporters. For the young mineworkers this was a new and invigorating experience. The more they expressed their pleasure, the more frightening it was for the mine officials who dutifully brought out the complaint sheets and destroyed them for all to see. This demonstration of defiance by Setona had a dramatic impact on the hostel residents. He led a march from No 9 shaft to No 10 shaft where his exploits were equally revered by the mineworkers. His demonstration of defiance was a catalyst for the union in the mine. When he called the next NUM meeting thousands attended and heard him declare that everyone in the two shafts had to join the NUM within two weeks or they would have to leave the mine. Most of them joined. In this unconventional way the union was established at Vaal Reefs and the foundation was laid for one of the three most militant branches in the union.

The Vaal Reefs branch displayed its greater strength by engaging in boycotts, go-slows and snap strikes early in 1985 until it was sufficiently confident to call the mine's entire black labour force of 42 000 out on strike for two days over a wages demand in February. This strike, however, revealed a sharp change in the attitude of management for it tightened the mine security measures and called the police in to force the strikers back to work. Thirty-five of the strikers were shot and injured. The police violence, however, did not deter the branch for in the following month it called a second two-day strike, but this time over safety and job reservation. These two issues had also been highlighted during the Third Congress of the NUM when they were included in the wages resolution. The union wanted to sit on the Chamber of Mines' Safety Committee and asked to be included in the talks with the white miners' unions over job reservation. It made no progress on either count with the Chamber of Mines. It remained, therefore, for the branches to take them up.

An opportunity arose at the end of April 1985, after the manager of Vaal Reefs had reached an agreement with the white miners in his mine to relax the colour bar so that he could deploy cheap black labour to do some of the work that had been reserved for whites under the Mines and Works Act. He then ordered 700 black mineworkers to do the semi-skilled jobs but without giving them either the necessary training for the work or the rate of pay received by the white miners. The union claimed that could endanger black lives. The mineworkers refused to be transferred and were instantly dismissed, on the grounds that they had disobeyed an order. This caused much resentment throughout the shafts. The NUM branch, led by Setona, called a strike in protest which was supported by 14 400 mineworkers. Anglo-American's response was to dismiss all of the strikers, making it the biggest case of mass dismissals since the 1946 strike. Despite this setback, Setona continued throughout 1985 to defy, belittle and attack the mine management in a manner that humiliated them. However, he paid dearly for demeaning white superiority because, towards the end of 1985, he was assassinated, allegedly by the South African Police, whilst visiting his home.

Other mining houses had begun to take a similarly hard line with strikers, no matter what the issues were. When, for instance, 1 800 mineworkers went on strike at the Rand Mines' Rietspruit Colliery in Witbank, in February 1985, to protest about defective safety measures after a mineworker had been crushed to death, they were assaulted by the mine security force and the SAP and ordered back to work at gun point. The union members, however, weathered the experience of armed attacks by the security forces. They began to protest about diverse issues and to vary their methods. Early in 1985, the NUM members in mines in the western Transvaal started boycotts of liquor outlets in attempts to break the monopoly of liquor sales by the mines. A boycott at the Hartebeesfontein gold mine in Klerksdorp was launched in March. After it had started, a group of about 80 members of the union branch marched through the compound shouting "*amandla*" and carrying a yellow banner with the slogans "Bury Job Reservation", "May Day - Workers' Day" and "NUM 85" written on it. The mine manager regarded this as provocation and called in the police to break up the march. The scene turned into a tragi-comedy when the police officer in charge accidentally fired his gun as he jumped off the back of his vehicle and, in doing so, wounded a mineworker. Treating the shot as a signal to fire, the police then attacked the demonstrators. Fighting also began between mineworkers as union members tried to force non-members to respect the boycott. One mineworker was killed and others were injured in the fracas while the bars in the hostels were damaged and fire extinguishers were destroyed. After the boycott there was a strike over the dismissal of four shaft stewards who had been involved in organising it. The police again intervened with tear gas, rubber bullets and birdshot while the manager played his customary role by dismissing all of the 2 400 strikers. In what could have been acts of retaliation, there were bomb blasts on 30 April at the head offices of Anglo-American and Anglovaal, the owners of Vaal Reefs and Hartebeesfontein respectively.

The Employers' Backlash

This uncompromising reaction to strike action by the various mining houses reflected a distaste of black trade unionism that was shared by mine managers in general. They were experiencing a new and defiant form of black trade unionism and did not like it. It was not at all like the tame, compliant type they had envisaged in 1982. So, wherever and whenever they could, they either got rid of union activists or simply dismissed all union members in one sweep. Rand Mines dismissed the entire branch leadership at the Rietspruit colliery during the strike over safety in February 1985. There was a similar reaction from Gold Fields when 11 300 mineworkers at its East Driefontein gold mine went on strike in the same month over a string of grievances about the *induna* system, the ill-treatment of black mineworkers by white miners, the defective quality of hostel food and the refusal of management to allow shaft stewards to represent workers at disciplinary hearings.[4] After two days of police violence, more than 145 mineworkers were injured, 813 were dismissed and 19 of the 22 shaft stewards were arrested, thus destroying the entire branch structure in the mine. The union took up the case of the arrested shaft stewards and the charges against them were dropped seven months after the strike, but none of them was re-employed, so the union had to build its organisation again from scratch. The Anglo-American Corporation was even more thorough in its endeavour to eliminate the NUM from Vaal Reefs. In a costly exercise after the mass dismissals in February, it transported all the dismissed mineworkers to their rural homes, re-interviewed them all and, provided they had not been union activists, gave them their jobs back and then returned them to the mine. About 80 per cent of them returned to the mine in that way.

The jobs of all union lay officials were at risk throughout the industry from the beginning of 1985. This did not, however, deter the union members for they extended their armoury of protest weapons to include snap strikes, sit-ins and, increasingly after the western Transvaal attempt to close the liquor outlets, boycotts. Indeed, they were so impressed with the success of consumer boycotts in the townships that they began to use them as an alternative to strike action over a range of issues concerning work and hostel conditions. As many as 60 000 mineworkers boycotted the concessionary stores and liquor outlets in the compounds during May 1985, mainly in protest against unwarranted price increases. But later their use was extended to cover assaults by white miners, the behaviour of hostel managers and the queuing system that discriminated against them. Mineworkers learned for the first time that through the boycott they could collectively challenge the provision and quality of the hostel services without unduly risking dismissal. Moreover, boycotts required little organisation to initiate and could be enforced fairly effectively. It was difficult for managers to identify the perpetrators. At the same time, however, the union learned that divisive management tactics could break boycotts. Some mineworkers were bribed to break them. There were instances where hostel managers reduced prices in the concessionary stores and liquor outlets simply in order to encourage black mineworkers to break boycotts. When that happened, there was usually violence as the union members attacked those who did not support them.

[4] *NUM News,* Vol 1, No 1, Oct/Nov 1985.

The 1985 Wage Negotiations

The wage negotiations with the Chamber of Mines, which overlapped these events, started for the union at its Third Congress in January when it listed its major demands. These were then examined at its Condition of Employment Workshop in February and finalised by the Central Committee of the NUM in March. The union submitted its list of demands, including one for a 40 per cent wage increase, in April. The Chamber of Mines responded with an offer of 14-17 per cent but the negotiations ended in deadlock, whereupon the union declared a dispute with it. The Minister of Manpower set up a Conciliation Board but it was unable to settle the issue. The NUM then suggested that both parties should call in the Independent Mediation Service, a private organisation, to arbitrate between the parties but the Chamber of Mines rejected the proposal.[5] Though both sides moderated their positions, with the NUM reducing its wage demand to 22 per cent and the Chamber of Mines slightly increasing the upper end of its offer to 19 per cent, they were not sufficiently close to reach an agreement.

At that point the union began to prepare its 200 000 members at 18 gold mines and 11 coal mines to take strike action. Rumours about a strike began to circulate and union members began to put pressure on non-members to support it. Hostel managers posted notices in each room explaining the consequences of a strike and promising protection to those who refused to support one. Late in June, TEBA, the mines' recruiting organisation, circulated copies of two short video films that were shown in each mine at regular intervals each day explaining the difference between a lawful and an unlawful strike, but emphasising that though a strike might be lawful, the strikers could still be dismissed for breaking their contracts. In some mines, the mine managers intimidated union members by refusing to give them permission to hold meetings, preventing union officials from communicating with their members and generally harassing union supporters to the point where many were afraid to strike.

Preparations for the strike took place within a situation of mounting tension in the townships as their residents reacted to police and army

[5]The Independent Mediation Service of South Africa became a significant factor in South African industrial relations during 1985. Most of its cases were on the Witwatersrand but it was beginning to extend to other centres. The arbitrators had no statutory authority but companies and unions agreed to accept their decisions, including the reinstatement of dismissed workers. (See Paul Benjamin, "Trade Unions and the Industrial Court" in *South African Review* 4, 1987.)

brutality by stoning them, burning municipal buildings, threatening the lives of collaborating blacks and driving them out of the townships. The affiliates of the UDF took up the ANC call to make South Africa ungovernable and create a dual power structure by forming organs of people's power. Street and People's Defence Committees were established in the townships that gradually extended their jurisdiction beyond organising protest activities to protecting participants from police repression. In some townships they undertook crime prevention measures and enforced black law and order. It was estimated that between March and August 1985, people's courts were set up in all of the major black townships.[6] These events had an instant radicalising effect on most black people for they illustrated the wider political context of their street activities. It became clear that the success of the police, with their incompetent brutality, was not in suppressing protests but in creating a stream of recruits for the ANC and Umkhonto we Sizwe.

Disunity Amongst the Employers

When the wage negotiations ended without agreement, the Chamber of Mines, as it had done two years earlier, announced that it was going to apply its own proposals from 1 July. It asserted that it could ignore the NUM because the union represented only a fraction of the mine labour force. Ramaphosa protested that as the Chamber had a recognition agreement with the union, this was an unfair labour practice. The Chamber, however, was unmoved and insisted that it would not respond to any pressure from the union. Within the Chamber of Mines, however, there were tensions about the union's demands. Gencor broke ranks first by paying two per cent less than the Chamber's rate in five of its gold mines where the NUM was not recognised.

As soon as the mineworkers at Gencor's Bracken and Leslie gold mines in Secunda heard this news they went on strike. The company responded immediately by issuing an ultimatum to the 7 600 strikers to return to work but 560 rejected it and they were dismissed. By the evening of Sunday, 30 June, the strike, without the intervention of the union, had spread to almost 20 000 mineworkers at Gencor's other gold mines at Kinross and Winkelhaak in Secunda and to its Beatrix gold

[6]Wilfried Scharf and Baba Ngcokoto, "Images of Punishment in the People's Courts of Cape Town," in Chabani Manganyi and André du Toit, eds, *Political Violence and the Struggle in South Africa*, 1990, p342.

mine in the Orange Free State. The police were called in to disperse the strikers in the compounds and break the strike. They killed one mineworker at Winkelhaak.[7] Altogether about 1 600 strikers were dismissed in the various mines. Gencor's management tried to play the hardheaded employer who knew the 'black mind'. "It is important," its executive director stated, "to remain consistent with black employees and not vacillate. Blacks have a different cultural approach. They want to know exactly where you stand, otherwise they lose confidence in you".[8] However, it was not pseudo-psychology that ended the strike but Gencor's use of force and its dismissal of union members and shaft stewards. Its mines played no further part in the wages struggle that year.

The conduct of the ballots, the continuing negotiations and the preparations for a possible strike were all interrupted and handicapped by the imposition of a state of emergency on 20 July in 36 magisterial districts around Johannesburg, Port Elizabeth and the eastern Cape under which the police were granted sweeping powers of arrest, detention, search and seizure. The emergency powers were later extended to Cape Town. Within three months more than 5 200 black activists, including many trade unionists, had been detained. Sixteen of the top leaders of the UDF were arrested and charged with treason on 5 August 1985.

Despite this handicap the NUM went ahead with the ballot of its members in the 18 gold mines and 11 collieries where it had recognition agreements. No ballots were held at the mines owned by Gold Fields, Rand Mines and Lonrho because they would not agree with the union on the procedures for holding them. The ballot results showed majorities in support of a strike ranging from 53 per cent to 100 per cent. A strike was then fixed for 25 August but as the negotiations had not been completed by that date, the union delayed the strike until 1 September. In the meantime the Anglo-American Corporation, JCI and Rand Mines broke away from the other mining houses to make improved offers that were accepted by the union. The strike was still scheduled to go ahead on 1 September but only at the mines owned by Gencor, Gold Fields and Anglovaal that had not been involved in the July strikes and where there had been no settlement.

Both the mineworkers and the mine managers made preparations for the strike. Mine managers alleged that mineworkers were arming

[7]*Business Day*, 2 July 1985 and the *Financial Mail*, 5 July 1985.
[8]*NUM News*, Oct/Nov 1985. The Manpower Director of Gencor, Naas Steenkamp, who had been a member of the Wiehahn Commission, gave a full account of the situation from Gencor's point of view in the *Financial Mail*, 2 May 1986.

themselves. For instance, at Hartebeesfontein it was claimed that mineworkers had stored various firearms and "a quantity of dangerous weapons including knives, sticks and swords". An Industrial Court was told later that an explosive device consisting of gelignite explosive sticks and detonators had been found on the top of a hostel perimeter wall and that information had been received that mineworkers "had been instructed to petrol bomb the hoist used to lower employees underground", but the manager provided no evidence to support the allegations.[9] For their part, mine managers liaised with the police and confronted the mineworkers with such force that in some instances, as at East and West Driefontein, the strike could not even start. In other mines, such as Kloof, all but a small minority were bludgeoned into submission. The mineworkers at Hartebeesfontein claimed that managers forced them around the compound at gunpoint. Everywhere, even in the best organised mines, as at the Marievale gold mine in the Witwatersrand, strikers were attacked with tear gas and rubber bullets, dismissed and arrested. As many as 50 Hippo armoured cars, carrying members of the Defence Force, surrounded the hostels at the Beatrix gold mine. Each room was searched for shaft stewards. The police then used guns and batons to force the other mineworkers to go underground. Twenty-eight thousand mineworkers went on strike, nonetheless, but during the course of two days of strike action almost 2 000 were dismissed, 80 were arrested and 30 were injured. The scale of management reprisals forced the union to call the strike off on its second day and to seek redress through the courts.[10]

Black mineworkers, for the second time in the short life of the NUM, questioned the wisdom of going through the mediation and conciliation processes set out by the law before calling a strike for when they complied they were treated with the same brutality as if they had struck illegally and were not protected from dismissal. Indeed, they found that by striking legally they advertised their strike plans and gave managers time to prepare their defences in conjunction with the army and the police. They discovered that their most effective strikes were in the unrecognised mines where no action had been planned and where, therefore, the managers were given no time to make anti-strike preparations.

[9]*Industrial Court Report*, 23 June 1987.

[10]See *ibid* for individual accounts of the struggles with the police in the compounds.

The Legal Basis of Mass Dismissals

The mine owners had always dismissed black mineworkers *en masse* at the slightest whim and without any redress for the victims. Under the terms of the Labour Relations Act an appeal could be made against the dismissals on the grounds that they constituted an unfair labour practice. This could only be done, however, where there was a union that had a recognition agreement with an employer. As most of the strikes in 1985 were in mines where the union was not recognised, the managers had a completely free hand to behave as they wished. Where the union was in a position to claim unfair dismissals, it had mixed fortunes.

After the September strike the NUM submitted appeals to the Industrial Court against unfair dismissals and eviction from the hostels at four gold mines and two coal mines. At Marievale, the only Gencor gold mine where the NUM was recognised, 88 per cent of the workers had voted for a strike but virtually the whole labour force of 1 200 had taken part in it. Gencor dismissed them all. In an out-of-court settlement on 3 September Gencor agreed not to evict the strikers from their hostels but did so, nonetheless, the very next day. The Industrial Court ruled in October 1985 that the mineworkers had been unfairly dismissed and that they should be reinstated. The parties to the dispute reached a settlement and the dismissed mineworkers were re-employed. The court found in favour of the union in four other cases, including Gencor's Beatrix gold mine where 52 mineworkers, amongst whom were six of the seven shaft stewards in the mine, were reinstated with three-and-a-half months' back pay.

The Marievale case was the first time in South Africa that a court had reinstated black workers who had been dismissed for participating in a lawful strike. The NUM regarded it as a landmark victory that had restored its faith in the court. "The judgment," it said, "will ensure that legally striking workers have some protection, and will encourage the use of the industrial dispute-resolving machinery."[11] The decision did, in fact, influence the subsequent behaviour of both unions and employers inside and outside the mining industry. Black unions increasingly referred disputes to conciliation and mediation and, where there was no agreement engaged in lawful strikes in the belief that their members would be protected against mass dismissals, while the employers became more hesitant about issuing ultimatums which led to mass dismissals.

[11]South African Institute of Race Relations, *A Survey of Race Relations in South Africa*, 1985, p200.

The effect in the mining industry, however, was limited because black mineworkers were employed on legally binding contracts and it was only in very exceptional circumstances that the employer's right to dismiss strikers for breach of contract was restricted. Those circumstances were described during the Marievale court hearing as being where the union's behaviour was blameless in that it had no option but to strike; where it had conducted the strike in an orderly manner without any violence or damage to property, and where the employer had broken all the rules. The court's judgment recognised that industrial action was an integral part of the collective bargaining process but only under those unique conditions.[12] When shortly after the Marievale case the Industrial Court heard the union's sixth claim, namely that the strikers at Hartebeesfontein had been unfairly dismissed, it rejected any comparison with Marievale and ruled that because of the aggressive and violent behaviour of the strikers, Anglovaal was completely justified in issuing an ultimatum to them to end their strike and then dismissing those who ignored it. This dissipated much of the union's euphoria with the Marievale case.[13] In any event, the mine owners retained the legal right to dismiss strikers for breach of contract and they continued to exercise it.

The Formation of COSATU

The intensified pressures on the black trade unions compelled them to speed up the moves to get greater unity between them. The NUM was a critical factor in this process. Ramaphosa's ideological transition had altered his attitude to unity in the trade union movement for he had learned that the racial fragmentation of the black trade union movement did not represent real divisions in the working class, but simply served the interests of the employers. He entered into discussions with Joe Foster, Chris Dlamini, Jay Naidoo, John Gomomo and some other trade union leaders, in an attempt to bring the various factions in the movement together into a single federation that was strong enough to confront both the government and organised employers.[14]

[12]See Benjamin, 1987, for an examination of the court's decision.

[13]The court on this occasion showed a clear bias towards management. It stated, for instance, that the union witnesses, under oath, "pretended to be uneducated and unable to understand English. Having sat with them over a period of more than 17 days, one could not help but notice how they listened to the evidence either in English or Afrikaans and even made notes of the evidence under oath. Counsel for Hartebeesfontein has described them as either mendacious or liars. I will content myself by saying that in my opinion they manifested little regard for the truth." (*Industrial Court*, 23 June 1987.)

[14]Joe Foster was the general secretary of FOSATU while Chris Dlamini was its president.

However, there were substantial institutional and ideological obstacles standing in their way. The wide ideological differences that existed between CUSA and FOSATU meant that they could rarely agree about objectives, let alone the means to achieve them. The group of community-based unions that were affiliated to the UDF were suspicious of both of the federations. Then there were a number of independent unions that adopted a pragmatic approach to industrial and political action. In Ramaphosa's view, however, these divisions simply reinforced the case for unity and he pressed for a single united federation, which combined the most progressive political and strategic elements of each group, such as the non-racial qualities of FOSATU, and the political engagement of the UDF affiliates.

Although CUSA was involved in the unity talks, Ramaphosa had been sceptical about its intentions from early in 1984 and his scepticism was reflected in the policy of the NUM. At its Third Congress in January 1985 the NUM, believing that CUSA was not serious about unity, had resolved to join a new federation even if it had to do so on its own. And when the unity issue was raised at the CUSA conference later that year, Elijah Barayi led the NUM delegation out in protest against its insistence on giving priority to the principle of black leadership. It was the disagreement over its insistence that only black workers, whom it regarded as the vanguard of the struggle in South Africa, had any hope of bringing about radical change in the country, that led to CUSA's eventual withdrawal from the unity talks in August 1985.[15]

During the period when the NUM was distancing itself from CUSA, Ramaphosa played an increasingly prominent role in the feasibility committee that was drafting the constitution for a new federation. Through this committee he became exposed to the influence of the South African Congress of Trade Unions (SACTU), the national federation that had operated in exile in support of the ANC since the early 1960s, and of the ANC itself, which were acting behind the scenes to shape a new democratic federation based on the principle of "One Industry, One Union". SACTU wanted all unions from all sections of the movement to

Jay Naidoo was the general secretary of the Sweet, Food and Allied Workers' Union. John Gomomo, a shop steward at the Volkswagen factory at Uitenhage, was prominent in the leadership of the Metal and Allied Workers' Union.

[15]This view about black advancement was expressed later in the year by the leader of the Azanian Confederation of Trade Unions when it merged with CUSA to form the National Council of Trade Unions (NACTU). (South African Institute of Race Relations, 1986, p232.)

join in the talks and it put pressure on the UDF affiliates, which were not in any federation, to take part. The feasibility committee agreed with this aim because it also wanted a broad representative organisation.

Four hundred delegates from a wide range of unions attended the final summit meeting to decide on the creation of a new federation at Ipelegeng in Soweto on 8-9 June 1985.[16] Ramaphosa chaired it in his first appearance in public as a national labour movement, as against miners', leader. The occasion gave him a new kind of prominence for he revealed a combination of diplomacy and firmness which gained him the respect even of those opposed to the creation of the federation. [17] His recognition was reflected in the fact that the meeting unanimously chose him as the convenor of the launching conference. But it was not simply Ramaphosa's abilities that impressed the delegates. The intimate involvement of the NUM was critical for the establishment of a new federation. It was by far the largest and fastest-growing union in South Africa. With 230 000 members it comprised 43 per cent of the total foundation membership of the proposed federation. After it had shown that it was serious in its intentions by disaffiliating from CUSA in June 1985, the NUM had a pervasive influence in all of the proceedings.

The Central Committee of the NUM, at its meeting on 2 November, gave its unanimous backing to the proposal for a new federation that had come from the final summit meeting in June. The proposal called for a tight, centralised constitution based on the five principles of non-racialism, one union one industry, workers' control, representation on the basis of paid-up membership and co-operation at national level. The Central Committee also decided to support a proposal to call the new federation, the Congress of South African Trade Unions (COSATU). Both proposals were endorsed by the launching conference in the sports hall of the University of Natal in Durban on 30 November 1985.[18] There were 733 delegates present from 33 different trade unions.

The proceedings of the conference were conducted solely by Ramaphosa who sat on the platform skirted only by two recording secretaries. In a short opening address, during which he said that "the struggle of workers on the shop floor cannot be separated from the wider political struggle for liberation in this country", he set out

[16]Jeremy Baskin, *Striking Back:. A History of Cosatu*, 1991, p45.

[17]*Ibid*, p49.

[18]The decision to use this name was influenced by the fact that by labelling itself 'Congress', the new federation was associated with the ANC in the minds of ordinary workers. (Baskin, 1991, p51.)

COSATU's political direction and, in doing so, indicated the role he envisaged for his own union. He went on to add, presaging the Tripartite Alliance with the ANC and the South African Communist Party, that "if workers are to lead the struggle for liberation we have to win the confidence of other sectors of society. But if we are to get into alliances with other progressive organisations, it must be on terms that are favourable to us as workers … When we do plunge into political activity we must make sure that the unions under COSATU have a strong shop floor base not only to take on the employers, but the state as well … What we have to make clear is that a giant has risen and will confront all that stands in its way…"[19] A number of the resolutions passed at the launch reflected this political emphasis. They called for the lifting of the state of emergency, the withdrawal of troops from the townships, the release of political prisoners, the unbanning of all restricted individuals and organisations and support for an international policy of sanctions.

The conference then elected the officials to lead the new federation. Jay Naidoo, the general secretary of the Sweet, Food and Allied Workers' Union, was elected general secretary, while Chris Dlamini, the president of the Federation of South African Trade Unions, became its first vice-president. The status of the NUM was reflected in the choice of Elijah Barayi, its vice-president, as the new president. Barayi, building on Ramaphosa's political lead, declared in his acceptance speech: "to the South African government, I say: your time is over … we do not apologise for being black. We are proud of it. As from today Mandela and all political prisoners should be released. P W Botha, you have failed in your duties to release Mandela."[20] Then, in a manner that became customary for him, he departed from his script and, in a rousing and memorable political peroration, gave the government an ultimatum to repeal the pass laws within six months or face the consequences.[21]

The NUM was the only former affiliate of CUSA to join COSATU.

[19]Quoted by Baskin, 1991, p54, from the original speech that was reproduced in the *South African Labour Bulletin*, 11 (3), January 1986.

[20]Baskin, 1991, pp59-60.

[21]This was the first, but not the last, threat to the government by Barayi which it appeared to take seriously. Within two months, the State President announced at the opening of Parliament that he intended to end the pass law system and was considering releasing Mandela on "humanitarian grounds". He made a further announcement on 21 April 1986, less than six months after Barayi's speech, that the pass laws were to be repealed and that those in prison for violating them were to be released. The appropriate legislation was then introduced. (cf Eileen Riley*, South Africa 1948-1990*, 1991, pp195-196.)

The remaining 12 CUSA affiliates and the 11 members of the Azanian Congress of Trade Unions (AZACTU), which also adhered to the black consciousness philosophy, perpetuated the divisions in the trade union movement by forming a separate federation called the National Council of Trade Unions (NACTU) almost a year later. Officially, NACTU was politically non-aligned, but it established a relationship with the Pan-Africanist Congress (PAC) and reflected a mixture of black consciousness and Africanist sentiments. Its formation was resented by COSATU, which had hoped to attract its affiliates, but it did not constitute a serious threat to it. NACTU had one mining affiliate, the Black Allied Mining and Construction Workers' Union, the membership of which was currently declining from a peak of about 75 000. This union had members in the asbestos mines in the eastern Transvaal but it did not challenge the growing authority of the NUM in the rest of the mining industry. In general, NACTU became the refuge for unions that did not want to be engaged in an alliance of class forces for liberation in the company of the non-racialist ANC.

The Impact of Mass Dismissals

A critical task of the national union secretariat at that time was constructing a functioning administration at regional, branch and shaft level from the small proportion of the members who had the educational standards necessary to enable them to operate an organisation and who, at the same time, were willing to face the risks that being a lay NUM official entailed. It had 3 800 shaft stewards in 85 mines in November 1985 and each one, without any prior experience of trade unionism, had to learn about union administration and organisation. The union attempted to make up for this deficiency through a Swedish-funded shaft steward training scheme, but it had barely started so the lay branch officials and shaft stewards had to acquire their skills through experience by trial and error. This educational task was made almost impossible by the fact that most of the strikes ended in mass dismissals that usually included the union activists. The dismissed mineworkers were normally only re-employed after being sifted through a process that identified and excluded the activists.

It was not coincidental that mine managers singled out the local union officials for dismissal whenever there was a strike for in a single act they could destroy the branch organisation of the union in their respective mines and, at the same time, damage the structure of the regions. On every such occasion the union had to rebuild its organisation with inexperienced people. The situation was worse, of course, where managers dismissed not only the officials but also all of the members of the union, for then the union had to begin the whole process of recruitment again, sometimes without having access to the compounds. The impact of the dismissals policy was greatest in the mines belonging to Gold Fields, Gencor and Anglovaal, which pursued the most aggressive anti-trade union policies, for once they had destroyed the union organisation in their respective compounds they took steps to ensure that it was not rebuilt. They were not always successful but the repercussions were felt in various negative ways. One problem was that having inexperienced, inadequate branch organisations, or even no organisation at all, did not prevent or even discourage mineworkers from striking but it did ensure that, when they did, the strikes were anarchic and marked by random violence. This gave the mine managers further justification, if they needed any, to bring in the police to try to destroy entirely any kind of collective action in their compounds. The government collaborated in the attacks on the union by arresting and detaining the activists and deporting those who were foreign migrants, including mineworkers from the 'homelands' as well as Lesotho. The courts had no difficulty in endorsing the deportations for once foreign mineworkers were dismissed, they automatically became illegal immigrants.

However, not all of the effects of the strikes on the union had negative consequences. Although mineworkers risked their jobs if they joined the union, they often reacted perversely to management repression. For instance, during the July 1985 strikes in the Gencor gold mines at Winkelhaak, Bracken, Leslie and Kinross in the far East Rand, where the union was poorly represented due largely to management repression, more than 10 000 mineworkers joined the union. On 17 August, 4 000 of them gathered in the nearby Embalenhle Township to launch Secunda as the 10th region of the NUM.[22] It was already clear, although mine managers failed to recognise it, that the idea of trade unionism had caught the imagination of the mineworkers sufficiently to make it impossible to suppress it by a show of force. This was further illustrated as the union expanded in other mines through the stimulus of successive strike defeats.

[22]It needed only two branches to be established in a geographical area to warrant creating a region. The regions at the end of 1985 were the Orange Free State, Klerksdorp, Secunda, Western Areas, Witwatersrand, Phalaborwa, Natal, Namaqualand, Northern Cape, Witbank and Carletonville.

This was particularly the case in Rustenburg and Bophuthatswana where the platinum mines were concentrated. Although platinum mining had expanded to become second only to the gold mining as an employer of black labour, the union had not been able to get established because the mine owners practised their anti-trade unionism under the cloak of the law of Bophuthatswana which stipulated that unions which were not based in Bophuthatswana were illegal. The platinum mineworkers, who had been unsuccessful in striking for recognition for the NUM in January 1984, resumed their struggle to organise exactly two years later when 30 000 of them struck at four Impala mines owned by Gencor. They had many complaints, including the management's refusal to provide facilities to the NUM, racial segregation and low wages, but Gencor rejected them all and, with the help of the armed Bophuthatswana riot police, dismissed 23 000 strikers after six days and forcibly removed them from the compounds, claiming that under the law it had no alternative. The management explained that mass dismissals were not a problem for Gencor because the high level of unemployment in Bophuthatswana made it easy to get replacements.[23] Despite this hostility, however, the union recruited sufficient members from those working on the South African side of the border to form a Rustenburg region in September 1986. In the same month, Lydenburg in the north-eastern Transvaal, with its mixture of minerals, also became a region of the union. The mineworkers there had originally been organised by the United Metal, Mining and Allied Workers' Union but they transferred their allegiance to the NUM after it had held an education seminar in the region which convinced them that there should be only one union in the mining industry.

The spread of new regions made it necessary to open additional regional offices and to define more precisely their functions within the overall administration of the union. By the end of 1985, only the Witwatersrand, Welkom, Klerksdorp, Phalaborwa and Namaqualand regions had their own offices. Ramaphosa wanted to make the regions more autonomous so as to reduce the burden on the national office and to make the administration more accountable to the membership. In the guidelines he issued in August 1985, he emphasised the principle that "all union offices effectively belong to the members and they should be

[23]Baskin, 1991, pp78-79. Ramaphosa described Gencor as an enemy company and pledged to launch an international campaign against it, including measures to prevent the shipment of its coal abroad. By that time Gencor, the second-largest mining house, rivalled Gold Fields in its anti-trade union activities. It was financed largely by Afrikaner capital and stood closest to the apartheid government.

controlled by the members themselves". He then listed the various administrative tasks the regions should undertake in their role as the link between the branches and the national office in order to give substance to that principle. They had, he advised, to be scrupulous in the collection of subscriptions and the use of regional funds and also be efficient at filing, letter-writing, completing stop-order forms, maintaining membership statistics and handling members' problems in general.[24]

These exhortations, however, presumed that the lay branch and regional officials were in office long enough to learn the basic tasks of administration. Unfortunately, due to the reprisal tactics of managers, that was rarely the case. The spread of the regional structures, in consequence, was formal, with little real substance. It made little visible impression on the decentralisation of the union's activities. The head office in Johannesburg, consisting of the secretariat and full-time staff, was the only section of the union organisation that the mine owners could not effectively destroy and which, therefore, had a continuing existence during which the acquisition of organisational skills was possible. This concentration of knowledge and expertise by the full-time, appointed staff, mostly with little or no experience of mining, gave them exceptional influence in managing the affairs of the union. The head office, in consequence, became the source of all vital decisions in the union and the main provider of its administrative services.

Disruption in the Head Office

The head office administration, however, operated within its own constraints and possessed its own contradictions that led to difficulties in the provision of services to members. There were two particular problems. First, due to the government's Bantu education policy there was a national shortage of black men and women with the education, training and experience to administer a rapidly expanding organisation. Ramaphosa, in consequence, found it hard to recruit and retain staff. Some he employed had no particular interest in trade unionism and regarded a job with the NUM as a stepping-stone to a more lucrative post. That alone would have been problematic in any administration, but the NUM's administration was not ordinary and this was the source

[24]NUM, Internal Memorandum re the Operation of Regional Offices, 12 August 1985.

of the second difficulty. In the absence of effective branch and regional administrations the ordinary members perceived the head office as the union and expected it to deliver services to them. They gave it credit when there were successes, but they also held it to be directly responsible for failures. When they were aggrieved they went directly to the head office in Johannesburg and demanded to see either Ramaphosa or Motlatsi.

In January 1986, the NUM's members were served by its head office at Lekton House in Wanderers Street, Johannesburg, with a staff of 32 full-time officials who were distributed, though not equally, between six departments. There were seven in the organising department but only one looking after health and safety, two dealing with finance, three with responsibility for research, information and publications, including a monthly *NUM News*, and three handling the ambitious education programme. Women were prominent on the staff doing legal, negotiations, clerical and administrative work. Computers had just been introduced in the office and Merriam Mashaba, who had previously worked at the Saaiplaas gold mine, near Virginia in the Orange Free State, was taken on as the union's first computer operator.

Ramaphosa recruited the union staff himself and both supervised and co-ordinated their activities. He also undertook a number of administrative tasks. He wrote his own speeches and memoranda, drafted his own letters and press statements, often acted as his own telephonist and gave a hand whenever there was pressure in the office. In addition, he consulted with and advised members, addressed meetings in various parts of the country, reconciled differences between the regions, activated the union's policy-making procedures and implemented policy decisions. He led the negotiations with the Chamber of Mines and intervened with the individual mining houses and the government over mine health and safety issues. He shared the international work of the union with Motlatsi and played his part as a roving ambassador for the union in the mining areas of the world. Increasingly, from the end of 1985, he liaised with other sections of the liberation movement, including COSATU and the UDF.

Ramaphosa's all-embracing approach to leadership did not encroach on the functions of James Motlatsi, with whom he had an agreed division of labour. Motlatsi, who had been consistently re-elected as the president of the union since 1982, was still employed as a personnel assistant at Western Deep Levels. He and Ramaphosa had become close friends even though they came from entirely different backgrounds. This friendship enabled them to support each other through many difficult spells in the union. But their relationship was more than that. They had contrasting but complementary qualities that served the union well during that period of its development. Ramaphosa, with a university background and without an intimate knowledge of mine work, adopted an analytical, but pragmatic, approach to its problems whereas Motlatsi, highly literate but without a formal higher education, responded to his own gut reactions resulting from his experience as an underground mineworker between 1970 and 1975 and his subsequent work as a personnel assistant at Western Deep Levels. Yet despite the fact that their understanding of situations was reached via different routes, they neatly dovetailed each other and they rarely disagreed over policy or strategy.

In practical terms they forged an effective division of labour. The control of the union administration and the conduct of negotiations were undertaken directly and solely by Ramaphosa. Motlatsi would be seen at the head office but would do little other than counter-sign cheques. He was present during negotiations but rarely made a vocal contribution. This was mainly because negotiating was an acknowledged part of Ramaphosa's brief. It was Motlatsi's responsibility to 'sell' the results of negotiations to the union members and to mobilise them when there were disagreements with the mine owners. He was also responsible for recruiting and extending the influence of the union in the mines. Motlatsi, of course, was restricted in his activities because he was a part-time official. He had to get permission from his immediate white superior at Western Deep Levels whenever he wanted time off though, he said, he was rarely refused.

This division of labour matched their respective abilities for Ramaphosa had, in fact, learned to be a consummate negotiator and administrator while Motlatsi was happiest when he was addressing mass meetings, recruiting and mobilising members. Any criticisms they had of each other in these areas were expressed in private and usually as banter. On one occasion, for instance, Ramaphosa asked Motlatsi, with reference to recruitment problems at Western Deep Levels: "What's the use of negotiating an agreement when you can't even get it recognised at your own mine?" At that time there were only 2 500 union members at Western Deep Levels out of a work force of about 18 000. On another occasion, Motlatsi quipped: "How do you expect me to get support for an agreement as bad as that?" But they took even implied criticisms of each other seriously. Ramaphosa had relatively little scope to improve on his agreements for, more than anything else, they resulted from the balance of power between the mine owners and the union which, at that time, was considerably in favour of the mine owners. Motlatsi, on the other hand, had much scope to respond. Within a few weeks of being 'ticked off' about Western Deep Levels he had organised a boycott of the liquor outlet, the concessionary store and the kitchen. On hearing a rumour that there was going to be a boycott, the hostel manager tried to subvert it by reducing the price of beer by 25 per cent but the boycott went ahead nonetheless and, in the heat of that struggle, 13 500 new members joined the union.

Early in 1986, however, Motlatsi and Ramaphosa did have a serious disagreement about an issue that disrupted the work of the head office and had potentially disastrous ramifications for the union as a whole. Its origin lay in the belief by ordinary members that the union was theirs and existed exclusively to serve them. It was stimulated by Ramaphosa's dictum that the union offices belonged to the ordinary members and which they took to heart. Whenever they wanted any kind of assistance they went straight to the head office and waited patiently to see Ramaphosa or Motlatsi. Sometimes they hung around for days, though they invariably went away each evening. It was a common sight to see them in small groups, wrapped in their blankets in the corners of the offices. For most of the time it was a manageable situation.

A survey conducted by the Human Resources Laboratory of the Chamber of Mines in the middle of 1985 revealed that the most widespread reason for the union's popularity amongst mineworkers was the belief that it could get them reinstated if they were dismissed.[25] This belief was boosted by the union's success in the courts in getting dismissed workers reinstated during the last quarter of 1985. When, therefore, Gencor dismissed 23 000 mineworkers at Rustenburg early in January 1986 many of them firmly believed that provided they could get to Johannesburg to give Ramaphosa the facts, they would get their jobs back. So a number of them travelled to the union's head office where they met Ramaphosa who, after hearing their story, launched an appeal against dismissal on their behalf.

There was a problem, however. They had no accommodation in Rustenburg for they had been forced out of their hostels in the compound and they did not want to return to their homes in the rural areas until they knew the outcome of the appeals. They decided they had no alternative but to remain in Johannesburg, though they had no money to pay for hotel accommodation or to buy food, drink and cigarettes. In their belief that the union was an all-providing institution, they demanded that it should finance them and they sat firmly in the head office until it did. This practice had been tolerable when a few

[25]Mine Hostel Managers' Association of South Africa, *Circular No 1/85*, 1 September 1985, p14.

people were involved but it became unmanageable when hundreds of them sat wrapped in their blankets, smoking and talking in every room of the office and insisting on staying there throughout the days and nights until their needs were satisfied.

Ramaphosa took two steps to try to ease the problem. First, he allowed 100 members to occupy one floor of the office where they could eat and sleep. Then he gave them money. The union had received finance from abroad during 1985 to support strikers, but it had not spent it all, so Ramaphosa proposed that the union should use the surplus to give R300 to each of the dismissed mineworkers in the hope that they would either return home or find their own accommodation. Motlatsi was opposed to this suggestion because it was an open-ended offer that might rebound on the union and possibly damage its finances if large numbers of disgruntled members decided to use it as a precedent to back their own claims. But he was over-ruled when Barayi and the national treasurer supported Ramaphosa.

Motlatsi proved to be right. The news spread that it was possible to get money from the head office and within about two months there were almost 400 dismissed mineworkers occupying the office and demanding cash. They wanted R450 each and refused to leave without it. Ramaphosa tried to compromise with an offer of R320, but they turned it down. They also refused to leave the office at the end of each working day, made it their home and cluttered it with their blankets and belongings. They became impatient and angry and threatened to assault members of the staff unless their demands were met. Merriam Mashaba, who was working in the office at the time, was told that she would be thrown out of the window unless she co-operated with them. Life became so unpleasant for the staff that they threatened to strike unless all of the mineworkers left, but their threat had no effect. The work of the office and, therefore, of the union, was so completely disrupted that some of the staff were convinced that the union would just "fade away".[26] This went on for about three weeks and cost the union approximately R15 000 a day in maintenance. It was already heavily in debt without that extra financial burden.

At this point, Ramaphosa, after some pressure from Motlatsi, left the office and returned to his home in Soweto. He was convinced that the problem had no peaceful solution yet he was opposed to the use of violence in the belief that it would create even more difficulties. He was in despair over the damage being done to the administration of the

[26]Conversation with Merriam Mashaba, 19 April 1995.

union. Motlatsi, on the other hand, believed in the dictum that every problem had a solution. "If there is no solution," he frequently said, "then there is no problem." Moreover, he was prepared to countenance violence if necessary. With Ramaphosa absent in Soweto, Motlatsi took the initiative. During the crisis he stayed at the Victoria Hotel where he discussed the problem with a small number of close associates in the union, including George Nkadimeng from the Witwatersrand region and Lebohang Hlaele, who was the chairperson of the nearby Grootvlei branch. They discussed various possible solutions but disagreed over whether or not they should use violence to evict the occupants. They did agree, however, without informing Ramaphosa, that Hlaele should go to Grootvlei and bring a group of his members back to the office in order to try to persuade them to leave.

The Grootvlei branch had a reputation for providing a heavy-handed security service for the union. Its members travelled around in red combis and were nicknamed '*Amabutho'*, or 'the army'. Hlaele, therefore, arranged for 30 of them to go to the union office in their combis in the middle of the night. As they neared Johannesburg, Hlaele decided that it would be impossible for them to persuade the occupants of the office, who were many times their number, to leave peacefully, so he planned to surprise them whilst they were sleeping and drive them out by force. The tactic was successful in one respect. All of the mineworkers were forced into the road and were dispersed. Some ran out without their clothes. Many were badly beaten, in some cases with broken arms or legs; at least one was seriously injured and one allegedly died from his injuries. The members of the Grootvlei branch disappeared into the night immediately after the incident. Hlaele then reported to Motlatsi that the office had been emptied.

When Motlatsi heard the news he believed that the problem had been resolved, so he did not immediately return to the office. In the meantime, however, some of the dispersed mineworkers had regrouped and, joined by others, had re-entered the building. They again detained the staff and threatened them with violence. During the day, members of the union's NEC, including Elijah Barayi, the union vice-president, turned up and discussed various ways of ending the siege. Motlatsi suggested getting the support of some local students who were not at school because they were supporting a school boycott. He thought that they could mingle with the mineworkers without being suspected and help to disarm them on a pre-arranged signal. Hlaele, however, contended that it was unrealistic to expect school children to overpower mineworkers. Barayi counselled caution and non-violence but the general view was that, as the occupants would not listen to reason, there was no alternative but to repeat the Grootvlei tactics. This time, members of the NEC were at the centre of the action. Some students were present but they did not play their assigned role. The protesting mineworkers were bundled out of the office for the second time. There was some rough handling, but less violence than before. The leader of the siege was a union organiser who had identified himself with the protest. Though he was armed, he was overpowered, forced into the boot of a car and driven away. He was never seen again, but it was rumoured that he was returned to his home in the rural areas.

At that point, the dismissed mineworkers gave up any hopes of getting more money from the union and they departed with only sufficient cash to pay their fares home. Although the tumult had taken place in public, no police were called so no-one was charged with causing violence. Motlatsi believed that this was because the state security forces had instigated the disruption in order to cripple the administration of the union. This was a plausible idea, but there was no evidence to support it. The press heard rumours about the incident but refrained from publishing reports about it. Motlatsi was glad of that because he did not want the union to be accused of organising an assault on its own members. Indeed, he wanted the disturbance to be kept entirely from public view. The union kept no record of it but news of it spread widely amongst the membership of the union.[27] Later, whenever retrenched mineworkers petitioned the head office for financial assistance, they generally dispersed at 6 pm because, they said, that was "Motlatsi's hour".

The disruption of the union's business had occurred at a difficult time for the NUM for it was struggling to assert itself against increasingly perverse and hostile managements while attempting to exercise control over its own exuberant, volatile and strike-prone members. It was also overwhelmed by debt and could survive only through overseas financial assistance. Both Motlatsi and Ramaphosa were trying to prevent the situation from getting beyond their control. The upsurge of unlawful strikes continued, however, and took on different forms, such as go-slows, sit-ins, short shifts and stay-aways. Many of them occurred without Ramaphosa and Motlatsi even knowing about them.[28] To add to the troubles, there had been such a dramatic increase in the incidence of mine violence from the second half of 1985 that the Gold Division

[27]There was an oblique reference in the NEC minutes to a problem in the head office, but it was written so as not to stimulate anyone's curiosity.

[28]South African Institute of Race Relations, 1986, p273.

of the Anglo-American Corporation had decided to co-operate with the NUM in an enquiry into its causes. The investigation had begun in November and ended with the publication of *Reaping the Whirlwind* in May 1986.[29] But even whilst the report was being compiled, there was a sharp escalation in the number of deaths and injuries from fighting in the compounds between groups and with the police and mine security.[30]

[29]A representative steering committee of four was established which included Cyril Ramaphosa and Bobby Godsell of Anglo-American. The research was undertaken by a small team directed by Professor James Leatt of the University of Cape Town and Paulus Zulu of the University of Natal.

[30]In the country as a whole, with the state of emergency still in force and the police and army operating on a free rein, there were 650 deaths from violence between January and November 1985.

CHAPTER SEVEN

APARTHEID IN CRISIS 1986-1987

Insurrection and Sanctions

The year 1986 was the high point in the insurrection in the townships that had begun two years earlier.[1] The mass protests through rent strikes, consumer boycotts, funerals, rallies, school boycotts, political 'stay-aways,' underscored by the ANC's intensified armed struggle, interacted with each other in large parts of urban black South Africa to make a reality of the ANC's call early in 1985 to "make the townships ungovernable". Rent strikes denied the local authorities the revenue they needed to operate. Consumer boycotts reduced the profitability of small firms and drove many of them into bankruptcy. The funerals and rallies enabled blacks to parade their defiance and their political allegiances through displaying the illegal flags and banners of the ANC, the South African Communist Party and the Soviet Union. The political 'stay-aways' involved hundreds of thousands of workers in identifying with the ANC, whose acts of armed insurrection in 'white' areas had not only increased from 45 in 1984 to 230 in 1986 but had spread to 'soft' targets where white people were likely to be killed. The school boycotts freed young people in the townships to attack all institutions connected with the government and all blacks who served in them. Black mayors, councillors, policemen and others defined as collaborators were driven from their homes or killed by 'necklacing' whereby the victim was beaten up and then killed by having a car tyre soaked in petrol hung around his or her neck and set alight until it clung to the skin like boiling tar. It was estimated that there were between 300 and 400 necklace murders in 1986. The strength of the insurrection was indicated by the assertion in the House of Assembly that the government had lost control of a number of black townships.[2] Government officials could not enter some parts of urban South Africa without substantial army protection and in many more areas they possessed little or no authority. The street committees, self-defence units, people's courts and

[1]See Robert M Price, *The Apartheid State in Crisis: Political Transformation in South Africa 1975-1990*, 1991, pp192-215 for an analysis of the insurrection between 1984 and 1986.

[2]South African Institute of Race Relations, 1986, p513 and p515.

committees that administered health and education filled these government authority vacuums.

The character of the insurrection and the harsh government reaction to it evoked quick responses from abroad as anti-apartheid bodies in the USA, Britain and other Western European countries put intensified pressure on their respective universities, churches, local and central government authorities and private institutions to introduce and extend economic sanctions against South Africa. Their reactions were tougher on apartheid than had been the case after both the Sharpeville massacre and the Soweto uprising. The officials of the NUM played a prominent role in promoting the policy of sanctions in the international trade union movement. They, along with other union representatives, became roving ambassadors for sanctions against South Africa as they attended foreign trade union gatherings.[3] For instance, Motlatsi and Ramaphosa received widespread publicity and a rapturous reception when they attended the Annual Conference of the British National Union of Mineworkers in South Wales in 1986 and urged the British to ban South African coal even though that would harm the job prospects of their own members. They met representatives of the British Trades Union Congress and the Labour Party to reinforce the case for sanctions. In a television interview in London, Motlatsi called for full, not partial, sanctions against South African gold, which riled the Chamber of Mines much more than demanding coal sanctions. When Motlatsi returned to South Africa, the Chamber of Mines demanded an explanation of his call for "sanctions against and disruption of the gold mining industry".[4] None was given because, a union statement explained, the NUM had persistently called for international economic pressure to be applied against the South African government "in a number of forums".[5]

A powerful campaign to withdraw investment from South Africa got underway. In the USA, President Reagan, who had been opposed to

[3]During 1986, they visited Egypt, Sweden, West Germany, Zimbabwe, France, Britain and Namibia.

[4]As a reprisal for that TV interview Anglo-American decided that Motlatsi could get time off for union work only if he was given permission by the mine manager at Western Deep Levels on each occasion. The Chamber of Mines continued to be bothered by the publicity given by Motlatsi and Ramaphosa in favour of sanctions. Mr Johann Liebenberg, its industrial relations adviser, told the author in February 1989 that he believed Motlatsi and Ramaphosa were doing immense damage to the mining industry by their international campaigning in favour of sanctions.

[5]Statement by Marcel Golding, who was then the union's press officer. (South African Institute of Race Relations, 1986, p241.)

sanctions and who supported a policy of 'constructive engagement' along with the British Prime Minister, Margaret Thatcher, had been compelled in 1985 to issue an executive order introducing some mild anti-South African measures in order to stall stronger measures from the US Congress. The pressure of anti-apartheid opinion forced Reagan to appoint an official advisory committee to examine the USA's South African policy. The European Economic Community (EEC) followed the US example and advocated a series of mild sanctions.

As the insurrection intensified through 1986, so the attitudes of foreign governments towards South Africa hardened. The Congress of the USA mustered a two-thirds majority to override the President's veto and pass the Comprehensive Anti-Apartheid Act in August 1986. This Act placed an embargo on the importation of a number of South African products and prohibited American firms from making new investments in South Africa and from giving bank loans to finance current production there. It was estimated that at least 48 US companies had sold up or shut down their operations in South Africa. Reagan's advisory committee reported that the 'constructive engagement' policy had failed.[6] The EEC again followed the US example but, under pressure from West Germany, on a limited and voluntary basis. West German firms, of which about 300 possessed South African connections, did not disinvest. The Scandinavian countries were the most steadfast in opposing apartheid and imposed legally enforced sanctions against investment and South African products. In the developed industrial world, only the British government consistently refused to penalise South Africa in any way for its apartheid policies though the anti-apartheid movement there was vigorously active. The British-based Barclays Bank, however, did finally withdraw from the country by selling its operations to the Anglo-American Corporation. Japan took steps in September 1986 to reduce the volume of its trade with South Africa, but it remained South Africa's biggest customer for coal and it left coal and iron ore imports unaltered. Some of the responses in all but the Scandinavian countries were symbolic gestures but they nonetheless carried a message of disapproval to the South African government.

The South African government and private companies responded to the sanctions on exports and embargoes on specified imports in various ways. Those companies and authorities that were determined to break

[6]Mr Herman Rebhan, general secretary of the International Metalworkers' Federation, claimed in April 1986 that "constructive engagement is dead and the American people have woken up to the dishonesty and equivocation of the administration's policy on southern Africa". (*Ibid*, p247).

the sanctions were assisted by the government, which established a unit to advise on evasion. In some instances, as with coal, ways were found to circumvent them. South African coal was exported to the port of Antwerp in Belgium where it was mixed with coal from other countries, lost its national identity, and was sold on the spot market without an attributed source. It entered Britain in this disguise. There were sanctions-busting companies in the west that provided South Africa with prohibited supplies of oil. The British government was active in this evasion process and provided a conduit for prohibited trade with South Africa. In general, South African firms were encouraged to divert trade to countries in Asia and South America where there were no sanctions policies.

However, the immediate practical effect of sanctions was to reduce the volume of trade between South Africa and the rest of the world. That led to a shortage of foreign exchange and created a balance of payments problem that, in turn, produced shortages of imports of vital materials for consumption and production. The white consumers of imported goods quickly felt the impact. The mining and manufacturing industries suffered a decline. The coal industry, for example, experienced a 25 per cent loss of exports. In manufacturing, it was the inability to get vital imported parts and materials that contributed most to its decline. Overall, there was a decline of almost two per cent in output per head of the population in 1986.

A fundamental and long-run effect of sanctions was that it generated a climate of disillusionment about the future. The unpredictable international environment of hostility towards South Africa made it difficult for individual business enterprises to anticipate the long term and lay plans for future development. They did not know what the cumulative demand for mandatory sanctions might produce. The deep gloom that was cast over South Africa's future led the leading conglomerates in the country, including the mining houses, to reduce the value of their domestic assets and invest abroad as a precautionary protective measure. Because they were multinational companies they were able to evade the government's exchange controls and illegally move funds to their foreign subsidiaries through transfer pricing and false invoicing.[7] In doing this they made the economy's future even more precarious. In so far as they did invest in South Africa it was through the acquisition of the subsidiaries that were being sold off by

[7]See Ben Fine and Zavareh Rustomjee, *The Political Economy of South Africa: From Minerals-Energy Complex to Industrialisation*, London, 1996, p11 and p247.

foreign firms under pressure from their respective governments and anti-apartheid organisations. The burden of acquiring these assets, however, was slight for it was usually done through financially painless arrangements that enabled repayments to be made out of profits over a long period.[8]

It was, however, the impact of sanctions on the capital market that made the deepest impression on the economy. South Africa was heavily dependent on the export and import trade. There was a high import content in its domestic production and consumption and in order to meet its import bill it needed foreign exchange either from exports or from international bank credits. The embargoes on those exports that generated significant amounts of foreign exchange resulted in a shortage of foreign exchange. This was made worse by the fall in the value of the rand which lowered the value of exports and made imports more expensive. But, equally significantly, South African firms were refused short-term trade credits to enable them to cover immediate currency demands and were not only denied loans in the international finance market but were threatened by the decisions of international banks to recall their outstanding short-term loans. Somehow they had to cut their dependency on imports and the only viable way of doing that whilst maintaining production was for the government to pursue a policy of self-sufficiency by investing in the production of substitutes for those commodities that were subject to import embargoes. Without success in that area the economy was likely to spiral into a terminal decline.

The combined effects of the international sanctions campaign and the township insurrection was like an impending death sentence on the life of the apartheid system. The government, more seriously than ever before, tried to avoid the sentence through the twin tracks of repression and reform. The purpose of repression was to eliminate the militant opposition and thereby reassure the white minority that its future was assured, while reform was intended to placate the international community and at the same time create empty spaces in the organisation of everyday life that black collaborators could fill. Through reform the government also aimed to destroy the base of the opposition in the townships by wooing the black population with promises of better living standards. In the process it hoped to create an apolitical black middle class buffer that would drain away the supply of educated political activists. Once that was underway it intended to return its collaborative

[8]See South African Institute of Race Relations, *Race Relations Survey, 1987/1988*, pp405-431 for data concerning the state of the economy. Also see Price, 1991, pp220-233.

black functionaries to their former positions in local administrations. Reform was about incorporating blacks into the white system of government without giving them any power.

Military Repression

Repression, however, was the government's first priority and one that it believed it could do well. It was not, of course, starting from scratch. Its security forces were extensive and well equipped. To circumvent the international arms embargo that had been imposed in the early 1960s, the South African government had established the Armaments Development and Production Corporation (ARMSCOR). Its primary task was to manufacture arms but in 1976 it was also made responsible for the purchase of arms abroad. The development of the armaments industry was a government priority that it projected as a success story. By the mid-1980s it was claiming that South Africa was almost self-sufficient in the supply of arms and had the 10th largest arms industry in the world.[9] It had the advantage over many foreign arms producers by being able to test and perfect its weapons in the townships and in the regional wars in Angola and Mozambique. This was one area of government activity where expense was no object.

Repression had always been a part of the daily lives of black people but following the imposition of the state of emergency in July 1985 it had been intensified to the point where no dissenter was safe. It mobilised the security forces and added to their ranks thousands of newly created 'special', euphemistically called 'instant', constables, who quickly gained a reputation for gratuitous brutality. The scope of the government's media restrictions already made it impossible to make an accurate assessment of the impact of the intensified repression from press reports. Those restrictions were extended after November 1985 to cover foreign and local journalists so that they could not "make, take, record, manufacture, reproduce, publish, broadcast or distribute … any film … or any photograph, drawing or other representation or any sound recording of any incident of unrest or any conduct of the security forces". By the end of 1986, subversion had been redefined to include involvement in any alternative administrative structures, such as street committees and people's courts. The security forces were able to search people, premises and vehicles, seal off areas, control traffic, close off

[9]Davies, O'Meara and Dlamini, 1985, pp104-105.

public and private places, control essential services, order curfews and arrest anyone they wished. Any member of the police, army, railway police and prisons department could order an arrest. And there was no legal redress against anything they did. Some townships were virtually occupied by the army. The government was especially concerned about the mobilising consequences of funerals. At the end of July in 1985 it had banned open-air funerals, allowed only one person to be buried at a time, insisted that only an ordained minister could speak, that no-one could mention politics or the emergency, that loudspeakers, banners and flags were forbidden and that mourners had to walk, not be driven, along designated routes.[10]

The scale of the emergency restrictions was reflected in the published crime statistics. During the course of 1986, there were 16 000 unrest-related incidents that resulted in 1 300 deaths. There were 11 000 arrests out of which about 7 700 people were charged. Almost 30 per cent of those who were charged with offences were under the age of 18 years. Apart from the arrests, there were detentions without trial. Almost 4 000 people were detained under the Internal Security Act of 1982 while there were as many as 25 000 emergency detentions, consisting mostly of school children, students and teachers. Many organisations were banned and their officials detained. The Minister of Law and Order, Mr Adriaan Vlok, stated that the ANC and the Communist Party had created about 1 000 organisations to mobilise opposition to the government and those were the ones he aimed to ban.[11]

Formal state repression was reinforced by vigilante activity where counter-revolutionary organisations, which were either supported by the state security services or were actually comprised of covert state security personnel, engaged in acts of harassment, disruption, the destruction of homes, plain indiscriminate murder and assassinations. Disguised white extremists or police personnel who disapproved of any concessions to black people led many of the vigilante groups. There were also wholly black vigilante groups that were made up of people who had lost status, influence, profits and even the limited power they had gained through collaborating with the apartheid system. They were local notables, mayors, councillors, black policemen, the owners of small businesses whose livelihoods had been destroyed and urban gangsters who had been pushed off the streets and punished by the

[10]See Roger Omond, *The Apartheid Handbook*, London, 1986, p254.
[11]South African Institute of Race Relations, 1987/88, pXL.

street committees.[12] Some of the people who had been hurt by the activities of the liberation forces wanted revenge and they sought it through vigilante activity. They mindlessly and indiscriminately attacked, burned and killed other black people, many of whom were innocent bystanders. The white authorities had willing helpers from amongst those who had suffered from the revolutionary implications of the insurrection.

There was much vigilante activity during 1986. In the early part of the year white vigilantes, sponsored by the white right-wing paramilitary group, the Afrikaner Weerstandsbeweging (AWB), were active in the west Rand in response to black consumer boycotts in the area and in retaliation for the killing of two white policemen during clashes with mineworkers in Krugersdorp. White men wearing balaclavas were identified in predominantly black vigilante groups during night attacks. In April, groups of men in unidentified uniforms entered Alexandra and assaulted residents and burned houses. Forty-five thousand people attended a rally there to protest against the violence of 'government-backed vigilantes'. Such violence was also common in the northern Transvaal, Natal and the squatter camps of the Western Cape and Soweto. Much of it arose out of the rivalry between Inkatha and the UDF. Some vigilante groups took on the name 'Inkatha' without actually being associated with the Zulu-based organisation. Insofar as the police were present, they played their usual supportive role.

Such was the scale of the repression after the reimposition of the state of emergency, on 12 June 1986, that the liberation groups in the townships began to suffer from exhaustion and lost their capacity to initiate action against the white authorities. They were depleted of their leaders and the UDF was in disarray. Many of the community organisations that had taken over the functions of the collaborating local authorities were destroyed. Indeed, the number of unrest-related incidents fell from 12 255 in the first half of the year to 3 701 in the second half. From the end of the year, the number of politically motivated deaths, including necklace murders, began to fall quite sharply. Not all township protests, however, were suppressed. Non-violent consumer boycotts continued but on a much reduced scale. There were also school boycotts, though fewer in number. Nonetheless, in Soweto there was no effective education in 54 schools. The use of rent boycotts as a non-violent means of protest that might avoid physical repression, however, increased. But so did the number of retaliatory

[12]See Price, 1991, p262.

evictions by the authorities.[13]

Legislative Reform

The parallel reform track had begun with constitutional changes in 1984 when the Tricameral Parliament was established in which Indians and coloureds, but not black Africans, had their own assemblies. Ultimate political power, however, rested with the President who possessed extraordinary executive powers and the President's Council that was appointed by the white assembly to advise him. This reform was accompanied by the repeal or weakening of some of the apartheid laws. An embargo was placed on the forced removals policy in February 1985 when the influx control sections of the Group Areas Act were eased. Then on 21 April 1986 President P W Botha announced that the numerous pass laws were to be repealed and replaced by a system of 'orderly urbanisation' through which black people could move to and between urban areas provided they could find houses and jobs. The Immorality Act and the Prohibition of Mixed Marriages Act that made interracial sexual relations illegal were repealed. The Prohibition of Political Interference Act, under which multiracial political parties were illegal and the involvement of one racial group in the affairs of another was prohibited, was also repealed.

Preparations were made to facilitate the emergence of a black middle class by extending freehold land ownership in designated areas to urban black residents and relaxing racial segregation in hotels, restaurants and public amenities. Black people were promised better education for their children. The President had declared in 1980 that the government would aim for parity in expenditure on education for black and white children but no progress was made on that score and in 1986 the policy was reaffirmed as part of a 10-year plan. In that year, 42 per cent of the state education budget was allocated to the education of black children who made up 76 per cent of the school population whereas 41 per cent of the budget was used for educating the 11 per cent who were white.[14]

[13]South African Institute of Race Relations, 1987/88, pXXXVIII. It was estimated that there were rent boycotts in 55 local authorities involving about 650 000 households (*ibid*, pp58-59). In general the boycotts were intended to force the resignation of town councillors, the dissolution of town councils, the lifting of the state of emergency and the release of all people who had been detained.

[14]*Ibid*, pL.

The Economic Crisis[15]

The fulfilment of the government's new policies and promises required a sustained economic growth rate of at least four per cent a year. It had already begun to restructure the economy through a policy of economic liberalism in an attempt to increase production. Following the pattern established in Britain, it embarked on a policy of privatisation and the deregulation of trade to attract private capital into productive activities that were previously controlled by the state. It provided tax incentives to encourage private domestic investment and gave favourable terms for private industry to engage in joint ventures with the government. It also set about encouraging the informal black sector of the economy both as a means of reducing black urban unemployment and of stimulating the growth of a politically compliant black middle class. These policies were combined with the contradictory one of increasing public expenditure through the government's import substitution programme that aimed to provide domestically produced substitutes for imports. The outcome was not a success, however, because the economy at that time was structurally incapable of responding to diverse real investment initiatives.

The South African economy in the 1980s was firmly under the control of large-scale capital. There was no longer a disjuncture between economic and political power caused by the Afrikaner dominance of political decision-making and the English ownership of the means of production. This had been reflected in the competition for state resources and protective legislation between Afrikaner farmers and the largely English mine owners. By and large, however, the mine owners had obtained the political services they required so that while the disjuncture remained on paper, it was not significant in reality.[16] The substantial acquisition of capital by Afrikaners during and after the 1960s had removed much of the friction between the owners of capital and the Afrikaner politicians. By the 1980s there was only white-owned capital. The contradictions then were between large-scale and small-scale capital, aggravated by the gradual but slow encroachment of black entrepreneurs into the terrain of small-scale capital through petty commodity production in the informal sector of the economy. The black entrepreneurs, however, carried no economic or political weight.

[15]See Fine and Rustomjee, 1996, for a detailed examination of the economic crisis in the apartheid system in the 1980s.
[16]See Allen, 1992, for the manner in which the mine owners were served by the Afrikaner political establishment between 1871 and 1948.

There were no obstacles to the adoption of an integrated economic strategy by the government in pursuit of economic self-sufficiency.

It was a case of a white-controlled government interacting with white-owned capital purportedly to preserve white supremacy but in reality to protect the interests of capitalists. Contrary to the general impression that the manufacturing industry dominated the economy, the six mining houses in alliance with the finance houses were still the controlling owners of capital. Through acquisitions and inter-locking directorships they were present in all sectors of the economy, often in a predominant position. The government, moreover, still served their core interests. The state resources were oriented in the 1980s, as they had been throughout the history of mining, to protecting and promoting the 'minerals-energy complex'. The focus of the government's investment was, in consequence, to develop mineral, heavy chemical and energy production but that was not a solution to the problems caused by sanctions.

Another critical factor in the crisis was the response of the mining houses to sanctions. They, with the finance houses, did not invest in production in an endeavour to rescue the economy, as they had done after the Sharpeville massacre, but to engage in a self-preservation exercise by concentrating on corporate restructuring, asset acquisitions and speculation. They acquired the off-loaded subsidiaries of the foreign companies that were quitting South Africa but that added nothing to the economy. Moreover, the capital they transferred abroad illegally was capital that was lost for investment in domestic production. It was estimated that South African firms accounted for the movement out of the country of as much as seven per cent of the gross domestic product. At the same time, hypocritically, they blamed the economy's poor performance on what they claimed to be unreasonable demands for real wage increases and on public expenditure programmes.[17] The mining and finance houses displayed no commitment to creating a vibrant and economically successful economy. The institutional consequences of their actions were equally negative for the country's financial infrastructure met their corporate needs by servicing financial speculation rather than facilitating the long-term provision of capital for industry.

The government, too, displayed the same negative qualities by "fostering unproductive investment and capital flight, both by the way it ... built up the financial system in South Africa as well as through its failure to develop coherent policies for industrialisation".[18] Nor could it

[17]Fine and Rustomjee, 1996, p11.
[18]*Ibid.*

significantly reduce South Africa's reliance on imports for the process of import substitution, which had taken place throughout the history of the minerals-energy complex, had virtually reached its limits. South Africa's dependence on mining for exports was firmly and structurally established and, given the authority of the mining houses, there was no possibility of disinvesting in that sector. In any case, the primary purpose of an import-substitution policy had already been largely satisfied for South Africa had the capacity to produce heavy capital goods. Insofar as the policy was aimed at increasing the investment in the production of consumer goods to ease the pressure on the balance of payments, it merely shifted the need for imports from consumption goods to the materials required for their production. There was no net gain. Every possible escape route was blocked.

The performance of the economy in the 1980s reflected its structural defects. Between 1980 and 1987 the gross domestic fixed investment fell, on average, by four per cent a year while the gross domestic product declined by an average amount of 1,1 per cent each year. Because the population growth was in the region of 2,2 per cent each year there was a dramatic fall in income per head. It was estimated that South Africans were poorer in 1987 than they had been in 1974.[19] They had to contend with an average annual rate of inflation of 14,6 per cent between 1980 and 1987. In 1986 it reached 18,6 per cent. The monthly average number of bankruptcies rose each year until it reached a high point in 1986.[20] Unemployment figures were difficult to calculate but all the indicators showed a sharp increase. The white population began to feel the direct economic effects of the crisis just as they were beginning to experience the disruption caused by black industrial action. They were no longer able to live as if in a white cocoon.

The Failure of the Government's Reform Measures

The effects of the interaction of the insurrection with the reforms and the repression began to emerge by the end of 1986. The constitutional changes were shown to be cosmetic. The UDF had organised a successful campaign to encourage Indians and coloureds to boycott the elections to the Tricameral Parliament in August 1984 and to refuse to become junior partners in a white-dominated government. The authorities

[19]South African Institute of Race Relations, 1987/88, p406.
[20]Price, 1991, p231.

tried to defeat the campaign by arresting its organisers and breaking up its meetings. Nonetheless, almost 83 per cent of the potential coloured electorate and more than 83 per cent of the potential Indian electorate refused to vote.[21] The legislative changes had varying consequences for black people. The abolition of the pass laws repealed the infamous Section 10 that made it an offence for a black person to be in a prescribed urban area for longer than 72 hours without permission. It removed the humiliation of having to produce a pass at the whim of a white policeman and made one less risk of being criminalised for being without a reference book, being an 'idle and undesirable' person or being without a permit of one kind or another. Between 1 January and 23 April in 1986, when the pass laws ceased to apply, 13 481 persons had been arrested for pass offences. But by the end of that period the rate of arrests was being run down. The 252 people who were either in prison or awaiting trial for pass offences in April 1986 were released.[22] But the freedom to travel to the urban areas was still constrained by the need to have a house available to occupy and a job vacancy to fill which made it a theoretical and not a real possibility in the vast majority of cases. Illegal urbanisation, therefore, continued. The embargo on forced removals that had been imposed by the government from February 1985 was a fiction for many black families. In the year before the embargo was imposed, the official number of people affected by forced removals was 23 000 but in the following year the number of black people who were 'assisted' to move from specified areas had risen to 60 000.[23] Moreover, the right to live in designated white areas, called 'grey' areas, was simply an acknowledgement of what was already taking place in at least 17 so-called white suburbs. For instance, in 1986, Hillbrow, in the centre of Johannesburg, was not a grey area but it had 45 000 black inhabitants. Apart from moving into the city centres, a small number of senior black employees and their families were beginning to infiltrate the more exclusive white suburbs of Johannesburg. This came about because their employers illegally allowed them to rent houses that they had legally bought. The government condemned this practice and a number of companies were warned to evict their black employees from their houses or face prosecution.[24]

The modifications in the apartheid system were overshadowed in the lives of ordinary black people by the impact of the economic crisis,

[21]Pampallis, 1991, p276.
[22]South African Institute of Race Relations, 1986, p345.
[23]*Ibid*, p489.
[24]South African Institute of Race Relations, 1987/88, pp488-489.

in terms of the loss of jobs and low wages, and on that score the government was able to offer nothing. The only escape route from destitution for many people was to move into the rapidly expanding informal sector of the economy, which, by 1986, constituted 20 per cent of the gross domestic product. But all that that achieved in most cases was simply to move the margin of impoverishment to a slightly less onerous level. All in all, the reforms failed, firstly, because the international community was not convinced by them and did not relax sanctions to ease the impact of the economic crisis and, secondly, because ordinary black people did not believe that, in terms of political freedom, reformed apartheid was more palatable than the unreformed type. There was less open resistance within the townships but that was due to the suppression of the township organisations during the state of emergency rather than to any moderation of black people's attitudes towards white rule. The decline in township resistance was the government's main achievement in that period but all it did, in effect, was to alter the balance of influence within the liberation movement by creating a vacuum which was filled by COSATU. The NUM played a leading role in that process.

CHAPTER EIGHT

VIOLENCE IN THE STRUGGLE FOR SOLIDARITY 1986-1987

Trade Unions in the Liberation Struggle

COSATU was unambiguously opposed to the apartheid regime and it played an increasingly political role both by reinforcing the United Democratic Front (UDF) and by contributing to the constant drip of economic pressures that was undermining the system. Through political strikes they added to the costs borne by employers through their endorsement of the apartheid state. Between August 1984 and the end of 1986 there were four times more political stay-aways than in the preceding three-and-a-half decades. The stay-aways, moreover, were bigger and more disruptive of production than ever before. 1986 had a record level of working days lost through strikes. Seventy-eight per cent of them were the result of political action and not dissatisfaction with industrial issues. This put pressure on many businesses that had been accomplices in the sustenance of the apartheid system to reconsider their respective positions. It was the combination of the subversive pressure on the economy from international capital with overt destabilising political protests from within the country that made the future of apartheid so uncertain.

Trade unions were privileged participants in the liberation struggle for they had forms of protection against suppression that were denied *ad hoc* community and street organisations. By the end of 1986 the majority of black unions were registered under the Labour Relations Act and had clearly defined legal rights that they could pursue through the courts, if necessary. The legal position of black trade unions under the Act was the same as that of white trade unions. The government could not discriminate against them or ban them without threatening their rights under the law. The Public Safety Act of 1953 under which the state of emergency was called insisted, moreover, that emergency restrictions should not be used to override the rights granted to unions under the Labour Relations Act. No actions arising from that Act, such as collective bargaining procedures, could be declared unlawful, nor could they fall under the ambit of restrictions relating to 'illegal' meetings and subversive statements.[1]

[1]Gerald Kraak, *Breaking the Chains*, London, 1993, p241.

This legal protective cover, however, did not prevent the security forces from harassing trade unionists. At least 2 700 trade unionists were detained during 1986 but always, the government claimed, for political and not trade union reasons. The police made arrests regardless of whether they were straightforward industrial disputes or not. They tried to ban trade union meetings and broke up those they could not ban. They seized documents and disrupted publicity campaign activities. Emergency restrictions were extended to cover the wearing of any garment, especially T-shirts, with identifiable trade union or political insignias. Landlords were encouraged to evict unions from their rented accommodation.[2] Vigilantes subjected some union offices to arson attacks. During the second state of emergency in 1986, the daily lives of active trade unionists were increasingly difficult and dangerous

But unions could no longer be destroyed through the harassment, detention or forced exile of their officials, as had been the case with SACTU.[3] They had existed long enough to acquire the institutional means for survival and were practised in recreating themselves. The NUM, in particular, had lost many of its officials and activists through selective and mass dismissals by the mine owners as well as through arrests and detentions by the security forces. In some instances its branch and regional organisations were totally wiped out by discriminatory dismissals and arrests. But its organisational structure was hierarchical and each level had the institutional means to fill gaps in other levels. Even if the whole of its leadership at national and regional level were eliminated and its authority structure was dismantled, the members had the constitutional means to fill the vacuums and repair the damage. The destruction that occurred at the base of the union was also reparable. The NUM had recovered from many periods of organisational damage and each time it had become more resilient and resistant to attacks.

During 1986, there was a rapid escalation in the involvement of the NUM in liberation politics. Ramaphosa had made his position clear at COSATU's founding conference where he had given his union's membership an indication of the political direction they should take. That direction was reflected in the resolutions from the regions to the Fourth Annual Congress of NUM in February 1986 that called for support for all kinds of international pressures on South Africa, an

[2]Baskin, 1991, p143.
[3]See Allen, 1992, chapter 7.

intensification of the internal struggle and greater resistance to the mining conglomerates. In his report to the congress, Ramaphosa stated that the "state of emergency is no remedy for the crisis. The military occupation of the townships will not restore 'order'... Only major economic and political changes in South Africa, which meet the aspirations of the oppressed and exploited working people, will end the current wave of unrest."[4] The decisions of the congress reflected its identification with the banned ANC. It called for the bans on all liberation organisations to be lifted and supported the campaign to secure the release of Nelson Mandela and other political prisoners. In an act of important symbolic significance it elected Nelson Mandela as the honorary life president of the union and organised a march on Pollsmoor Prison in Cape Town, to where Mandela had been transferred from Robben Island, in order to attempt to present him with his union membership card.

The NUM extended its political involvement to the international sphere by becoming active in the international trade union movement. It resolved at its congress in 1986 not to affiliate to any one international union federation but to work with "all international working class organisations that are fighting for the rights and freedoms of the working class". But in practice it was catholic in its choice of international allies and sided with whoever would give it financial support. It condemned the International Confederation of Free Trade Unions (ICFTU) for promoting 'imperialism and capitalism' but retained its affiliation to the MIF, an international trade secretariat linked to the ICFTU. The MIF, however, provided a valuable service to the NUM by channelling funds to it from the Scandinavian unions. The international mining situation, however, was complicated after the MIF had split in 1985 over its hostility to the communist miners' unions. The British and Australian miners' unions disaffiliated from it and joined with the miners' section of the French Confederation General du Travail (CGT) and miners' unions from the Soviet Union, other socialist countries and some Third World non-aligned countries, to form the International Miners' Organisation (IMO).[5] Despite its affiliation to the MIF, the NUM reserved its freedom to forge links with the newly formed IMO that it regarded as its political ally. When Motlatsi and Ramaphosa

[4]NUM, Fourth National Congress, Soweto, 14-16 February 1986.

[5]The CGT and the miners' unions from the Soviet Union and the other socialist countries were affiliated to the Miners' Trade Union International, an international trade secretariat that was part of the World Federation of Trade Unions. The ICFTU had been formed as a breakaway from the WFTU in 1949.

visited Europe in June 1986 they went to Paris specifically to talk with Alain Simon, the general secretary of the IMO, about the NUM's relationship with it.

The NUM's international links had been extended to the USA late in 1985. The Rietspruit opencast coal mine in the eastern Transvaal, that was owned in equal shares by Shell South Africa and Rand Mines, fired four shaft stewards for walking off the job for two hours to attend a memorial service for a colleague who had been killed at work. Eighty-six mineworkers then went on strike for the reinstatement of the four stewards but they, too, were dismissed. The NUM claimed that the dismissals were part of a company policy of intimidation against union members. Although it had signed a recognition agreement, the company refused to allow union meetings in the compound and barred the shaft stewards from contacting the members. The NUM appealed to the international trade union movement to put pressure on the parent company of Shell. A major response came from the USA where, early in 1986, the United Mine Workers of America joined with the United Automobile Workers and the American Federation of Labour and Congress of Industrial Organisations in declaring a boycott of all Shell products. They asked their members to boycott the 11 000 Shell service stations in the USA and to cut in half and mail to the AFL/CIO their Shell credit cards. Richard Trumpka, the president of the United Mine Workers, gave an additional reason for supporting the NUM. The low-paid South African mineworkers, he said, were undercutting American miners and making it more difficult to sell American coal.

The Spread of Unrest in the Mines

Events in the mine compound began to resemble those in the townships as the volume of collective protests, repression and counter-violence in them increased to an unprecedented level from the beginning of 1986. All of the mining houses felt the effects but the Anglo-American Corporation, the largest mine employer and with the highest proportion of union members among its black employees, felt them most. The extent and the intensity of the disputes in the Anglo-American gold mines in 1986 were to be described by Advocate J F Myburgh, the counsel for the corporation, in an affidavit during an arbitration dispute following the national mining strike the following year. The details he related were not complete and they made superficial presumptions about the causes of industrial unrest but nonetheless they presented a disturbing picture for both the corporation and the NUM.[6] He stated:

"At Western Deeps North Division ethnic conflict resulted in two deaths. At Free State Geduld and Freddies there was a boycott of the liquor outlets. As you know the liquor outlets are managed by the mines and are found in each hostel complex - there was a boycott of that. In March 1986, at Vaal Reefs, there was a continued period of short shifts and violence. Following the February incident where the team leaders were killed at President Brand. Senior mess was disrupted and team leaders intimidated ... that was another festering sore throughout this period, where senior employees were allowed their own mess and the union leaders did not want that institution and a lot of intimidation and conflict arose as a result of that.

"Then, in April 1986, at President Steyn, there was ethnic conflict: at Western Deeps in May there was a boycott of a concession store; in June 1986 at Western Deep Levels North, there was a work stoppage following a dismissal - a work stoppage by 700 workers; at Erfdeel there was a strike by 775 workers on June 16; at Western Holdings there was an attack following a NUM meeting by a mob on workers drinking in the liquor outlet. This occurred twice in June and was due to NUM supporters objecting to fellow workers not attending NUM meetings; at Freddies, 3 000 workers were involved in a work stoppage in an attempt to force the re-employment of a shaft steward; in July; short shifts took place at Vaal Reefs, Erfdeel, Free State Geduld; at Western Holdings there was an enforced boycott of a senior mess; increased numbers of assaults on team leaders eventually resulting in four team leaders being executed, publicly, before a mass of people. [7]

"... In August 1986 the problems at Free State Geduld continued, workers demanding the dismissal of a mine overseer. This eventually led to the dismissal of 500 workers who were on strike; in September at Vaal Reefs East, a senior black employee was kidnapped and assaulted by a group of NUM supporters. He was then forced to march round the hostel carrying the NUM flag; at Western Holdings 2 097 workers refused to go underground and demanded the reinstatement of four employees who were dismissed for their involvement in the murders in July; in October there was a stay-away; at Vaal Reefs East there were problems underground in sympathy with those workers who were disciplined for their involvement in the assault of a senior black employee; the liquor outlet was petrol-bombed; boycotts of liquor outlets started again; in November at Vaal Reefs East there were clashes between groups in favour of the boycotts and those opposed to the boycotts, resulting in the death of 13 employees; December, Vaal Reefs, 10 additional deaths occur; at Vaal Reefs South the conflict spreads - another nine workers die ..."

Trouble had started at JCI's Randfontein Estates gold mine in Westonaria in December 1985 when a group of mineworkers were

[6]Stated by Advocate J F Myburgh in a submission to the Industrial Court, March 1988, on behalf of the Anglo-American Corporation. The submission was reprinted in *Patterns of Workplace Behaviour by Officials and Members of the National Union of Mineworkers*, Memorandum by the Anglo-American Corporation, 1988, pp3-5.
[7]See pp216-222 below for details of this incident.

arrested for fighting at Bekkersdal, a nearby black township noted for its support for the liberation struggle. The arrests were followed by a protest, then a strike, at Cooke No 2 and No 3 hostels during which grievances about living and working conditions were raised. At that point the issues were taken up by the NUM, which, though not recognised, claimed about 8 000 members out of a labour force of just over 17 000. The union promised to approach the management and report back. Early in January 1986, the NUM wanted to hold a mass meeting to discuss the abolition of the *induna* system and the behaviour of the mine security force. The mine manager refused to give it permission to meet within the grounds of the hostel so the venue was switched to the open veld in between the mine and Bekkersdal. That, however, was illegal under the state of emergency regulations and the police were called to break up the meeting.

The police patrol, comprising a white sergeant and white constable, tried to disperse the crowd by using a loudhailer but it had the reverse effect for the mineworkers charged at the van, attacked and killed the two policemen and confiscated their arms, which they then used against the police when police reinforcements arrived. The mineworkers moved to attack the police but this time there were many more policemen and they were heavily armed. Police fire killed seven mineworkers and injured 40, two of whom died later in hospital. The remainder ran into the veld towards Bekkersdal. The police sealed off the township and arrested several hundred mineworkers for questioning. None of the weapons was recovered.

On previous occasions when mineworkers had been confronted by armed police they had retreated, so why, the police wanted to know, did they advance this time? They were given the answer during the subsequent interrogations. Before the December strike a small number of mineworkers had been treated with a special *muti* by a Bekkersdal witchdoctor which, he claimed, would turn bullets to water.[8] The *muti* consisted of receiving small razor incisions on the neck, forehead and other parts of the body that were smeared with a special ointment. The witchdoctor insisted that no-one with this treatment could be harmed. The course of the December strike had indicated to many mineworkers that his claim was justified. What had happened then was that in order to scatter the strikers the mine security guards had fired over their

[8]A *muti* was a magic recipe or potion. The details of its use on this occasion are to be found in Anthony Hocking, *Randfontein Estates*: *The First Hundred Years*, South Africa, 1986, pp248-249. See also South African Institute of Race Relations, 1986, pp756-757 for information about the killings.

heads. The *muti*-treated ones had stood firm so the guards had turned on them but, instead of aiming to kill, they had turned a water cannon on them. The survival of the strikers with *muti* convinced hundreds of mineworkers that they should also buy the treatment. They were the ones who attacked the police van and amongst whom were the fatalities and casualties. The mine manager ended the strike by dismissing 571 mineworkers, mostly from the Transkei, while the police charged 11 with murder and 76 with public violence. In a public demonstration of sympathy with those who had been killed, a convoy of buses containing hundreds of mourners joined the funeral procession to the cemetery.

Later, on 28 January, another seven mineworkers were killed in an inter-group clash involving about 2 000 people at Gold Field's Kloof gold mine in Westonaria. The Blyvooruitzicht gold mine in Carletonville was also the scene for violence in March when six were killed and 118 were injured at the end of a week-long strike over production bonuses that had been introduced in some shafts and not in others. This strike had an unusual characteristic in that the mineworkers went underground as if on a normal shift and then refused to work. They returned to the surface at the end of the shift and were replaced by the next shift that also refused to work. When, after two days, the managers refused to allow any more workers to enter the shaft those who were already underground decided to stay there. They remained underground without food for 36 hours. The violence occurred when, after they had returned to the surface, the police tried to arrest the organisers of the strike. The strikers reacted by stoning the police who opened fire and killed four of them. Two others died in fighting between groups. The police arrested a union organiser and 54 strikers and charged them with a range of offences from murder to attending unlawful gatherings.[9]

White Miners Under Threat

White miners, most of whom were Afrikaners, perceived themselves to be the custodians of the natural supremacy of the white race in their dealings with black mineworkers whom they regarded as a motley group of hapless, ineducable individuals who depended on the innate superiority of the whites for their survival. The subjugation, and even

[9]Baskin, 1991, p81.

violence, against black mineworkers was perceived as a natural consequence of their inferiority. It was not seen as calculated racial discrimination. The emergence of the NUM began to change that situation by politicising acts of discrimination through exposing and constantly questioning white privileges. The union's presence, in consequence, tended to inhibit white miners from acting in their usual abusive, and often violent, manner. At the same time, black mineworkers became increasingly sensitive and intolerant of their treatment by white miners.

It was the idea of having to deal with unionised black mineworkers underground, however, that began to disturb the white miners most for they knew that they were vastly outnumbered and were under the constant scrutiny of the union. The mine authorities had always been fearful of black uprisings in the compounds.[10] The presence of armed security forces in each mine compound was a reflection of that fear. But it had never before extended to underground work. White miners had always believed that although they often worked with gangs of black mineworkers in isolated roadways and stopes and frequently without any means of immediate communication with the surface, they were in complete control. Indeed, they were usually sufficiently confident of this to assault individuals in their gangs without any fear of retaliation.

The extension of fear to underground work made relations between white miners and black mineworkers particularly fragile in a number of Witbank collieries and incited the white miners there to accuse their mine managers of being lax in their discipline of black labour. In February 1986, after a strike by the entire black workforce at Rand Mines' Douglas colliery in support of a demand that the white hostel manager should be suspended for carrying a gun, the atmosphere became so tense that the white miners threatened to arm themselves at work. P J Paulus, the general secretary of the Mine Workers' Union, declared: "Often one white miner works with between 30 and 40 black workers. If there is intimidation of white miners we will do one of two things. We will carry firearms or demand to be given protection by management."[11] The NUM insisted that Paulus should retract his statement and that the Chamber of Mines should make its

[10]The basis for their fear had been described in 1914 by the Native Grievances Inquiry when it stated that black mineworkers were "scattered over 50 miles of country in blocks of from 1 000 to 5 000 in each compound. They can mobilise themselves in a few minutes, armed with such weapons as assegais, jumpers, axes, etc … Most of them are savages, whose only idea of reform is violence … It is not necessary to emphasise the possibilities of mischief which are latent in such a condition of affairs." Quoted in Allen, 1992, p256.

[11]South African Institute of Race Relations, 1986, p277.

position clear over the question of carrying arms. It added that if management permitted "white workers to arm themselves, labour relations in the mines would be detrimentally affected".[12] The Chamber of Mines responded by stating that any employee, whether black or white, would be disciplined if caught carrying unauthorised weapons on mine property. But that did not remove the tensions because the reality of power relations underground remained unchanged. White miners continued to assault black mineworkers who began to retaliate with strike action. At the beginning of March, just a day after the exchange of statements between the NUM and the Chamber of Mines, there was a strike at Duvha opencast mine, owned by Rand Mines, after a white supervisor had assaulted a black mineworker. This was followed the next month by a strike over the same issue at Anglo-American's Kriel colliery in Witbank. The NUM forewarned the mine owners that it would respond with strikes whenever such cases occurred. By September there was concern amongst mine managers that the increased polarisation over the issue of race was leading to a confrontation between the black and white unions in the mines. It had become a matter to be dealt with by the mine security forces.

Most of the mines possessed their own armed security forces and used them to underpin the authority of management. The Vaal Reefs east division gold mine, for instance, had a security force of 162 officers and other ranks who were issued with weapons and tear-gas and had at their disposal one large rescue vehicle or Casspir, two smaller armoured vehicles and 13 'soft-skinned' vehicles. All of the 24 senior officers had been members of the South African Police. Its duty was to maintain law and order in the compounds and to deal with any incident that might involve the loss of production, damage to mine property or a threat to human life. The security force had no legal basis to justify it acting in the role of the SAP but it did so on the instructions of the mine manager. In practice it worked in close collaboration with the SAP and sometimes referred cases to it.

As if to prepare for a confrontation between white miners and black mineworkers, while at the same time intensifying the tension that could lead to it, an increasing number of white miners, supervisors and managers enrolled as part-time members of the voluntary Emergency Protection Units (EPU) that were formed in each mine under the control of a Superintendent of Mine Security to assist the mine security forces in emergency situations. Their sizes varied according to the

[12] *Ibid.*

number of black mineworkers in a mine. The one at the large Vaal Reefs east division, for instance, had about 100 members and so considerably supplemented the security force itself. The units were intended to protect the nearby white townships and vital mine equipment and to operate security control centres in collaboration with the security forces but they were not supposed to be used in confrontations with mineworkers. Nonetheless, their members were issued with shotguns on their operations and were allowed to use the smaller patrol vehicles of the mine security forces. All of the members were white though black personnel were not officially precluded from joining them. It would, however, have been difficult for any black persons to have participated because, in the main, the volunteers were white mine employees who were eager to pursue a common agenda to protect white supremacy. In general, contrary to their officially declared purpose, the EPUs were used during disputes with black mineworkers and, it was alleged, they behaved with 'vindictive ruthlessness' towards them. They singled out those they regarded as problems in order to settle personal scores arising from work and they meted out special repressive treatment to those ethnic groups whom they considered as being particularly strong supporters of the NUM and the ANC.[13] This was in addition to their usual strike-breaking activities whenever black mineworkers went on strike. Whereas black mineworkers were reluctant to injure members of the security forces in confrontations because, in some sense, they regarded their work as legitimate, they did not act with the same restraint when the EPUs were involved. The black mineworkers' hostility towards those bodies became a part of their liberation struggle. Central to the issue of black and white relations in the mines, though no longer confined to it, was the campaign by the NUM to remove the privileges attached to job reservation. Indeed, NUM's Annual Congress had resolved in February 1986 to fight for its abolition before the end of the year. In order to fulfil this resolution, the union acted at different levels and whenever opportunities arose. The union, however, not only wanted to remove the colour bar but to prevent it from being relaxed surreptitiously by managers without giving any training or financial compensation to those black mineworkers who were ordered to do the 'white-only' jobs, as was happening in some gold mines. In most cases, groups of mineworkers reacted spontaneously. That had been the case the year before when 14 400 mineworkers at Vaal Reefs had been dismissed for striking in support

[13]AAC and the NUM, *Reaping the Whirlwind?,* May 1986, p20.

of the 700 mineworkers who had refused to undertake white semi-skilled jobs without adequate training and pay.[14] There was another instance at Vaal Reefs in February 1986 when a shaft steward was dismissed for trying to prevent a white miner from getting into a cage reserved for whites. The black mineworkers waiting at the shaft with him reacted by going on a four-hour strike against the dismissal but the management ignored their protest. The NUM, then, appealed against the dismissal at the Industrial Court but the court found that it was justified.

For different reasons, the NUM and the Chamber of Mines were broadly in agreement about removing the colour bar but they disagreed about practical ways of bringing that about. In consequence, they were slow in making real progress. The government took the first practical step when, in August 1986, it tabled a Bill to delete the offending parts of Section 12 of the Mines and Works Act which enshrined the colour bar in law. Instead, a clause was inserted in the Act that obliged the Minister to appoint advisory committees, in consultation with the white miners' unions, to devise an alternative standard comprising practical work experience, a command of the English or Afrikaans language, educational qualifications, physical health and age. Neither the NUM nor the Chamber of Mines approved of the change because, in effect, it gave the white miners' unions the right to continue controlling black advancement. The NUM had an additional reason for disapproving of the new clause because it involved black mineworkers taking a command of language test that very few of them could pass. The vast majority of black mineworkers were not fluent in either of the official languages. Nor could they reach the minimal educational standards that were likely to be set. Under the new arrangements, therefore, the vast majority would be ruled ineligible to apply for blasting certificates irrespective of their competence at work and knowledge of mining. The Bill, that was placed on the Parliamentary order paper for 1987, was intended to remove the colour bar but without removing job reservation which favoured white miners. Nonetheless the Mine Workers' Union saw it as a threat, engineered by the NUM, and opposed it.

The Changing Occupational Character of NUM

Many of the original members of the NUM had been the relatively highly paid personnel assistants, clerks and team leaders and they had

[14]See Chapter 6, p155

formed the core of its active membership during the first two years of its existence. However, once the union's organisation had begun to take shape its main constituency was the mass of unskilled workers who gradually moved into branch positions. As they did so they took with them the concerns of the unskilled, namely wage differentials and the privileged treatment accorded to team leaders and other senior black employees by mine managers in a typical divide-and-rule strategy. This did not necessarily provoke challenges to those senior black mineworkers who were active in the union but it did result in bitter compound disputes about privileged pay and conditions in general.

The issue was the extent to which team leaders, *indunas* and other senior black employees had been granted separate and privileged hostel accommodation, including canteens, that set them apart from the rest of the mineworkers. Dissatisfaction over wage differentials was also a continuing part of this situation but both elements developed into the bitter question of loyalty to the union. The team leaders, who traditionally had stood between the white supervisors and black mineworkers, and had in some ways acted as a buffer between them, began to be identified with management and, in consequence, be regarded as part of the enemy camp. Some of them encouraged this perception by directly collaborating with managers in order to protect their privileges. Moreover, as the NUM branches came under the control of the unskilled mineworkers and became the catalysts for the protests, some team leaders began to ingratiate themselves even further with management by adopting anti-union positions. They opposed union-inspired boycotts and strikes. In some cases they arranged for supplies of scab labour. The *indunas* did much the same things but for a different reason. They regarded the creeping influence of the shaft stewards as a threat to their own authority in the compounds and, in order to contain it, they also engaged with managers in trying to undermine the union. Thus disputes that were basically about material conditions became a matter of solidarity with the union at a critical time in its development. The issue was further complicated by the fact that the bulk of the unskilled workers were Xhosas who had traditionally worked on short-term contracts and, therefore, had not acquired the skills and experience necessary for promotion to senior black positions. The disputes, therefore, took on an overt ethnic character. The arguments over privileges, that became issues about solidarity with the union, ended as faction fights as each side sought and obtained support from their respective tribal groups in the compounds.

This swing in the balance of influence in the union towards the Xhosas was accelerated from January 1986 by the defection of some

Sotho mineworkers from the NUM following the overthrow of Chief Leabua Jonathan's government in a military coup that had been assisted by the South African government. The military government which it established, led by Major-General Metsing Lekhanya, set about repaying its debt by establishing friendly relations with South Africa. As a sign of its good intentions it expelled the ANC from its territory and opposed sanctions. In April, Lekhanya urged Lesotho mineworkers not to support NUM activities on the grounds that to do so would result in their dismissal, jeopardise their livelihoods in South Africa and spoil the friendly relations between the two countries. Some Sotho mineworkers, conscious that they would be deported if they were dismissed and aware of the dangers from the Lesotho security forces if they were forced to return home because of strike action, responded to the exhortation by dropping out of the union and refusing to support union-inspired boycotts and strikes.

The compounds were in such a fragile and volatile state as the NUM expanded and the anti-union forces forged alliances against it that even a vibration in the balance of power between the union and management was capable of destabilising them. Such a vibration occurred after 8 October 1986, when the government declared that no new Mozambican migrant workers would be allowed into the country and that those already employed could complete their contracts but they would not be permitted to return. At the time of the ban there were more than 56 000 Mozambican mineworkers, making up almost 12 per cent of the total mine labour force. Many of them held senior positions as team leaders and *indunas* and, unlike the Sothos, they represented a powerful pro-management pressure group within the compounds. Their disappearance, therefore, would weaken management's ability to control the hostels. Relatively few of them had joined the NUM either because they regarded its preoccupation with South African affairs as having no relevance for them or because their government had openly discouraged them from doing anything that offended the Chamber of Mines in case the flow of deferred payments, the country's main source of foreign exchange, was disrupted.

The background to the government's ban had nothing to do with the situation in the mines but concerned the failed attempts by the South African government to neutralise the borderline states in its war against the ANC. The government insisted that the catalyst for township unrest came from external sources and that, if it could block them, its internal task would be simplified. At various times it made allegations that Botswana, Lesotho, Zimbabwe, Zambia, Angola and Mozambique were either directly supporting the ANC or ignoring its presence on their territories. It engaged in destabilising tactics, including clandestine armed border raids by the South African Defence Force supported by vigilante attacks, to dissuade them from sustaining the ANC. The brunt of its military assaults, however, were experienced by Angola and Mozambique where, either directly through armed intervention or indirectly through military support for the rebel forces of UNITA and the MNR, their respective economies and social infrastructures were virtually obliterated. In a further measure in 1985, the South African government had threatened to expel all nationals of the neighbouring states who were working in South Africa as a punishment for their support of sanctions. On that occasion, the NUM had announced that it would call a national strike as a retaliatory measure because it would lose up to 40 per cent of its membership if the government carried out its threat. The government took no action and there was no strike but the NUM repeated its threat in August 1986, exactly a year later, when the United Nations was debating the implementation of mandatory sanctions on South Africa. Then President P W Botha threatened to punish the front-line states by expelling all of their one-and-a-half million migrants from South Africa. He had second thoughts, however, about tackling such a daunting administrative task and chose, instead, to expel only the Mozambicans as a lesson to the rest.

The ban was a powerful economic lever on Mozambique because its economy was highly dependent on the flow of foreign exchange that came from the earnings of their migrant workers in South Africa. It was particularly dependent on the compulsory deferred pay agreement with the Chamber of Mines, whereby 60 per cent of the monthly earnings of every Mozambican mineworker was deposited in South African currency with the Mozambican government until such time as the mineworkers returned home after completing their contracts.[15] The repatriation of all Mozambican mineworkers would mean the loss in foreign exchange equivalent to more than the total value of all of Mozambique's exports for 1985. The immediate and direct sufferers of the ban, however, were the estimated one million dependants of the mineworkers in the southern Mozambican provinces of Gaza and Inhambane.[16]

[15]Under the Mozambique Convention of 1928, 60 per cent of the pay of Mozambican mineworkers was paid in gold bullion at the official price. This arrangement enabled the Spanish government through the Mozambique authorities to make a substantial profit as well as having a continuous supply of foreign exchange. The convention was cancelled in April 1978 and, thereafter, the deferred pay was transferred to Mozambique in South African currency. (See Allen, 2003, volume II, chapter 14.)

[16]*The Star*, 13 October 1986.

The economies of those provinces were already on the brink of collapse through drought, floods, the ravages of war and their destructive impact on the government administration. It was clear that Mozambique could not afford to break its link with the South African mines.

The South African government had earlier tried to neutralise Mozambique through the Nkomati Accord of Non-Aggression and Good Neighbourliness that they had signed in March 1984.[17] It had aimed to persuade Mozambique to refuse transit facilities to the ANC and to mould its economy in harmony with that of South Africa through a policy of 'good neighbourliness'. Mozambique, for its part, hoped to end the supply of arms from South Africa to the rebel MNR. The supply of Mozambicans to the mines came into the equation through the Joint Economic Working Group that was established two months after the Accord was signed to discuss the terms and conditions under which the recruitment of mineworkers would continue taking place.[18] The Mozambican government wanted the Chamber of Mines to cover the cost of flying the migrant workers to and from the mines, to raise their number from just over 44 000 to 80 000 fairly quickly and then, subsequently, up to 120 000 and, finally, to increase the proportion of mineworkers' pay that was compulsorily deferred. The Chamber of Mines, for its part, wanted concessions relating to recruitment facilities and the health and safety of recruits. The end result was that the Mozambican government allowed TEBA agents greater scope by lifting locally imposed restrictions on recruitment while the Chamber of Mines increased the number of recruits by 12 000 in a year.

The Accord turned out to be a paper one. Seventeen months after it was signed, Zimbabwean troops found incontrovertible documentary evidence of the continuing presence of South African government and army officials in Mozambique when, in support of the Mozambican Liberation Front (FRELIMO), which was the legitimate government, they stormed the rebel MNR headquarters in Casa Banana in central Mozambique. The rebel forces, obviously sustained by external assistance, were showing no signs of weakening and were successfully challenging FRELIMO over wide areas of the country. On the other hand, the South African government claimed that Mozambique was not honouring the Accord because, it insisted, the ANC was still present on its territory and had organised 23 incidents from there since the beginning of the year.[19] Then, on Monday, 6 October 1986, there was a land mine blast near the

[17]See Davies, O'Meara and Dlamini, 1985, ppP5-P11, for an analysis of the Accord.
[18]See Crush, Jeeves and Yudelman, 1991, pp112-113.
[19]*The Star*, 11 October 1986.

Mozambican border in which six members of the South African Defence Force were injured. The ban on the employment of Mozambican labour in South Africa was imposed two days later.

There was an immediate reaction to the ban from the farmers of the eastern Transvaal and the Chamber of Mines against the government's interference in their affairs. The NUM added its voice when Ramaphosa described the government's decision as an "outrageous and unwarranted retaliatory action which the miners would not take lying down". But though there was press speculation that the NUM might activate the decision to strike it had taken in August 1985, Ramaphosa made no mention of it. The government made immediate concessions to the eastern Transvaal farmers by exempting from the ban the 17 000 Mozambicans they employed. It also allowed in the 750 Mozambican mine recruits who had already signed their contracts. By December, the Chamber of Mines had negotiated a 'temporary deal' which exempted from the order all semi-skilled and skilled Mozambicans, namely those employed in categories 4-8, plus all unskilled Mozambicans with more than seven years' experience. It was estimated that between 40 per cent and 50 per cent of the total number of Mozambican mineworkers were included in these categories. Thus the senior black employees, who were closest to management and critical of the union, were not seriously affected.

Weeding out the Collaborators

Those who did not join in the struggle to extend the jurisdiction of the NUM were regarded and treated as traitors. In the circumstances of South Africa in 1986 being a traitor involved more than taking management's side. It implied collaboration with the apartheid government and, in the townships, black collaborators were executed, often by the gruesome method of 'necklacing'. This intolerance towards people who identified with the apartheid system by accepting government jobs spread to the mines. Team leaders and *indunas*, simply by virtue of their positions, were sometimes accused of being '*impimpis*' or 'snakes', 'informers' or 'sell-outs' and were either arbitrarily killed or punished by improvised people's courts in the compounds.

The suspicions about collaborators were fuelled both by the management's use of *agents provocateur* within the compounds to foment faction fights and by the government's use of vigilantes outside and inside the compounds in order to attack union officials and members and their families. The treatment of trade union opponents of the regime by the security forces was as brutal as anything meted out to collaborators in the townships. Many trade union activists were killed or wounded in their homes at night by unidentifiable vigilante groups, not only in KwaZulu-Natal but also throughout the country.[20] Whole families including children and other innocent members were massacred. Members of the NUM had justifiable reasons for being suspicious of the activities of those who refused to support their struggles in the compounds.

The issue, however, was much wider than being hostile towards collaborators in particular disputes. Members of the NUM, for the first time in their lives, were experiencing the exhilaration and excitement of democratic activity whereby they were able to determine decisions in important areas of their lives and follow their own collective instincts in defiance of management. It had become incomprehensible to them that any black mineworker would want to act otherwise and defy the wishes of the majority. There were always some, however, who fell in that category either because it was in their personal interests to collaborate with management or because they were foreigners who regarded the union's preoccupation with South African politics as being of no concern to them. Both could be the source of mineworkers who were willing to undermine strikes and ignore boycotts and who, therefore, failed the all-important test of solidarity with the union whenever there was a dispute.

Those who failed that test were treated with a hostility that became more violent and uncompromising as the struggles between the union and the mine owners intensified. The bitter compound disputes that resulted, however, were not always clear-cut ones between the supporters and opponents of the union. They also became vehicles for expressing tensions and settling scores caused by the changing ethnic and occupational structure of the labour force. This mixed pattern of causality was graphically and cruelly illustrated through the conflicts at Vaal Reefs and Western Holdings during the course of 1986.

The Turmoil at Vaal Reefs

The management of the Vaal Reefs gold mine, in contrast to the publicised approach of the senior executives of its parent company, the Anglo-American Corporation, was reluctant to co-operate with the NUM

[20]See South African Institute of Race Relations, 1986, pp254-257, for a list of the victims.

and regarded the spread of its influence as a threat to its own authority.[21] Its tendency to counter strike action with mass dismissals, including shaft stewards, had meant that the union branches had continually to rebuild their organisations. The managers, moreover, used every trick they could think of to undermine the authority of the local union officials. They constantly tried to belittle them by insisting that the union had only half of the membership it claimed and they used that to justify their refusal to consult or negotiate with the union representatives. They had not always succeeded as the confrontations with Lira Setona in 1985 had illustrated but they did everything they could to ensure that whatever gains the union made were short-lived. Every dispute became a trial of strength. In that situation, the union believed that it had no alternative but to demand unquestioned solidarity from all black mineworkers in the compound. That became clear during 1986.

The disputes at Vaal Reefs in 1986 began during February and March in No 1 hostel. Most of the hostel dwellers came from either the Transkei or Lesotho and there was no history of animosity between them. The NUM was generally popular with both groups and there was a close connection between its members and the comrades in the townships of Kanana and Jouberton. Indeed, important NUM conferences, including its Inaugural Congress, had been held in Jouberton. Some of the shaft stewards were themselves participants in the militant township groups. In that way, the rising tension and escalation of violence in the townships politicised the mineworkers in the hostels. The state of emergency under which many trade union leaders had been detained was still in force. The period was described as "one of the most troubled in South Africa's history".[22]

The mineworkers who lived in No 1 hostel had participated in that troubled history from the beginning of 1985 through a series of strikes, boycotts, sit-ins and sporadic attempts to defy the colour bar, all of which invariably provoked retaliatory measures from the white authorities. Their living conditions were a constant source of grievance. More than 4 000 of them were accommodated in 430 rooms, each of which held from 12-16 people. The occupants of every 14 rooms shared ablution facilities, consisting of toilets and showers. In common with hostels in other mines, there was much over-crowding and a lack of privacy in the ablutions. Each room was provided with steel beds, a central stove and

[21]The Vaal Reefs mine was an extensive and complex organisation. It had 10 shafts situated within an area of 67 square miles.
[22]AAC and the NUM, 1986, p1.

lockers. The mineworkers compared the rooms with prison cells. As the occupants sometimes worked on different shifts, they slept at different times. It was often the case that friends were being entertained when others were trying to sleep. There was, in consequence, no privacy even in the rooms. There was no refuse disposal system in the hostel and, as the sanitary pipes leaked, it was pervaded by an overwhelming stench of "moisten filth".[23]

It was the NUM's official policy to abolish the hostel system but in practice its attitude was ambivalent because the hostels were essential units for recruitment and the organisation of the union. Once the NUM had access to them it had an unequalled opportunity to agitate and recruit for 24 hours a day. In industry at large, the interface between workers and employers was at the point of production but in mining it also occurred in the hostels for the hostel managers regulated and controlled the after-work life of the mineworkers and thus extended the area of potential conflict between workers and management. What actually happened in the hostels, therefore, was of direct concern to the union. This made the presence of the union in the hostels doubly important and gave rise to the union demand for them to be democratically controlled. Indeed, the NUM's declared strategy for 1986 was for mineworkers to "take control of the hostels". This was a threat to management but it spread alarm amongst the *indunas* whose power in the hostels was derived solely from management.

At Vaal Reefs about 1 000 team leaders, clerks and *indunas* were accommodated with their families in a small township apart from the hostels. The rest lived in special rooms in the hostels where they were provided with separate eating facilities that were superior to those given to the ordinary mineworkers. These privileges were targeted by the union as the first step in its campaign for hostel democracy at Vaal Reefs. During the course of the campaign, however, as well as attacking management for providing the privileges, animosity was directed against those who benefited from them. It was alleged that the chairman of the union branch encouraged this animosity after he had returned from leave in January 1986. At a succession of union meetings it was said that he described team leaders as *impimpis* because they acted as informers for the white managers. They were, he was reported as saying, "tale-bearers for the white men". The damning allegation against him

[23]National Union of Mineworkers' Representations. Report of the Commission of Enquiry Into Violence at Vaal Reefs Mine, No1 Shaft Hostel, by Advocate D A Bregman, Johannesburg, 12 May 1989 (*Bregman Commission Report*).

was that he was said to have spoken of how *impimpis* were punished in the townships by being set alight with tyres and that the team leaders should be treated in the same way.[24] The audiences responded by chanting: "Yes, they are *impimpis*, they are *impimpis*" and the feeling of hostility towards the team leaders intensified.

During February a rumour circulated in the hostel that the shaft stewards were going to attack and kill the team leaders who, in response, armed themselves. When the shaft stewards heard what they were doing they, too, collected weapons so that both sides were in a state of armed readiness. When the branch chairman learned that the team leaders had hoarded weapons in their rooms he asked the hostel manager to investigate, but he refused. The shaft stewards, therefore, decided to take matters into their own hands and, on the night of 18 February, a number of them, led by the branch chairman, moved around the hostel searching for team leaders. The group expanded from 10 to about 100 members as it went from room to room in one block after another, dancing and singing "*impimpi*, what are we going to do to you today, you devil? Today you are going to die!" and: "Team leader is an *impimpi*. This *impimp*i will be unable to do anything, the devil will burn!" They assaulted any team leaders they encountered. The climax came when they approached room 12 in block No 1, in which six team leaders lived. As they entered the room, they sang: "What are you, *impimpi*? You devil, today you are going to burn … today you are going to learn a lesson!" Then they attacked the occupants, reportedly killing four of them before setting the rôom alight.

The scene was subsequently described in court in the following way:

> "At about 8pm on the night of 18 February 1986 a number of men launched a vicious, violent attack on room 12 and its occupants at the time. As far as the room itself is concerned, photographs taken of it shortly after the arrival of the police at about 10pm depict a scene of havoc and devastation. Viewed from the front, every single pane in he window was broken. No door is visible in the doorway. Marks on the wall showwhere stones, pieces of concrete and other missiles had missed their mark. At the back, many, if not all, of the window frames had also been broken. Inside … the room was a complete shambles. Metal bedsteads were overturned and up-ended, the

[24]All of the details relating to this incident, including the allegations made by witnesses, are taken from the *Bregman Commission Report* and the submissions by the NUM and the Anglo-American Corporation. The Bregman Commission, which was convened in January 1987, was a private enquiry established by the Anglo-American Corporation and, as a consequence, it had no powers of subpoena or sanction. Another view of the events at Vaal Reefs is presented by T Dunbar Moodie in "Ethnic Violence on South African Gold Mines", typescript, n/d, South African Church Leadership Programme. See also T Dunbar Moodie with Vivienne Ndatshe, *Going for Gold*, 1994, pp201-210.

chairs were broken and overturned. The floor was littered with an almost unbelievable quantity of stones, rocks, bricks, pieces of concrete, pieces of metal, and innumerable pieces of debris of various kinds. Also, a large fire had been set alight inside the room. The police evidence was that the inside walls of the room were blackened by smoke and that there was a strong smell of petrol in the room along the tops of the windows at the front and at the back."[25]

Nine of the mineworkers involved in the attack were arrested and five of them were subsequently charged with murder. All but the branch chairman were acquitted. He was found guilty on two counts of murder and one count of attempted murder and was sentenced to death.[26]

The shaft stewards reacted to the arrests by convening meetings in all of the shafts. Each one reached a unanimous decision to strike on 25 February in support of the arrested mineworkers. The Sotho mineworkers in general supported the strike decisions but there were some who, whilst they did not register their opposition to the strike, were reluctant to take part in it because the team leaders who had been killed were Sothos and it seemed to them wrong to support a strike in favour of the men who had allegedly killed them. A group of Sothos led by 'Seven' Mojakisane, a gang supervisor from Lesotho, attempted to break the strike after a Sotho shaft steward had tried to persuade him to support it. He took a group of about 20 new Lesotho recruits to work dressed in underground clothes but with blankets draped around them to hide weapons of various sorts. It was alleged that they had been threatened with dismissal if they did not accompany him. The hostel manager allowed them to enter the shaft through his office so they would not be seen carrying illegal swords, spears and sticks underground. They returned in the same way without hindrance and repeated their performance the following day. In the meantime, the

[25]An extract from the report of the Appeal Court in the case of the branch chairman's appeal against conviction. (Myburgh, *Memorandum by the Anglo-American Corporation*, 1988.)

[26]The branch chairman appealed against his conviction but his appeal was dismissed. He then went to death row in Pretoria prison to await his execution. Whilst there his case was raised by James Motlatsi in his presidential address to the Sixth National Congress in 1989. Motlatsi said: "We have seen how the courts of law have been used to sentence our comrades to death. Here, comrades, I want you to give a thought to ... (the former chairman of Vaal Reefs who)... was sentenced to death by a judge who was once detained during the Second World War for his activities which were aimed at undermining the struggle against the Nazis. He was a committed member of the Ossewabrandwag, a pro-Nazi organisation. How can a man with such a background be expected to dispense justice?" In May 1989, P W Botha, the state president, granted the convicted mineworker a reprieve and commuted his sentence to 20 years in prison. He benefited from a subsequent amnesty and returned to work as an official in the NUM.

mine security officers made a number of announcements over the loudspeakers promising protection for those who wanted to work, emphasising that those who did not end their strike would be dismissed and, allegedly, even shot. This prompted about half of the strikers to return to work. After hearing the mine manager repeat that all those who failed to go back to work would be dismissed, the shaft stewards recommended everyone to return and the strike ended. The mine manager took no action against 'Seven' Mojakisane for taking weapons underground. Instead he gave him protection by allowing him to have 30 days' leave.

The strike in favour of the arrested men ended after two days but five days later the shaft stewards met again and agreed to organise a series of short shift strikes in protest against the arrests. The workers in No 1 shaft started on 5 March by working four-hour shifts instead of the usual eight-to nine-hour ones. At the end of four hours they left their workplaces singing, dancing and shouting slogans. That evening the violence against the property of senior black personnel continued when explosives damaged the cars of the Sotho *induna*, and the Swazi and Tswana *indunas* who supported him. The Xhosa *induna* did not possess a car. When the strikers returned to the surface on the second day they found that a number of clerks in the departmental personnel office were still working whereupon they forced some of them to run around the hostel, singing and dancing and carrying the NUM flag. Others had to *toyi-toyi* before a growing crowd of hostel residents.

On the third day of the action, Anglo-American obtained a court interdict restraining the NUM and all those working in No 1 shaft from working short shifts. It announced through a loudhailer that no-one should report for work that day but should, instead, go to the shaft office to sign a document reaffirming their conditions of employment, their desire to continue working on the mine and their willingness to abide by the company's rules and regulations. Armed security officers, supplemented by reinforcements from Carletonville, controlled the mineworkers on their way to the shaft office against a background of singing and shouting. They were present when the document was signed. The Bregman Commission, however, contended that the armed security force was simply restraining "a stampede of workers who wanted to demonstrate their desire to continue working". The union regarded their presence as intimidation.[27] The mineworkers

[27] *Bregman Commission Report*, 1989, para 4.5.7.

were not given an opportunity to discuss the terms of the document beforehand nor was a union official permitted to be present at the signing. They were given two options - sign or be dismissed. Only 11 refused to sign. They were dismissed and provided with transport to take them home.

The signing of the document did not diffuse the situation. The day after the signing, the house and car of the chief *induna*, James Mpalinyana, who was a Sotho, were bombed. After this incident the management allowed the main Sotho officials to have armed protection even though it was contrary to the mine regulations. The mine security had warned Mpalinyana that the Xhosas were going to attack him but they took no steps to prevent it happening. A normal shift was worked on Monday, 10 March, but there was much tension in the compound. The Sothos were prepared for a fight and in the evening they milled around carrying weapons under their blankets but, although two Sotho policemen signalled them to fight by blowing their whistles, they did not respond and nothing happened until the following night.

On that day, 11 March, the Sothos and Xhosa representatives asked the hostel manager for permission to hold a meeting to discuss the whistle-blowing of the night before, but he refused. Both groups, therefore, resigned themselves to the possibility of a clash that night and began to collect arms. The tensions in the mixed gangs underground increased when a white shift boss reported, falsely, that fighting had started in the hostel. Before returning to the surface, therefore, the Xhosa and Sotho mineworkers began to make weapons out of roof bolts, but they did so in a co-operative way. They took the weapons into the cage and then to the crush office where, contrary to common practice, they were separated into tribal groups by the mine security. The Sothos were fed and allowed to keep their weapons. The Xhosas, on the other hand, were disarmed and sent to their rooms.

At about 8 o'clock that evening, the two Sotho policemen who had blown whistles the night before blew them again and the two groups lined up to fight. The Sothos stood in their blankets while the Xhosas were draped in white sheets. Only a minority of the Sothos was there, however, for many of them did not want to fight at all. Some refused to leave their rooms and others ran out of the compound. Even those who were lined up were confused about why they were there. The mine police, firmly behind the Sotho team leaders, tried to rally the rest of the Sothos by promising to support them but there were still insufficient to confront the Xhosas who forced them out of the compound where they remained until the police, true to their promise, turned out with four armoured vehicles to back them up. The police fired rubber bullets at the white-sheeted Xhosas and gave protection to an advancing stone-throwing Sotho group, until the Xhosas were forced into one corner of the compound. The Sothos then had the freedom to maraud around the hostel, assaulting the Xhosas, sacking their rooms and looting their property. There was chaos with mineworkers seeking refuge wherever they could, irrespective of their tribal identities. Wherever the mine police found Sothos and Xhosas together they allowed the Sothos to stay in the rooms but ordered the Xhosas to be taken outside where they were beaten. On one occasion, the Sothos invaded a room and killed its Xhosa occupant while the police stood by and watched. There was no precise count of the casualties though the mine manager estimated that two were killed and many were injured.

The dispute ended following a series of meetings in the Thembisa stadium between the management and the representatives of both the Sothos and Xhosas. They commenced on 12 March and continued intermittently until 17 April and, at different stages, involved official delegations from Lesotho and the Transkei. But though the fighting was stopped, not all of the main issues were satisfactorily settled. There was continued hostility towards the union from the Sothos and a lingering grievance among the Xhosa-speakers, the Pondos in particular, at not receiving any compensation for their property that had been looted. The mine manager had assured them that action would be taken against those culprits whose identities were known. They were asked to present lists of stolen items and told that if possible the stolen property would be recovered and that where that was not done, compensation would be paid. The manager, however, took none of those steps. The Pondo Mtshongolo Dance Group, that had lost its equipment, was especially unhappy about the lack of compensation. It sought its revenge later in the year.

There were differing views about the cause of the dispute. Mlamli Botha, a Xhosa shaft steward, reckoned that neither side actually wanted to fight.[28] His view was reinforced by the evidence to the Bregman Commission which showed that though there was hostility between the Sotho and Xhosa *indunas* there would have been no tribal conflict without the active encouragement of the mine security. This reinforced the general view of the Xhosas that the management had deliberately fomented a faction fight in order to undermine the union. The delegation

[28]See Moodie and Ndatshe, 1994, p207.

from the Lesotho government absolved the mine security but saw the dispute as a conflict between a small group of Sotho *indunas* and team leaders whose authority was being undermined by the NUM on the one hand and the NUM branch, including the Sotho shaft stewards, on the other hand. This was roughly the opinion of those Sothos who were engaged in the fighting.

The management had its own version of events. Both Kenneth Dicks, the manager of No 1 shaft, and Hendrick Wilden, the hostel manager, insisted that the fighting stemmed from the rivalry between the Sotho and Xhosa *indunas* and would have ended if they had agreed to shake hands. There was an element of truth in this in that both of the *indunas* were active participants in the dispute and were targets for mutual recrimination. But according to the evidence submitted to the Bregman Commission they were embroiled in an attempt by the management to destroy the union by facilitating ethnic polarisation. The management believed firstly that the Xhosas were the mainstay of the union and were responsible for the strikes and violence and secondly, that all of the Sothos, not just the strike-breakers, were hostile to the union, and, therefore, to the Xhosas. As a consequence of this thinking the mine security consistently encouraged the Sothos to fight, assured them they would get police support if they did, allowed them to collect and carry illegal weapons, refused to intervene when the Xhosas were attacked and allowed a Sotho gang to go on a looting spree around the compound. Not surprisingly, the Xhosa mineworkers began to regard the Sothos in general and not just the Sotho *indunas* and team leaders as their enemies. It was clear that there would be further trouble in the compound.

The Executions at Western Holdings

While the fighting at Vaal Reefs was taking place the Anglo-American Corporation and the NUM, both deeply and equally worried about the intensity of mine violence, were conducting a joint investigation into its causes.[29] On 21 May 1986, not long after the settlement at Vaal Reefs, they published their report, *Reaping the Whirlwind?* It was based on an examination of five mines amongst which the Vaal Reefs and Western Holdings gold mines represented the extremes. Vaal Reefs appeared to contain all the elements that produced factional violence

[29]See chapter 6 below for details of the composition of the research team.

whereas Western Holdings was a model mine with no experience of violence since 1974. The report identified the underlying causes of violence in the work situation and the hostels and made recommendations based on the principle that all approaches to mine violence should be joint union/management ones. There was general agreement that ethnicity was not the cause of violence but was the conduit through which the tensions and conflicts with management were expressed. It stated that the roots of the violence lay in the system of labour migration which should eventually be scrapped but that, in the meantime, other measures had to be taken such as removing workplace grievances, providing more housing accommodation, making hostel life more congenial and basing hostel management on worker and management representation. The question of ethnicity figured prominently in the report. The Anglo-American Corporation had been moving from segregated to integrated ethnic accommodation since the mid-1970s but its mines still contained ethnically segregated blocks and rooms.[30] The report suggested that all ethnically-based accommodation should be abandoned, that recruitment along tribal lines should be ended, that the induction courses should be tribally mixed and that the tribal *induna* system of authority in the hostels should be abolished. Anglo-American agreed to work along those lines but before it could even start there was unprecedented violence at its model mine, Western Holdings.

Western Holdings had been at the hub of the activities that had led to the formation of the NUM and it had retained a strong union presence. In 1986, 64 per cent of its labour force of about 20 000 were members of the union and were represented by more than 100 shaft stewards. In some shafts the degree of organisation was much higher. In No 6 shaft, for example, 73 per cent of its 2-3 000-strong black labour force had joined the union. This high level of commitment to collective action was accompanied by an exceptional intolerance towards those who refused to support the NUM.

The difficulties in No 6 shaft began when a hostel manager with an anti-union reputation was transferred there from No 1 shaft. He had been seen actually shooting at strikers in support of the police during a strike at No 1 shaft in 1985. The union's No 6 shaft committee regarded his transfer as an attempt to subdue the union in their shaft and it refused to co-operate with him. He could speak Sotho well and he

[30]A sample of mineworkers questioned at the five mines believed that the process of ethnic integration had been slowed down after the NUM was formed.

The hostel at no 6 shaft in Western Holdings.

chose his allies from among those Sotho team leaders who were opposed to the union. It was alleged that when they met him at night they wore balaclavas over their faces. The Xhosas saw this as confirmation that the Sotho team leaders were management spies with the result that the campaign against giving them preferential treatment was extended and intensified.

The shaft committee began its campaign by organising a boycott of the kitchens and dining rooms early in July 1986 and demanding equal and uniform eating facilities, the return of belongings that had been confiscated in a raid by the mine security and the removal from the shaft of one particular person whom they regarded as an *impimpi*. The committee presented its demands to the hostel manager on 4 July 1986 but he rejected them out of hand and dismissed four of the shaft stewards, including the chairperson and vice-chairperson of the shaft committee, whom he blamed for organising the boycott. The chairperson, Zolile Gongge, had worked at the mine for more than 20 years. As he and another of the dismissed shaft stewards came from the Ciskei, which was a designated homeland, dismissal meant that they would lose their right to remain in South Africa and their work in mining would end.

The mineworkers on all the levels of No 6 shaft protested against the dismissals by refusing to return to the surface on Monday, 7 July. They had gone to work that day obviously prepared for a stay-down strike for they had taken extra supplies of jabulani with them.[31] This type of action caused much consternation amongst the managers because it imposed a complete halt on production. They were never particularly worried by ordinary strikes because they were able to counter them by arresting the strikers and organising strike-breakers. They could do neither of those things when the mineworkers stayed underground. So they called in Patrick Phatoli, the chairperson of the Western Holdings branch, to tell him that they would "sincerely offer" to discuss the demands of the strikers with him provided they returned to the surface. The union agreed to permit two of the dismissed shaft stewards, Gongge and Joloza, to go underground to explain the management offer to the strikers. The strikers were persuaded to return to the surface that evening but they left the cage in a militant mood and were chanting union slogans. They immediately occupied the kitchen and held a meeting with the four dismissed shaft stewards to discuss what retaliatory measures they could take. It was an angry meeting but their anger was directed at collaborators and not management. Joloza

[31]Jabulani was powdered mealie-meal which had to be mixed with water.

The NUM office at Western Holdings.

presented a list of the team leaders who deserved to be "severely punished" while another of the shaft stewards identified an attendant in one of the boycotted bars as an informer and allegedly called for him to be killed. They decided to resume the boycott of the kitchen and dining rooms two days later.[32]

This time the anti-union team leaders organised their opposition to the boycott. During the evening of Wednesday, 9 July, 15 of them, with weapons hidden under their blankets and blowing whistles to muster support from other Sothos, went into the hostel in search of the members of the shaft stewards' committee. They injured two stewards, one seriously, and drove most of the rest out of the hostel. The next day the shaft stewards asked the hostel manager, Attie Roux, for protection but he refused to give it to them. He simply questioned one person involved in the assault but did nothing to parade and discipline either him or any of the others even though their identities were known. This led the shaft stewards to organise their own protection and, on the principle that attack was the best form of defence, they went out that evening in search of the assailants of the previous night. One was caught and injured. Tit-for-tat violence had begun but nothing further happened until the following weekend.

Many mineworkers, including shaft stewards and union officials, went on leave on Friday evening and did not return until Sunday night. Those who remained spent most of their free time on Saturday drinking in the township. When they returned to the hostel that night a number of them began to search for the team leaders who were on Joloza's punishment list. They found only one, Chemist Mbalo, a Xhosa team leader, who had openly supported the anti-union group. They took him away for questioning and then executed him by beating him to death before a crowd estimated by different witnesses to be between 80 and 400. One of those involved in the execution then read out the names of the others who were alleged to be management informers and threatened to kill them. There were three more executions the following night. Four team leaders whose names were on the punishment list were intercepted at the compound gate as they returned from their weekend leave on Sunday evening and were taken to the dance arena. There they were made to stand on a table before about 2 000 mineworkers until

[32]Most of the information about this incident has come from discussions with mineworkers who were working at Western Holdings at the time and from the report of D A Bregman who was appointed by the mine manager to investigate it on 30 July 1986. Bregman was asked specifically to adjudicate over the dismissal of the four shaft stewards. He endorsed the dismissals.

they confessed not only that they were *impimpis* but also that the manager had supplied them with guns to be used against union members. One of the four jumped from the table, escaped from the arena and subsequently gave evidence at the inquiry into the events. The other three, Mosa Posholi, Ramarumo Libenyane and Bofelo Martini, were beaten to death with sharpened steel bars. No-one at the time was arrested and charged with causing their deaths.[33]

When about 60 other team leaders, who believed their names were on the punishment list, returned to the compound after their weekend break and heard the news of the executions they ran straight from the gate to the compound security barracks for protection and remained there without pay for many weeks. They were ostracised by other mineworkers. Those in No 6 shaft refused to have any dealings of any kind with them. Those who were members of the NUM asked to have Cyril Ramaphosa as a mediator but management failed to pass the request on to him. Some were transferred to Nos 1 and 2 shafts at Western Holdings or to other Anglo-American mines, such as Free State Saaiplaas, where they became anonymous figures. Others, in fear of their lives, left the mine altogether and never returned. A few waited until the tensions had eased and then returned to their former jobs in No 6 shaft. At the end of July, the union branch and the management agreed to appoint Advocate D A Bregman, who had investigated the violence at Vaal Reefs, to inquire into the dismissals. He endorsed the management's actions. Though the union branch accepted his report, the bitterness generated by the dismissals did not subside. The question of reinstating the four shaft stewards who had been dismissed in July 1986 was raised again in June the following year during a particularly violent dispute.

[33]The police announced in May 1989 that no-one had been arrested for the executions. A shaft steward, who was said by two witnesses to have been present when one of them occurred, was already in prison for murders committed in 1987. The following month the President of the Chamber of Mines, in his Annual Report, deplored the fact that no person had been punished for them. But even before he spoke the police were moving towards making arrests. Almost three-and-a-half years after the executions 18 mineworkers were arrested and charged with the murder of the four Sotho team leaders. This came about following a special report on the case by the editor of *Business Day*, Ken Owen, in 1989. Owen was subpoenaed in May of that year to hand over information. Afterwards, the police confiscated tapes of the proceedings of the Bregman Commission which Anglo-American had presumed were lost or stolen but which, in fact, were in a special vault at Western Holdings. Information on the tapes led to the arrests. (*Business Day*, 20 November 1989.)

The Stay-aways and the Second State of Emergency

The first major intervention of COSATU, in its move to become the focus for action in the liberation struggle, was when it joined with CUSA and the Azanian Confederation of Trade Unions (AZACTU) in organising a national stay-away on 1 May 1986 in support of their demand for May Day to be declared a public holiday. This was in accordance with one of COSATU's foundation aims. May Day had been celebrated sporadically as International Labour Day for many years and had been recognised as a paid holiday in some industrial council agreements. In 1961, however, the government had passed a law excluding it from such agreements and, thereby, had made it a contentious issue for the whole trade union movement.

The NUM had raised the question of declaring May Day as a paid holiday for mineworkers in its 1985 wage demand to the Chamber of Mines but the issue had been shelved, to be dealt with later. When the NUM announced it would be supporting the 1986 May Day stay-away the Chamber of Mines stated that as the NUM demand had not been resolved, the union would be acting illegally if it went on strike and sought an injunction in the Rand Supreme Court to prevent it from "instigating, calling for, supporting or organising" a May Day strike. The court refused to grant the injunction on the grounds that as the NUM had tabled its demand the previous year it had won a right to strike over the issue which had not lapsed over time. Approximately a quarter of a million mineworkers, representing almost all of those from South Africa and more than half of the total mine labour force, were part of the one-and-a-half million workers who took part in the stay-away.[34] It was a new experience for black mineworkers to participate in a stay-away on such a large scale.

The May Day stay-away was held during an interregnum between one state of emergency and another but that did not deter the police from making arrests. Three days after he had participated in the strike, James Motlatsi was detained with four other NUM members whilst attending a funeral at the Khutsong Township in Carletonville but they were released within two days after union members in Carletonville, Klerksdorp and the Orange Free State had threatened to strike for their release. This defiant reaction to detentions was repeated in relation to other mine incidents and spread to other industries after the state of

[34]South African Institute of Race Relations, 1986, p285. See Baskin, 1991, pp120-128 for a description of the May Day stay-away.

emergency was reimposed on 12 June, just four days before a second large stay-away was held to celebrate the 10 anniversary of the Soweto uprising. The second stay-away was as large as that on May Day but on this occasion only 150 000 mineworkers supported it.

The reimposed state of emergency was directed primarily at trade unions for they were the only opponents of apartheid that were left to be visibly active. Their members were summarily detained or attacked by vigilantes, unidentified assailants and Inkatha groups. There were large-scale arrests of workers for acting unlawfully by striking against detentions. The arrest of union leaders led to 100 protest strikes in the retail industry alone during the first two weeks of the emergency. The NUM was less affected by police repression than were the other unions, nonetheless, whenever arrests did occur it protested in various ways. In the first week of July about 18 000 mineworkers from various gold, coal and diamond mines went on strike in protest at the detention of their union officials. Mineworkers at the four De Beers diamond mines in Kimberley and at its nearby Finsch diamond mine were on strike from 3 July for six days until the company agreed to campaign for the release of all the members of their branch executive who had been detained. According to NUM, there was even a go-slow protest against the detentions by a small group of mineworkers at the Gold Fields' Kloof and Libanon gold mines in Westonaria, where the union was relatively weak. At the Free State Geduld gold mine in the Orange Free State, mineworkers went underground but refused to go to their workstations. There were token two- and four-hour stoppages at various other gold and coal mines. The NUM nationally organised a further protest against the detentions with a widespread boycott of mine concession stores and liquor outlets on 1 July. The mine compounds were no longer isolated from the mainstream black protest movements.

The state of emergency came at a critical stage for wage negotiation. The NUM had submitted a claim for a 45 per cent wage increase 'across the board' and a number of fringe benefit improvements in the third week of May. The Chamber of Mines made a final offer of between 12 and 20 per cent on 9 June so the government clamp-down on union activity came just as the NUM was due to hold report-back meetings with its members. The government banned the NUM, with a number of other unions, from holding any indoor meetings in the magisterial districts of Johannesburg and Roodepoort at the beginning of July, although the ban was lifted after two days when it was appealed against in court.[35]

[35]The unions protested about the government ban and on 9 July it was lifted, allegedly

Local magistrates prevented the union from holding at least six report-back meetings, including those at Vaal Reefs, Kloof, West Driefontein and two Witbank collieries. The actual negotiations, moreover, were hampered because at least 10 of the negotiating team, including the NUM vice-president, Elijah Barayi, were in detention and an unknown number were in hiding. Both Motlatsi and Ramaphosa had illegally left the country to attend the British NUM's Annual Conference in Wales in the first week of July and they fully expected to be arrested on their return.[36] The Chamber of Mines, recognising the damage being done to industrial relations in the industry by the detention of so many union negotiators, protested to the government about it.[37] In the main, the mining houses adopted an enlightened policy towards the detained mineworkers. The Anglo-American Corporation stated that their employment rights would not be prejudiced at all while Gencor guaranteed to give them back their jobs and lost pay. Gold Fields, on the other hand, stated that it would guarantee the jobs but not the pay of detained mineworkers.

COSATU continued with its opposition to the government by organising a third national stay-away on 14 July. This time the initiative and planning of the stay-away was left to its individual affiliates with the result that there was a mixture of responses ranging from strikes to lunch-time protest meetings and symbolic sit-downs. In general, the stay-away was described as a failure. It was successful only in the towns of the Eastern Cape. In the Western Cape it was called off altogether whilst elsewhere the rate of participation varied from between eight and 24 per cent of the workforce. There were various reasons for the low turnout. The employers had tolerated the first two stay-aways but were hostile to the third and said they would take disciplinary actions against those who took part in it. Many of the COSATU officials who had organised the earlier stay-aways were unable to participate because they were in detention. Eighty-eight per cent of the 245 trade union leaders arrested by the security forces were COSATU members. There were some unions, including the NUM, that were unenthusiastic about embarking on a third protest in quick succession. The NUM estimated that only about eight per cent of the mine labour force supported it with any kind of action. There were

because it contained "certain errors" and because trade unions should not have been included in it anyway. (South African Institute of Race Relations, 1986, p252.)

[36]When they arrived back at Jan Smuts Airport in Johannesburg they were met by British diplomats who enabled them to return safely to their homes.

[37]*The Star*, 2 July 1986.

partial and patchy strikes at only four collieries and in one small section of Anglo-American's Erfdeel gold mine near Welkom.

This was not, however, a sign of strike-tiredness among mineworkers but a reluctance to make too many symbolic gestures. They were preparing for a showdown over wages with the Chamber of Mines at the time. The NUM had lowered its wage demand from 45 per cent to 30 per cent but the final offer from the Chamber of Mines came nowhere near that figure. So, whilst further rounds of negotiations were being held, strike ballots were organised in the mines where the union was recognised. However, while these were being held the union compromised by accepting an offer which ranged from 19,5 per cent to 23,5 per cent for its members in the gold mines. Gold Fields refused to pay the agreed rates and precipitated a strike at its Kloof, Libanon, Deelkraal and Doornfontein gold mines. The strikers returned to work after three days but the union remained in dispute with the company. The NUM accepted a lower offer of 21,6 per cent for its members in the collieries. The Rand Mines' collieries refused to pay this so the NUM remained in dispute with that company. Neither issue was settled before the 1987 wage round began so that the negotiating process for that year started with some important unsettled business.

The Resurgence of Violence at Vaal Reefs

The problems at the massive Vaal Reefs gold mine, like those at Western Holdings, had been shelved rather than resolved by the intervention of Advocate Bregman. Hostel relationships continued to be threatened by the simmering hostility between the Xhosas and the Sothos, the unresolved antagonism of the Pondos towards the shaft stewards and the grievance of the Pondos at having had their property looted the previous year. But underlying these, and in some ways fuelling them, was the perpetual conflict between the management and the NUM which Bregman barely acknowledged. The union members were determined to extend and consolidate their membership in defiance of management and as a base from which to challenge its authority. The management, on the other hand, recognised this threat for what it was and did all it could to contain, control and, if possible, destroy the union.[38] Its attitude was illustrated by the manner in which it tried to regulate the business and conduct of the Vaal Reefs union branch.

[38]AAC and the NUM, 1986, p10.

The branch had to apply to the hostel manager for permission to hold union meetings and with its application it had to name the chairman, provide a copy of the agenda and name the speakers. If he disapproved of a speaker or an item on the agenda then he could, and did, refuse permission for a meeting to be held. But whilst the branch officials were compelled to follow these procedures they nonetheless tried to retain some control over the actual proceedings of the meetings. This led the hostel manager to intervene in their conduct. For instance, informers in the branch reported to the hostel manager during August 1986 that at the beginning of each meeting there was a ceremony in which the union flag was paraded and the members sang union songs. Although the mineworkers had always been encouraged to engage in tribal dancing and singing in order to support the tribal system of authority in the compound, the hostel manager tried to prevent such activities from being used to support the union. Indeed, he used tribal dancing as a device to counter the influence of the NUM and gave the Pondo Dance Group housing privileges in order to enhance its popularity.

All attempts to suppress the use of singing and dancing as means for mobilising and politicising workers failed for they became part of a ritual whenever black workers gathered together. This was especially so in the NUM. The mineworkers became renowned throughout the labour movement for their singing and dancing. In their union meetings, they would *toyi-toyi*, either on the spot or moving along in unison, with the steps and the gestures varying with the lyrics of the song, but always with a rhythm and energy that were constantly mobilising. The dance was infectious and when it started everyone, including non-members, participated. It was not a war dance but an enjoyable expression of solidarity. This, by itself, was not perceived as a threat by management or the state. It was its combination with anti-apartheid song lyrics that converted it into a seditious activity and made it, therefore, potentially illegal.

In the four years since the union had been formed a litany of songs had been written in praise of the union, the various sections of the liberation movement and their leaders. They danced and sang, for instance, to the lyrics: "Ramaphosa, we will die in prison, Ramaphosa" and so on in a similar vein. They flouted the law by praising Nelson Mandela, Oliver Tambo and Joe Slovo, the leaders of the banned organisations, and by intermittently shouting: "*Viva* the ANC, *viva*! *Viva* the South African Communist Party, *viva*! *Viva* COSATU, *viva*!"[39]

[39]In 1986, Oliver Tambo was the President of the ANC and Joe Slovo was the General

At either the beginning or the end of their meetings they sang the ANC national anthem, *Nkosi Sikeleli'i Afrika* (God Bless Africa), which was also illegal. A feature of NUM meetings was that singing and dancing constantly interrupted each one. A speaker had only to pause to take a breath to give someone the opportunity to rouse the meeting with the lyrics of a liberation song in which everyone joined. All of these expressions of feelings were sufficiently subversive to warrant the intervention of the police and the mine security forces in the compounds but the interventions simply made the mineworkers more determined to carry on. There was no question, therefore, of the NUM branch at Vaal Reefs acceding to the hostel manager's demand to stop displaying its flag and singing at its meetings. The mine security force periodically raided them but no arrests were ever made because someone would tip off the union that there was going to be a raid and by the time the mine police arrived the members would have dispersed.

There were rumblings of discontent at Vaal Reefs for about six months following the March fighting. White miners, as well as management, began to feel threatened by the spread of black militancy and some of them reacted by wearing the stickers of the white racist organisation, AWB, on their helmets as a sign of their white supremacist beliefs.[40] Black mineworkers, for their part, showed their allegiances by putting up UDF and COSATU posters and by resisting white racism at work. There was, for example, a spontaneous sit-down strike on 7 July when a white mine overseer called Porky Rheeders told a gang of black mineworkers at the end of a shift that they would have to wait for a cage to hoist them to the surface because the ones in use at the time were not for 'kaffirs'. This was consistent with the prevailing practice in the mines. On this occasion, however, the black mineworkers refused to enter empty cages that were sent for them and sent a message to the mine manager stating that they were going to sit-in until he sent down a 'kaffir lift' for them. The irony was lost on the manager who sent three shaft stewards to persuade them to return. He then instituted an enquiry into the affair which resulted in the white mine overseer being reprimanded for failing to control the situation.

During August, just at the time the hostel manager was complaining about the union meetings, the branch committee organised a short-lived boycott of shops in the compound and intimidated those who ignored it by confiscating the groceries they had bought and scattering them on

Secretary of the South African Communist Party.

[40]AWB was the acronym for the Afrikaner Weerstandsbeweging, a white supremacist semi-terrorist organisation formed in 1979.

the floor. This was followed by trouble during the Transkeian general election on 25 September. The Vaal Reefs management had agreed to allow the Transkeian authorities to use the compound recreation hall as a polling station, to provide hostel staff to assist with the arrangements and to allow a Pondo personnel assistant called Raymond Matyaba to act as the electoral officer. These were provocative acts because the union had rejected the government's 'homelands' policy and was opposed to the elections. Matyaba, moreover, had been in conflict with the union the year before due to his association with strike-breakers. His house and car had been blown up. As a result of the union's opposition, only 44 out of an electorate of about 1 500 had turned out to vote by mid-afternoon. Just at about that time 100 mineworkers, singing, dancing and waving the union banner, entered the electoral hall and disrupted the proceedings by tearing up the electoral rolls and throwing the furniture about. Those who were conducting the elections ran away. Matyaba tried to get to his car but he was caught, assaulted and made to run around the hostel carrying the union banner and singing union songs. Three union officials were arrested and charged with assaulting Matyaba but they were subsequently acquitted of all the charges by the Klerksdorp magistrate.

Ever since the NUM had been formed, hostel managers had infiltrated the hostels and the union branches with informers who, aided by 'in-house' security officials, kept them informed about the union's intentions. Without this constant flow of information they would have been unable to maintain their control of the hostels. But as the union members expressed their contempt of informers by punishing them, sometimes seriously, it became increasingly difficult for the managements to keep the system going. Their sources of relatively reliable information dried up and were replaced by rumour. At Vaal Reefs, for instance, it was rumoured on 8 October that mineworkers at level 65 in No 1 shaft were staging a sit-in though no-one knew whether it was true or, if it were, why it was happening. Nonetheless, two senior managers went to level 65 with a security force and, without asking questions, drove the mineworkers out of the mine with batons and hose pipes. The mineworkers were dismissed when they reached the surface. The rumour was never substantiated. Such events kept the tensions in the compound alive.

There were also constant rumours about armed strangers wandering around the compound. They were described as 'vagrants' by management who reported during October that they were involved in a number of serious incidents. Others gave them different names and allegiances. The Transkeian delegation believed they were 'comrades' from the townships who mixed with the shaft stewards, wore union T-shirts but used balaclavas to hide their identities. They allegedly injected politics into hostel complaints, led the dancing, singing and flag-waving and encouraged the killings. The union described them as '*Amarussia*' or vigilantes recruited from criminal gangs in the townships who were brought in by the mine security, fed by the mine, given mine overalls for camouflage and allowed to move around freely. They could, the union alleged, blow a whistle to mobilise the Sothos or a bugle to mobilise the Xhosa. They were adept at fomenting fights. Various witnesses to the Bregman Inquiry stated that vigilantes were party to the violence that had occurred in November and December 1986. Mine managers acknowledged their existence when they stated to the Anglo-American/NUM investigation into violence:

> "The 'Russians' were a reality, a Mafia-type organisation that used all kinds of weapons including explosives and they could be hired by anybody. The possibility existed that segments within the mining context did use them. Also, workers could have justifiably suspected collusion between the 'Russians' and management as the gang leader 'Kimberley' did offer his services to Vaal Reefs management as a 'mediator' in recent conflicts, but management had officially rejected the offer."[41]

The shaft stewards' committee renewed its campaign after a meeting was held with the hostel residents on Tuesday, 4 November. They discussed bar price increases, the discriminatory effects of the production bonus and the arrest of three people in connection with the disruption of the Transkeian elections. The meeting unanimously agreed to implement an immediate boycott of the liquor outlet. By the following day no-one was using the outlet but the day after that, 6 November, an armed group of Pondos draped in white sheets entered the liquor outlet to break the boycott and, allegedly, to provoke a fight so that they could have an opportunity to loot as compensation for the goods they lost during the fighting in March. The shaft stewards did not want a confrontation with the Pondos at that time so they assured the 'boycott-busters' that there would be no reprisals against them and that the boycott would be called off.

The hostel was quiet until 18 November when a shaft stewards' meeting decided to re-impose the boycott from the next day. The union branch draped its banner across the entrance to the liquor outlet to remind potential drinkers of the risk they took if they went in. The boycott was maintained until 21 November when, in the late afternoon, a group of armed white-sheeted Pondos again invaded the liquor outlet. While

[41]AAC and the NUM, 1986, pp17-18.

they were there a petrol bomb was thrown into the bar. The union alleged that the thrower was a white man but others claimed that a shaft steward was responsible. In any event, the bombing triggered off a renewal of violence. The beer-drinking Pondos, armed with pangas, sharpened iron bars and sticks, went from the bar to the recreation hall and chased a number of people who were watching a film out of the building and over a high wall. Forty were injured. They then went in search of the shaft stewards in the hostel and the black married quarters. They caught and twice assaulted Ernest Kgang, the chairman of the shaft stewards' committee. No shaft stewards slept in the hostel that night. In the following two days the Pondos killed Patrick Masilo, who was a union member but not a shaft steward, and both killed and mutilated Alton Mhlamvu, who was a shaft steward. Up till that time it was a matter of Xhosa-speaking Pondos chasing and assaulting largely Xhosa shaft stewards in order to destroy the union's authority in the shaft. Then the ethnic complexity of the situation became more evident.

On 23 November about five groups, each consisting of from 10 to 15 armed Pondos, moved from room to room in the hostel searching for shaft stewards but warning all Xhosa-speakers to get out of the hostel and join with them or they would also be assaulted. An unknown number of Xhosas ignored this threat but the general effect of it was to revive the animosity between the Xhosa-speakers and the Sothos and to give the violence a predominantly ethnic element. The Sothos, who were worried about the quantity of arms being brought into the hostel, armed themselves but also asked the hostel manager to call in the mine security for protection. He did nothing to defuse the situation. The mine security allowed the Pondos to go with their weapons from room to room in the hostel and from house to house in the black married quarters. At no stage did they try to disarm the Pondos who openly boasted that in their hunt for shaft stewards they had the support of the mine security and, therefore, of the management. This view was shared by the Transkei delegation which stated in its evidence to the Bregman Commission that the Pondos were given "an open licence to … hunt for and kill shaft stewards 'within six days' after a petrol bomb had been thrown into the outlet…" It added that the "search for shaft stewards was common knowledge to the hostel inmates, mine security, hostel police and, presumably, to mine management" but the "group of Pondos living in No 1 shaft hostel … were never disarmed or their possessions searched for dangerous weapons".[42] This failure to disarm

[42]Bregman Commission, *Transkei Delegation Report*, pp16-19.

what the Transkei Delegation called a "rampant group of Pondos" led, it believed, to a faction fight in the evening which was totally unrelated to the initial anti-union campaign and seriously implicated management in the causes of unrest at the mine.

When the Pondos returned to the hostel after attempting to clear out the Xhosas they were confronted at the main gate by a large number of armed Sothos and a smaller number of Xhosas who repelled them. Two of the Xhosas were killed. The Sothos and Xhosas then regrouped to guard the two gates into the hostel and prevent the Pondos from getting in. However, a security policeman guarding another entrance into the liquor outlet allowed the Pondos through so they could attack the Sothos from the rear. In the fighting which followed the Sothos were driven out of the hostel. Just then members of the mine security force arrived in three Hippo vehicles. They made their intentions clear when one of them allegedly said to the Pondos: "We have not come to stop your job. We have come to take the dead ones."[43] The mine security then separated the factions by keeping the Sothos outside and allowing the Pondos to remain in the hostel where they looted the Sothos' lockers, thus getting their revenge for their loss of goods in March. The mine security force entered the hostel later but only to collect the bodies of the six Sothos, four Xhosas and one Pondo who had been killed.

Early the following morning the management arranged for representatives of the Xhosas and Sothos to meet in the Thembisa Stadium for peace talks and the laying down of weapons. The two tribal groups, each heavily armed, sat on opposite sides of the stadium. The process of reconciliation began when first the Xhosas and then the Sothos put down their weapons. Each side then talked to the hostel manager but before joint discussions could get underway there was a disturbance when some of the Xhosas grabbed their weapons and chased after a group of Sothos who had decided to leave the stadium. The police used tear gas to disperse them. As each side gave its version of events neither of them could understand why they had fought each other when they regarded the union shaft stewards as their common adversaries. After four hours of discussions they agreed to make peace. They put their weapons down and requested that the mine security should patrol the hostel for two days, check it for weapons and search every person who entered it.

The peace accord, however, was tenuous. Within two days another clash was threatened and more weapons were confiscated. It appeared

[43] *Bregman Commission Report*, p107.

that not all the weapons had been handed in during the meeting at the stadium so, on 27 November, the mine security undertook a full weapons search which involved cutting the padlocks on private lockers though it was illegal to do so. Truckloads of weapons were removed yet it was clear that they continued to be taken into the hostel at No 1 shaft. This confounded both the management and the NUM branch and national officials who had met to discuss the situation. No-one with authority could understand how and why the weapons were being taken into the hostel or the real cause of the tensions.

The question of looted property remained a sore point. An argument about it between two mineworkers in the liquor outlet on 6 December led to an attack by the Pondos and Xhosas on the Sothos. By the time the mine security force had arrived nine people had been killed and 14 injured. Almost 700 mineworkers from various tribes, though mostly from Lesotho, decided the next day to leave the mine rather than face further risk to their lives. Voluntary resignation was a dramatic act for them as they were not given a Return Certificate entitling them to return to the mine. In effect, it meant that they could no longer work in the mines. This incident resulted in an agreement for the first time between the union branch and the mine manager when they united to enforce a strict policy of dismissing anyone caught carrying weapons, of disciplining those guilty of intimidation and of meeting unofficial worker violence with official mine security violence.

Despite this display of unity the mine manager continued to side-step the NUM. When he met the branch officials on 11 December he suggested that the tribal chiefs should be invited to try and pacify the situation. Thabiso Lenka for the NUM disagreed and said that if the chiefs were invited, the NUM would not meet them.[44] He proposed instead that Ramaphosa should be invited to the mine. The mine manager agreed and three days later Ramaphosa and Motlatsi met with the managers of the East, West and South Divisions of Vaal Reefs. By that time there was hardly a section of the mine that did not have problems. Firstly, in No 9 shaft a Pondo unit supervisor's flat was petrol bombed, the liquor outlet was boycotted and the mobile NUM office was bombed. Then there was a faction fight at No 2 shaft in which the Pondos took on everyone else in the hostel and caused a number of deaths and injuries. Thirdly, at No 6 shaft the trouble was of a different order for it was about the refusal of underground workers to work for a particular white shift boss. Lastly, a number of mineworkers resigned

[44]Not long after this meeting, Thabiso Lenka was detained under the state of emergency.

from Nos 3, 4 and 5 shafts due to their unease about outbreaks of violence. The managers reported on the actions they had taken since their last meeting with the branch committee. They had, they said, continued with weapons searches and had improved security at the gates. They had disciplined three people out of seven who had hidden caches of weapons in the black married quarters and had told the concession stores to stop selling pangas and swords to mineworkers. Lastly, they had allowed normal trade union activity to resume.

Ramaphosa and Motlatsi had the dual task of supporting the union's activities as well as suggesting a solution to the problems in Vaal Reefs. It was made more difficult by the fact that both the Anglo-American Corporation and Advocate Bregman had accused the union branch of being responsible for the boycotts, strikes and violence at Vaal Reefs. Ramaphosa's response was unequivocal. It was to support the branch committee and the shaft stewards in all actions they had taken in the name of the union. That was the end of the matter so far as he was concerned. But he was prepared to talk about the incidence of violence at Vaal Reefs even though he thought that it was not the right place for a discussion of its general causes. This led him to ask about the mine's attitude to the tribal integration of the hostels. The mine manager said it was the mine's policy to mix the tribes together in the hostels except that tribal dancers were allowed to live together as ethnic groups in order to foster a team spirit. One of the hostels, however, had resisted integration. When the discussion turned to the immediate circumstances of Vaal Reefs, Ramaphosa suggested that there was an urgent need for the NUM to hold a mass meeting so that everyone could be brought together, that defence committees with tribally integrated members could be formed in the hostels, that workers could be told what had been done with confiscated weapons and that the management could meet regularly with the branch committee. The manager, however, was not prepared to go that far. He accepted in principle the proposal to establish defence committees and agreed to allow shaft stewards to witness the destruction of weapons but in return he wanted the NUM to desist from tribally divisive boycotts and to agree on procedures for conflict resolution.

The situation at Vaal Reefs and Western Holdings illustrated the difficulty confronting those seeking to understand and explain the general problem of mine violence. During the course of 1986, disputes that had been about building union solidarity in attempts to win concessions from management, and which had been countered by management through initiating, aiding and abetting anti-union forces, were transformed into what appeared as straightforward faction fights between ethnic groups. Other mines had similar experiences and equivalent levels of violence. There were many deaths through seemingly ethnic disputes at the Kloof and President Steyn gold mines during November and December. This transformation of and emphasis on the ethnic character of mine disputes both relieved the mining houses of any responsibility for the troublesome situations in their mines and hid the impact of the rapid spread of trade unionism on management authority.

Both the Anglo-American Corporation, which had borne the brunt of that impact, and the NUM publicised their respective versions of the year's events in the press, early in 1987. The Anglo-American version appeared in *The Star* on 5 January 1987. It stated:

> "The last major ethnic violence occurred on an AAC-administered mine in 1974. A thorough investigation resulted in the restructuring of hostel residential arrangements to accord with work teams and to eliminate the emphasis on tribal groupings. This exercise proved successful and a decade of relative peace followed. Tension in the mines, much of which exists by virtue of the single-sex hostels and migrant labour system, has been heightened over recent months by the state of emergency and general unrest in the country. Concern and anxiety over reports of repatriation of foreign workers have also contributed. A further cause of tension arises from the frustration of better-educated workers being blocked in their career advancement by job reservation restrictions. This has been exacerbated on numerous occasions by irresponsible action, including 'trials' and executions, general intimidation, forced boycotts of liquor outlets and mine stores during the last three weeks … The deaths have occurred in spite of a strong mine security presence in the hostels and strenuous efforts to restore order … We shall … continue to press to prevent illegal work stoppages and the coercion of workers to join them. In many instances union elements at local branch level have ignored the established procedures and channels for conflict resolution and, despite strenuous efforts by management, violence has resulted. There was also an element of outside interference in the hostels prior to the incidents … In relation to unions the Corporation has long encouraged the growth of representative trade unions amongst the workers. In fact, mines administered by the Corporation are the most unionised in the country … However, recent disturbing trends of behaviour on the part of the NUM give causes for concern and need to be addressed by management and this union if sound industrial relations are to persist."[45]

The NUM responded with its own advertisement in the *Weekly Mail* on 10 January 1987 in which it stated:

> "Let it be known once and for all that the source of conflict is rooted in the institutions of oppression and exploitation that exist in the mining industry. The hostel system, migrant labour and *induna* system were pioneered at the turn of the century by mineworkers' lives. It is from this brutal and draconian system that the

[45] *The Star*, 5 January 1987.

AAC has benefited. Over time these structures have been refined but kept intact. AAC has identified and acknowledges some of the issues which have caused the tensions. But what has it done? AAC wants industrial relations to be sound and orderly yet it is not prepared to remove the archaic structures which are the source of conflict. It wants to publicly articulate its liberal views and distance itself from the deaths and violence, when the very cause of the problems emanate from the institutions it has created ... Its empire has been built with the blood, sweat and toil of black workers .. Together with its trusted partner, the apartheid regime, it has wrecked families and maintained people in conditions of enslavement ... How many workers must die, and how many 'independent' commissions of enquiry must be held before the hostels, migrant labour and *induna* systems are dismantled as the first phase in the organisation of labour relations in the mines?"[46]

In the majority of disputes in the gold and coal mines in 1986, the presence of the NUM was a significant factor and in some cases it was the central one. It did not record every dispute during that year but all of the 35 it listed between January and September arose out of attempts by its members to redress grievances against management and the security forces.[47] In the background to them all was the expansion of the union and the hostility shown to it by mine managers irrespective of the public stance of the different mining houses. The Anglo-American Corporation, that had initially supported the formation of the NUM and which, in consequence, had the greatest number of NUM members in its work force, changed its policy on the ground to one of hostility after 1985 when it discovered that the NUM was not, after all, a pliant 'sweetheart' union. From then onwards, virtually all mine managers from each of the mining houses had unfettered freedom to attack and undermine the union. Yet despite this opposition the NUM increased its signed-up membership from 250 000 to over 344 000, making it the largest trade union in South African history. Not all of those who signed up were paid-up members but their numbers had increased by 128 per cent to 228 000 during 1986. The union had more than 5 000 shaft stewards operating in many of the 178 companies it was endeavouring to organise.[48] This represented unprecedented collective power in South Africa. It encouraged the leaders of NUM to contemplate challenging the Chamber of Mines in a head-on struggle.

The presence of the NUM had exposed the violent context in which the freedom of association was being achieved in the mines. Because it was a representative organisation it was able to bring together numerous disparate incidents and show the scale of the violence in terms of

[46] *Weekly Mail*, 9-15 January 1987.
[47] NUM Fifth National Congress, *Annual Report*, 1987.
[48] *Ibid.*

deaths, injuries and dismissals. In its overall picture it showed that most of the 133 mineworkers who had been killed in compound disputes during the year had died in struggles to establish the union and that virtually all of the thousands who had been dismissed had been casualties of management retaliation against union action. The scale of the consequences of union action politicised the union's officials and its members but that was only part of the picture. The rest of it was made up from their experiences underground, including the daily treatment of black mineworkers by their white superiors at work. It was difficult, however, for the union to present this aspect on a single and simply understood scale that would expose and add to the stark reality of being a black mineworker in a politically meaningful way. But, before the year was out, what the union found difficult to do was achieved through a massive, collective tragedy when a fire swept along an underground roadway at the Kinross gold mine in the east Rand. What happened there, from the start of the fire to the completion of the investigation into its causes and the subsequent struggle for compensation for the victims, was a graphic public illustration of the treatment of black mineworkers at work and in death. It completed the case for a strong NUM, as is shown in the details of the following chapter.

CHAPTER NINE

CRIMINAL NEGLIGENCE

THE KINROSS GOLD MINE DISASTER 16 SEPTEMBER 1986

Between Disasters

After the Hlobane disaster, Ramaphosa had had to collect safety advisers around him as quickly and effectively as he could. He had informed his National Executive Committee in November 1983 that he had found attorneys to represent the union at the Joint Inquest and Inquiry into the disaster and that he had approached several international organisations for financial assistance. At his suggestion the executive agreed to the establishment of a health and safety department and he began a search for someone to head it. In his report to the NUM's Second Congress on 10 December 1983, Ramaphosa included a health and safety department in his discussion of the internal structure of the union but it had no designated staff. Nonetheless, the congress gave it work to do by passing a resolution calling on the Chamber of Mines to involve the union in all of its committees dealing with safety and for the mines to negotiate safety agreements with it. It also asked the government to institute a Commission of Inquiry into mine safety, as Ramaphosa had done shortly after the Hlobane disaster. The government ignored the request while the mine owners steadfastly refused to discuss safety issues with the union at any level. Nonetheless, there was much work for a department to tackle for within three months of the Hlobane disaster, the NUM had been involved in disputes over safety in four mines.

When Ramaphosa started his search for a co-ordinator to set up a health and safety department, he was still influenced by the ideology of the black consciousness movement in that he perceived the union to be an organisation run by black people for black workers. This severely limited the pool of possible candidates for the job for he could not turn to the small circle of white people who supported the black liberation movement for assistance and there were very few black people who had had the opportunity to acquire knowledge about mine health and safety issues. When Paul Benjamin, a young white lawyer, who understood the laws relating to mining, offered to work for the union in February 1984, Ramaphosa turned him down because, he said, employing a white person involved a "major policy problem" for the union. The union's NEC, however, agreed to use Benjamin on a retainer basis for any issues that required the expertise of a lawyer.[1] In that way, Benjamin began a long and fruitful association with the NUM. At the same time as Benjamin's case was being discussed, the union agreed to collaborate with two other white intellectual supporters, namely Jean Leger and Eddie Webster, from the Sociology Department of the University of the Witwatersrand, in conducting a research project on mine safety. That, too, was beneficial to the union for it resulted in a relationship that produced a constant flow of critical data about mine safety for the union for a decade.[2] Thus indirectly the union adopted a non-racial position with regard to people who were willing to undertake work for it. Its search for a black head of the health and safety department continued until July 1985 when Howard Gabriels was appointed. He was, however, its only member. He had little time to construct safety structures because from the time of his appointment until the Fourth National Congress of the union in February 1986, he was involved in inquests and inquiries in 10 mines. He also organised legal representation for the dependants of mine fatalities and endeavoured to look after the interests of the injured.

Various factors influenced Gabriels in his work. He had access to an ample supply of funds from overseas donations given specifically for health and safety purposes. He continued, however, to have no staff. The main restriction on his work stemmed from the opposition of the Chamber of Mines and the Government Mining Engineer to allowing the NUM to play any part in their health and safety deliberations. Its complete exclusion meant that its activities were confined to events outside the jurisdiction of the mines such as inquests, inquiries and trials. But even that was too much for the mine owners because they had discovered that wherever the black unions had legal representatives with the right ask questions in public, they could generate adverse publicity about them. This had been their experience at Hlobane. It had been repeated at the inquiry following a methane gas explosion at Middelbult Colliery in Secunda in August 1985, when 30 mineworkers were killed.[3] The Government Mining Engineer, therefore, in support of their interests, embarked on a damage limitation exercise. Fortunately for them, those areas of investigation were not beyond his influence.

[1]NUM, National Executive Committee Minutes, 12 February 1984.

[2]From then on, Jean Leger became South Africa's most prominent expert on mine safety.

[3]The Middelbult Colliery was owned by the South African Coal, Oil and Gas Corporation (SASOL). Its black employees, including mineworkers, were organised by the Chemical

In the short time since the NUM had been intervening in health and safety matters, the Government Mining Engineer had learned that there were distinct differences between inquests and inquiries, on the one hand, and trials, on the other. From experience, the least damaging option from the point of view of the mine owners was litigation. A court case was less likely than an inquiry to probe into the detailed operation of the mine both because its concern about facts was limited to proving or disproving guilt and because it precluded the democratic participation of the appropriate trade union with its own witnesses and experts. In trials, veracity was the least important quality in the struggle to achieve or avoid convictions. Indeed, defence counsels usually determined what the cross-examinations should be about in trials about mine safety so that the central issues could be avoided. This was not the case in inquests and inquiries for they were governed by rules that precluded cross-examinations and did not permit counsels to approach witnesses on the presumption that they were guilty of offences. Their proceedings were expected to be investigatory and, in consequence, they presented fairly detailed pictures of the actual events. That did not always suit the mine owners. In consequence, the Government Mining Engineer's preference was usually for litigation and, whenever he could, he made sure that that course was taken. Although the Mines and Works Act stipulated that the Government Mining Engineer had to initiate an investigation whenever a disaster occurred, he claimed that he had some discretion in deciding whether to hold an inquiry or refer the matter to the Attorney General with the possible intention of initiating litigation. His interference on behalf of the mine owners did not stop there. He took a further step to curtail the involvement of unions in accident inquiries during 1986 when he ordered the Mines Inspectorate to cease publishing its own inquiry findings, thus adding to the secrecy surrounding accidents and disasters. This blocked an authoritative source of information from the public about the causes of accidents, whether regulations had been breached and whether any person or persons were responsible. It also made it more difficult for unions to

Workers' Industrial Union. The investigation into the disaster was entrusted to the Government Mining Engineer. Fifteen months later, when the disaster had largely been forgotten, the Attorney General levied a charge of culpable homicide against the owner and manager of the colliery. In the subsequent trial the accused were acquitted of that charge and found guilty of minor offences involving infringements of some of the regulations of the Mines and Works Act. When another disaster occurred at the Middelbult Colliery in May 1994, resulting in the deaths of 53 mineworkers, the legal proceedings arising from the 1985 disaster had still not been completed satisfactorily.

pursue compensation claims as a result of accidents because they were not in full possession of the facts.

Early in 1986, Ramaphosa attempted to inject some new life into the union's health and safety department by moving Howard Gabriels, still its only occupant, to the education department and putting Hazzy Sibanyoni in his place. Sibanyoni had worked for the Industrial Aid Society that assisted workers who were not members of trade unions and had gained experience in organising safety workshops. Gabriels had recruited him to work for the NUM. He immediately set about trying to build organisational structures at branch, regional and national level to deal with safety. His successes, at first, were confined to the national level. He organised two safety conferences. A hundred delegates attended the second one in May from most of the union's regions. It endorsed the Bill of Rights that Ramaphosa had demanded soon after the Hlobane disaster and elected a National Safety Committee that met twice in 1986, though it never functioned effectively. The problem was that without some degree of co-operation from the Chamber of Mines and the government, the union's representative safety bodies were talking to themselves. It seemed that the only way to avoid that knot and move towards collective discussions about safety policies was to engage in public rhetoric about disasters. It was a way Ramaphosa was skilled at taking. However, it was a matter of bitter concern to him that the incisiveness of his rhetoric depended on the scale of the disaster. Many people had to be killed in order to make a telling point about the need for democratic safety procedures. This was painfully illustrated on 16 September 1986 when the largest gold mining disaster in South African history occurred.

The Kinross Gold Mine

Kinross Mines Limited had been floated in November 1963 to exploit the Evander gold field at Secunda, about 130 kilometres east of Johannesburg. Shaft-sinking had begun almost a year later and it had gone into full production in January 1968. At that time it was a wholly-owned subsidiary of Union Corporation that merged with the General Mining and Finance Corporation in 1980 to form Gencor, an Afrikaner-owned mining house. By 1986 the mine employed 7 000 black mineworkers. It was one of a cluster of four gold mines in the region, belonging to Gencor, a strongly anti-union company that reinforced the government's policy of repression of dissent amongst black workers. Nonetheless, the NUM at Kinross had organised about

60 per cent of the mine's black labour force and had applied for recognition. It was well established, though, apart from being allowed to exist, the union branch had no rights in any matters that were vital to the well-being of its members.

The mine had started production about 1 500 metres below the surface. By 1986, it had reached a depth of 1 900 metres by extending its operations to 18 levels. The maze of tunnels that served those levels extended to a total of 314 kilometres in length. One of them alone, the cross-cut tramming haulage on level 15, was three-and-a-half kilometres long. The mine, then, extended over a wide and complex area of intersecting tunnels in which more than 5 000 out of its total black labour force would be working during shift time.

In the early days of the mine, when the main haulage cross-cut north on level 15 was being prepared for production, all of the tunnel was roped, laced and covered with wire mesh and then, except for about 600 metres, was concrete-blasted in order to protect the rails from falling rocks. After that the overhead electric power supply was installed throughout the length of the tunnel in order to power the ore-carrying electric locomotives. No-one seemed to know why the whole length of the tunnel was not concrete-blasted in a single operation. In any event, when the last remaining stretch came to be completed in 1980 there were problems because the mine was in full production and the haulage was in constant use. The job could not be done without stopping production because concrete-blasting required heavy vehicles to be moved along a track that was already being used by the ore-carrying locomotives.

In planning the completion of the tunnel's insulation, the mine manager's primary consideration was to avoid the disruption of production. He decided, therefore, to line the remaining stretch with polyurethane foam that could be sprayed on without seriously interfering with the flow of ore traffic. The initial outlay was higher than for concrete blasting but it was cheaper when all the other costs were taken into account. Polyurethane foam, moreover, appeared to have technical advantages over the other options. It was, for instance, more powerfully elastic than concrete in that it could cover cracks in moving strata without cracking itself. It could also protect the steel lacing and wire mesh from corrosion and was an effective insulator because it prevented heat from penetrating through the lining from the outside. Its technical qualities were known because it was already being used in the mine. It was used for brattice walls to control the flow of air along the tunnels and sheets of it were placed between the rails to prevent the sludge that dripped from the ore-carrying hoppers from seeping into the ballast that carried the sleepers and thereby disturbing them. There seemed to be no question to the mine management that polyurethane foam was the right material for the job on hand.

The Morning of 16 September

The gang responsible for pipes, tracks and ventilation, which operated as the emergency repair force, was washing and examining tracks along the main tramming haulage cross-cut north on level 15 on the morning of Tuesday, 16 September 1986. Its members had been told that they were going to be visited by the general manager and for that reason they had started work at 4 am, which was earlier than usual. Cross-cut north was the main production artery for the mine and it was the gang's responsibility to ensure that the traffic flowed freely along it. It was used only by electric locomotives, powered by the electric cable running overhead for the length of the haulage, to carry ore from No 1 to No 2 shaft. The haulage was never used to transport people. Nor did many people ever work along its length.

There was contradictory evidence about what happened that morning. In effect there were black and white versions of the events. The black team leader, Isisi Mbuthuma, who was interviewed on British television and gave evidence at a subsequent trial, stated that some time after the gang had started work he had been informed that a rail had buckled at a point about 1 000 metres south of No 2 shaft.[4] He said that he telephoned the foreman's office on the surface to pass this information to Frederick Viviers, the white miner in charge of the gang, as only he was allowed to weld the rail. He was told that Viviers was on his way down so he instructed the gang to collect its tools and take the welding equipment and replacement rail to the place where the rail had buckled. He said that Viviers arrived at about 9 am, fastened his cutting torch to the acetylene cylinder and began to do the repair work. After Viviers had finished cutting the bolts holding the fishplates to the track he switched his torch off and, as he did, Mbuthuma heard a small explosion and saw a flame flashback burst from the valve in the neck of the cylinder and set it on fire. Mbuthuma then said that Viviers dropped his torch, shouted to him to put the fire out and then ran towards No 1 shaft,

[4]See Regional Court for the Regional Division of the Northern Transvaal at Witbank, "The State *v* Kinross Mines, Limited and Others", *Official Transcript*, 21 September 1987 and subsequent dates. Also Laurie Flynn, "Death is Part of the Process", *World in Action*, Granada Television, Manchester, UK, 13 October 1986.

followed by four members of the gang. Left by himself, he tried to contain the flames by pushing the cylinder on its side but it got too hot to handle so he also ran for safety towards the No 1 shaft. Running against the flow of ventilation saved his life because, by that time, the tunnel was on fire and the fumes were spreading. He found a safe exit before reaching the shaft where he met the underground manager from the No 1 shaft and told him about the fire.

The white miner, Frederick Viviers, gave a different account that put him in a more positive light. He was one of the accused in the subsequent trial but he did not give evidence or make any public statements. He did, however, describe events as he saw them to an Inspector of Mines on the day following the fire and in the presence of an official of the white Mine Workers' Union.[5] He said that he was working on level 14 when his team leader, Mbuthuma, told him about the buckled rail on level 15. He ordered Mbuthuma to take the flame-cutting equipment and a replacement rail to where it was needed and followed him to do the job. When he had cut through the fishplate bolts he switched his torch off and, as he did so, there was a bang behind him. He turned round and saw flames coming out of the spindle of the acetylene cylinder. His immediate reaction, Viviers said, was to try and close the valve on the spindle in order to stop the supply of oxygen but the shut-out key turned without effect as if the threads were worn. He realised that he could not put the flame out when the cylinder got too hot to handle so he went for help and, on the way, met the underground manager. His movements after that were not recorded.

The Spread of the Fire

Neither Isisi Mbuthuma nor Frederick Viviers could explain why the cylinder had caught fire. Nor did they know why the flame from the cylinder had so quickly and easily set the walls and ceiling of the tunnel on fire, as well as the insulated cables and pipes that ran through it. Since 1971 Kinross had been classified as a 'fiery mine' and was, therefore, covered by special regulations concerning tests for methane gas. No-one had conducted a test before the welding began so it was possible that gas might have contributed to the ignition. But this did not account for the intensity with which the fire burned and the speed with

[5]The statement was not written but was memorised by the Inspector of Mines and accepted as evidence in the trial. It was a rather patchy recollection. See the Regional Court for the Regional Division of the Northern Transvaal, 21 September 1987, pp376-378.

which it got out of control, giving off thick black smoke and noxious fumes that quickly spread throughout the ventilation system in the direction of No 2 shaft until it affected about half of the total area of the mine.

At that stage there was little any one of them could have done to extinguish the fire. There were no fire extinguishers near at hand though there were a water tap and five empty paint tins about 170 metres away. There were also at least 400 metres of hose pipe in the platelayer's store some distance further away. The tunnel was wet and a lot of water lay on the track in between the lines. In normal circumstances there would have been sufficient water around to extinguish a fire, but the circumstances were not normal. Moreover, the live electric cable running along the length of the tunnel made it dangerous to use water.

As the polyurethane lining burned and smouldered it gave off a thick black smoke containing a lethal combination of carbon monoxide, cyanide and some other gases that were difficult to identify. Carbon monoxide was the main element. It killed by causing either *hypoxia*, which meant there was too little oxygen or *anoxia*, meaning there was no oxygen, thus causing asphyxiation by depriving the tissues of oxygen. Death from carbon monoxide poisoning came very quickly. There was none of it in the atmosphere naturally and as little as 0,2 per cent, breathed for a few seconds, was enough to kill.

The fumes spread with the flow of the ventilation from No 1 shaft to No 2 shaft, permeated throughout levels 15 and 14 and even penetrated levels 17 and 18. They caused havoc, panic and death wherever they went but their greatest impact was amongst those working in the area of No 2 shaft who were caught completely unawares. The mineworkers there tried to escape but they did not know where the fumes were coming from and, in any case, because of the density of the black smoke, they could not see to where they were running. They became breathless and had to hold on to the walls to orient themselves in order to find their way to the shaft. Only seven of those working on level 15 managed to escape for the noxious gases penetrated all the routes to level 14 and then along that level too. Some died more quickly than others so that escaping mineworkers stumbled and collapsed on to others who had just died until their dead bodies, with blood flowing from their mouths, were piled in heaps along the tracks.

There were about 5 200 mineworkers in the mine when the fire started and 2 200 of them were in areas directly affected by the fumes. When an Inspector of Mines went underground at lunchtime there were still 800 people unaccounted for. By 8 o'clock that evening the last mineworker to leave the mine alive reached the surface and an hour later the fire was finally extinguished. The mine management held a press conference at 8 o'clock the following morning and stated that 170 had died. The death toll rose during the day and was reported to be 181 by the evening but four mineworkers who were thought to be missing had actually escaped from the shaft shortly after the fire began. The final death toll was 177, making it the worst disaster in South Africa's gold mining history and the third worst in terms of the number of deaths in any mine disaster in South Africa. Five of the dead were white. At least 228 black mineworkers and two white miners were detained in hospital with injuries of various kinds. Black families were bereaved in every main recruitment region in southern Africa. Forty-five of those who died were Sothos, 29 of whom had come from Lesotho. Twenty-nine were Xhosas from the Transkei, 21 were Shangaans from Mozambique, 20 were Pondos, 15 were from Malawi, 14 were Tswanas, eight were Swazis, another eight were Zulus, six were Hlubis, while the rest were either Vendas or Pedis.

The age distribution of the dead resembled that of the mine as a whole. Most of those killed were young mineworkers. Almost 66 per cent were under the age of 30 years while, even more tragically, just over 43 per cent were between the ages of 20 and 25 years. The youngest to die was 20-year-old Joshua Tiro, a Tswana from Molopo. There were a few old mineworkers amongst them. One of them, Fernando Jacob Bie, a Shangaan, was 67 years of age; another, Magungqu Mahotyana, who was a Pondo, was 60 years old while Michael Jumbe, a Malawian, was 59 years old. These must have been men of great experience for the mines only retained old black mineworkers if they were highly skilled. Altogether, 33 of those killed were over 40 years of age. Two-thirds of all of the Malawians who died were in that age group and it was most likely that they were team leaders. Most of the young mineworkers who were killed were far away from their families in the Transkei and the Ciskei. There were no records of the number of widows who were left or of children who became fatherless. Of the five whites who died, only three were working down the mine at the time. The other two, a surveyor and a shift boss, were members of rescue teams.

Abuse After Death

A racially segregated flow of ambulances and mortuary vans continued to move bodies throughout the night. The management had ordered the same number of ambulances for white as for black mineworkers even though the black casualties outnumbered the rest by more than 30 to one. This meant that the black mineworkers who were rescued, numbering more than 2 000, had to be ferried by mine buses and lorries to the Winkelhaak mine hospital some distance away to be treated for shock or acute respiratory distress. The Winkelhaak hospital dealt only with black patients. Its facilities were limited and inadequate to cope with the demands made on them. Some hundreds of injured mineworkers had to sit on the grass outside the hospital waiting for treatment. Urgent medical supplies were quickly exhausted and replacements had to be obtained from Johannesburg. Though there was much spare capacity at the spacious white hospital nearby at Evander, where the 52 white miners who were injured were being treated in comfort, no black casualties were transferred there. All but two of the white miners were discharged by the next day.

After the casualties had been examined, Gencor made a statement that no-one had been injured by rock falls or collapsed walls or by fire itself though some of the mineworkers brought to the surface were seen to be so badly burnt that their faces were unrecognisable.[6] One white miner who escaped added that he saw some black mineworkers panic and fall down the shaft after the cage had come down and had not stopped to pick them up.[7] It soon became clear to the mine managers and Inspectors of Mines, however, that most of the victims had either died or suffered from some form of gas poisoning.

The final decision as to the precise causes of death could not be made until after the post-mortems had been held. In order for those to be done properly it was necessary for the location of each body to be marked and for blood and tissue samples to be taken on the spot. The corpses then needed to be kept separately in a properly refrigerated mortuary and to be examined as soon as possible after death. If those conditions were not fulfilled then the information available to the pathologists would be both misleading and insufficient to make a valid prognosis. There was a large state mortuary at Diepkloof, near to Soweto in the south-west region of Johannesburg on the freeway from the Secunda area. It could have been reached from Kinross quickly and without difficulty. It had modern facilities to conduct autopsies on up to 600 bodies but it was reserved exclusively for white corpses. There was, then, no difficulty in conducting prompt and proper autopsies on

[6]*Business Day*, 18 September 1986.
[7]*The Star*, 17 September 1986.

the bodies of the five white miners. The first of their funerals, in consequence, was held within three days of the disaster.

It was a different matter for the deceased black mineworkers. The nearest state mortuary where they were allowed to be admitted was at Springs but it had refrigeration facilities for only 24 bodies. The remainder of the bodies were distributed between various locations where there were no refrigeration facilities. Even at the Springs mortuary there was no proper equipment for the autopsies. They had an agricultural bow saw to open skulls rather than an electric one while industrial bolt cutters, rather than cartilage shears, were used to open the thorax for the extraction of the lungs. The scales for weighing the various organs of the bodies were obsolete and broken. The one used for weighing the bodies was also antiquated. It was calibrated in pounds and ounces and of the type normally used for weighing bags of grain. They had no specimen containers for carrying blood or tissue samples and there was insufficient protective clothing for the district surgeons and members of the mortuary staff.[8]

The results of the official autopsies were inconclusive and contradictory. Two district surgeons came from Springs and Standerton on 19 September and examined the bodies of 134 of the black victims and found them with no signs of trauma or abnormalities. Dr J A Greyling from Springs thought that the main cause of death was most probably carbon monoxide poisoning but he could not be sure. Dr D C Grobler from Standerton, on the other hand, much to the gratification of Kinross Mines Limited, refused to commit himself to any explanation. Greyling saw the cherry pink tell-tale colour in the dead bodies, indicating carbon monoxide poisoning, while Grobler saw nothing of the kind.

The NUM, in anticipation of such a situation, had hired Dr J B C Botha, the State Pathologist of Namibia, to conduct autopsies on its behalf. He was appalled by the conditions at the Springs mortuary when he arrived there on 22 September. He found that the bodies he was meant to examine were stacked on top of each other in four piles on the floor and were in such a state of decomposition that some of them had stuck together, making it difficult to distinguish between the injuries caused by the disaster and those created by neglect afterwards. Ten days later he sent a highly critical assessment of what he had seen to the NUM, which then sent a copy to the Department of National Health and Population. Dr Botha wrote:

[8]Dr J B C Botha, State Pathologist for Namibia, *Report: Autopsies: Kinross Mine Disaster*, 30 September 1986.

"The Kinross Mine accident is of national importance and may have far-reaching implications for the South African mining industry; I therefore find it inexplicable that the post-mortem examinations of the victims were not performed with more thought and co-ordination. In my opinion, the following should have been done:

(i) All the bodies should have been transferred to a central point with adequate facilities for the performance of a large number of autopsies, eg Diepkloof State Mortuary.

(ii) A State Pathologist should have been appointed to supervise and co-ordinate the autopsies.

(iii) The prosecutors should have been provided with detailed information regarding the circumstances of the accident and the positions of the bodies when found (I believe the latter information was recorded by the rescue teams).

(iv) The autopsies should have been attended by representatives from the National Centre of Occupational Health and possibly the Department of Labour.

(v) Blood samples should have been obtained from all the deceased and not just a random few so that a systematic analysis of the toxic gases in selected victims from different parts of the shaft could be performed; the differing morphologic features at autopsy suggest that the concentration of toxic substances present may have varied from place to place within the shaft.

(vi) Routine tissue specimens should have been obtained from all the cadavers in case detailed histological examination was required at a later stage."[9]

The NUM generated much controversy by releasing the report to the press before the government had responded to its contents. The headline in *The Star* stated: "Kinross autopsy report sparks storm of protest." Gencor reacted strongly to it and described the allegations as "blatant untruths", adding that it found it "distasteful that anyone should go to such lengths to sensationalise this tragic accident and cause further distress to the families of the victims". Gencor, however, later conceded that it had been misinformed about the autopsies though it did not apologise to the NUM.[10]

Dr Botha's second report was based on 33 post-mortems he performed at the Springs mortuary. He submitted an autopsy report for each of the bodies but stated that he was only able to remove specimens for a histological examination from 16 of them because the only specimen containers available were those he had brought himself. This factor, plus the deteriorated state of the bodies due to neglect by the authorities, meant, he said, that he had insufficient and inadequate information on which to base any sound conclusions. He wrote:

"The features observed suggest that death was due to the inhalation of toxic gas which caused asphyxia and/or intense irritation of the respiratory tract; some of the

[9] *Ibid.*

[10] *The Star*, 16 and 17 October 1986. See also Flynn, 1992, pp27-30. Dr Botha's allegations were publicised by Flynn, 13 October 1986.

> gases which may have been inhaled include carbon monoxide, nitrogen oxide, hydrogen chloride, hydrogen cyanide, phosgene and methylene di-isocyanate. The microscopic features observed in 16 of the cases revealed features consistent with the inhalation of a respiratory tract irritant but the identification of the responsible agent(s) was not possible by histological evaluation alone. As the same spectrum of changes were not observed in all cases, it was not possible at autopsy to indicate with any degree of certainty which gas or gases had been inhaled and which had caused the death of any one individual; such analysis requires blood samples from each body to be subjected to gas chromatography and mass spectrography. In the absence of such investigations, the cause of death in each of these 33 cases remains speculative."[11]

The effect, though not necessarily the intention, of abusing the black victims in death was to destroy the conclusive evidence about the cause of their deaths. But it did not eliminate polyurethane foam as the prime agent of death. Too much was known about the lethal consequences of burning polyurethane foam in enclosed places for any other agent to be suspected.

The Polyurethane Scandal

All of the initial reports of the disaster mentioned toxic gas as the primary cause of the deaths. When Kobus Olivier, the general manager of Kinross, was first interviewed, he believed that the fumes had been caused by burning an anti-corrosive sealant called rigor seal which had been applied to the walls to make them waterproof and airtight. By the following afternoon, however, the Government Mining Engineer had informed him that no rigor seal had been used in the tunnel where the fire started and that the fumes had come from burning polyurethane foam. The subsequent investigations and controversies involving the NUM and the press centred on the advisability of using polyurethane for underground mining though a number of other factors, which made the disaster more likely and less easy to control, were also raised.

Olivier consistently claimed that he had not known that polyurethane foam lining was inflammable until the day after the disaster and that he had never believed there was any need to check on it when it was installed. This alleged ignorance formed the basis of the mine's defence against charges of negligence in subsequent investigations. In fact, when Olivier was questioned about it at the trial in October 1987 he said that "no timber or other combustible or inflammable material

[11] J B C Botha, *Kinross Mine Accident*, Report to the NUM, 2 November 1986.

were present in 15 cross-cut north" where the fire occurred.[12] He disputed that the polyurethane lining was combustible or inflammable. The Chamber of Mines endorsed his view by stating: "Polyurethane and other products used in the mines have until now been regarded by mines as safe."[13] On the basis of this belief the South African mine owners had never bothered to test the product. Nor did the Mines' Inspectorate have an opinion about polyurethane. The Government Mining Engineer had never issued warnings or regulations about its use and no inspectors had ever instructed a mine to remove it. An Inspector of Mines who visited Kinross the day before the disaster said nothing about the hazardous nature of the polyurethane foam in the mine.

However, once it was recognised that the deaths and destruction were caused by polyurethane fumes, the NUM discovered that there was a wide body of international knowledge concerning its uses and dangers about which the individual mining houses, the Chamber of Mines and the Mines' Inspectorate could not have been ignorant. The first important disclosure had come on 9 September 1967 when an underground fire at Michael Colliery in Scotland resulted in the death of nine miners. Just five weeks before that fire a 2,54 cm coating of polyurethane foam had been applied as a lining along about 46 metres of roadway to stop air leakages in case of fire. It was then common practice in British mines to use polyurethane as a protection against underground fires. In total, about 1 610 kilometres of roadway had been lined with it.

The fire in Scotland had been started through the spontaneous combustion of coal behind the lining. It had spread rapidly and with intense heat until the polyurethane disintegrated into dust or remained like charred blisters on the wall. When Ramaphosa went to the site of the fire at Kinross he saw similar charred and blackened remains. The fire at Michael Colliery had produced white smoke that turned grey or a blue haze and then became pitch black. It was so dense that a miner could not even see if anyone was standing alongside him. No-one lived to describe the evolution of the colours of the lethal smoke at Kinross. The survivors there could only describe its blackness and density.

The public inquiry into the disaster in Scotland, ordered by the British Ministry of Power within two months of it occurring, was presided over by the Chief Inspector of Mines. His report, published in

[12]See Regional Court for the Regional Division of Northern Transvaal, 5 October 1987, p13.
[13]*Sunday Star*, 21 September 1986.

June 1968, described the experiments with burning polyurethane foam that had been conducted at the government's Safety in Mines Research Establishment. They provided the first scientifically based information about polyurethane and they were devastating. It was found that within two minutes of ignition, the whole area containing polyurethane foam lining was on fire, giving off volumes of dense black smoke and generating a temperature in excess of 1 000ºC. This was roughly the heat of the core of the flame of an acetylene torch when it cut through steel.

In the experiments, the flame was found to spread at a phenomenal speed. An underground fire involving ordinary untreated timber usually advanced at the rate of about one-third of a metre every minute. With polyurethane, however, the flame moved at the rate of between 36 metres and 46 metres every minute, which was roughly 137 times faster. The fire, moreover, increased in its intensity as it moved from the source of ignition. At such a high temperature it destroyed everything that lay behind the foam and killed or devastated everything that got in its way. Indeed, the fire itself was so intense that it was capable of consuming all the oxygen in the air, thus making life impossible. The black smoke, however, moved even faster than the flames and raced with the flow of the ventilation at 70 metres a minute. It was quite obvious, then, that mineworkers in the vicinity of the fire at Kinross, who were choking from a lack of oxygen, unable to see because of the black smoke and unfamiliar with the details of the mine layout, would quickly be overtaken and overwhelmed by the flames and the smoke.

The conclusive evidence from the British experiments that mineworkers could not survive a polyurethane fire was provided by the measurements of the quality of the air in the wake of the fire. In ordinary circumstances, about 21 per cent of the air consists of oxygen but, due to the fire in the underground roadway, that amount fell to seven per cent or less. Carbon dioxide is present in the air in a tiny constant quantity amounting to 0,03 per cent but after the fire it comprised 13 per cent of the air, which was 433 times more than the normal amount, and it took the place of the oxygen. Carbon monoxide, which is a poisonous, tasteless, colourless gas without smell, is not normally present in the air but as a result of the burning polyurethane foam it made up more than one per cent of it. This combination of gases would produce instant asphyxiation. The scientific officer who conducted the experiments stated in his conclusions: "I understand that almost a mile of roadways have been treated with polyurethane foam but, notwithstanding the difficulties involved, I am sure that the fire hazard warrants its complete removal or sealing off as soon as practicable."[14]

The British Chief Inspector of Mines had no hesitation in attributing the deaths at Michael Colliery to the fumes from the combustion of two coats of polyurethane foam which had been applied just a month before the fire as lining in one of the main belt conveyor roadways. A flame inhibitor had been incorporated into the polyurethane but it had little, if any, effect on its flame-resisting qualities. It had not been treated with any fire-retardant material but even if it had been, it is unlikely that the outcome would have been any different. The inquiry's conclusion was unequivocal and straightforward.[15] It stated:

> "(a) Polyurethane foam as at present constituted, whether or not treated with a fire-retardant coating, should not be used underground at any coal mine or at the surface of any mine in such a situation that the products of its combustion would be likely to enter the mine;
>
> (b) Any foam which has been placed in moving ground or in a situation where there is a spontaneous combustion hazard shall be removed immediately, or sealed off, or be given a fire-retardant coating and be removed within 12 months;
>
> (c) Any foam which has been placed in a situation other than in (b) above shall be given an adequate fire-retardant coating and be removed as soon as practicable or sealed off."

The British National Coal Board reacted before the report was even published by formally banning the use of polyurethane foam in all British mines. In addition, although the question of self-rescue masks was not raised at the inquiry, the British National Coal Board made a submission that it had decided to issue them to all underground workers. It would be almost 20 years before the Government Mining Engineer in South Africa decided to do likewise. The results of the British experiments were published in a batch of scientific papers that were made available throughout the mining world, including South Africa where the Chamber of Mines Research Department received and filed them and then summarised the first one in a circular, dated 21 October 1969, which it sent to all members, including those mining houses which eventually merged to form Gencor.[16]

Yet another warning about the danger of using polyurethane foam in mines had come from the USA. There had been a series of 10 fires

[14]HMSO, *Report: Fire at Michael Colliery, Fife*, Cmnd 3657, June 1968, para 104, p25.
[15]*Ibid*, para 118, p29. Also see the Minutes of the Safety and Health Sub-Committee of the National Executive Committee of the National Union of Mineworkers, London, 23 January 1969.
[16]Flynn, 1992, p258.

involving polyurethane in American mines. The most devastating of them was at the Sunshine silver mine in Idaho where, in May 1972, 91 miners were killed by fumes that were made particularly lethal by the combustion of a stretch of outgoing airway lined with rigid polyurethane foam.[17] These fires prompted the United States National Materials Advisory Board to report on the hazard of burning polyurethane with the consequence that the substance was banned in US mines from 1980.[18] As in the case of the Michael Colliery disaster, the fire at Sunshine Mine was widely publicised and discussed at successive international mining conferences where South African mining experts were invariably present.[19] It was difficult to imagine that any South African mining engineers could have overlooked the fact that polyurethane was banned in two of the largest mining countries in the world.

Then there was the evidence from South Africa itself. There had been at least three occasions before the Kinross fire when mineworkers were killed by fumes from burning polyurethane in South African mines. At the Buffelsfontein gold mine, also owned by Gencor, two mineworkers were killed by polyurethane fumes in 1977 while a further 12 died in the same mine and from the same cause in 1982. A year later there was a tragedy at the Anglo-American gold mine at Vaal Reefs when 10 mineworkers died from polyurethane fumes. On that occasion the NUM raised the polyurethane foam issue with the management and, as a result, the Chief Inspector of Mines held a Joint Inquest and Inquiry at the end of January 1984. The situation resembled that at Kinross. A fire-retarded polyurethane foam had been applied in a roadway and a flame from welding equipment had set it alight. Johannes Vellema, the production manager who gave evidence at the inquiry, wrote a report after visiting the scene of the fire in which he emphasised the "dangers of using this material underground ... In particular he warned that from his experience it was vital to assume that all claims for fire-retardant qualities were completely false unless cast-iron proof to the contrary was provided."[20] It was clear that the Chamber of Mines, the Mines Inspectorate and Gencor knew all about the danger of using polyurethane underground.

[17]*Ibid*, pp258-259. Flynn reported that because the American owners of Sunshine Mine had ignored the warning arising from the British Michael Colliery disaster, it had to pay punitive damages to the widows and orphans totalling US$6 million.
[18]*Sunday Star*, 21 September 1986.
[19]Flynn, 1992, p260.
[20]*Ibid*, p266.

Within a week of the Kinross disaster the Chamber of Mines distanced itself from Gencor when it announced that it not only knew about the possible danger of using polyurethane foam, but also had repeatedly warned the mining houses about it. It stated that it had circulated information to its affiliated members in 1968, 1969, 1970, 1977 and 1982. In 1970 it had distributed an article called *The Dangers of Polyurethane Foam* while seven years later it had circulated information about Gencor's Buffelsfontein mine fire. David Gaynor, who was the technical adviser to the Chamber of Mines at that time, said that discussions about the use of polyurethane and its dangers were held with the safety, environmental and ventilation committees of the Chamber of Mines. It was, he added, the function of the Chamber to keep the industry informed about the perils of the foam but that the responsibility for taking action lay with the mine managers.[21] Some of the mining houses had taken notice of the warnings and had restricted the use of polyurethane in the early 1980s but in general it continued to be fairly widely used in South African mines.[22]

Eventually, Gencor officials admitted that they had known about the dangers. In fact, they revealed that the engineers in Gencor's coal division, Trans-Natal Coal Corporation, had distributed a circular to all of the company's coal mines before the Kinross disaster stating: "Polyurethane is liable to be a fire hazard ... As a result it has been banned in collieries."[23] The head of Gencor's mining, metallurgical and medical services, Dr Con Fauconnier, informed the *Sunday Star* that he had known of the risks relating to polyurethane foam and for that reason the mine had treated it with flame-retardant chemicals. Unfortunately, he said, the chemicals must have deteriorated over time.[24] It emerged that the Kinross managers had been warned in February 1981 that this deterioration was likely to take place and would, thereby, make the polyurethane they had installed a fire risk.[25] Furthermore, both the general manager and the manager of mining at Kinross had been told of the dangers of using polyurethane underground shortly after it had been installed on level 15.

Henry Rose, a Union Corporation ventilation engineer, learned that

[21]*Business Day*, 23 September 1986.

[22]The Chamber of Mines' Press Conference, *Post Kinross*, 26 September 1986 and *The Star*, 23 September 1986.

[23]Flynn, 1992, p259.

[24]*Sunday Star*, 21 September 1986.

[25]Minutes of Meeting of Gencor Group Ventilation Engineers, quoted in Regional Court for the Regional Division of the Northern Transvaal, September 1987, p634.

polyurethane was being used in the mine when he visited it late in December 1980. He was immediately concerned for he had read the report of the Scottish Michael Colliery disaster, knew about the deaths at the American Sunshine Mine and had come across polyurethane whilst working in a copper mine in Zambia in the early 1970s. That same day he warned the mine manager at Kinross about the possible consequences of using it. He raised the matter at a meeting of Gencor ventilation engineers on 25 February 1981 where he emphasised that polyurethane foam should be used only for fire-fighting purposes and not to line roadways.[26] The minutes of that meeting stated that Rose had commented that "it was alarming that a major intake airway was being lined with the material and the implications were that a large number of persons would be exposed to noxious gases in the event of the foam catching fire".[27] These minutes were sent to the general manager of Kinross, as were those of a similar meeting held on 27 May at which the matter was raised again. Rose claimed that he had been told by his superior, Carl Netscher, the Chief Consulting Engineer for Gencor, to instruct Kinross's general manager to stop putting the foam in the mine and to take out any that had already been installed. He stated that he had passed this instruction on. Netscher, however, denied ever having given the instruction because, he said, it was a function of line management to make a decision of that kind and Rose, as a staff member, had no authority even to pass it on.[28] He did not deny, however, that the issue had been raised with the Kinross management.

There was, therefore, incontrovertible evidence that the Kinross management had been warned of the dire risks involved in using polyurethane foam since December 1980 but the warnings were negated by the ambivalence of managers in Gencor who were superior to Rose. One of them, for instance, had appended a note to a letter from Rose, stating that he thought the issue concerning polyurethane had been overstated and that he was not convinced its use should be prohibited underground.[29] As a result, the mine ceased installing it but did not remove the foam lining already in place on level 15. The last communication to the general manager of Kinross about the matter had

[26]As a means of helping to prevent fires it was used to make airtight seals, to seal off a fire area, to establish fresh-air bases, to prevent oxygen from migrating into a fire area and to stop carbon monoxide from escaping from it.

[27]Regional Court for the Regional Division of the Northern Transvaal, September 1987, p635.

[28]*Ibid*, pp991-996.

[29]*Ibid*, p678.

been on 25 June 1981. After that no-one in Gencor raised it again in relation to Kinross. Rose himself made many visits to Kinross between 1981 and 1986 without asking whether the polyurethane had been taken out or not. The mine manager retired and a new one was appointed on 1 January 1985 while just over a year later a new general manager was appointed. Neither of these newcomers had been informed about the history of events concerning the installation of the polyurethane foam in the Kinross mine.

The missing persons in this distressing saga were the Inspectors of Mines. They acted as if they were unaware of either the installation of the foam or of the risks involved. They were not present at the meetings of the ventilation engineers, when Rose raised the issue, nor did they read the minutes relating to them. Indeed, an Inspector of Mines with 25 years' experience who was investigating the Kinross disaster had not known that polyurethane had been involved in the fires at the Buffelsfontein mine in 1977 and 1982 and he could not explain why the material had not been banned after the Vaal Reefs fire in 1983, except to say that he was not a part of the Government Mining Engineer's policy-making team and that it was his job to carry out policy in his particular region and not to formulate it.[30] The Government Mining Engineer himself, however, could not use this excuse, for he had been involved in the Vaal Reef Inquiry and was responsible for formulating the government's safety policy.

The Embargo on Information

The disaster had occurred at between 9 o'clock and 10 o'clock in the morning but Gencor made no announcement about it until 5 o'clock that evening, even though the company had a long-standing agreement with the South African media's Conference of Editors to provide immediate details of mining disasters. By 5 pm, however, the rescue and fire-fighting teams, which had entered the mine within 40 minutes of the start of the fire, had almost got it under control. The whole mine had been evacuated; some of the dead had been recovered and many of the injured had been ferried to the hospitals. It was not until that evening that James Motlatsi, the president of the NUM, learned about the disaster from a news item on his radio which stated that 13 were known to be dead and about 150 were missing. He did not know

[30] *Ibid*, p490.

whether Ramaphosa had also heard the news so, as Ramaphosa did not have a telephone at his home in Soweto, he drove there from Johannesburg to consult with him. Ramaphosa was not at home when he arrived so it was late that evening before both of them became directly involved in making representations on the union's behalf. By that time they knew much more about the scale of the disaster and that at least 60 NUM members had been killed.

The withholding of news about the disaster for eight hours was the beginning of Gencor's endeavour to control the flow of information about it. In this respect it mirrored the government's approach to the freedom of the press.[31] Gencor tried to treat the disaster as its private affair. It resented any kind of adverse publicity and tried to stifle criticisms at birth. It took steps to prevent the large number of media representatives who converged on Kinross from gaining information but it did so incompetently. A big contingent of journalists was told at a press conference on the night of the disaster that they would be admitted to No 2 shaft but when they got there they were refused entry.[32] Then reporters from *The Star* were given permission by the Winkelhaak Mine Hospital to interview patients but, just as they were about to enter the wards, they were stopped because Gencor had issued a directive stating that patients were not allowed to speak to the press in order to "avoid conflicting reports".[33]

This practice of issuing contradictory messages caused the journalists to become increasingly impatient and critical of Gencor. They continued to arrive at the mine but when they tried to interview mineworkers they were restrained by armed security guards and told that if they persisted the police would be called to remove them from the mine.[34] No press photographers were allowed underground to record the conditions after the fire though managers themselves took photographs. The embargo on news was extended to other organisations with an interest in safety. The head of Evander's Protective Services and Civil Defence, who was asked by the Evander Hospital to assist with the care of patients and with handling "a massive number of corpses", was also refused access to information. "The mine management," he said "has totally clamped

[31]David Webster, "Repression and the State of Emergency", *South African Review* 4, Section 2: States of Emergency, 1987, pp141-172. The government already had more than 100 statutes to restrain the media in its reporting of unrest in the townships and of the measures it was taking to repress it, but it went to new lengths in censorship under the new state of emergency which it imposed on 12 June 1986.

[32]*Business Day*, 17 September 1986.

[33]*The Star*, 17 September 1986.

[34]*Business Day*, 18 September 1986.

up about the accident. They don't want to speak to us. In fact, they don't want to speak to anybody."[35]

Gencor could not, in quite the same way, refuse to speak to the NUM which, in consequence, became the principal conduit for information to the public. The management refused to allow Hazzy Sibanyoni, the union safety officer, to enter the shaft but it could not deny Ramaphosa access to it. He was the only person, apart from the mine managers, various Inspectors of Mines and Danie Steyn, the Minister for Mineral and Energy Affairs, to see the immediate aftermath of the disaster. He spent three hours down the mine the day after it had occurred and then went on to talk with patients at the Winkelhaak Mine Hospital.

Ramaphosa made a quick assessment of the situation. He stated: "The first thing that struck us was that the entire tunnel ... was covered in soot from roof to bottom. It was dark, pitch black, not a single light shone, except for our headlamps. There was a deadly silence. There were no injured or dead, all had been removed from the area. The occasional rock fell from the overhanging roof. It seemed we were in some alien environment hostile to all life forms. We still choked on the toxic gases that had claimed so many lives, 24 hours after the accident. For 500 metres, everything had been reduced to charcoal. The polyurethane that covered the walls simply flaked off, as did the covering on what once were electric cables. Aluminium supports, which had earlier held up lamps, were now tangled lumps of metal, a testimony to the intensity of the heat generated by the fire. The intense heat had even cracked water pipes. Metal rods supporting overhanging rocks had given way. A stark contrast confronted us when we reached the end of the area that was previously covered in the polyurethane lining. Everything beyond that point was intact – nothing had burned. It then became apparent that the wall lining had fuelled the fire..."[36] He noted that the conditions where the fire broke out were "highly irregular". There was no fire extinguisher near the welder who was using the flame torch and there were highly inflammable materials lining the walls. He said that the gas could have been checked. It was, in his opinion, an accident that could have been avoided.

After the Hlobane disaster in 1983 the NUM had suggested to the Government Mining Engineer that all mineworkers should carry self-rescue units that gave them a supply of oxygen for 30 minutes. In

[35]*Ibid*, 17 September 1986.
[36]*New Nation*, 25 September 1986.

some circumstances where there was an excess of carbon monoxide that amount of time was sufficient to enable mineworkers to get to relative safety. The Government Mining Engineer had agreed with the union and suggested that the units should be issued in the coal mines in 1985 and in the gold mines by 1986. This, however, was only a recommendation and at the time of the Kinross disaster none had been issued. The Chamber of Mines claimed that the delay was because they had to invent their own self-rescuer to satisfy the unique conditions in South Africa's gold mines. Their self-rescuer prototype was still in the testing stage when the fire broke out.[37]

Ten days after the disaster the Chamber of Mines called a press conference to clarify its position with regard to the self-rescue units.[38] It displayed a small self-contained unit that could be worn wherever carbon monoxide was present and which was in use in overseas coal mines. It had provided 1 400 units to rescue teams but was unwilling to give them to all mineworkers because of the potential cost. The price of each unit was high, a member of its executive committee declared, and there were 500 000 mineworkers who would need them. But it was clear from the demonstration that if they had been issued at Kinross, then many of the victims would have escaped death. The Government Mining Engineer eventually issued a regulation in October 1986 compelling the mines to issue a self-contained self-rescuer to every person who, for whatever reason, went underground.[39]

The sensitivity of the Chamber of Mines to the criticisms levelled against it by the NUM and the press over its approach to polyurethane and self-rescuers resulted in its establishing two new research groups within days of the disaster. One was called the Hazardous Materials Unit that was given the task of identifying and assessing the extent of

[37]*The Star*, 20 September 1986.

[38]Chamber of Mines Press Conference: *Post Kinross*, 26 September 1986. The member of the executive committee who addressed the conference was K W Maxwell, Chairman of the Gold and Uranium Division of JCI. The regulation introduced by the Government Mining Engineer as a reaction to the Kinross disaster recommended that body-worn self-rescuers should be issued.

[39]For a description of the problems faced in developing a self-contained self-rescuer in South Africa, see H J M Rose, "The Introduction of Self-contained Self-rescuers Into the South African Mining Industry: An Update", *Mine Safety and Health Congress Proceedings*, Chamber of Mines, 1988. Rose, who had warned Gencor about polyurethane in 1981, was sceptical about the effectiveness of self-rescuers due to the fact that under the conditions which might be found in the event of a fire they would not last for much more than 15 minutes. He was keener on the installation of refuge bays where mineworkers could collect in safety from fires and fumes. The October 1986 Regulation also insisted on the provision of refuge bays.

the hazards associated with all materials used underground. It was expected to provide a central database of flammability and toxic properties and to collect and analyse all information relating to underground fires. The second group was called the Expert Multi-Disciplinary Task Force whose remit was to investigate fires and explosions and make recommendations to improve underground safety.

The Memorial Services

Under the state of emergency those active members of the NUM who were urging militant action over safety issues were as likely to be the subject of police repression as the young township activists who were advocating violence. The mineworkers' reactions to the disaster, therefore, have to be seen against this hostile background. Immediately after the disaster the mine manager said that he intended to close No 2 shaft for at least a week in respect for the dead. This news was welcomed by the union, and by those who had been working there, not only out of respect for the dead but also because of their fears that it was unsafe to do otherwise. Within three days, however, the manager broke his word and mineworkers were forced to go underground. The collective grief of the black mineworkers quickly turned to anger and 2 000 of them refused to work. They were hounded and harassed by mine security officials and threatened with having their pay docked as well as by being disciplined. Gencor's explanation was that though they intended to halt production for a week, workers still had to go underground to clear the debris and make the area ready for production.

There was further trouble over the memorial services. An emergency session of the union's NEC decided to hold a memorial service in Secunda on 24 September and to call for a memorial work stoppage for a day on 1 October on all mines affiliated to the Chamber of Mines. The mine management, however, held its own service with Gencor officials, including Derek Keys, Gencor's general manager, at No 2 shaft on 22 September, to which neither union representatives nor individual black mineworkers were invited. The mineworkers attended anyway. A large number of them gathered some distance from the service. One of them addressed the crowd with a hand-held loudspeaker and said: "We have never prayed with whites and we don't intend to start today."[40] When the prayer service began in Afrikaans some mineworkers

[40]*The Star*, 23 September 1986. Also *The Citizen*, 23 September 1986.

NUM's poster reminding mineworkers of the Kinross disaster.

shouted: "Speak Fanagalo!" Then about 3 000 of them, stamping their feet, singing, waving branches and their hard hats and led by members of the NUM wearing T-shirts carrying the slogan "One Industry, One Union", formed a vast column and moved through the area where the management service was being held, circled it a number of times and drowned out the voice of the minister of religion who was conducting the service. They then stopped and held their own impromptu service, after which a number of them broke away and, as Gencor's representatives hurriedly left at the end of their ceremony, moved into the service area, toppled the pulpit with its flower-coloured backdrop and swarmed over the refreshment tables to take the sandwiches, sausages and drink that had been left behind. They were admonished later by the NUM safety officer for behaving as if they did not care for their brothers who had died underground.

The NUM's own memorial service in the stadium at Embalenhle Township, near Secunda, was a political occasion despite the state of emergency. The police were present with 20 armed vehicles. Police helicopters circled overhead. Ramaphosa said that about 15 000 mineworkers attended it and that there would have been more if the mine security at Kinross had not forced some, who wanted to attend, to go to work at gunpoint. The mourners carried posters urging "Safety before profits" and "We want decent working conditions". Two of the main speakers, Ramaphosa and Jay Naidoo, the general secretary of COSATU, both said that the only means of making the mines safe was for the workers to take control of them. From that day, Ramaphosa added, the union would not allow mineworkers to be killed as at Kinross. The proceedings were constantly interrupted as the mineworkers sang ANC songs praising Oliver Tambo and Nelson Mandela. The principal event of the service, however, was Winnie Mandela, the only non-union speaker, who was mobbed when she arrived in the stadium. She promised them the support of the black townships and said that the day of mourning the following week would be the beginning of the long battle against 'racists' in Pretoria. "For far too long," she said, "you have been locked in these ghetto compounds. For far too long we have been separated from one another by influx controls ... Now we shall forever be together." At the end of the service the police gave the mourners five minutes to leave the stadium.[41]

Both COSATU and the UDF supported the day of mourning on 1 October but allowed their respective affiliates, the individual unions,

[41] *The Star*, 24 and 28 September 1986.

factories, regions and associations to decide for themselves on how best to take part. Gencor also realistically agreed to it but the Chamber of Mines stated that the other mining houses would only allow five minutes' silence for the dead miners, though it was the view of Motlatsi that mineworkers "from Springbok to Messina" would take the day off. The action became a test of union strength. Estimates of the numbers who supported the stoppage varied, but all were large and the biggest in South African mining history. The NUM claimed that 325 000 mineworkers out of a total of about 450 000 took the entire day off and that 275 000 members of COSATU-affiliated trade unions also participated in stoppages of varying duration.[42] In some mines, such as the Anglo-American mine at Vaal Reefs and the Gold Fields mine at Kloof, there were reports that mine security forces had used guns to force mineworkers to go underground so as to prevent them from participating in the stoppage.[43] Apart from those incidents, however, there were surprisingly few arrests arising from it.

The Aftermath

Once the identities of the dead mineworkers had been established and the post-mortems were completed, the bodies were returned to their homes for burial. That, however, was not the end of the matter for the dependants. For most of them it was the beginning of a phase of more intensive poverty. Very few of the bereaved families were able to sustain themselves without the pay of their migrant menfolk. The cost of returning the bodies and holding the funerals was covered by the death benefit scheme set up by the Chamber of Mines in 1981. The compensation to provide an income for the dependants, however, was meagre. The widows received a pension equivalent to 30 per cent of a mineworker's monthly earnings, which, at the current monthly rate of pay of R244, amounted to about R73. Gencor admitted that this was insufficient and announced that it had set up a trust fund of R1,7 million to supplement the compensation with an average payment of R10 845 for each family.[44] The responsibility for distributing the money was given to the depots of TEBA but many families did not

[42]The independent Labour Monitoring Group reported that between 250 000 and 275 000 mineworkers stayed off work. (*The Star*, 2 October 1986.)
[43]*The Star*, 2 October 1986.
[44]*Ibid.*

receive any.[45] Some of the victims' dependants in Mozambique were never located and the compensation payments they should have received were returned to the South African government.

The mineworkers who survived the fire also had their worries. There was always apprehension, a sense of impending doom, when mineworkers descended the shafts but it became almost unbearable for those who had to return to work in the aftermath of a disaster. This was heightened at Kinross by two unrelated fires that occurred on level 15 almost a month after the disaster. The fires were minor ones in a short section of the rubber-coated overhead electric locomotive line and their causes were unknown. A small amount of smoke, but enough to remind them of the disaster, entered some of the underground workings and 2 000 mineworkers in No 2 shaft were evacuated. The NUM called for the shaft to be closed until an inquiry into the disaster had been completed and reiterated its demand for a judicial commission of inquiry into the conditions at Kinross "with the participation of international safety experts as proposed by the European Community's foreign ministers". Gencor, however, replied that the mineworkers had been in no danger and that there was no reason why operations should not proceed almost immediately.[46] Gencor's view prevailed, so the men had to stifle their fears.

Hiding the Facts

Gencor continued to do everything it could to cover its tracks. It obstructed Dr Herbert Eisner, who had been invited from Britain by the NUM as a foreign expert. When Dr Eisner, who had previously investigated the Hlobane disaster for the union, arrived he was refused permission to go underground to examine how far the flames went from their origin, the ferocity of the fire and the amount of polyurethane that had been used, because, Gencor stated, its own technical consultant was not available to accompany him. As if Eisner were an ordinary member of the public, it added, offensively, that it "didn't want people tramping around while they were clearing jackhammers".[47] Eisner was scheduled to be in South Africa for one week only but it was two weeks before he and three NUM officials and advisers were allowed down the mine and

[45]Eighteen months after the disaster only 108 families had received any payments. (Leger, *From Fatalism to Mass Action*, July 1988.)
[46]*Business Day*, 16 October 1986.
[47]*Citizen*, 24 September 1986.

even then it was as part of a delegation of about 60 people from other mining houses, white trade unions, the Inkatha union and trade unions that had no members at all at Kinross. The occasion was not to investigate the site of the fire but to escort visitors around the mine, as if on an 'open day'. Eisner was given no time to investigate and take measurements. He just had to walk around as if on a guided tour.

This treatment made a mockery of the invitation to foreign experts by the State President, P W Botha, in response to a demand by the ICFTU for an independent inquiry into the disaster. Botha, in his reply, praised the mining houses and attacked their critics. He stated: "South Africa has over the years developed an outstanding system guarding the health and safety of our mineworkers. This system is jealously preserved by both the public and private sectors concerned with mining ... Responsible representatives of all foreign governments and mining industries are welcome to visit our mining industry and ascertain for themselves what our safety standards, precautions and mechanisms entail ... The statistics and records will speak for themselves."[48] The South African Foreign Minister, Pik Botha, reacted similarly when the West German Foreign Minister, Hans Dietrich Genscher, called for an international commission of experts to assess the conditions at Kinross. Genscher, Botha responded, was ignorant of the safety standards of South African mines.[49]

The standards to which both of the government ministers referred were those set by the 'International Safety Rating' system which had been criticised by the NUM after the Hlobane disaster.[50] The Chamber of Mines had reacted to those early criticisms by refining and adapting the system to satisfy local conditions but it remained controversial and was at the centre of the argument about safety standards after the Kinross fire. Kinross had first implemented the system in 1980 when it was awarded three stars. By 1982 it had four stars and remained at that level until June 1984 when it received five stars, the highest accolade for safety for any mine. Just over three months before the disaster, however, it was downgraded to three stars due to shortcomings in the planned inspections of operations such as welding and fire control. But from its introduction until 1986 the 'International Safety Rating System' had failed to identify the use of polyurethane foam as a major lapse in safety standards.

[48]*Ibid*, 23 September 1986.

[49]*City Press*, 21 September 1986.

[50]See chapter 5 for a discussion of the origins of the International Safety Rating System and its early use in South Africa.

Whilst Gencor did not want to speak to the press about the disaster it was very willing to provide it with details of Kinross's safety record. Outside of No 2 shaft was a large notice declaring "Fatality-free shifts: 133 000 fatality-free shifts". Its last fatality had been a month previously when a fall of ground had killed one mineworker. The mine had achieved one million fatality-free shifts on three occasions during the previous 10 years and claimed that its safety record was well above the average.[51] A mine spokesperson said that from 1976 to 1985 there had been a 57 per cent improvement in the fatality rate and a 26 per cent improvement in reportable injuries.[52] Gencor claimed that it was a safety-conscious company, "ahead of all the mining houses".[53]

Insofar as these statistics indicated real improvements at Kinross, the union welcomed them. Unfortunately, according to the NUM safety officer, the figures used by the mine were misleading and acted to cover over the reality of working underground. The figures, he claimed, were incomplete, as many accidents were not counted. None was included that resulted in absences from work of less than 14 days. Moreover, they did not show important short-term fluctuations as, for instance, during the 12 months prior to the disaster when the mine's average fatality rate rose higher than that for the industry as a whole. The injury rate just before the disaster was the highest for 10 years.[54] Moreover, even if it were the case that the improvements were real, they simply took the mine from a level of unacceptably high death and injury rates to lower, but equally unacceptable rates. There was a long distance to travel before the mine could be described as safe.

The chairman of Gencor, Derek Keys, admitted that, in the light of the disaster, his company could not remain satisfied with its past safety practices. "It's start again and think again time in Gencor safety management," he said.[55] Yet the company was remarkably unwilling to answer the questions about Kinross raised by the NUM and the press. They wanted to know: (a) Why was polyurethane used when so much was known about its noxious qualities? (b) Why was gunite, a spray cement, not applied when it was both effective and safe? (c) Why, in any event, was the foam lining installed without following the manufacturer's instructions which required a coat of cement plaster and firebreaks of 12 metres long for just over every nine metres of

[51] *Business Day*, 18 September 1986.
[52] *The Star*, 17 September 1986.
[53] *Business Day*, 18 September 1986.
[54] *City Press*, 21 September 1986.
[55] *The Star*, 20 September 1986.

polyurethane coating? (d) Why did the underground manager give written permission for the welder to use an acetylene torch with a naked flame in such a potentially dangerous area? (e) Why, contrary to the Mines and Works Act Regulations, were there no fire extinguishers on hand when the welding was taking place? Why, finally, had the Mines Inspectorate taken no action over the matter since 1980?[56]

These questions were central to the investigation into the disaster yet, despite the government and Gencor's much publicised concern about safety, they were unwilling to have a public inquiry to answer them. Neither of them wanted to provide Ramaphosa and Motlatsi with the opportunity to attack their behaviour in public. Kobus Olivier, the general manager of Kinross, and the Government Mining Engineer decided, therefore, to avoid a public debate by pressing for a summary trial. They had reached their decision as soon as they had met on the day of the disaster, for the very next day, when reporters questioned Olivier about the use of inflammable materials underground, he said he could not reply because it was *sub judice*. When he was asked why that was so, he replied: "There is a legal case going on, absolutely, it will be in a legal court ... The GME will make certain recommendations to the public prosecutor and a magistrate will sit."[57]

G P Badenhorst, the Government Mining Engineer, referred the case to the Attorney General, as Olivier had forecast, and this prompted the NUM to seek a meeting with him on 4 May 1987 to ask when an inquiry was going to be held.[58] Ramaphosa, who led the union delegation, was surprised to be told that there had already been an inquiry in terms of Section 5(1) of the Mines and Works Act and that the union had been represented at it by one member of his staff and Clive Thompson, the union's legal representative. The inquiry turned out to have been the brief investigation Badenhorst had conducted soon after the disaster when he had simply taken statements from some of those present at the fire. When Ramaphosa asked why the usual procedure of cross-examination with a Presiding Inspector submitting a report was not followed, Badenhorst said that no cross-examination took place because no witnesses were called. He insisted that there was no difference between that investigation and an inquiry because the taking of statements constituted an inquiry. He added that he had the

[56]Regulation 11 (3) (4) of the Act stated: "No welding, flame-cutting or flame-heating shall take place unless adequate means are available to extinguish any fire which may result in such operation."

[57]Transcript of press interview, 17 September 1986.

[58]NUM, minutes of meeting with the Government Mining Engineer, 4 May 1987.

discretion to decide on the form of an inquiry and that, in any event, the NUM would not have had the right to participate because no NUM members were held responsible for the disaster. Badenhorst concluded that as far he was concerned, there would be no further inquiry.

The NUM continued to insist that a formal inquiry should be held and told Badenhorst that it would challenge his decision in the courts. On hearing this, he referred the matter to the state legal advisors who advised him that the NUM was correct in its interpretation of the Mines and Works Act and that an inquiry still had to be held. This was communicated to the NUM on 10 December 1987, but by that time the trial arising from the disaster was underway and the Government Mining Engineer had to wait until it was over before taking any further action.

The Trial

The Attorney General brought a charge of culpable homicide, namely the negligent killing of 177 people, and 12 other charges involving infringements of the Mines and Works Act against Kinross Mines Limited and six employees, including Kobus Olivier, the general manager of the mine. The court case began on 21 September 1987 in the Regional Court of the Regional Division at Witbank just over a year after the disaster. The main charge of culpable homicide consisted of two parts. It stated that the accused had failed to:

(a) Identify the presence of flammable substances used in 15 cross-cut north and the dangers involved in the use of such flammable substances.
(b) Take adequate precautions to prevent such flammable substances from igniting and/or burning out of control.

The other charges comprised a variety of infringements of the Mines and Works Act, such as failing to provide an efficient means of conveying water for fire-fighting purposes, failing to supply fire extinguishers where welding was taking place and failing to test for methane gas.

A magistrate conducted the trial proceedings. There was one state prosecutor and three counsels for the defence. Thirty-seven witnesses testified, including a number of experts. The NUM was not represented in any form and no representative expression of black opinion about the disaster was allowed. The only view that reflected the NUM's position was that put by Herbert Eisner who was called by the prosecution as an expert witness. But he was severely restrained in what he could say by the court rules that allowed witnesses to be questioned only about their own direct involvement in events. This meant that although Eisner, as the Director of the British Explosion and Flame Laboratory, had initiated and supervised many experiments into the use of polyurethane, he could not comment on the results of those where he had not actually been present or on those about which he had read in the scientific journals and conference papers. Evidence, moreover, that fell outside of Eisner's immediate expertise was discounted. These restrictions applied to all expert witnesses. Thus only a qualified chemist was allowed to comment, for example, on the chemical composition of polyurethane even though other witnesses may have acquired much more relevant knowledge about it from experience.

The counsel for Kinross Mines, Attorney C Plewman, dominated the court proceedings. He set the limits to the issues discussed and the manner in which they were pursued through his cross-examinations and challenges to the magistrate over admissible evidence. Plewman, moreover, had a cosy relationship with the magistrate. Throughout the trial they shared jokes and teased each other. In most instances the magistrate followed his advice. It was clear that the question as to what was and what was not admissible evidence pervaded and restricted the proceedings of the trial. It severely limited the exchange of information and illustrated the inadequacy of trials, compared with inquiries, as a means of identifying causes.

Each of the accused submitted a brief affidavit denying the charges but none was called to give evidence. The only black witnesses were four members of the pipes, track and ventilation gang that was working at the scene of the accident. Frederick Viviers, the white welder who was in charge of the gang and whose equipment started the fire, was not called to give evidence. None of the six accused individuals had been employed at Kinross when the polyurethane was installed in 1980. Indeed, two of them had only relatively recently taken up their positions. The director who represented the company was an accountant who, on his own admission, knew nothing about mining practices. One of the under managers on trial had actually been on holiday when the fire occurred. Those who had made the critical decisions had retired and were not before the court. Some of the witnesses had been connected in some way with the installation of polyurethane in 1980 but none of them had possessed any decision-making authority in the mine at that time.

The trial began on 21 September 1987 but the actual proceedings proper did not start until 5 October. It continued for 12 days until 22 October when it was adjourned until 14 March 1988. It went on for four more days and ended on 24 March 1988. At the outset the prosecution dropped the main counts on four charges but the principal one of culpable homicide remained. After the brief affidavits from each of the defendants denying the charges had been read to the court, details were given about the causes of death. The trial was not about the causes of the tragedy but why Kinross Mines and the six other defendants had nothing to do with it. There was no sense of drama or awe about what had happened. There was no-one there to express the grief and suffering the deaths had caused and to show graphically why the defendants were in court at all. There was, in consequence, no-one with the incentive to pursue the main questions until all the answers became clear. The prosecuting counsel was satisfied simply to put the case. Not surprisingly, the trial ended with no answers.

Rather than reflecting its enormity, the proceedings of the trial trivialised the disaster. It was as if it were a point-scoring game for clever white attorneys. For instance, when the medical report was read describing the main cause of death as carbon monoxide poisoning, the defence counsel tried to diminish the significance of the presence of polyurethane foam by showing that, as any fire involving any organic material produced carbon monoxide, there was nothing unusual about the deaths. He contended that there was just a fire, unfortunately in an enclosed space that made matters worse. The game was also racist. All witnesses for the prosecution were harangued or disparaged by the defence counsel by belittling their competence. But the black ones suffered most. The questions to them assumed that they were ignorant about the simple, basic features of their work situations. For instance, one black witness with almost 20 years' experience of mining was asked if he knew what a box spanner was and whether, as he never did any blasting, he knew why water was used afterwards. There was one particular example of discrimination when Isisi Mbuthuma, the team leader who had remained behind by himself to try and extinguish the flame in the cylinder, was harangued by the defence counsel for his inability to move the burning cylinder, yet that counsel made no mention of the actions and whereabouts of the white welder who had allegedly run from the scene. That the court proceedings were conducted to favour the white defendants was shown by the fact that they were conducted in their languages, namely English and Afrikaans, and that no provision was made to translate them into any of the indigenous African languages. When a black witness did not fully understand the questions, the defence counsel became impatient and irritable and on one occasion even stopped his cross-examination.

The Judgment

The meaning of the trial was encapsulated in the magistrate's final judgment. For him the main issue before the court was 'foreseeability'. That which was not foreseen could not have been anticipated and therefore carried no blame. He believed that none of the six employees knew of the presence of polyurethane foam in the haulage and, therefore, could not have foreseen the consequences of burning it. The company knew that the haulage had been lined with a foam but believed, he stated, that it was a fire-retarded isocyanurate and not polyurethane. He contended that the company was possibly misled by the Rock Grout Construction Company that had supplied the material, as the containers in which it had been brought were labelled 'isocyanurate'. Or perhaps, he surmised, the correct substance had been supplied but its flame-retardant qualities had deteriorated over time. In any event, in his view, the company carried no blame because, from the tests it had conducted, it believed that the foam would not burn. Moreover, it could not have foreseen the other events that led to the fire.

In his conclusions the magistrate emphasised that the foam alone did not constitute a danger because, by itself, it had not caused the fire. There had been foam lining along the haulage for six years without mishap. It seemed, he said, "that three eventualities interacted to cause this grave calamity. There was the absence of the metal washer in the valve of the cylinder. Secondly, the unexplained combustion of the leaking gas and thirdly, the inflammability of the foam along the walls of the tunnel."[59] Thus, in his view, the contact between the foam and the flame from a leaking oxy-acetylene cylinder caused the disaster. But this line of reasoning downgraded the significance of the polyurethane for if there had been no polyurethane, there could have been a flame from a leaking cylinder without causing a fire, or a fire that did not streak along the tunnel spreading toxic fumes. It was the foam alone that had made a confinable event into an uncontrollable tragedy.

So, having dismissed the foam as having any special significance, the court was left looking at leaky valves and providential acts of

[59]Regional Court for the Regional Division of the Northern Transvaal, September 1987, p935.

combustion. The leak occurred, it was discovered, because a metal washer was missing in the valve at the head of the cylinder. But, the magistrate said, neither the welder nor the other accused could have known about that. Only the person "who was so careless as to overlook the installation of the missing washer" was responsible. "In the long chain of causation," he continued, "one could seek for culprits far beyond the perimeters of the present inquiry."[60] In other words, no-one in the trial carried any blame for the leak. As for the actual ignition of the leaking gas cylinder, that, he concluded, was and remained a mystery.

The magistrate did raise one highly significant issue in words that indicated he believed it was important. "What the court finds most astounding," he said, "is the fact that the result of Dr Eisner's tests that were conducted in Britain had not resulted in the immediate banning of polyurethane or related substances in the mines of the Republic. Could it be that the results of these tests never became known to the mining fraternity in the Republic or to the manufacturer of the chemicals? It seems highly unlikely that through the passage of many years since the tests were done in Britain and the material banned there, it never reached the eyes or ears of those concerned with the matter in the Republic, or was the information perhaps deliberately suppressed?" But he stopped there and the matter was dropped. No reference was made to it during the trial except that the defence counsel implied that the questions it raised were irrelevant because there was virtually nothing South Africa could learn from European mining expertise. Nothing was allowed to flaw the characters of those responsible for importing, installing and using polyurethane even though a senior engineer with the South African Bureau of Standards informed the court that foams had been bought from overseas without being properly evaluated and sometimes with misleading instructions. He gave an example of how the words "*leicht brennbar*", meaning "easy to ignite", were translated from a polyurethane instruction booklet in German into English as "slightly combustible".[61]

The magistrate found that, except in the case of two minor infringements of the Mines and Works Act by the welder, Frederick Viviers, all of the accused were either excusably unaware of the dangers or had acted in a reasonable manner in the circumstances. Thus, he did not think it was negligent not having fire extinguishers on hand because so few people travelled along the haulage. Nor did he believe it was a

[60] *Ibid.*
[61] *Ibid,* p615.

mistake not testing for gas before the disaster because the regulation stipulated only that the test should be completed sometime during the shift and it could have been the intention to test later in the morning. No-one, in the end, was really responsible for anything. In his confusion about the responsibility for the events the magistrate even asked whether providence might have played a role.[62]

As in the Coalbrook trial, where the defendants pleaded not guilty, they were found not guilty and where they pleaded guilty, they were found guilty. Only the welder, who carried the least responsibility for any of the misdemeanours, pleaded guilty to anything. His plea was that he had infringed two minor regulations. All the cases against everyone else, including the other charges against the welder, were dismissed. Viviers was fined R50 on each charge so the total penalty for causing the deaths of 177 men was R100.

The Gagging Order

The NUM reacted angrily to the verdict. Cyril Ramaphosa suggested that the union should send a cheque for R100 to the convicted white miner, as he was not the person who was responsible for the disaster. He said the verdict was "offensive to our members' sense of fairness and justice" and that it implied "that the fire was an act of God, like an earthquake or flood, which could not have been prevented". He could not understand, given the evidence showing that the Kinross Mine was aware of the dangers of polyurethane foam, why the company "was not criticised or censured or brought to book in any way by the court."[63] He contended that if the 172 black victims had been white there would have been an entirely different outcome.

Three days after the trial verdict, the Government Mining Engineer, acting on the advice given by his legal advisers the previous December that a proper inquiry into the disaster still had to be held, ordered one to start on 28 June 1988. It was scheduled to last for three days. At its inception, Paul Benjamin, who represented the NUM, asked the Presiding Inspector, H Liebenberg, whether he could cross-examine witnesses. Liebenberg replied that he could question a witness only if it were shown that NUM members would be implicated or blamed by that witness. As the NUM members were victims of the disaster and were

[62]*Ibid*, p935.
[63]Flynn, 1992, pp264-265.

in no sense implicated in causing it, the ruling effectively excluded them from the inquiry and prevented Paul Benjamin, who had come prepared to unravel the background to the disaster, from asking Gencor and the mine manager a single question.

The Presiding Inspector then went on to ask each of the trial defendants to read out the statements they had made soon after the disaster but he allowed them to make amendments in the light of evidence submitted at the trial. They each made exactly the same alterations, emphasising that the mine had ordered one type of foam but had been supplied with another and that they had been unaware that the foam was not self-extinguishing when in contact with fire. The witnesses were not cross-examined to elicit more information about their behaviour either by the Presiding Inspector or the lawyers who represented Gencor. In the end, the inquiry lasted for only three hours, including breaks. The exclusion of the NUM from the inquiry was consistent with the general policy of the Government Mining Engineer of preventing the union from making any meaningful contribution to inquiries into disasters. His ruling acted like a gagging order.

CHAPTER TEN

THE GROWING MOOD FOR CONFRONTATION JANUARY-AUGUST 1987

The Adoption of the Freedom Charter

Once the NUM leaders had decided to engage the union in political action they set about clarifying and exposing their attitude and position in relation to the most vital political questions of the time, namely the liberation struggle and the kind of society they wanted to emerge out of it. They gave notice of their vision of a future society in the tail-end of the union's reply to Anglo-American's advertisement about mine violence in January 1987. "The NUM believes," it stated, "that only in a society where workers are in control of all aspects of their lives, including their lives on the mines, will conflict be eliminated. The AAC, the entire mining industry and those in power should take note that the NUM intends to make the achievement of this its agenda for 1987."[1] But it was the Fifth National Congress of the union in February 1987, held in a period of acute state repression, which clarified and publicised its political focus. Ramaphosa set the stage for the debate with an analysis of the political climate in his Annual Report. He wrote:

> "The state of emergency has failed to restore law and order, hence the regime's refusal to lift it; the townships remain ungovernable because the masses have evolved alternative structures of governing themselves; the tricameral parliament has been exposed for the fraud it is; the mass detention of activists has not solved the country's problems and the economic crisis is still with us.
>
> "The regime's inability to govern our country has resulted in immense hardship for the ordinary people; rising unemployment, brutal deaths at the hands of the SA Defence Force, mass detentions of children and other innocent people, an unaccepable rate of inflation and rising food prices. We have seen the country's resources being squandered and diverted to the homelands and the military.The intensification of internal mass resistance and the application of international pressure against the government continue to bewilder the regime.
>
> "It is clear that the continuous resistance, which started in August 1984, is not a passing phenomenon. It holds out the promise of some really basic transformation The factors that usually combine to set the scene for a major national transformation are beginning to come together. Firstly, the ruling class has virtually conceded that it has failed to govern in the same way it has for years. Secondly, it is clear that the masses are no longer prepared to be governed in the same old way. Thirdly, there is the tremendous growth of mass support for the liberation movement which heads the people's onslaught against the regime. As we move towards this dramatic transformation the interest of the working class is championed, amongst other forces, by COSATU and its affiliates..."

[1] *Weekly Mail*, 9-15 January 1987.

In his presidential address, James Motlatsi extended the debate with a description of the meaning of transformation for black mineworkers. "The struggle," he stated, "must be taken to a new phase. We must take control of our lives in every front ... We must take control of the hostels ... Together with the compounds and *induna* system developed over a century ago, the mine owners have kept us in conditions of wage slavery and oppression ... 1987 is the year NUM will begin to change this. We will fight every inch of the way to destroy the hostels, migrant labour and the *induna* system ... Comrades, the answer to our problem on the mines is clear: we need to take control. 1987 is the year we march in that direction." Motlatsi warned the mining houses that either they started negotiations "with the NUM to begin the process of dismantling the migrant labour system and establish workers' control of the hostels, or the mineworkers seize control without their involvement". This threat was encapsulated in the union's slogan, which designated 1987 as "The Year Mineworkers Take Control". There was a close similarity between this and the new slogan of the ANC. Seven weeks previously, the ANC had celebrated its 75th anniversary by declaring that 1987 would be the "Year of Advance to People's Power".

The political sentiments expressed by Motlatsi and Ramaphosa were reflected in the resolutions submitted to the congress by the regions for consideration as the policy of the union as a whole. It had been obvious at the various regional conferences held just prior to the congress that the members had become conscious of the need for the union to define and adopt a common political strategy. They had been politicised by their experiences of struggles during 1986 and by the state of emergency, which, Ramaphosa stated, had "brought about a level of dedication and commitment amongst mineworkers that had never been experienced before".[2] The delegates, however, found it difficult to translate their newly found solidarity into specific political terms. The regions made various suggestions but none amounted to a political statement that satisfied the majority of the delegates. Then, at a meeting trying to construct a compromise resolution, someone suggested that the *Freedom Charter*, as formulated by the Congress of the People at Kliptown on 26 June 1955, should be adopted as the most cogent expression of their political aspirations.[3] The *Freedom Charter* was fresh in their minds because the ANC had recently reaffirmed its

[2]NUM, Fifth National Congress, *Annual Report*, 1987, p1.

[3]See Allen, 2003, volume II, chapter seven, Appendix A, for a copy of the *Freedom Charter.*

commitment to it and its adoption had been discussed within the UDF. It was, nonetheless, a political adventure for the NUM to consider adopting it. The meeting eventually agreed that this should be done and a resolution was submitted to the congress that "the *Freedom Charter,* as adopted by the Congress of the People on 26 June 1955, is hereby recognised as a guiding document in our struggle from national oppression and economic exploitation ... All mineworkers should be united under the constitution and political policy of the NUM and COSATU ... The NUM and the organised working class in COSATU must play their historic role in the struggle for a non-racial and democratic South Africa."[4] The resolution was endorsed by the congress and became an integral part of the NUM's policy.

This move closely reflected the trend in black public opinion. Following the uprising in the townships between 1984 and 1987, the ANC had become recognised as the leader of the liberation struggle.[5] This had come about as support amongst black urban residents for direct action, and even for the armed struggle, as advocated by the ANC, had escalated while, by the same process, support for middle-of-the-road policies to oppose apartheid, as advocated by the Inkatha Freedom Party (IFP) in open opposition to the ANC, had declined.[6]

[4]NUM, Fifth National Congress, *Resolutions*, 1987, Resolution 1.

[5]See Mark Swilling, "The United Democratic Front and the Township Revolt", in W Cobbett and Robin Cohen (eds), *Popular Struggles in South Africa*, 1988, and Jeremy Seekings, *The UDF: A History of the United Democratic Front in South Africa, 1983-1991*, 2000, for accounts of the uprising in the townships.

[6]Inkatha was formed in the late 1920s as a Zulu nationalist movement to bolster support for the monarchy. Chief Mangosutho Gatsha Buthelezi revived it as the Inkatha Freedom Party in 1975 to give himself a political base and to project himself outside of KwaZulu as a national liberation figure. But it remained essentially a Zulu organisation with its support in the rural areas of Natal. Initially it was supported by the ANC, as was Buthelezi's decision for Inkatha to assume the government of the KwaZulu bantustan. The black trade unions in Natal also endorsed Buthelezi's decision. By 1980, however, there was a considerable political distance between Buthelezi and the ANC and Buthelezi's popularity amongst blacks in general began to decline. From then onwards, Buthelezi was marginalised until his power base was mainly confined to the peasants and petty bourgeois in Natal. Inkatha controlled the government of KwaZulu and its leading officials doubled up as members of the government. Subsequently, a fierce competition developed between Inkatha and the ANC-led liberation movement for the allegiance of black people in KwaZulu that was, in fact, a struggle for the political survival of Buthelezi and Inkatha. The tensions between Inkatha and ANC supporters developed into open violence after the formation of the UDF in August 1983. Inkatha claimed that the UDF was a front for the ANC and, as a consequence, its organisers attacked the UDF activities such as school boycotts, student protests and trade union stay-aways. This led the National Education Crisis Committee that directed the campaign against the educational policies of the apartheid government to resolve in March 1986 to

One survey showed that between 1977 and mid-1985, the nation-wide support for the IFP had fallen from 44 per cent to eight per cent. Even in Natal, only one-third of the Zulu-speaking residents backed it.[7] The adoption of the *Freedom Charter*, therefore, both set the NUM in pursuit of broad policies with socialist implications and placed it firmly behind the ANC in the liberation struggle. In consequence, it moved closer to the leaders of the township insurrection. It was a hazardous, even dangerous, course for a legally recognised organisation to take.

NUM's move set off a chain reaction in the trade union movement. The question as to whether the *Freedom Charter* should express their political aims was put on the agendas of a number of trade unions in 1987 but it was not a clear-cut issue for it exposed sharp political perspectives. The National Union of Metal Workers of South Africa (NUMSA), the second-largest union in COSATU with 130 000 paid-up members, adopted the *Freedom Charter* but with some reservations and with a more overtly political purpose than the NUM for it made socialism, rather than a non-racial democratic state, its primary and immediate objective.[8] The Chemical Workers' Industrial Union, with almost 30 000 members, followed the NUM's example after a prolonged debate.[9] COSATU itself then accepted a NUM resolution to adopt the *Freedom Charter* at its Second National Congress in July 1987. Gwede Mantashe, representing the NUM from the Matla Colliery in Witbank, moved the resolution. In doing so, he called for disciplined alliances with democratic non-racial organisations in the tradition of the ANC to "promote the leading role of the working class in a united

expose, isolate and fight against Inkatha. Thereafter, there was open warfare between Inkatha and all organisations, including the NUM and other trade unions, which identified themselves with the ANC. For details about the formation of Inkatha and its struggle with the ANC, see Davies, O'Meara and Dlamini, 1984, volume two, pp387-395; Baskin, 1991, chapter 21; Gerhard Maré and Georgina Hamilton, *An Appetite for Power: Buthelezi's Inkatha and South Africa*, Johannesburg, 1987; Mzala, *Gatsha Buthelezi: Chief with a Double Agenda*, 1988 and Cheadle Thompson, "The Natal Violence: South Africa's Killing Fields", an unpublished paper prepared for the International Defence and Aid Fund, nd.

[7]Price, 1991, pp197-198.

[8]See Baskin, 1991, pp214-217, for a discussion of the trade union debates about the *Freedom Charter*.

[9]The delegates at the Congress of the Chemical Workers' Industrial Union could not agree when the issue was first raised in 1987 so another congress was convened three months later when it was decided to adopt the *Freedom Charter* as a set of basic demands to escape from national oppression and economic exploitation. (CWIU, *Struggle for Workers' Rights: A History of the Chemical Workers' Industrial Union*, Durban, 1994, pp50-52.)

front alliance". He added: "We cannot divorce the struggle for national liberation from the struggle against capitalist exploitation … we cannot have our struggle demarcated into a workers' struggle and a community struggle."[10] The political audacity articulated in the speeches about the *Freedom Charter* at these various trade union gatherings raised the spectre of the Defiance Campaign of the 1950s and indicated a new political brashness in opposition to the government and its emergency powers. It also evoked cruder than ever forms of repression from the government's security services.

NUM's Approach to the Migrant Labour System

NUM's resolution on the migrant labour system at its Fifth National Congress matched the tone of Motlatsi's speech. It described it as "an evil system which constitutes a crime against humanity and … a gross violation of human rights". It asked the mine owners to make a clear statement of intent to dismantle it and, in doing so, destroy the compound system "within a reasonable specified time period". It demanded the involvement of the NUM with the mining houses in planning a more acceptable means of accommodating mineworkers. For the interim period it suggested various ways in which the hostels could be improved, including the transfer of control from management to democratically elected representatives. The sting in the tail of the resolution was the demand that if the mine bosses did not respond positively by 30 March 1987, then the union's National Executive Committee would declare a dispute and call a national strike. In preparation for such a contingency every branch committee was instructed to form a defence committee. Ramaphosa, however, did not act as if the issue were urgent because he failed to inform the mining houses of the terms of the resolution, including the strike threat, until 20 March, which was three weeks after it had been passed and only 10 days before the deadline.

Anglo-American responded in two parts. G W H Relly, the chairperson of the Corporation, replied in general terms on 10 April while E P Gush, the head of its Gold and Uranium Division, replied in greater detail on 8 May. Relly spelt out the commonly known position of Anglo-American, namely that it was in favour of destroying the migrant labour system and of settling its employees in permanent family housing as

[10]Baskin, 1991, p219.

"an essential step towards a more stable and orderly community and work life for all South Africans". He added that he believed that mineworkers should be involved in "shaping the housing schemes that are being planned for an orderly transition towards permanent housing options". The sticking point, however, was the legislative barrier the government had created to prevent black mineworkers from exercising their choice as to whether or not they wished to live in permanent mining accommodation. In law they were required to be migrant workers. In the case of foreign mineworkers that law was absolutely inflexible for the issue of permanent accommodation raised the question of nationality. Relly referred to this when he stated that the "residential choices available to foreign workers are governed by legislation in South Africa and the country of their origin. The more restricted choices available to foreign workers and the implications of demanding married accommodation for them could, we believe, prejudice their continued employment opportunities in South Africa. This matter will require careful thought by the respective governments, employers, employees and their leaders."[11] As almost 43 per cent of Anglo-American's black labour force was foreign it was not prepared to jeopardise its future labour supply by defying the government and prompting it to restrict, or even ban, the recruitment of foreign labour.

The second part of Anglo-American's response, namely the reply to the union from E P Gush, on behalf of the Gold and Uranium Division, was itself in two parts. The first part reprimanded Ramaphosa for not attending a meeting arranged for him in February to discuss Anglo-American's revised plans for housing all of its employees and for not suggesting an alternative date. It added that the Regional General Manager of Vaal Reefs had asked to meet him nine months before to discuss proposals for housing and hostels in his mine but he had not even received a reply. The second part answered Ramaphosa's letter, point by point. It stated that Anglo-American had been engaged for a decade in upgrading its hostel accommodation by officially dismantling the tribal *induna* system and enabling mineworkers to participate in the administration of the hostels. Gush emphasised that the Corporation had made progress by ceasing to finance tribal dance groups that were often the focus for tribal enmities and by allowing mineworkers to choose with whom they wanted to share accommodation. The gist of the response was that the resolution did not apply to Anglo-American mines because they were already doing what it demanded. The

[11]Correspondence from G W H Relly to the General Secretary of NUM, 10 April 1987.

delegates at the congress, who were primarily from those mines, thought otherwise.

When Ramaphosa reported to NUM's Central Committee in June he had already had two meetings with mining houses about the migrant labour situation. The question of implementing the strike threat, however, was never raised even though none of the mining houses had responded by 30 March and two of them, Gold Fields and Anglovaal, had not replied at all. But that was not the end of the matter. The formal approach to the mining houses was part of a wider, unco-ordinated and often spontaneous campaign by NUM members in their branches to wrest control of the hostels from management. They had taken action over a range of issues in attempts to undermine the legitimacy of management's authority. They had boycotted concessionary stores and liquor outlets and had gone on strike to have the *indunas* removed and replaced by elected representatives. Their protests against the provision of special facilities for team leaders and for the replacement of liaison committees by joint union-management committees were part of their campaign.

Dissatisfaction with hostel conditions, rather than opposition to the migrant labour system itself, was usually expressed through boycotts and attacks on concessionary stores and liquor outlets. For instance, such grievances had led to a café block being burned down at the Vaal Reefs gold mine in January 1987 and for the liquor outlet and concessionary store being damaged at the Saaiplaas gold mine in the Orange Free State in the following May. A strike over the quality of hostel food by 5 000 mineworkers at the West Rand Consolidated gold mine near Krugersdorp, also in May, resulted in them being attacked by the security forces with tear gas and rubber bullets. Trouble returned to Vaal Reefs in June when its liquor outlet was bombed.

There were two cases, in particular, where changes in hostel practice occurred as a result of local action. In April 1987, about 600 NUM members in Anglo-American coal mines in Witbank in the eastern Transvaal had taken their wives and children to live with them in the hostels in defiance of management. Then, a pilot scheme was initiated at Western Deep Levels whereby the *indunas* were removed and union representatives were assigned the task of implementing the union's resolution on hostel control. This achievement was undermined, however, by the fact that the Western Deep Levels' management had reported the union branch under the Labour Relations Act in the same month for a number of breaches of its recognition agreement. The Industrial Court subsequently found the branch guilty of 16 unfair labour practices, including a refusal to dissociate itself from threats of murder made by local union officials.[12] There were some other setbacks. JCI took two retrograde steps, firstly by reintroducing ethnic segregation in its hostels in retaliation for a strike in 1985 and, secondly, by refusing to stop its managers from making racist statements as a means of dividing, and, therefore, controlling their respective labour forces.[13] Gold Fields continued to act as before by simply ignoring the union and being impervious to any demands for reform.

NUM Members Set the Agenda

NUM members followed their customary practice of taking industrial action as the first step in pursuit of a settlement of a grievance. This may have been their preferred choice but they had no feasible alternative because most mines had no negotiated grievance procedures to which branch officials could resort. Strike action was the only sure way of communicating messages to management. And as black mineworkers were becoming increasingly conscious of messages they wanted to communicate, the tempo of strike action rose and spread.

During the first seven months of 1987 there were disputes in platinum, zinc and chrome mines as well as in the gold and coal mines. They occurred in small independent mines as well as in those belonging to the big mining houses and the parastatal producers of steel, electricity and chemicals. The issues in contention included anti-trade union activity by managers, retrenchments, Inkatha activities in the mines and white privileges at work. There were strikes against retrenchments in a number of coal mines as a result of the fall in the demand for coal by ESKOM and in the international market. The retrenchment of 37 union members at the Finsch diamond mine near Kimberley provoked a strike there in May but the largest of such disputes occurred at the Randfontein Estates and Western Areas gold mines in April after JCI, the owner of the mines, introduced labour-saving long-haul dumpers to scoop up broken rock. The dumpers, that were new to South Africa, displaced 2 310 mineworkers out of a combined labour force of 24 000. All of the black mineworkers in both mines went on strike but returned to work after three days when the police detained their union officials.

The pressures of the previous year that had focused the attention of

[12]The Industrial Court issued its judgment in November 1988.

[13]The management at the Randfontein Estates gold mine, for example, described the NUM as a Sotho union in order to create hostility towards it by other ethnic groups.

mineworkers on the need for solidarity and which had resulted in so much violence, continued into 1987. Some disputes were carried over into the New Year. The President Steyn gold mine in the Orange Free State, for example, which had been the scene of much faction fighting in mid-December 1986, continued to be disrupted until February 1987. Its union branch was expanding and was already the fifth-largest in the union. The branch officials alleged that the management was trying to destroy its influence by allowing an anti-trade union body to create animosity between ethnic groups in the mine. The animosity became violent at the end of January when two days of faction fighting resulted in the deaths of seven mineworkers. Almost the entire black work force of 19 000 went on strike against the continuing violence early in February. The mine manager issued an ultimatum to all the strikers to return to work and dismissed the 2 200 who defied it.[14] There were other indications of the fact that black mineworkers, contrary to the employers' view that faction fighting was an inherent feature of black lives, were intolerant of the violence and wanted either to end it or escape from it. Almost half of the labour force of 8 000 left Gencor's Beatrix gold mine in the Orange Free State and returned to their homes in January because of faction fights that had left eight of them dead and 53 injured. In the same month there was a seven-day strike at the Goedehoop Colliery in Witbank to get rid of an *induna* because it was alleged that he caused trouble by fomenting ethnic divisions.

The black mineworkers continued to protest about 'petty apartheid' practices, such as segregated facilities and job discrimination. Fifteen thousand mineworkers at the Randfontein Estates went on strike at the end of January in sympathy with a small number of black computer operatives who had been forced to leave the mine after complaining about the racist behaviour of their white supervisor. Then, at the beginning of April, there was a dispute in Freddies gold mine in the Orange Free State over the refusal of black mineworkers to act as 'piccanins'. The 'piccanin' system, in which black mineworkers were compelled to look after the welfare of white miners and supervisors, mainly through carrying their tools and bags, was a long-established practice in the mines. Ramaphosa described a 'piccanin' as a black who

[14]*Weekly Mail*, 6 February 1987. At first Anglo-American insisted that the mineworkers had left voluntarily but within a few days it admitted that they had been dismissed because of "the prospect of renewed violence and the consequent loss of life". Anglo-American rejected the union's allegation that anti-trade union activity had been the cause of the violence and blamed it on tribal tensions between Xhosa and Sotho mineworkers.

carried his "master's" satchel containing "food, clothes, newspapers and comics".[15] The Anglo-American Corporation, which owned the mine, rejected Ramaphosa's demeaning description of the contents of the satchels and claimed that blacks were only required to carry bags containing explosives. All black mineworkers in the mine, not only the 'piccanins', knew that this was not true and they waged a go-slow campaign against the practice for five days. The management claimed that the campaign was making mining unsafe, so it locked out the workforce of 8 000 and would not allow them back until the NUM gave an assurance that "its members would return to normal, safe working practices".[16] The union eventually gave the required assurance.

This context of bitter and violent struggles in the mines preoccupied the interests of the ordinary union members and was not propitious for building permanent union structures that required a commitment to routine administrative processes. Yet that was what the union needed, particularly in the area of health and safety. Although since the Hlobane disaster, Ramaphosa and Motlatsi had been able to mobilise support within the union for mass demonstrations after mine disasters this had not resulted in the generation of a continuing pro-active concern about health and safety issues amongst the ordinary members. The branches had shown little interest in demanding health and safety committees that could negotiate draft mine agreements covering relatively mundane issues such as the provision of time off for health and safety training without loss of pay and facilities to caucus to discuss safety problems. This was so even though the memory of Kinross was frequently rekindled.

The mine managers had contributed to this situation through their constant refusal to discuss safety issues with the branches. Safety, therefore, had not become the focus of sufficient branch activity to cause reports and resolutions to be sent to the regional committees. In the absence of a flow of information about safety, accompanied by resolutions demanding action, the regional safety committees were unable to function properly and made no demands on the health and safety department. There was little, therefore, apart from its own initiatives, for the national safety committee to co-ordinate. Insofar as a health and safety structure had been formed, it had been established from the top down, without any real basis in the branches. As the union's new health and safety co-ordinator, Sibanyoni's prime task, in

[15] *Weekly Mail*, 3 April 1987.

[16] *The Star*, 3 April 1987.

consequence, was to galvanise a concern about mine safety amongst the ordinary members that did not depend on the shock horrors of disasters. He had some small successes in that branch health and safety committees were formed in a few mines but they were distributed unevenly within and between the regions and were unequally developed. The committees resulted from agreements with mine managers that allowed for the election of safety stewards who were required to make regular inspections and to follow agreed safety procedures where there were dangerous conditions. They were vulnerable, as they were soon to discover, to repressive action by the employers and could readily be ignored by the Government Mining Engineer. The union was ill-equipped to forestall events that led to accidents or confront the issues that arose from them. Too much of what it did still depended on the personal interventions by Ramaphosa and they were all invariably made after the events.

Sibanyoni had many incidents to contend with. About six months after the Kinross fire, whilst the union was still grappling with its consequences, there was a cluster of three serious accidents within a week. The first, early in April 1987, was a methane gas explosion in a Secunda colliery through which 16 mineworkers were injured. A few days later, on 11 April, 34 mineworkers died and 17 were injured in a methane gas explosion at Gencor's Trans-Natal Ermelo colliery in the eastern Transvaal. On that occasion the Government Mining Engineer, in an attempt to avoid the adverse publicity he had received over his behaviour at Kinross, allowed an Inquest and Inquiry to be held. The NUM did not organise the majority of mineworkers in the mine and did not, therefore, have a recognition agreement with the management. Nonetheless Ramaphosa spoke on behalf of the mineworkers and condemned the mine, saying that as methane gas could be effectively dispersed by ventilation there was no reason why any should be present in the mine. It was there, he said, because of human error. The managing director of Gencor's Trans-Natal Coal Corporation replied that of course the gas should have been swept away by the ventilation system, but "it can go undetected between inspections, it can be released into the air by rock falls, or it can be released during controlled collapses".[17] The clear implication was that, in his view, it was nobody's fault.[18] Then, within the same week, a fall of ground at the JCI Randfontein Estates gold mine in Westonaria killed seven mineworkers and injured three others. This mine had been

[17]South African Institute of Race Relations, 1987/8, p342.

[18]Minimal charges were eventually laid against the mine and four of its officials and, in November 1988, they were each found guilty of negligence.

disrupted by large-scale strikes in protest at a retrenchment programme and there was much recrimination between the NUM and the management. As soon as the Mines' Inspectorate declared it would hold an inquiry into the accident there were instant contradictory explanations about its cause. The management claimed that it happened when 10 machine operators were drilling holes to install roof supports. The NUM, on the other hand, asserted that they were not installing roof supports at all but were drilling to blast out an excavation to store equipment. Neither side was able to substantiate its assertion but, in the climate of 1987, the important factor was that the mine owners no longer controlled the flow of information.

Recurring Problems at Western Holdings

The causes of the dramatic events at No 6 shaft in Western Holdings which had led to the public executions the previous July had not been resolved and the resentment over the dismissal of the four shaft stewards who had been involved in them had continued to simmer. In an act of defiance against the management, the union members elected two new shaft stewards whom the management described as instigators of the execution proceedings. In April 1987 these two, along with three other mineworkers, were dismissed. The whole of No 6 shaft demonstrated for their immediate reinstatement but added the four who had been dismissed the previous year to the list. The mine manager insisted that they should wait and follow the procedures laid down for an appeal against dismissal. This they did but the matter had not been resolved by the beginning of June, so about 700 night and early-morning shift workers demonstrated outside the section manager's office. The NUM insisted that they were simply holding a meeting but the manager alleged that they were planning to attack the office. In any event, the mine's Emergency Protection Unit was summoned to assist the mine security in its efforts to break up the demonstration and two of its members, a security officer and a section engineer, were stabbed and beaten to death. Five of the demonstrators were shot and killed and up to 20 were injured.

The deaths of the two white workers prompted an immediate management reaction that was rarely displayed when black mineworkers were killed. Indeed, most black deaths in the mines were not the subject of police investigations. On this occasion, the shaft was closed and the mine security services moved into the compound in force. Union meetings were completely banned and no meetings of any kind involving more than 10 people were allowed. Informers were strategically placed in the mine and the compound, and their evidence alone was sufficient to warrant arrest and dismissal. There were many dismissals and although there was a dismissals appeals procedure, no-one was allowed to use it. The entire branch leadership was amongst the 243 mineworkers who were arrested and detained until December. There were some injured mineworkers amongst those arrested but they received only cursory medical attention. As a result of the police action the union organisation in No 6 shaft was destroyed. The workers became disillusioned with the union and not only refused to be members, but resented its presence.

Nineteen of the detained mineworkers were charged with the murder of the two white officials and nine were found guilty. In the case of eight of those convicted, the judge found extenuating circumstances and sentenced them to prison for periods ranging from 18 months to 10 years. The ninth convicted person was given two death sentences and the right to appeal against only one of them. He was just one of the 143 people who had been refused the right to appeal, out of 248 who had been sentenced to death in 1987. He was, therefore, dispatched to take his place on 'death row' in a year during which there were more executions and fewer reprieves than in any other year since the National Party had come to power in 1948. In 1987 there were 51 people waiting to be hanged for politically related crimes. Altogether, in that year there were above one-third more executions than in 1986 and almost four times more death sentences for unrest-related incidents.[19]

Inkatha Enters the Mines

Despite the emergence of a collective consciousness amongst black mineworkers, they often perceived themselves first as members of different tribes, nationals of different countries or as workers in distinct mining sectors. It was not difficult to exploit these perceptions and create divisions within the union that aimed to organise them all. There was also a perception that numbers conferred status and superiority and

[19]See South African Institute of Race Relations, 1988/1989, pp532-533. There were many petitions for the convicted mineworker's release whilst he was on 'death row'. Eventually his sentence was commuted to life imprisonment and he was freed under the government's amnesty in 1992. During the demonstration that had led to the deaths he had been shot in the leg by the mine security and taken to hospital, but the police had whisked him away to jail before he could be treated and he remained permanently crippled.

that, therefore, the biggest occupational or industrial group was the most important sector in the union. This particular perception influenced the attitude of those who worked in the gold mines, who comprised about 85 per cent of the union's membership, towards those from the smaller mining segments. At the 1984 Annual Congress, for instance, delegates from the gold mine regions had ridiculed those from the coal mines and had described them as weak and effeminate. The effect on the coal mineworkers had been to generate solidarity amongst them and encourage militancy but it had also sowed the seeds of separatism. Late in 1986, the chairperson of the coal mining Witbank region, influenced by the fact that most overseas mining unions were exclusively coal mining ones, tried to establish a breakaway union for his coal region but he was opposed by those loyal to the NUM, led by the regional secretary, Gwede Mantashe, and the unity of the NUM was maintained.

Mine managers knew these weaknesses and were keen to exploit them. In particular, they encouraged tribal divisions and whenever the opportunity arose for them to foster a rival union, they took it. Their first significant opportunity arose through the actions of the United Workers' Union of South Africa (UWUSA) that was linked to Inkatha in Natal. From early in 1987, UWUSA began to infiltrate collieries, mines and refineries, always with the consent of the local managers and well beyond the confines of Natal. Whenever it found it difficult to penetrate any of these establishments it resorted to a policy of burning union property and assaulting or murdering union officials and activists.[20] In order to make sense of these destructive activities in the mines, it is necessary to understand something about UWUSA's origins.

UWUSA was the creation of the IFP. There had been a possibility for some time that the IFP would establish its own trade union

[20]Gerald Kraak, *Breaking the Chains: Labour in South Africa in the 1970s and 1980s*, p148. The violence had spread well beyond the confines of work situations into any communities that were identified with the UDF. The violence was open and public with the intention of intimidating people at large. Insofar as it was perpetrated by Inkatha groups, it was under the control of self-styled 'warlords' with the connivance of the police which had become, in effect, the armed wing of the IFP. There was much symbolic brutality in the violence. Victims were not clinically executed but mutilated. And though the IFP had a death list that contained the names of union and political activists, the killings were indiscriminate and included babies, small children, the sick, infirm and the relatives and friends of activists. The monthly death rate was more than 160. Though the press at the time described the violence as tit-for-tat, the vast majority of aggressors were IFP members and most of the victims came from organisations and communities sympathetic to the UDF.

federation to compete with and challenge the unions that were linked, however remotely, with the liberation movement and, at the same time, appeal to Zulu workers as an ethnic group. It was possible to do that in KwaZulu because, contrary to the situation in all of the other bantustans, it had legalised trade unions even though the bantustans were exempt from the findings of the Wiehahn Commission. This followed from Chief Buthelezi's posture of 'collaborative opposition' whereby he used some of the rhetoric of the ANC, yet opposed majority rule in favour of a regional power-sharing arrangement and gave the Pretoria government practical support. He disagreed with sanctions and disinvestments as means of putting pressure on the government. Within this mixture of contradictory policies he had endorsed the existence of trade unions. The KwaZulu legislature, in consequence, had passed the Industrial Conciliation Amendment Act No 10 in 1981 that gave unions the right to register and receive the benefits of conciliation and arbitration that hitherto had been the exclusive preserve of white unions. But it went further than the Wiehahn proposals in one important respect by allowing black unions in KwaZulu to affiliate to a political party. This enabled the IFP to establish its own trade union.

The IFP, however, had had little incentive to take such a step at that time. Before the formation of COSATU the trade unions operating in KwaZulu had gone out of their way to avoid antagonising the IFP by trying not to challenge its hegemony. But after the Natal unions had begun to expand and the UDF extended its strikes and boycotts to KwaZulu, ruining many small Zulu businesses in the process, the relationship deteriorated. Antagonisms developed into outright conflict, almost on a war footing, after COSATU began to organise stay-aways in Natal contrary to the wishes of the IFP.

Buthelezi's concerns about trade unions took place against this background. There had been discussions within the IFP since at least 1983 about the need for a closer relationship between it and the trade unions but always on the assumption that the IFP would be in the dominant position. In order to reach that end, Buthelezi, in 1983, had asked for funds from the American AFL/CIO to employ someone who would devise ways of "establishing a working relationship between Inkatha and the trade union movement" but he had no success even though, late that year, the American CIA had been accused of working through the AFL/CIO to assist the South African government "in a deliberate programme to undermine the independent unions in South Africa".[21]

[21]Maré and Hamilton, 1987, p129.

At first, Buthelezi had simply wanted an Inkatha presence in the trade union decision-making bodies. Only when that appeared to be impossible did the IFP talk, just a week before the COSATU launch, of forming its own unions. The launch was a catalyst for Buthelezi. He was hostile to COSATU because he believed it was an ANC front but his hostility grew more intense after Elijah Barayi, the newly elected President of COSATU, had ridiculed him, called him a "tribalist", described him as a puppet of the apartheid government and promised to "bury" him.[22] The IFP responded quickly. It set up a working committee to organise a new union and planned its launch for May Day, 1986.

UWUSA was launched as the trade union arm of Inkatha at a May Day rally in Natal, addressed by Chief Buthelezi in his capacity as both the President of the IFP and the Chief Minister of KwaZulu. It was a lavishly expensive launch, financed "with considerable covert funding from the SAP's Security Branch".[23] The local chiefs sent busloads of supporters to the gathering. Seventeen special trains were laid on to transport others. The Natal Rugby Union Football stadium was hired as the venue. Members of the KwaZulu legislature were present, as were many members of the Natal business community. It was estimated that between 60 000 and 70 000 people were present, representing all sections of KwaZulu society. The governments of both South Africa and of the KwaZulu bantustan supported the launch. It was in no sense a trade union rally though many workers were present. To add glamour to the event, Chief Buthelezi arrived in a helicopter with the name of the new union emblazoned on its side.[24]

The aim of UWUSA had been to establish an industry-based union in every sector where it was able to recruit 5 000 members and, in that way, transform itself into a federation comparable with COSATU. It decided at the outset to concentrate its efforts in the mining and manufacturing industries where many Zulus worked. The NUM, therefore, was a prime target for a violent, though limited, competition for members. Unlike any other competitive union, UWUSA had built-in qualities that enabled it to mount an immediate campaign against the NUM even before it possessed any members. It was backed by the resources of the IFP and received financial support from the

[22]Barayi's comments were made at a workers' rally on 1 December 1985, the second day of the launch of COSATU.

[23]Truth and Reconciliation Commission of South Africa, *Report*, Volume Two, 1998, p469.

[24]Maré and Hamilton, 1987, pp132-133 describe the launch. The attendance contrasted with the 10 000 who were present when COSATU was launched.

administrative structures of the state, including the security services. Its president, general secretary and treasurer were businessmen who brought with them the backing of Natal's petty bourgeoisie. Only two members of its executive possessed any previous trade union experience. Within a week of its launch, UWUSA had appealed for "large-scale assistance" to the African American Labour Centre through which the American State Department channelled funds for the purpose of buying the allegiance of African unions. It also asked for personnel from American unions to train its officials, as well as help from Israel and West Germany.

A month after COSATU had appeared on the scene, it had published a 45-page dossier containing affidavits and statements alleging incidents, threats and assaults on its members by Inkatha supporters, adding that "the link between Inkatha and UWUSA is so interrelated that it is not always possible to separate them".[25] UWUSA lost no time in entering the mine compounds and was welcomed by the mine managers for its opposition to strikes, its support for 'liberal' capitalism and its emphatic rejection of any act with a socialist flavour. It appeared as the mine managers' dream of a trade union and they used it wherever they could to break strikes, divide the labour force and undermine the NUM. But in entering the compounds they invited retaliatory violence. As was being shown at Western Holdings and Vaal Reefs, the members of the NUM had no tolerance for workers who collaborated with management.

The first major incident of violence between the NUM and UWUSA occurred at the Hlobane colliery, near Vryheid in Natal, shortly after the state of emergency was reimposed in June 1986. The NUM had members in the colliery but the manager had refused to grant it recognition and had done all he could to obstruct the union's activities. When the union organiser was refused access into the compound, the NUM called a strike. UWUSA had been allowed into the compound shortly after its formation and its members opposed the strike. But that was not all, for the mine manager allowed non-mine UWUSA supporters, alleged by the NUM to be Inkatha vigilantes, to enter the colliery compound and assist in breaking the strike. In the ensuing fighting between the two factions, 11 NUM supporters were killed and more than 100 were injured. UWUSA insisted that the NUM shaft stewards, who threatened to 'necklace' any mineworkers who attempted to break the strike, caused the violence. A similar situation was

[25]South African Institute of Race Relations, 1986, p246.

threatened at the nearby Anglo-American Vryheid colliery when the manager allowed UWUSA into the colliery in order to enlist members in competition with the NUM. On that occasion, however, the NUM went to court and successfully obtained an injunction from the Natal Provisional Division of the Supreme Court in May 1987 ordering UWUSA not to assault, threaten or intimidate any of its members. But by that time factional violence had become endemic throughout Natal.

Because of its determination to exclude the NUM entirely from its mines, Gold Fields was one of the most enthusiastic supporters of UWUSA. It had refused to grant recognition to the NUM for any unskilled categories at its seven gold mines. It was alleged that it employed a private police force to suppress dissent in its mines but its main ploy was the generation of ethnic tensions that destroyed the solidarity of the workforce.[26] It used UWUSA for this purpose early in 1987 at Zincor, its zinc refinery plant near Springs. There were a number of assaults on NUM members in March during which unidentified gunmen killed three and injured five. Then, one night in April, six armed gunmen, wearing balaclavas, burst into a Zincor hostel at midnight and opened fire on the sleeping workers. Five NUM members were later admitted to hospital with gunshot wounds. Eighteen bullets had hit one of them. The NUM claimed that UWUSA members were responsible for the shootings on both occasions. Six hundred workers at the refinery went on strike over the violence, claiming that the collusion between Gold Fields' security and UWUSA had turned the plant into a killing field.[27]

The Gold Fields management maintained its opposition to the NUM throughout the dispute. It sealed off the compound and refused to allow workers to go in or out. Nor would it allow NUM officials to enter the compound to discuss the incidents with their members. It turned down a request by Ramaphosa for an independent commission into the violence on the grounds that it was already addressing the underlying problems. When the NUM requested permission to hold a memorial service for those killed in the March shootings it had to give a written undertaking that the gathering would hold only a memorial service and that nothing would be said about allocating responsibility for the deaths. But as all black funerals had become opportunities for symbolic protests, the NUM was unable to abide by its undertaking. As a punishment, the Zincor manager refused to allow the NUM to hold any

[26]See South African Institute of Race Relations, 1987/1988, p346 for a report on Gold Fields' private police force which, a mineworker declared, had tortured him.
[27]*Weekly Mail*, 16 April 1987 and *The Star*, 16 April 1987.

meetings unless they were recorded and the tapes were handed over to him. This condition remained in force until after the April shootings but Gold Fields insisted that it was refusing to permit meetings to be held only because "mass meetings held at a time when emotions are running high … could lead to an exacerbation of the situation".[28] The strike ended after three days when the Zincor manager assured the workers of their personal safety in the plant and hostel. Six days later they were allowed to hold a meeting without any stipulations about its conduct. This incident was the beginning of a series of violent activities by UWUSA in mines outside of Natal. Its practices were common. It required only a small group of supporters to undertake systematic disruption. In some situations, one was enough. This was the case at one of Gencor's Impala Platinum refineries in Springs, late in June, when an UWUSA member was suspended for a month for his "disruptive activities". When he returned to work at the end of July 1987, virtually the whole of the black labour force at the plant went out on strike because, the NUM stated, he was an obstacle to harmonious relations there.[29]

The May Day Stay-Away

Meanwhile, the NUM endeavoured to carry out its obligations as an affiliate of COSATU by responding to its calls for mass action on May Day 1987 and during the whites-only general election on 6 May. However, it faced a special problem concerning the May Day stay-away. Some black unions had commemorated May Day with a stay-away for a number of years. In 1986 COSATU had entered the trade union scene with such a highly successful May Day stay-away that some employers had agreed to regard 1 May as a paid holiday and had urged the government to look into the question of including it as a paid public holiday. The state president responded by announcing in March 1987 that in future the first Friday in May would be a paid public holiday and would be called 'Workers' Day'.

It so happened that 1 May in 1987 fell on a Friday, so the stay-away was consistent with Botha's announcement. The celebrations went ahead with few difficulties but in anticipation of future years both COSATU and NACTU insisted that the commemoration of Labour Day on 1 May was not negotiable. This view reflected the opinion of black

[28]*Business Day*, 22 April 1987.
[29]*Ibid*, 30 July 1987.

unions in general for 47 out of 56 trade unions interviewed about the matter supported 1 May and no other day as Labour Day.[30] A further problem arising from the state president's statement was that it did not apply to the manufacturing and mining sectors. He, therefore, appealed to employers to be "accommodating towards workers" who were not covered by his decision. Some in manufacturing did respond positively but this was not the case with the Chamber of Mines.

A difficulty arose because the NUM had been in dispute with the Chamber of Mines since the 1985 round of negotiations over its refusal to grant mineworkers a paid holiday on 1 May. The matter had come to a head in April the following year when the Chamber of Mines applied to the Rand Supreme Court for an injunction to prevent the NUM from calling a May Day stay-away on the grounds that it was illegal. The court's decision on 28 April 1986 had been in favour of the union. It declared that the NUM had a legal right to strike on that day because it was officially in dispute with the Chamber of Mines over the issue. The judge said that the legal right to strike, which the NUM had obtained in 1985, had not lapsed. The ruling had given the mineworkers a big morale boost with the result that they supported the 1986 May Day stay-away in large numbers. Afterwards, the NUM returned to the negotiations with the Chamber of Mines that carried on until April 1987 and then collapsed. The cause of the collapse was the insistence by the Chamber of Mines that all mineworkers who supported a stay-away should notify management of their intention to do so. Ramaphosa replied that this would expose them to intimidation and victimisation by white miners and supervisors whose earnings would be reduced because they were based on bonus payments. This, he said, had already happened. Thus, there was no agreement to permit black mineworkers to observe May Day as a paid public holiday, as was the case with many other workers throughout the country. Those who participated in the 1987 May Day stay-away lost a day's pay. The anticipation of that markedly reduced the mineworkers' support for it.

The COSATU stay-away against the whites-only general election was spread over two days, 5 and 6 May 1987. It was estimated by the Labour Monitoring Group that 520 000 black workers participated in it on the first day and that 630 000 took part on the second day.[31] It was said that only about 20 000 of these were mineworkers and that most of them came from Anglo-American and Gencor mines. Black mineworkers,

[30]South African Institute of Race Relations, 1987/1988, p685.
[31]See Baskin, 1991, p189.

however, had always disliked political stay-aways for, unlike township workers, they could not simply stay at home and become anonymous amongst their families and friends. When they were not at work they were visible in the hostels, under the control of the white hostel managers and exposed to coercion from the mine security forces. Their participation, therefore, depended on the extent to which they were able to relate the stay-aways to their own mine work interests. The issues had to be close and immediate to justify risking physical coercion or dismissal. For this reason, commemorative stay-aways were relatively low down the list of mineworkers' priorities. For instance, the anniversary of the Soweto uprising on 16 June, celebrated in the townships, was mainly an ordinary working day in the mines.

The Changing Character of Strike Action

As the NUM branch protests were taking place there was a growing mood of militancy in other industries that culminated in relatively prolonged national strikes. It was clear that the state of emergency was not successful in repressing strikes over legitimate industrial grievances even though the state made no distinction between strikes for political and industrial purposes and damned them both. In the first half of 1987, there were 1,1 million strike days compared with 1,3 million for the whole of the previous year. The vast majority of strikes were in manufacturing, followed by the trade and servicing industries. They were not only more frequent but were longer. Some of them, moreover, were highly successful. Lastly, because they concentrated on industrial relations issues and targeted their respective employers strictly according to the labour law, they were able to maintain their status as legally recognised organisations. In practice, however, many of the strikes, indirectly, inadvertently, and sometimes as an after-thought, were directed at the apartheid system. It was, as Gwede Mantashe suggested in the union debate over the *Freedom Charter*, a fiction to separate worker from community issues or industrial from political ones. This overlap of purposes aided the security services for it gave them a justification to intervene, whatever the stated purpose of the strike. In practice, they rarely made any kind of distinction and attacked them all.

An indication of the changing consciousness of black workers was reflected in the characteristics of the strikes in retail distribution, the postal services and the South African Transport Services.[33] Each of these

[32]See the *South African Labour Bulletin*, Volume 12, No 3, March/April 1987.

strikes lasted for some weeks, had a relatively successful outcome and was led by a union that was strengthened in the process. The survival of black trade unions after relatively long strikes was a new phenomenon in South Africa. Another distinctive quality of these strikes was that they intruded into and disrupted the lives of the white communities at the same time as they attacked apartheid practices. The first of the strikes was over wages and working conditions at the retail chain, OK Bazaars, an Anglo-American subsidiary. The strike, which involved almost half of OK Bazaars' labour force of 23 000, started on 18 December 1986, lasted for 10 weeks and ended with a mediated settlement.[33] The strike had extended over the Christmas period so that many shoppers in different parts of the country had felt its disruptive effects. These did not disturb many of the black customers for about 60 per cent of them boycotted the shops in support of the strikers.

A series of strikes occurred in the postal services in the regions of the Witwatersrand, the north and eastern Transvaal, East London and Port Elizabeth at different times between April and September 1987, over a variety of petty apartheid issues. The Post and Telecommunication Workers' Association (POTWA), that led the strikes, had demanded, amongst other things, "an end to the arrogance by management during meetings with union representatives ... the scrapping of apartheid which denied African workers the right to use lavatories, telephones and canteens designated for whites ... (and) ... the scrapping of the quota system whereby management hired on racial lines and promoted other racial groups to supervise African workers..."[34] There was a partial settlement of the Witwatersrand strike after 28 days when the management agreed to soften the attitudes of white managers during negotiations and to relax the enforcement of segregated facilities. The central issue in the postal strike in the eastern Cape, which began in June 1987, was a demand for wages parity between black and white workers. Despite threats and an ultimatum, the strike lasted until early in September when the government agreed to work towards wages parity. As in the case of the OK Bazaars strike, the union was strengthened in the eyes of the workers by showing that it could both survive a relatively long strike and wring concessions from the employer. Its reward in the Witwatersrand was that it signed up 90 per cent of its potential membership.

The last of these influential strikes started in a small way and

[33]Baskin, 1991, pp168-171. OK Bazaars was owned by the South African Breweries Group which was a subsidiary of the Anglo-American Corporation.
[34]South African Institute of Race Relations, 1987/1988, pp675-678.

finished dramatically. It clearly revealed the complexity of the environment within which trade unions were operating. On 13 March 1987, about 500 black workers in the City Deep railway depot in Johannesburg protested against the dismissal of a colleague.[35] Other depots joined in until about 20 000 black workers employed by the South African Transport Services (SATS) were involved. In the process of striking, the workers' demands escalated in the belief that once on strike they might as well clean their slate of grievances, of which there were many. SATS, a state-owned company, was unashamedly racist. The majority of its 206 000 employees were Afrikaners who had obtained their jobs through state patronage, primarily on the basis of being both white and Afrikaners. In order to get permanent tenure they were simply required to serve two years' probationary service. All the black railway workers, on the other hand, were migrants with no security of tenure. They received poverty-level wages, no fringe benefits and were housed in squalid single-sex hostels. Like many other black people who had to work alongside white manual workers, they were subjected to humiliating and degrading racist abuse. The central issues during their strike were their migrant status, job discrimination and racial abuse.

The mood of the strikers hardened during the second week and they vented their anger on railway property by burning 136 railway carriages, 30 of them beyond repair. Many black commuters refused to pay for their tickets and rode free, thus threatening the financial viability of the company. Rail travel during peak times in the Witwatersrand was chaotic. The ANC, in its assessment of the strike, claimed that the burning of the coaches had spread fire from the townships into the white areas for the first time. White people were affected in another way for SATS tried to break the strike by taking on unemployed white workers and white school children as scab labour. The children were given work as station cleaners and were paid well above the rates given to black workers. The company's basic strategy to break the strike, however, was repression. After just over a month, it gave the strikers two days to return to work or be dismissed. It also tried to evict them from their hostels. The deadline was extended to 22 April when the dismissals were enforced and the strikers were given notice to vacate the hostels. The strikers rejected these deadlines, in turn. In the background, the police raided the homes of union officials and physically attacked them. Strike meetings were disrupted and dispersed

[35]For a fuller description of this strike, see *ibid*, p670-675, and Baskin, 1991, pp171-181.

by the police through the use of sjamboks, tear gas and, on occasions, live ammunition. Many strikers were arrested under the emergency regulations.

After the rejection of the SATS ultimatums on 22 April the strikers met at various locations to discuss their future tactics. The police forced their way into one such meeting in the Germiston office of COSATU and killed three strikers and injured many more. The workers scattered and some of them, with their injured comrades, made their way to the head office of COSATU where thousands of strikers were already gathered. After the meeting had been told of the Germiston police raid it sent a delegation to assist those who remained there but the police intercepted it at the Doornfontein railway station and shot and killed a further three strikers. The delegation, with its dead and injured members, returned to COSATU House.

At the beginning, the South African Railways and Harbour Workers' Union (SARHWU) that was leading the strike had very few members and was not recognised by the company. Though, during the course of the strike, the majority of strikers joined the union it was not allowed into the negotiations. On the contrary, SATS went out of its way to try to discredit it by showing documentaries on prime-time television to illustrate that it was affiliated to an organisation that supported the ANC and was, therefore, controlled by the Communist Party under directions from Moscow.[36] The SATS propaganda, however, rebounded on it because any association with the ANC and its allies encouraged, rather than discouraged, recruitment and solidarity.

The government's intention had been to use the strike as a pretext to attack COSATU House itself. Within a few hours of the return of the delegation and its casualties, 200 heavily armed police cordoned the building off and then entered it. This gave them access to the offices of a number of COSATU affiliates that were located there. When COSATU was first formed it was given a room at the head office of the NUM in Lekton House, Wanderers Street in Johannesburg. At its first executive meeting, however, it had decided to establish its own headquarters as soon as possible and rented a large building at the intersection of Jeppe and End Streets with sufficient office space to accommodate some of the affiliated unions as well as acting as a centre for the whole trade union movement. The NUM moved its own head office into three floors of the building. A number of other unions, including SARHWU, took up residence there. A raid on SARHWU,

[36]Baskin, 1991, p173.

therefore, gave the security services the opportunity to disrupt the work of COSATU and some of its affiliates, including the NUM.

COSATU House was raided twice in April 1987. During the first raid on 22 April, though the police detained up to 400 SARHWU members and officials, they did not confine their attention to the strikers in the main hall or to that union's offices. They went through every room in the building destroying "property, union records and equipment. Papers were thrown out of windows, expensive machinery was vandalised or rendered unusable, food and coffee were laced with unknown substances, money was stolen and locked cupboards and safes forced open".[37] Ramaphosa reported to his executive that NUM's offices had been entered and damaged, money had been stolen and members of his staff had been trapped inside. That was only the beginning.

The police raids, however, did not cause SARHWU's biggest problem during the strike. That resulted from the continuous supply of black and white strike-breakers. Under the terms of the state of emergency, peaceful picketing was impossible so the strikers had no legal means of preventing workers from crossing their picket lines. They were not allowed even to talk to them in an attempt to dissuade them from working. The strikers, therefore, decided to use means consistent with the methods of the police. On 28 April they accosted five strike-breakers and took them to COSATU House where they intended to beat them as an example to others. Some of the strikers, however, thought otherwise and demanded that the strike-breakers should be killed. The five men, therefore, were taken to a piece of wasteland. One of them broke away and escaped. The remaining four were necklaced and killed.

The police discovered the dead bodies early the next day and, during the afternoon, threw a massive cordon around COSATU House. "Armoured vehicles and trucks surrounded the building. Police placed the swastika of the extreme right-wing Afrikaner Weerstandsbeweging (AWB) on the bonnet of at least one of their vehicles ... Scores of heavily armed police forced their way into the building ... They surrounded it adopting combat positions. Others took up positions on the roof of the neighbouring building. Their weapons were aimed at the entrance to the hall and at those trying to watch the scene from the upper floors of COSATU House."[38] In the search for the killers of the four strike-

[37] *Ibid*, p177.
[38] *Ibid*, p178.

breakers, "every one of the many thousands of people present in the building was slowly led out at gunpoint. They had to pass, one by one, in front of a video camera and informers wearing balaclavas. The informers pointed out certain individuals, and 11 were arrested, including a 12-year-old child."[39] Subsequently, 18 people were arrested and tried for murder. Eight were found guilty and four of them were sentenced to death.

Just over a week later, on the night of 7 May, two bomb blasts virtually destroyed COSATU House and caused R2 million worth of damage. Jay Naidoo, the general secretary of COSATU, claimed that experts had done the bombing. "The saboteurs," he said, "had apparently cut their way into the building by scaling a five-metre wall and cutting iron bars…" without attracting the attention of the security forces who had it under surveillance at the time.[40] Naidoo's intuition was later confirmed by Adriaan Vlok, the Minister of Law and Order, and General J V van der Merwe, the chief of the South African Security Police, who admitted that they had ordered the destruction of COSATU's head office because it was the centre for organising violence during the national railway strike and was being used by underground ANC members to train MK soldiers in attacking the police. They believed that they would not be able to close it by legal means and that if it were destroyed it "would cause so much disruption that it would give us a breathing space".[41] They passed the responsibility for planning and executing the plan to other police departments. The bombs they used were described as "the most powerful ever detonated on the Reef".[42] All union activities in the building were terminated. The Johannesburg City Council declared it unsafe and no-one was allowed into it even to collect documents. The five national unions with offices in the building lost their equipment and many of their records and had to find alternative accommodation. As a short-term measure, the NUM rented rooms in the Victoria Hotel in Johannesburg but was compelled to vacate them almost immediately because an anonymous caller threatened to blow up the building. It moved, therefore, into its Witwatersrand regional office in Johannesburg.

The railway workers' strike ended on 5 June, exactly three months

[39]*Ibid.*

[40]South African Institute of Race Relations, 1987/1988, pp626-627.

[41]In their evidence to the Truth and Reconciliation Commission after they had applied for an amnesty for the bombing. (Truth and Reconciliation Commission, Volume Two, 1998, pp289-290.)

[42]*Ibid*, p289.

after it had started, with a settlement that satisfied many of the workers' demands, including a guarantee that none of the strikers would suffer from discrimination. All of them, including the railway worker whose dismissal had prompted the strike but excluding those who had been arrested, were to be reinstated. Those in detention were promised their jobs back after they were released. No workers were to forfeit any benefit rights because of the strike. In addition, the hostel conditions were to be improved and the black employees were to be given the right to elect their own representatives to negotiate with management. Moreover, like white employees, they were to be entitled to permanent jobs after two years' probationary service. Lastly, all of the casual workers, including white ones, who had been taken on during the strike were to be dismissed.

Within the context of the current apartheid situation, the strike, in terms of its length, solidarity and achievement, was historic. It exposed what was achievable despite both the government's determination to destroy any black representative organisations that were associated with the liberation movement and the hardening of the white electorate's opposition towards black emancipation. In the general election for the exclusively white House of Assembly on 6 May, held whilst the strike was in progress, the Conservative Party, which was in favour of entrenching the apartheid system, increased its share of the white vote and displaced the more liberal Progressive Federal Party as the official opposition party in Parliament. It was clear that although the government's reform rescue operation was limited and cosmetic, it went too far for many white voters.

The 1987 NUM Wages Claim

Such, therefore, was the evolving mood for confrontation whilst the NUM conducted its negotiations with the Chamber of Mines over wages and conditions of work during 1987. It had been decided at the Fifth National Congress in February that the union, as well as campaigning to end the migrant labour system, should present the Chamber of Mines with a number of demands that represented an indictment of the everyday conditions of work for black mineworkers. The minimum monthly wage for underground workers in the gold mines after the 1986 wages settlement was R228. This was less than half of that received by black workers in the manufacturing industry. The underground workers in the coal mines received about nine per cent less than the gold mine rate. The Fifth National Congress instructed the union negotiators to press for a wage increase of 55 per cent for all workers in job categories one to four on the surface and underground and for a 40 per cent increase for those in job categories five to eight and including all monthly-paid black staff. The congress also wanted to reduce working hours from the prevailing level of between 92 and 102 hours for every two-week period, down to 80 hours. It demanded an extension of the annual leave from an average of 18 to 44 days with a 100 per cent holiday leave allowance; no compulsory overtime; an end to the deferred payments system; the recognition of 16 June as a paid holiday and the payment of danger money to black mineworkers that was equivalent to the safety incentive bonus paid to white miners. Finally, it wanted to increase the benefit paid to the beneficiaries of deceased mineworkers from two to five times the mineworker's annual earnings.

The list was presented to the Chamber of Mines on 23 March but it had been leaked to the press well before then. There was the usual spell of silence before the Chamber of Mines responded. Negotiations eventually began on 15 May with an offer of a 12,5 per cent increase on minimum rates for gold and coal mineworkers. This amounted to a basic monthly increase of R28,50 compared with that of R125,40 demanded by the union.[43] The gap narrowed during the two days of negotiations. The NUM lowered its across-the-board wage demand, first to 40 per cent and then to 30 per cent while the Chamber of Mines raised its offer for gold mineworkers to 23 per cent for all surface and underground mineworkers in the lowest wage categories and 17 per cent for the highest-paid categories. The offer for colliery workers ranged between 15 per cent and 23,4 per cent.

The NUM rejected the terms offered by the Chamber of Mines on all the wages and fringe benefits issues and declared a dispute under the Labour Relations Act. The first step in that process was the appointment of a Conciliation Board. Two meetings of the board failed to bridge the gap, so Ramaphosa proposed that they submit their respective cases to mediation and arbitration in order to break the deadlock. In a clear sign that it was prepared to call what it perceived as the NUM's bluff, the Chamber of Mines rejected the proposal and imposed its own wage increases on 1 July, just as it had done each year for decades before, in the belief, presumably, that the balance of power in the industry had been unaffected by the rise of the NUM. It raised the basic rate for underground workers in the gold mines by 15,4 per cent

[43]The offer for gold mineworkers ranged between 12,5 per cent and 17,9 per cent.

which gave them a monthly wage that was just over 42 per cent of the average wage given to black workers in manufacturing after their 1987 round of wage negotiations.[44]

The union believed, in consequence, that it had no serious option but to proceed to the next stage and hold a ballot for strike action in each of the mines where it had a recognition agreement. On 13 July, the balloting began in 28 gold mines and 18 collieries that mainly belonged to two mining houses. Thirteen of the gold mines and nine of the collieries were owned by the Anglo-American Corporation while nine gold mines and five collieries belonged to Gencor. The only Gold Fields mine where a ballot was held was at its New Clydesdale Colliery in Witbank. The other mining houses were only marginally involved in the balloting process. Altogether, 210 000 mineworkers were balloted.

There were clear and early indications that the Chamber of Mines was prepared to avoid an agreed settlement in order to confront the union. There was, first, its refusal to go to arbitration and then its unilateral imposition of its own terms on 1 July. But the clearest sign was that there was no attempt amongst the mining houses to break ranks as they had done previously. The offer in money terms was the highest the Chamber of Mines had ever made and it insisted, despite contrary claims by the NUM, that the majority of the mineworkers were satisfied with it. It repeated its allegation not only that the NUM spoke only for a small minority of mineworkers but that it did not even represent the views of its claimed membership which, in any event, it believed was inflated. This policy of trying to sideline the union by repeatedly understating its influence annoyed Ramaphosa, who began to see an all-out national strike as the only way of proving that the Chamber of Mines was wrong. Thus the attitudes of both sides ensured that they were on a collision course.

The ballot results showed that 200 000 black mineworkers, representing more than 95 per cent of all those eligible to vote, were in favour of a strike. The union's national executive committee was summoned to a meeting on Sunday, 2 August to receive the results and, on the recommendation of the secretariat, decided to call a national strike in a week's time, starting with the night shift on 9 August. Ramaphosa announced the decision to a press conference the next day when he also explained the situation from the point of view of the union. He said:

> "Thank you for coming. I am sure all of you have expected that we have come to inform you all of the date our strike begins. But before we do that we just want to

[44]See South African Institute of Race Relations, 1987/88, p332 and p337.

sketch a few things and announce a number of developments that have been taking place on the various mines. We have been in dispute with the Chamber of Mines for some time now. After reaching deadlock at the conciliation board we proposed to the Chamber in the usual reasonable way that the dispute should be referred to mediation and arbitration. The Chamber of Mines rejected both proposals. We then proceeded with the strike ballot which received overwhelming support from our members Almost 200 000 members voted in favour of strike action and clearly showed their determination to proceed with strike action to pursue their demands.

"The Chamber of Mines, on the other hand, believes that it has offered the highest increase ever in the mining industry in money terms. It says and it argues that workers are quite satisfied with the wage offers. The strike ballot was a clear proof that our members are not satisfied. Our national executive committee met yesterday and decided that strike action should proceed this coming Sunday, that the night shift should not proceed underground and the day shift the following day until we have won our demands.

"The issues that are in dispute, as you might recall, are wages, where our union is demanding a 30 per cent increase, holiday leave of 30 days, June 16 as a paid holiday and danger pay for workers who are working in dangerous or hazardous areas, as well as an increase in compensation for death benefits to be increased to five years' amount of earnings per year."[45]

In answer to questions from the journalists, Ramaphosa said that he had not had any talks with the Chamber of Mines since the strike ballot and he did not expect to have any before the strike started. He then proceeded to describe how he anticipated the government, the mine managers and the police would react and what precautions the union was taking to protect its members against them. There were no attempts, he stated, to achieve an 11th-hour settlement.

[45]NUM, Strike Press Conference, *Transcript*, 3 August 1987.

CHAPTER ELEVEN

THE GREAT NATIONAL STRIKE[1] AUGUST 1987[2]

Pre-Strike Preparations by the NUM

During the six days between Ramaphosa's strike announcement and its planned commencement, the government's state of emergency was in full force so that in all matters and activities planned and occurring outside the private property of the mining houses, the NUM was subject to the strict controls imposed by its regulations. There it could do nothing without the permission of the magistrates and the police. It was not free to organise meetings outside the compounds, to solicit sympathetic strike action from other unions or to mobilise opinion in the townships without the risk of repressive retaliation by the government. Even its ability to project its case or highlight its treatment by the mine owners was limited by the censorship regulations that governed the media. This context of rigid control was sometimes defied but the union leaders could never ignore it when they were planning their activities. Nor, at any time, was it relaxed.

On the other hand, the political pressures on the NUM were seemingly eased by a number of contradictory features in the country at that time. There was the proclaimed neutrality of the government in relation to the dispute because the union was pursuing it strictly in accordance with the terms laid down by the Labour Relations Act. The union, in consequence, did not anticipate large-scale repression. Foreign countries were closely scrutinising the government's responses to guide them in enforcing international trade sanctions. The international media was on hand to film and report to critical foreign audiences every deviation from established western norms. Cracks were appearing in the broad opposition to the liberation movement as the ANC gained some legitimacy with business, the churches and the Afrikaner intelligentsia. There was even speculation that some of

[1]In his report to the Sixth National Congress in April 1989, Ramaphosa referred to it as the "Great Miners' Strike".

[2]The data for this chapter has been obtained from NUM documents, interviews with participants in the strike, newspaper reports and the following descriptions of the strike: South African Institute of Race Relations, 1987/1988, pp678-684; *South African Labour Bulletin*, November 1987 and Bettina von Lieres, *The 1987 NUM Strike*, a dissertation submitted in partial fulfilment of the requirements for the degree of Bachelor of Arts with Honours, University of Witwatersrand, Johannesburg, March 1988.

the Robben Island prisoners might be released. These factors were indicative of the attitude of the government but none of them influenced the mine owners. The consequence was that no matter what restrictions were imposed on the union's ability to plan its strike, the mine owners ensured that there would be a scenario of bitter conflict from the outset.

The NUM followed the procedures laid down by the Labour Relations Act before calling a strike so that its members would receive limited legal protection in the event of being dismissed for the act of striking. This process had given each side notice that there might be a strike some weeks before the actual decision to call one had been taken and, therefore, had provided each of them with the opportunity to prepare for it. The mine owners would have ample time to take their usual precautions of building up stockpiles of ore or coal, preparing their security forces for conflict, recruiting vigilantes and ensuring that there were adequate supplies of strike-breakers. The NUM, on the other hand, was able to mobilise its members who were scattered over large areas of the country and who had to be consulted through constitutionally prescribed union structures. It was also able to make tentative strike preparations but these were reactive steps for what the union needed to do largely depended on the policies pursued by the mine managers.

How each side used the time leading up to the strike depended on their respective perceptions of their opponents in conflict. The mine owners had a low opinion of the union and did not believe that it had the capacity to organise and sustain a lengthy national strike. Its previous strikes had been largely disorganised, short-lived affairs. Indeed, Johann Liebenberg, the chief negotiator for the Chamber of Mines, said that he expected the strike to last for only two to five days and warned that if it lasted beyond that period the mines might apply a policy of mass dismissals.[3] Ramaphosa, already with much bitter experience of the behaviour of the mine owners during strikes, believed that they were relying on this dispute lasting no longer than 24 hours and that they "were going to use any means to crush it in that time".[4] He anticipated repression and violence, as he indicated at his press conference on Monday, 3 August when he said:

> "During this week our union's regional and branch Strike Committees will be meeting to prepare and finalise strike details. This will include informing workers of the various strike rules and the procedures that have to be followed during the strike. Our members are fully prepared but there have been a number of developments on

[3] *New York Herald Tribune*, 10 August 1987.

[4] *The Guardian*, 3 August 1987.

managements' side where we have seen them building up their own mobilisation against our union and against workers in general. Managements have refused to grant our union meeting facilities on the various mines to enable us to communicate properly with our members. They have refused to let us distribute pamphlets to our members; in certain cases they have confiscated those pamphlets. They have threatened to dismiss workers. They have threatened to close certain mines. They have threatened to call the police during strike action and they have threatened to use their securities. But what is more disturbing to us and to our members is that mine managements have threatened to starve our members and, as you all know, our members are akin to captive labour in that they are in a compound and they rely on managements' good-heartedness to be fed and so forth. Mine management have now threatened that they will not be fed during strike action. On certain mines they have stated that the workers will have to pay R5 per day for food and lodging while they are on strike. This is clearly a breach of agreement because it affects conditions of employment which they have never negotiated with us. We are at present taking legal advice on the matter and will pursue it if necessary.

"The state of emergency has also made matters worse. Each time our Branch Committees and Regional Committees have approached the magistrates for permission to hold meetings, they have been refused. The total number of mines that will be affected are 28 gold mines around the country and 18 collieries. Our union is at the same time in dispute with a number of other companies where strike action could take place at any time should the various mine managements refuse to accede to our members' demands. Those companies are … all the mines under De Beers' administration … Samancor, De Wall Dubatse Chrome, Impala Refinery, NUFCOR, Winterveld Coal Mine, Rietvlei Opencast Colliery, Consolidated Metallurgical Industries, Consolidated Murchison as well as Ergo."[5]

Three days after the press conference Ramaphosa sent the following letter to the Chamber of Mines.

We believe that some or all of your members intend to deprive our members of food during the strike or charge them for the cost of food and accommodation. We believe this would be a breach of the applicable contracts and also an unfair labour practice. We also know that in previous cases, attempts have been made to break strikes by violent, unlawful or improper means. These methods include acts of physical violence by mine security personnel and the authorities, threats and intimidation by mine management. We are very concerned that there will be a repetition of this conduct in the forthcoming strike with disastrous consequences.

So we are calling upon your members each to give us the following undertakings:

1. that they will provide our members with food of the usual standard and accommodation for the duration of the strike and that no charge whatever will be made for this;
2. that there will be no restrictions on the normal freedom of movement of our members, and in particular that no roadblocks will be erected on the premises;
3. that union officials will be allowed their normal rights of access throughout the strike;

[5]NUM Strike Press Conference, *Transcript*, 3 August 1987.

> 4. that mine security personnel will not enter the hostels unless accompanied by a person nominated by the union head office;
> 5. that the South African Police or Security Forces will not be invited onto the mine premises without prior consultation with and the consent of the union head office. If the South African Police or Security Forces should enter the premises on their own initiative, then the mine management will notify the union head office immediately...
>
> Because of the urgency of this matter we require the respective undertakings to be given by not later than 8 pm 6 August 1987.[6]

Ramaphosa had already stated that if there were no agreement and the strikers were attacked by the mine security forces or by the police, they had been instructed to defend themselves as best as they could.[7] In order to put external pressure on the mine owners, and much to their consternation, he released his letter to the press. But when a reply had not been received from the Chamber of Mines by the deadline on the following day he decided that, in order to avoid the possibility of violence, the mineworkers should leave their hostels and return to their homes. The General Manager of the Chamber of Mines was informed of his decision in a letter marked "extremely urgent". Ramaphosa stated: "In view of the potential for bloodshed … and in order to minimise the risk of harm to our members and to diffuse potential conflict, we are advising our members for their safety, during the course of the strike, to vacate their accommodation on mine premises and return to their homes or to occupy alternative accommodation outside mine premises until a settlement is reached on the dispute." Ramaphosa emphasised, however, that by leaving the hostels the mineworkers were neither abandoning the strike nor deserting their employment. The strike committee at each mine, he added, would remain in the compound to meet with management when necessary and to act as a liaison between it and the National Executive Committee of the union.[8]

The Chamber of Mines replied to the first letter the day after the deadline. It criticised the union for persisting with its intention to strike and rebutted the implied accusation that the mines were the main instigators of violence. "Experience for some considerable time," it stated, "has shown that it is the union officials and members who have been the source of violence on the mines … If the mines concerned were to give you the undertakings which you request in your letter … they would be failing to discharge their responsibility to take all steps

[6]Correspondence, NUM to Chamber of Mines, 6 August 1987.
[7]NUM Press Conference, 3 August 1987.
[8]Correspondence, NUM to the Chamber of Mines, 7 August 1987.

necessary to protect their own employees from injury." It then dealt with what were described as the specifics of Ramaphosa's letter:

> 1. Mine managements and mine security personnel do not initiate violence and do not use violence to prevent or break strikes.
> 2. Mine security personnel are called in only when there is a threat of injury or intimidation to employees or damage to property.
> 3. The South African Police are not called in by mine managements to prevent or break strikes. They are called in only as a last resort and where violence has erupted and the mine security personnel assess it to be beyond their capabilities to control.
> 4. Large numbers of employees have intimated to mine managements that they are satisfied with the wage increases that they received from 1 July and that they do not wish to participate in the strike. They have appealed to managements to protect them against intimidation by union members who they believe are going to use violence to prevent them from going to work. Mine managements will not tolerate violence being used to intimidate those who do wish to work and will protect them.
> 5. Your union seeks a number of undertakings without giving any undertaking itself that its members will not resort to intimidation and violence. Whilst the Chamber remains willing to evolve a code of conduct with the union, there will clearly have to be both a commitment from the union to work for peace and order and some means of ensuring that the code will not be a dead letter through the union merely disclaiming responsibility for the actions of its individual members.

In response to the fear that food and accommodation might be denied the strikers, the Chamber of Mines declared that the mines were entitled to recover the cost of these provisions since they were part of a contractual arrangement which striking mineworkers were breaking. It added, however, that the mining houses concerned were not suggesting withdrawing food and accommodation from striking members of the union.

The NUM had no further contact with the Chamber of Mines in the few days before the strike was due to start, so no agreed guidelines were laid down for its conduct. It did, however, formulate its own rules for the guidance of strikers. In the special strike edition of its newspaper, *NUM News,* under the heading "Strike Rules", it stated:

> We have reached the stage where we must move towards united action. All regions and branch strike committees must fully understand the strike rules and ensure that strike stewards carry them out. Our strike starts on Sunday night, 9 August 1987. For the strike to be effective, we must have all workers on strike. We must be disciplined in our strike action.
>
> 1. The strike will continue until we have won our demands or until the union's National Executive Committee calls off the strike.
> 2. Only strike committee members will be responsible for matters affecting workers during the strike. Workers should listen to the strike committee.
> 3. During the strike, all workers should stay in their rooms and only come out when the shaft stewards or strike committee members tell them to.
> 4. Workers are reminded not to drink alcohol or to fight among each other or intimidate

each other. Any worker who fights or drinks must be taken to the strike committee.
5. Should the management refuse to give food and water to workers, the strike committee will meet.
6. We must be prepared to go on strike for many weeks, like the railway workers and OK workers, because the Chamber is refusing to give in to our demand for a living wage.[9]

The union's organisational centre was crudely and inadequately equipped for the task of co-ordinating a widely spread national strike. The head office staff had to work in temporary premises after its offices in COSATU House had been bombed three months before. They were crammed into an open-plan office in Darragh House in Wanderers Street in Johannesburg. But, nonetheless, they managed to take essential co-ordinating measures. Every branch was provided with an Authorisation Form that had to be signed by 30 members and deposited with either the legal department in the head office or a designated attorney. This gave the union authorisation to act on behalf of its members in any legal proceedings arising from the strike. At the same time it distributed an Attorneys List containing the names and telephone numbers of attorneys who were available to respond promptly to requests for legal assistance in each of the regions. The strike committees were supplied with a list of contact numbers in the head office for advice and emergencies. They were also given report forms to complete showing the statistical support for the strike and the manner of its organisation in their respective mines so that the secretariat could keep records of the numbers on strike and the extent of strike-breaking.

The responsibility for preparing for the strike at mine level lay with the regional and branch committees but most of them were reluctant to take any action before the strike deadline in case it gave managers a justification for taking repressive action against them. A shaft steward at the President Steyn gold mine described how they coped with this situation. He said:

We did not make any preparations before the strike because we did not want to give management the chance to disrupt our preparations. We only wanted to make sure that the strike was legal. We had a meeting with management two days before the strike to confirm that our strike was legal. We agreed that hospital and kitchen staff and one boilermaker per shift would be exempt from the strike. The organisation of the strike was done with the threat of management disruption in our mind. The shaft stewards were told not to shoulder all the responsibility for the strike because they were well known by management and might be detained. We have a big hostel at our

[8]*NUM News*. Strike Special, "Now is the Hour", 9 August 1987.

James Motlatsi after addressing a pre-strike rally in Secunda on 9 August 1987.

> shaft. It is divided into blocks and each room in a block is represented by a room steward. The strategy was to avoid that shaft stewards be labelled as intimidators The room stewards would be given a message and go to another room to deliver that same message. My role as a shaft steward was only to open a meeting with *Nkosi Sikelela* and say a few words. The whole job was not carried out in the union offices. It was masterminded in the rooms.[10]

As a result of the widespread fear of management repression most of the pre-strike activity took place during the weekend the strike was due to start. Regional Shaft Stewards' Councils were summoned on 8 and 9 August followed by branch meetings to discuss how to maximise support for the strike and what they should do about strike-breakers. A number of branches resolved to punish strike-breakers. Two of them, Randfontein Estates and Blinkpan Colliery, decided to impose the death penalty for strike-breaking after the strike was over. On the day the strike was due to start, the union organised rallies in strategic sections of the industry to mobilise support for it. James Motlatsi addressed the biggest one, in the Secunda region, where there were both gold and coal mines and where there had been outspoken criticisms of the national leadership. He expected opposition but the very large rally gave its unanimous backing to his call to maintain discipline, reject intimidation and be prepared for a lengthy strike. There were similar expressions of solidarity in other regions. After the rallies the union activists in the compounds returned to put their strike organisations into place. They knew that the initial response to the strike call could determine its outcome.

The primary responsibility for organising and conducting the strike in the compounds was delegated by the National Executive Committee to mine strike committees which were elected by mass meetings convened by the branches. The composition of the committees varied between the mines and collieries. They were not exact copies of branch committees. They invariably included branch officers who were usually given designated tasks but other union activists who held no union positions were also elected. At the Optimum colliery, for instance, 12 people were elected at a mass meeting and incorporated into the shaft stewards' committee. In other instances, the committees displaced branch committees that were suspended for the duration of the strike. Strike committees were not mentioned in the union's constitution and only had the authority given to them by the NEC. They had powers

[10]From one of many revealing interviews conducted by Bettina von Lieres shortly after the strike had ended. See Von Lieres, March 1988, p64.

to do whatever they thought best in pursuit of victory, including disciplining strikers and non-strikers, but they had no authority to end the strike. They were responsible for maintaining contact with management, mobilising support and keeping mineworkers informed about strike developments. They functioned as defence committees with "the duty to protect workers against management-inspired faction fights and mine security involvement…"[11] They were required to report regularly to mass meetings of strikers from whom they received instructions concerning the conduct of the strike but within the union hierarchy they were responsible only to the National Executive Committee which had created them. In general, strike action for a number of mineworkers was an exciting and stimulating venture and provided leadership opportunities that brought out hitherto hidden qualities. This made strike committees adventurous, sometimes reckless and certainly different from each other. There was, therefore, a considerable diversity in the ways the strike was prosecuted at mine level.

In the week leading up to the strike, supporters and opponents of it made their respective positions clear. In the discussion with journalists following his press statement on 3 August, Ramaphosa said that he did not expect the government to intervene because the mining industry was not covered by Industrial Council regulations and, therefore, the Minister would need additional legislation if he wished to do so. The government, however, was looking for ways to prevent unions in general from confronting employers and on 7 August the Minister of Manpower, Pietie du Plessis, said he was preparing legislation to bring order to the trade union movement and reduce, thereby, its burgeoning power. Both COSATU and the UDF pledged their support for the NUM. COSATU, fearing as Ramaphosa did, that the police would be called in to try and crush the strike, announced that if that occurred it would call sympathy strikes in key industries. "We see the NUM strike," it stated, "along with the wave of strikes in other sectors, as a high point in COSATU's living wage campaign."[12] The UDF conveyed the unqualified support of black public opinion for the mineworkers. But that was something which was beyond the comprehension of the Chamber of Mines for when it rebuked Ramaphosa for ignoring "a public call … to reconsider" the decision to go ahead with the strike, it obviously had only white opinion in mind.[13]

The union received pledges of support from the International

[11]*Ibid*, p61.

[12]*Financial Times*, 10 August 1987.

[13]Correspondence from the Chamber of Mines to the NUM, 7 August 1987.

Miners' Organisation, the Miners' International Federation, the British NUM, the German and American miners' unions and from miners' unions in Peru, Chile, Bolivia, Algeria, Zambia and Zimbabwe. But when Ramaphosa was asked about the kind of financial support the NUM might obtain from abroad, he said: "We are very touchy on the question of financial matters… It is something we cannot disclose quite happily."[14] It was known that the State President, P W Botha, wanted to restrict the flow of international aid to trade unions in general and Ramaphosa did not want the NUM to give him the excuse to introduce legislation for that purpose.

The Mine Owners' Strike Precautions

The mine owners, too, completed their strike-breaking plans. There was a belief amongst some mine managers that a short strike might be necessary to teach the union a lesson but the mine owners, in general, were fundamentally opposed to one because they saw it as a threat to their authority as controllers of the mining industry. They were particularly apprehensive about a long strike because of the damage it might cause to the underground conditions as well as its effect on production and, thereby, profits. From the outset, both Peter Gush, the chairman of Anglo-American's Gold and Uranium Division, and Johann Liebenberg, the senior general manager of the Chamber of Mines responsible for industrial relations, urged mineworkers to ignore the strike call and promised protection for those who wished to keep on working. Both were confident that many mines would be able to maintain production. Different types of protective measures were taken. In the coal mines, they were not especially concerned about a short strike because of the slump in the international trade for coal though they, nonetheless, took the precaution of stockpiling coal at Richard's Bay terminal. In the gold mines, most of the contingency plans were laid at mine level. Some managers concentrated their resources on the maintenance of immediate production by moving workers from development to production and by transferring white miners from supervision to production work. They stockpiled ore in surface dumps and drew up plans to confine production to high-grade ores so that the value of their gold output could be maintained. They were silent over the amount they had stockpiled but one estimate put it

[14]NUM Press Conference, 3 August 1987.

at a month's supply.[15] It was, in any event, more than the NUM had anticipated.

Security strategies to counter the strike were also left in the hands of the mine managers so that, as with the strike committees, there was much diversity in mine responses. There was, however, one common element. Each of the gold mines had its own private armed security force that included one or two paramilitary vehicles of which one was usually large enough to carry up to 28 armed personnel equipped with tear gas canisters, rubber bullets and live ammunition. Each security force also had its own dog squad comprising from 20 to 50 dogs. In one mine there was a horse patrol. They all had the option of the occasional use of helicopters. The coal mines had similar security provisions, except that one belonging to Anglo-American which did not possess a security force and borrowed from other mines when the need arose. In situations where mine managers could not rely on the loyalty of the black members of their forces or where they wished to supplement them, then they were able to draw on the support of the white voluntary Emergency Protection Units and even the security personnel of other mines. Overall the control of black mineworkers in the compounds was more tightly repressive than that in the townships under emergency regulations.[16]

The mine owners, moreover, were able to call on the South African Police for help whenever they wished. Although mine managers constantly denied that they deployed the SAP, the police often had unrestricted access to the mine compounds and were widely and indiscriminately deployed to attack strikers in their rooms and meeting places and were present in large numbers on the roads leading to the mines. The Reuters correspondent, Christopher Wilson, described the security situation at Carletonville just after the strike had started. He wrote, under the heading: "Gold Mines Resemble Military Compounds", that "South Africa's strike-bound Western Transvaal gold mines this week looked more like military installations than the major

[15] *Weekly Mail*, 21 August 1987.

[16] Bobby Godsell, the Anglo-American Corporation Group consultant in industrial relations, who was a frequent spokesperson for the corporation, told the British Broadcasting Corporation on the first day: "Our security forces have only one role to play in a strike situation and that is to simply maintain order and prevent violence … we emphatically will not use force to break the strike." (Interview with Peter Hobday. *Today* programme, Radio 4, BBC, 10 August 1987.) He reiterated this in a subsequent interview with the *Weekly Mail* when he stated that "we've now given a categorical undertaking from the Anglo-American group that we will in no way and in no circumstances use force to force people to work" (14 August 1987).

profit centres that form the backbone of the economy. Most of the big mines dotted along a rich gold reef that stretches in an arc from Johannesburg to the Transvaal's south-western border were sealed off by heavily armed mine security men backed by plainclothes police. Armoured vehicles crammed with steel-helmeted soldiers and police patrolled the main roads on the gold fields."[17]

An array of anti-union tactics was deployed that often bore little relation to the policies declared by the mining houses. There were some mines where restrictions were imposed on union meetings, roadblocks were erected to enable mineworkers' movements to be controlled and monitored and where union officials were barred from entering the compounds.[18] Union activists at JCI mines were searched as they entered their hostels so they hid pamphlets in their shoes as they went through the hostel gates. All those caught carrying strike-related materials had their identity cards confiscated. It was reported that in the mines around Klerksdorp, the mine managers distributed their own pamphlet bearing the logo of the NUM which stated: "The national executive committee regrets to inform all workers that it has decided not to go ahead with the proposed strike."[19] The manager at Randfontein Estates, with two shafts, tried to stop the strike by telling the mineworkers in each of the shafts, incorrectly, that those in the other one did not intend to join the strike. As the mineworkers in both shafts knew that was not true, none of them believed him. The manager at the President Brand gold mine in the Orange Free State used the same trick when he circulated a message on the first day of the strike stating that the mineworkers in most other mines had reported for work when it was already known that all the other mines in the Anglo-American group had been closed by the strike. The message stated:

A MESSAGE TO ALL STRIKING EMPLOYEES

1. The management of this mine regrets that many employees took the decision to participate in the strike, whilst employees on most other mines have reported for work

2. Management wishes to advise all employees who are not at work of the following consequences of their action:

2.1 All employees on strike have now breached their contracts and could be

[17]Reuters, *Report*, 13 August 1987.

[18]Bobby Godsell, on behalf of the Anglo-American Corporation, denied this. He said: "It's absolute nonsense to say the mines have been sealed off to union officials. To the best of my knowledge, NUM officials have exactly the same kind of access now as they've had at any other time." (*Weekly Mail*, 14 August 1987.)

[19]*Irish Times*, 10 August 1987. The Chamber of Mines denied any knowledge of the pamphlet.

dismissed at any stage.
2.2 Your shift for today has now been **cancelled** and you will continue to receive no pay whilst on strike
2.3 Your leave due date has now been adjusted in view of your action today and will be further adversely affected, should you not return to work immediately.
2.4 Your work record has been endorsed and this could affect your future job advancement on the mine.
2.5 You have been charged for board and lodging which will be deducted from your wages.
3. You are only harming yourselves by not working as your actions will not bring you more money – only more losses.
4. The shaft cages will be running normally and you are urged in your own interests to return to work immediately
5. Management wishes to advise all employees that security is there for their protection. Any incident of intimidation must be reported to the hostel manager immediately.

The managers of the Gold Fields gold mines took the most hostile and rigorous pre-emptive action to prevent the strike from taking place, though there was no possibility of a legal strike being called in their mines because none of them had recognition agreements. They were afraid that the union activists in each of their large mines would call strikes despite the law. Gold Fields was confronted with a problem of its own making for through its anti-unionism it had stimulated an interest in the union which was most probably greater than it would otherwise have been. At Deelkraal, Kloof, Doornfontein and East Driefontein, between 50 and 60 per cent of the black mineworkers were members of the union, supported the strike and were keen to join it. If they had done so they would have wreaked revenge on Gold Fields for denying them union recognition but a strike would also have had a dramatic impact on the total level of South African gold production because their mines exploited the richest reefs.

The Gold Fields managers made sure that did not happen. They refused to allow union officials to have access to the compounds for some weeks beforehand and monitored union discussions and activities through a network of informers. They also set up roadblocks to restrict movements whenever there was any union activity in the area.[20] The union members at Doornfontein alleged that the manager had ordered up to 2 000 union membership forms to be torn up prior to the strike to prevent the membership reaching the level where the union could claim recognition. But unrestrained and indiscriminate violence in the hostels was Gold Fields' principal means of crushing support for

[20]See *South African Labour Bulletin*, "The 1987 Mineworkers' Strike", November 1987 for a description of the activities of Gold Fields in relation to the strike.

the strike. Large groups of armed vigilantes entered the hostels at Deelkraal, Kloof and Doornfontein just before the strike was due to start on Sunday, 9 August and began a campaign of terror. They threatened the lives of those who wanted to strike and created such a climate of fear in the hostels that the strike never started at those mines. A union member at Doornfontein described the events in his hostel that evening. He said:

> The majority of our members were willing to join the strike. On the evening of 9 August Doornfontein workers became notoriously difficult to organise ... due to the behaviour of a group of workers, especially Pondos, who ... managed to break the strike ... This group of workers had pangas, knives and *knobkieries* in their possession. We were also secretly informed that they were given firearms to cause terrible faction fights when approached by us. Some workers were even offered ... money ... to help in the harassment of shaft stewards.[21]

A spokesperson for Gold Fields made no reference to the violence and simply reported that the night shifts at its mines had gone to work normally at 9 pm.[22]

The Strike Committees

The strike committees succeeded in getting support for the strike at all of the 26 gold mines and 18 collieries where the NUM had recognition agreements though the Chamber of Mines disputed this and declared that only 16 of its 44 gold mines and 13 of its 55 coal mines were affected. The strike was also supported at a number of mines and collieries that did not have recognition agreements and where, therefore, strike action was illegal. In all of such cases, as at the Loraine gold mine owned by Anglovaal, the union was meticulous in advising the strikers to return to work.[23] The union could have jeopardised the continuation of the legal strikes if it had become involved in illegal ones. It was estimated that 150 000 mineworkers in the gold and coal mines affiliated to the Chamber of Mines refused to enter the shafts at the start of the Sunday evening shift and that by midday on Monday, 10 August, the number had risen to 280 000.[24] The vast majority of

[21]*Ibid.*

[22]*Business Day*, 12 August 1987.

[23]Three unrecognised collieries belonging to JCI, namely Tavistock, Phoenix and South Witbank, came out on strike on 13 August and returned to work the next day.

[24]This is a NUM estimate. The Chamber of Mines estimated that between 220 000 and 240 000 were on strike by the second day and it did not revise these figures throughout the strike.

those working in Anglo-American gold and coal mines went on strike immediately. They were accompanied by almost half of the 70 000 mineworkers in Gencor's gold and coal mines. Despite the management-sponsored violence there were a few short-lived stoppages at some shafts in mines belonging to Gold Fields. Douglas Colliery was the only mine in the Rand Mines group to join the strike. Ramaphosa claimed that, by the second day, 340 000 black mineworkers, including 80 000 who were not members of the union, had joined the strike.[25] Altogether, these represented 72 per cent of the total black labour force in the gold and coal mines. Though a quarter of the strikers were not members of NUM, they were all subject to the discipline of NUM's strike committees.

The immediate and high level of response to the strike call was, in large part, a reflection of the work of branch organisations in generating solidarity amongst mineworkers during the struggles over the previous two years. It was as if they had been organising 'dress rehearsals' for a national strike. The strike committees, however, had to sustain that solidarity in the face of determined management opposition. They discovered that solidarity was not simply a product of an emotional bond with the union and that mineworkers had to be constantly reminded of their loyalties and priorities. In some instances, the waverers could only be held in line by discipline and when that was infringed there were penalties. In effect, the strike committees treated strike-breaking as an act of treachery.

The hostels were the places where the issue of solidarity was resolved in one way or the other. They were, therefore, the focus of attention from both sides. The vast majority of mineworkers remained in their hostels for the duration of the strike for only a few thousand from Randfontein Estates in the west Rand had heeded Ramaphosa's pre-strike call to board buses and taxis and return to their rural homes. For 24 hours a day, therefore, the mineworkers were exposed to the competitive moral and physical pressures from the local strike leaders, on the one hand, and the managers and security forces on the other hand. The strike leaders needed to be able to move freely in and between the hostels because the solidarity of the strikers depended on the extent to which the union could sustain their morale during trying circumstances, prevent strike-breaking by forces from inside and outside the mines and provide physical protection from attacks by the mine security forces and the SAP. These conditions were satisfied

[25] *The Scotsman*, 11 August 1987.

effectively only where the strike committees had control of the hostels. This was revealed during the course of the strike for, in every hostel where the strike committee had control, support for it was virtually complete, whereas where the management remained in charge there was little, if any, strike activity. It was not surprising, therefore, that the hostels were the scenes of many bitter battles between the strikers and mine security forces.

The Anglo-American Corporation had asked Ramaphosa on the first day of the strike to stop shaft stewards from attempting to take charge of the hostels by putting up roadblocks and controlling the hostel gates in order to regulate access to them. It alleged that mineworkers were prevented from exercising their choice as to whether to work or not and that the delivery of food and other supplies was being obstructed. Indeed, in six gold mines and four collieries the strike committees took immediate and full control of their hostels and the hostel kitchens. In other mines, the strike committees exercised varying degrees of control. In every case, however, the strike committees were in a precarious situation because the mine security forces and the SAP were constantly searching for weak links in their defences, waiting for lapses in their vigilance and using any excuse to put on displays of armed strength in order to disarm and dispirit the strikers.

A major problem, common to all strike committees, was preventing their organisations from being penetrated by management informers. Informing for management, however, involved risks, which relatively few ordinary mineworkers were prepared to take, so the most likely sought-out culprits were the *indunas* and the team leaders.[26] However, it could never be assumed that informers were not present, even in the hostels controlled by the strikers. Strike decisions, therefore, were rarely, if ever, taken at general meetings of strikers. In order to overcome this difficulty, the union proposed that the committees should adopt, in one form or another, the communications system used by its Education Department called the E-Plan. Its intention was to organise educational groups in the hostels in defiance of management. It was a scheme similar to that drawn up by Nelson Mandela in 1953 to prepare the ANC for an underground existence in the eventuality that it was declared illegal.[27]

[26]In an unusual situation, five of the *indunas* at the Elandsrand gold mine were shaft stewards so it was difficult for the management to use tribal authority to usurp that of the NUM.

[27]See Thomas Karis and Gwendolen M Carter, (eds), *From Protest to Challenge: A Documentary History of African Politics in South Africa, 1882-1964*, volume 3, Thomas Karis and Gail M Gerhart, *Challenge and Violence, 1953-1964*, USA, 1977, pp35-40 for a description of Mandela's plan.

It matched the needs of the NUM completely. The E-Plan required that the activities of the sub-structure of the strike committees had to be based on the maintenance of secrecy. Strike units, comprising one or more people, were organised in each of the hostel rooms on a cell basis with the responsibility for carrying out the decisions of the strike committees. The identities of their members was kept secret as far as possible so that few people knew exactly who was responsible for spreading information about the strike from other mines and regions and for mobilising support for strike committee decisions.[28]

The Witbank collieries were the most tightly organised sector of the strike. They had the advantage over the gold mines in having relatively small hostels that accommodated 1 000 mineworkers or less compared with many thousands in the gold mine hostels. The union branch committees in the collieries were retained but were superseded by volunteer strike committees for the duration of the strike. Each of the volunteers was given a specific task such as guarding roadblocks, supervising the hostels for cleanliness, identifying drunkenness and preventing all forms of anti-strike activity. There were variations in practice amongst the committees. At Matla and Optimum, where there was already a tradition of militancy based on frequent confrontations with the security forces, they were highly organised. They operated on a 24-hour basis and assumed the duties of the mine security forces, though within the perimeter fences of the hostels. Others engaged in various degrees of collaboration with management. For instance, the strike committee at Douglas colliery, though it had serious disagreements with the manager over the issue of food coupons, agreed to share the task of guarding the gates with the mine security. The Wolwekrans and Vandyksdrift collieries recruited scabs yet their strike committees co-operated with the mine security forces and even agreed to allow them to supervise the roadblocks and the gates unheeded.

The strike committees in the gold mines followed the general pattern stipulated by the union secretariat and were a mixture of volunteers and branch committee members. Some of them, such as those at Western Deep Levels, Vaal Reefs and Western Holdings, started from a high level of militancy and immediately began to push back the boundaries of management jurisdiction, whereas others, though not collaborative, had to be satisfied with lesser

[28]See *Inqaba Ya Basebenzi*, issue No 24/25, October 1987, where a striker from Secunda related how the E-Plan worked in his mine. This issue was devoted to the strike. The journal was published by a faction that called itself the Marxist Workers' Tendency of the African National Congress.

achievements.[29] On the whole, however, the atmosphere in the gold mines was more tense, unpredictable and potentially explosive than in any other mines. Their managers appeared to be more aggressively determined to break the strike than were the colliery managers and this gave rise to a greater determination by the strikers in the gold mines to preserve their unity.

The situation at Western Deep Levels was not typical because it was one of the largest mines and was relatively highly organised but it illustrated what the strike committees in general were aiming to achieve. At the outset, a union mass meeting in the mine suspended the branch committee for the duration of the strike and elected a strike committee on which the branch officials, along with volunteers, were given specific jobs. The strike committee members then introduced themselves to the manager and demanded that the control of the hostels should be handed over to them. They talked for six hours but departed with different and confused views about what had been agreed. The mine manager believed they had agreed to allow the union to control the strike but that the management would retain control of the hostels. He insisted that while strikers could stand at the gates they would not have the right to search persons coming in. The strike committee, on the other hand, believed that it had been given control of the hostels and the conduct of the strike. In a separate set of negotiations a union organiser had obtained a written agreement giving him the right to mobilise in the hostels.

The position was clarified during the first week. The strike committee mobilised support for the strike on a room-by-room basis whilst at the same time taking over the control of the dining rooms, kitchens, change rooms, liquor outlets and the walkways to the shafts. It controlled the hostel gates from the inside while the mine security stood outside. The management personnel did not have free access to the hostels. Indeed, the hostel manager had to get permission from the strike committee to visit the kitchens. On one occasion, when a mine security official entered the hostel gates to find a toilet, he was arrested and detained for two days. Some of the strikers wanted to kill him. In

[29]This was the case in all but one of the shafts at Western Holdings. No 6 shaft had been the most militant in 1986 and had led the protests in June 1987 against the management reprisals after the killing of two white officials. But the scale of arrests, dismissals and repression in general against mineworkers in the shaft was so great that they became disillusioned with industrial action and turned against the NUM itself. Indeed, such was the feeling of resentment towards the NUM that, although all the other shafts supported the national strike, only approximately 50 to 100 joined it from No 6 shaft. The non-strikers ostracised this small group and forced them out of the hostel.

order to uncover anti-strike activities at Western Deep Levels the strike committee established its own counter-espionage measures. No meetings of more than five people were permitted in the hostels and an intelligence network of union agents was established in the rooms to inform about possible anti-strike moves and to identify management informers. The committee reacted quickly when it received any such information. When, for instance, it learned that an *induna* was holding a meeting in the hostel manager's office in order to instigate a faction fight to divide the workers, it broke the meeting up and assaulted the *induna*. It became difficult for the mine manager to communicate with the strikers so he had to rely on the public address system to get messages into the rooms and, on occasions, on the use of helicopters to shower the compound with propaganda material.

The committees in general tried to follow the strike rules laid down by the national secretariat. Apart from establishing intelligence networks they were concerned about the social behaviour of the strikers. They insisted that the rooms were kept clean and tidy though as the strikers spent most of their time in them playing cards, reading or drinking, they had a difficult task. It was especially difficult to control the consumption of alcohol for it was often the only means the strikers had to escape from the tensions caused by being confined in hostels surrounded by armed security forces. The manager of Gencor's Stilfontein gold mine in Klerksdorp raised the issue of drink with Ramaphosa on 19 August when he complained that some of the strikers were drinking so much liquor that they were becoming violent in the hostel. At Western Deep Levels no liquor was allowed to be drunk anywhere in the hostel. If a mineworker left the compound and returned drunk he was not allowed back in again.

The strike rules were generally strictly enforced. Anyone caught infringing them was taken before the strike committee, questioned and then punished according to the nature of the offence. Those guilty of lesser offences were usually given hostel-cleaning duties for the duration of the strike.[30] The paramount rule of the strike committees, however, was that no-one was allowed to do anything that might break or weaken the solidarity of the strike. Anyone who infringed that rule was tried and punished privately by the appropriate cell without any discussion amongst the strikers in the mine or shaft about the nature of the charges or the severity of the punishments. It was estimated that

[30]Interview with Sazi Jonas, who had been the secretary of the branch committee but during the strike was the co-ordinator between the strike committee, the regional committee and the head office. (4 January 1994.)

through this process up to 10 mineworkers were killed for anti-strike activities and that a large but indefinite number were assaulted and injured.[31]

The NUM members at Gold Fields' mines and collieries faced exceptional difficulties when trying to organise strike support for they had to overcome the unrelenting and uncompromising hostility of the managers. Shaft stewards' committees were not allowed to meet regularly nor, except in secret, could they meet with mineworkers. Mine security officials regularly searched the hostel rooms on information supplied by management-appointed room representatives. Armoured vehicles patrolled the compounds and SAP helicopters circled above them. Union activists were constantly harassed to the point where ordinary mineworkers were afraid to talk with them or become identified with the union in any way.

The hard-line attitude towards the NUM in the Gold Fields mines drove union activists to seek unusual solutions. Two shaft stewards from the Libanon gold mine went to a witchdoctor on 20 August to get *muti* from him that they believed would protect them against bullets. With this, they convinced a meeting of about 250 union members that they could safely confront the mine security and take part in the strike. The witchdoctor advised them that once they possessed the *muti* they should not retreat when confronted by the guns of the mine security forces and the SAP.[32]

The mineworkers returned to the compound from their meeting and as they entered the main gate they were stopped by the mine security that had been reinforced by members of the security forces from other Gold Fields mines. According to Nqeleni Marawa, one of the shaft stewards who had initially approached the witchdoctor, a hippo drove behind them and a helicopter circled above. He said that when the first group from the meeting arrived at the gate the SAP gave them five minutes to disperse but they did not respond because they were waiting for the rest of the mineworkers to arrive so that they could all gather together and tell the night shift not to go underground. The police started to fire well before the five minutes had expired. Nonetheless, following the advice of the witchdoctor the group advanced on the police and, indeed, caused them to retreat. However, one mineworker was killed and 27 were seriously injured with shotgun and handgun wounds. Marawa, who believed that the *muti* would protect him, was shot in his

[31]Conversation with Gwede Mantashe, 4 January 1994. Mantashe, who worked at Matla colliery, was the secretary of the Witbank Regional Committee during the strike.
[32]Von Lieres, 1988, pp62-63.

right and left shoulders and in his right arm. The morning after the shootings the police conducted body searches of all the mineworkers involved in the incident and arrested all but one of those who suffered from bullet wounds because, in line with past police practice, they regarded the wounds as evidence of complicity in the violence. They were charged with public violence offences but those who were not wounded were deported to their homes in the neighbouring states. Marawa managed to avoid arrest by slipping unseen out of the hospital.

Gold Fields telexed to the South African Press Association a different version of the incident, which stated:

> One employee of Libanon Gold Mining Company Limited was killed and 20 others seriously injured in an attack on the mine's hostel gate by an armed mob of approximately 250 people at 7.20 pm last night (20 August 1987).
>
> This unfortunate and disturbing incident occurred at the hostel gate after a large group of about 100 armed people had initially forcibly gained access to the hostel grounds. A further group then stormed the gate, at which stage mine security was forced to fire rubber bullets at the mob which was armed with sharpened sticks, metal bars fitted with bolts and also with *knobkieries*. After approximately 40 minutes, security staff eventually managed to bring the fighting under control.
>
> At the commencement of the attack, the SA Police was notified and arrived on the scene at approximately 7.40 pm. When they arrived order had partially been restored by mine security and thus the SA Police did not actively participate in controlling the mob. They remained on the scene until the night shift went underground at 9.pm.
>
> At the request of the employees, who feared for their personal safety, they were escorted to the shafts where they went underground as normal.
>
> From reports it is clear that the employees involved in the attack were under the influence of narcotics and had been incited by a witchdoctor (*sangoma*), who scarred them by making skin incisions on their chests and foreheads, thereafter dabbing these injuries with blood. It appears that this attack is similar to the one near Randfontein approximately a year ago, when two policemen were killed by a mob which then attacked a mine's hostel.
>
> The 20 injured were immediately evacuated to the mine hospital. Some of the injured were later transferred to the Rand Mutual Hospital in Johannesburg.
>
> At present the situation is calm and all the shifts at the mine are working as normal.

The postscript to the event was acted out in court. The Libanon Gold Mining Company made an urgent application to the Rand Supreme Court to dismiss 41 mineworkers who had been at the meeting, including those in hospital, to evict them from the hostel complex and to prohibit them from being anywhere on the property of the mine. The NUM was not mentioned in the application though the mine manager stated that he was conducting investigations to see whether it was implicated. The application was granted on 26 August with the rider that if the mineworkers opposed the application they would have to pay the costs of the action.[33]

That marked the end of the attempts to set up strike committees in Gold Fields' mines and to bring its 75 000 black employees into the strike. It was a major setback for the NUM.

Making and Breaking Solidarity in the Compounds

Once the members of the NUM had begun to refuse to go underground, sometimes in opposition to black mineworkers who declined to support the strike and always contrary to the wishes of management, there was an expectation of confrontation and violence. There were low levels of tolerance on all sides. Mine managers acted quickly. On the day before the strike, the manager at Gencor's Beatrix gold mine in the Orange Free State sent *indunas* into the hostels to advise those intending to work to arm themselves, as they were likely to be attacked by strikers. During the first day, mineworkers at the Harmony gold mine in the Orange Free State claimed that they were forced at gunpoint to go underground. However, when they got there they protested by staging a sit-in strike. The manager denied that force was used but agreed that there had been a brief sit-in strike. At the Bracken gold mine in Secunda, the mine security went through the hostel rooms with guns to force mineworkers to go to work. They injured five with rubber bullets in the process. There was a similar incident in Anglovaal's Loraine gold mine in the Orange Free State, where the union was not recognised. There the police fired at the strikers with birdshot and tear gas as well as rubber bullets. The NUM alleged that strike-breakers in the mines at Welkom in the Orange Free State had attacked strikers with machetes and had caused an unspecified number of casualties.[34] Also on the first day of the strike, the NUM accused the manager at Vaal Reefs of colluding with the SAP to force mineworkers to go underground.[35] As a prelude to providing more intensive support for the mine owners, the police tried to destroy the NUM's strike organisation at Klerksdorp and Westonaria by surrounding, then raiding their offices and preventing union officials from entering them.

[33]*The Citizen*, 27 August 1987.
[34]*New York Herald Tribune*, 12 August 1987.
[35]*The Guardian*, 13 August 1987.

The strike got off to a violent start at Vaal Reefs when the security force injured six mineworkers with rubber bullets in three clashes in attempts to stop them assaulting strike-breakers. They made three arrests. The hostility towards strike-breakers continued into the second day when, during the evening, a small group of strikers armed with swords, knives and *knobkieries* abducted three of them from the dining hall in No 1 hostel and took them to the union office where they were told to sit on chairs on one side of the room in front of about 40 strikers who were sitting on the floor, armed with weapons. The three men, who were interrogated separately and repeatedly that evening and throughout the night, were threatened, one of them alleged, with being necklaced. While they were held, they were refused food and water. Eventually, they were told that the shaft stewards at No 8 shaft would decide their case the following afternoon. But before that happened, members of the mine security force, acting on information provided by an informer, forced their way into the union office and released the captives. Long after the strike had ended, an official of the Chamber of Mines claimed that, on that evening, two of the strike-breakers had been "necklaced" for the 'crime' of refusing to join the strike.[36]

There were other early instances of the intolerance shown by the strikers to those who went to work. At the Kinross gold mine in Secunda, seven arrests were made when strikers attacked 750 strike-breakers as they returned from work and refused to allow them to enter the hostel. The first alleged fatality of the strike was at Gencor's Blinkpan colliery where one of six strike-breakers was found strangled in bed following the first night shift after they had been involved in a fight with the strikers. As a result, five mineworkers were arrested but released the same day. At Trans-Natal Coal colliery two cooks were arrested for allegedly poisoning the food provided not only for the strike-breakers but also for the colliery manager. It was estimated that 30 black mineworkers and at least three mine security officers were injured during the first two days.[37]

The familiar pattern of provocation and assault by the mine authorities was repeated at Western Deep Levels where there was a mine security force comprising 50 men, housed in military-style barracks, with a helicopter and two Casspir armoured personnel vehicles at their disposal. Its armoury included automatic shotguns, rifles and pistols as well as riot control equipment such as tear gas guns and rubber bullets.

[36]South African Institute of Race Relations, 1987/88, pp23-24.

[37]*Yorkshire Post*, 12 August 1987.

An ammunition store was filled high with wooden boxes of bullets, shotgun shells, tear gas canisters and masks.[38] The SAP stood at the main gate on the first day preventing taxis from entering and intimidating and assaulting mineworkers as they left. On the following day they assaulted one person so badly that the union attorney complained to the mine manager about it. In another incident, the mine security at the President Steyn gold mine called a meeting over the public address system on the day the strike started and then fired at the mineworkers as they walked to it.

Police action was intensified on the third day, when, according to the union, 177 arrests were made in a co-ordinated attempt by the police in all of the regions to remove the strike leadership from active participation in the strike. Shaft stewards and strike committee members were detained or arrested in Carletonville, Westonaria, Witbank, the Orange Free State and Klerksdorp.[39] "We are seeing," Ramaphosa stated, "a whole pattern of repression that is beginning to evolve to try and destroy our strike. It was very well calculated by the Chamber of Mines as well as the government to crush the strike by arresting the leaders and leaving the strikers leaderless."[40] The police, moreover, had begun to attack and arrest strikers for the most trivial of offences. In many cases the offence was simply being present in the hostel or in a meeting or in just being a mineworker. The catalogue of arrests on that day, Wednesday, 12 August, was a long one. At the Randfontein Estates gold mine 15 strikers were arrested as they waited at a taxi rank because, it was alleged, they were obstructing the road. Six strikers escaped to the Orange Free State regional office with gunshot wounds after being shot by mine security in an ambush whilst going to a public meeting. The SAP then occupied the office and detained the injured men. Sixteen shaft stewards from the Deelkraal and Doornfontein gold mines in Carletonville were arrested when they returned from attending a NUM meeting in Johannesburg the previous night even though there were no strikes at their mines. At another of Gold Fields' mines, Venterspost in Westonaria, the SAP arrested 24 mineworkers as they stood outside their hostel within the precincts of the mine. Another 22 arrests were made at the Libanon gold mine but, on that occasion, by the mine security. They were released but were told they would be charged later.

The police raided the Klerksdorp regional office for the second time

[38]Reuters, 13 August 1987.
[39]*South African Labour Bulletin*, November 1987.
[40]*Wall Street Journal*, 13 August 1987.

in three days when, on 12 August, between 300 and 400 mineworkers met there to discuss the progress of the strike. This was the most ambitious police operation of the day. Ten armed policemen occupied the room and arrested 78 people, including the whole of the regional committee and two union officials. Nineteen of those arrested were shaft stewards at Vaal Reefs.[41] Some were charged under the Internal Security Act while others were charged with subversion and a conspiracy to murder. The conspiracy to murder charge arose, the police alleged, because, during the meeting, in response to a claim that the strike was failing, someone had said, "radical steps had to be taken to prevent miners returning to work". Ramaphosa believed that the conspiracy to murder charge was "just one of their smear campaigns" and would be rebutted in court if ever the case reached there. In any case, believing that the union offices were bugged, he added that "they would have been pretty stupid to meet in a union office to discuss the murder of anybody".[42] A union attorney travelled to Klerksdorp and, after some difficulty, secured the strikers' release on bail for R68 000.[43] They were not able to resume their strike duties, however, so an interim strike committee was elected until such time as they were freed, if at all, by the court. The harassment at the Klerksdorp office continued the next day when it had to be evacuated due to a bomb scare after two suspicious-looking parcels, one connected to a battery device, were found. The police bomb squad took away the parcels but did not report back to the union about their contents. Indeed, when journalists inquired about the incident the public relations officer of the police told them that nothing had happened, that the police had not been called to the Klerksdorp office and, in any event, they had not taken any suspected parcels away.

Despite this violence and intimidation, the strike numbers increased. When the NUM held its press conference on 13 August it claimed that 10 extra mines and plants employing about 10 000 workers had joined the strike during the previous two days. There were a few sympathetic strikes in mines not connected to the wage dispute, such as at the Rand Mines Rietspruit opencast coal mine in Witbank, but most of the additional strikers came from mines where the balloting had not taken place until after the strike had started. They included the Rand Refinery owned by the Chamber of Mines, through which all the South

[41] *The Guardian*, 13 August 1987.

[42] *The Independent*, 13 August 1987.

[43] The NUM had to raise substantial sums of money for bail in a number of other cases also.

African gold was refined, Consolidated Murchison, a highly organised antimony mine in the north near Phalaborwa, and the East Rand Gold and Uranium Company (ERGO), a gold and uranium reclamation plant in Springs, owned by the Anglo-American Corporation.[44]

The most significant of those strikes in terms of its influence on the course of the strike in general was at ERGO. There was no history of militancy there but the NUM's wage dispute with it was particularly bitter. The NUM went through the normal pre-strike procedures with the government's conciliation service and ERGO but the conciliation board stage was not reached until after the national strike had started. When the conciliation board failed to resolve the issue on 10 August the NUM asked the company for facilities to hold a ballot of union members who then comprised less than half of the work force. This was agreed and the ballot was completed on 12 August. The result was overwhelmingly in favour of a strike and 1 022 workers, more than the total number of union members, began it immediately.

Before the ballot had been completed, and in anticipation of a successful result for the union, the management had written to Ramaphosa stating that all strikers would have to leave the premises of the company for the duration of the strike. The letter added that "the company will not in any circumstances expose itself to the risks for the safety of persons and property which are inherent in the presence on company premises of employees on strike ... the company will not hesitate to protect its interests, and persons and property, in the event of any failure or refusal to observe our request..."[45] A copy of the letter was sent to the shop stewards' committee. The company also distributed a circular to all of its black employees making the same point but adding that those who went on strike would not receive backdated wage increases that the company had unilaterally imposed as from 1 June that year. A letter of 'Acceptance and Undertaking' was enclosed for each worker to sign, agreeing to accept the wage offer and to refrain from taking or supporting any form of industrial action. The union advised its members not to sign the document but to leave the plant as requested.

The strikers, however, had their own plans and engaged in a sit-in. The manager then demanded that they either evacuate the plant

[44]Rand Refinery employed 195 white workers and 195 black workers of whom 150 were members of NUM. One hundred and twenty-eight of the union members voted to go on strike. The relatively high number of white workers in the production process enabled the refinery to continue operating despite the strike.
[45]NUM Correspondence, 11 August 1987.

voluntarily or face forced evictions. They ignored this demand so the company went to the Supreme Court and obtained an order to evict the strikers. The company's argument before the court was that the strikers had committed 14 acts of sabotage in the plant. It stated that valves, which should have been closed, were found open; a volt breaker was switched off and one tank had overflowed because of unauthorised adjustments to the equipment. It claimed that seven tons of sulphuric acid had been released. The union denied the claims and insisted that inexperienced workers, including management staff, brought in to keep the plant working, did the damage. The plant had to operate continuously in order to protect the equipment from serious damage and to make this possible the NUM had agreed that ERGO should be allowed to obtain replacement labour during strike. This ERGO had done. Journalists who asked to be allowed into ERGO to see for themselves what damage had been done by the spillage of sulphuric acid were refused access because, the chairman of Anglo's Gold and Uranium Division, Peter Gush, stated, the ERGO officials were "tied up" and there were bigger problems "elsewhere".[46]

The strikers still continued to refuse to leave the plant and met together to seek advice from their shop stewards about the eviction order. The manager, however, refused to allow the shop stewards to address the meeting and told the mine security to order the strikers to disperse. There were disputed versions about what happened subsequently. The manager claimed that the strikers were still obstructive so he ordered the gates to be locked and called in the SAP. Just at that moment, at about 4 pm on Thursday, 13 August, George Nkadimeng, who was the union representative in the plant, was negotiating with the management for a peaceful evacuation of the strikers. He and the strike committee had been negotiating with management all day, he said, but no-one had mentioned the question of sabotage. Nor had he been told about the court eviction order until 3.30 pm that day. He asked the manager for time to telephone the union head office but before he was able to reach the phone, "the place was swarming with police and dogs".[47] The dogs were set loose while the police fired tear gas and rubber bullets and used their *sjamboks* to beat up the strikers. Those who wanted to leave peacefully were unable to escape the violence because of the locked gates. The strikers were not armed or threatening. Many were injured, seven of them seriously. Twenty-three were arrested

[46]*The Star*, 14 August 1987.
[47]*Ibid*, 16 August 1987.

on charges of trespassing and contempt of court. The following day, Ramaphosa called a press conference to discuss the question of violence during the strike and to meet injured mineworkers. About 200 had been injured during that week. Scores of them arrived in Johannesburg "from different mines by the vanload".[48] They had missing eyes, closed eyes, missing teeth, fractured arms, swollen faces, *sjambok* weals across their faces, bruised and swollen bodies, injured legs and savaged ears. The 20 who came from ERGO had the most serious injuries.

The Anglo-American Corporation won the struggle at ERGO but it, and the other mine owners, knew that in terms of their overall strategy, victories at places such as ERGO were relatively insignificant and that in order to make a serious impact on the solidarity of the strikers they had to attack and break the citadels of union power in the large gold mines and the militant collieries. This meant, in particular, breaking the spirit of the strikers at Vaal Reefs and Western Deep Levels, two of the largest and most aggressive gold mines, on the one hand, and at the defiant Matla and Optimum collieries on the other.[49] Anglo-American, with Vaal Reefs and Western Deep Levels in mind, complained that its compounds were being turned into military bases patrolled by armed strikers and that the NUM was encouraging the "adoption of a military base-type attitude by hostel members".[50] But it was far from being a one-sided affair. In many incidents the reality of compound life was that the police and mine security, with their vastly superior arms, were adopting a pro-active role in relation to the clashes with strikers.

From Tuesday to Saturday of the first week, the Anglo-American Corporation continued its attack on the strikers at Vaal Reefs and Western Deep Levels. Firstly, on Tuesday, it successfully appealed to the Bloemfontein Supreme Court for an interdict restraining 12 shaft stewards and four ordinary strikers at Vaal Reefs from arming strikers, abducting strike-breakers, setting up 'kangaroo courts' and threatening to necklace those they pronounced guilty.[51] The corporation, moreover, endeavoured to penalise the Vaal Reefs strikers in all but one of the Vaal Reefs shafts, by depriving them of medical treatment at the medical centre, cutting off the power for cooking in the

[48]*Ibid*, 15 August 1987.

[49]Vaal Reefs and Western Deep Levels had a combined labour force in the region of 65 000. Each comprised a number of shafts that had their own distinctive approaches to the union and the strike.

[50]*Sunday Times*, 16 August 1987.

[51]*The Star*, 12 August 1987.

kitchen and by refusing to allow the TEBA bank to issue money to them. The situation was relatively quiet until Saturday, 15 August, when a large mine security force, accompanied, the union alleged, by members of the SAP disguised as mine officials, moved into No 9 hostel with firearms and tear gas and ordered its inmates to return to work immediately. Their methods of persuasion left 24 strikers injured, three of whom had to be taken to hospital. When asked by reporters why the security officials needed to force their way into the hostel, the manager gave two inconsistent answers. One was that they went to rescue a clerk who was badly beaten by strikers. The other was that they went to investigate a claim that hostages were being held there. It was reported that they found neither a badly beaten clerk nor any hostages.[52]

There was renewed repression at Western Deep Levels on Thursday evening, 13 August, when the SAP and mine security arrived at the No 1 shaft hostel in seven armoured vehicles to confront about 700 chanting strikers who, the management claimed, were armed with pangas, petrol bombs and stones, which they threw at the security forces whenever they were asked to disperse. James Motlatsi, however, who worked in No 3 shaft, denied that the strikers were armed, riotous or unruly or had caused any damage to property. The police gave the strikers five minutes to disperse but after one minute, as the strikers began to move back, they fired rubber bullets and tear gas at them. The strikers were then driven into their rooms where the assaults continued. Some were cornered in the changing rooms and the kitchens and were fired at there. There were many casualties. Three of them had their eyes damaged and were in danger of losing their sight. A fourth, Seluku Tladi, described how he was injured.[53] He was shot, he said, in his lower left leg by a rubber bullet. Then two policemen grabbed him while a third put a gun to his head and threatened to blast it off. The policeman, he said, was serious and pulled the trigger but, just as he did, Tladi jerked his head so that the bullet ripped his upper gum away and damaged his tongue. The NUM doctor who tended him later said he needed extensive plastic surgery to his mouth. Ten of the mineworkers had been shot in the head with rubber bullets. Others had wounds in their legs and backs. Altogether, 14 of the 78 injured strikers were

[52]See *City Press*, 16 August 1987 and *The Citizen*, 17 August 1987.

[53]Seluku Tladi was 57 years of age. He came from Mafikeng in the north-western Transvaal and had seen his family, on average, for only 20 days a year, at the end of each annual contract, during the 30 years he had worked in the mines. When he was interviewed he had been in hospital for a week and could barely talk because of his mouth injury. His family had not been informed that he was injured and was in hospital.

in a serious condition.[54]

The ordeal of the strikers at Western Deep Levels did not end there. Shortly after the shooting had stopped, the mine manager announced over the intercom system that all those requiring medical attention should go immediately to the medical station in No 1 hostel. Many of them, including at least four with quite serious injuries, responded and somehow made their way to the medical station. When they got there, however, they were greeted not by medical attendants but by the police who were waiting to arrest them on the assumption that if they were wounded then they must have been active participants in the fighting. Fifty-three of the injured strikers were arrested on the spot and charged with public violence. They were taken to Potchefstroom prison where they were remanded over the weekend until their bail applications were heard on the following Monday. Thirty-eight of the prisoners had rubber bullet wounds. None of them received any medical treatment from the time they were arrested early on the morning of Friday, 14 August, until after their bail of R200 each had been paid on the following Monday and they were released from prison. Their requests for pain-killing tablets were ignored. Nor, during that time, did they receive any food or drink.[55] The serious cases were sent from the clinic to a hospital but union officials were not allowed to visit them there. That was not the end or even the climax of repression at Western Deep Levels. Three days after the injured strikers had been released from prison another 37 strikers were arrested and detained in cells at Fochville police station.

Ramaphosa's contention that the mining houses were determined to seek a quick end to the strike by using force was borne out by the manner in which the intensive and large-scale confrontations conducted by the security forces and the police were synchronised between Wednesday and Saturday in the first week. This period covered the five-day time-span that Johann Liebenberg of the Chamber of Mines had predicted the strike would last. The concentration of police violence, however, was not confined to the large gold mines. As the strikers at Vaal Reefs and Western Deep Levels were being attacked, so were strikers in the most strategic coal mines at Witbank. At the time of the strike, 16 collieries in the union's Witbank region had an

[54] *The Star*, 15 August 1987.

[55] This information was recorded by NUM attorneys when they bailed the strikers from prison and took them to the Alexandra Clinic where they were treated by a team of eight doctors. It was compiled into a memorandum by Mono Badela for the NUM at the time but it was not dated.

Injured strikers from Western Deep Levels on 14 August 1987.

average 68 per cent rate of union organisation compared with 58 per cent at Vaal Reefs and 56 per cent at Western Deep Levels. Whilst the whole region was well organised, the Matla and Optimum collieries, both owned by the Gencor Group, provided the strike leadership in the region by their unwillingness to make concessions and their intolerance of strike-breakers. Both collieries became the focus for violent retaliatory action by the mine managers.

The NUM had organised almost 70 per cent of Optimum's black work force so it had little difficulty in getting strong support for the strike. The mine manager, however, questioned whether the support was genuine and voluntary so he brought in the police to interrogate each of the strikers. The SAP entered the hostel, which had about 1 000 residents, on the fifth day of the strike and ordered the strikers to return to their rooms to be questioned about their involvement in, and, presumably, their commitment to the strike. The strikers defied the order so the police set about enforcing it by firing at them with both rubber bullets and live ammunition. Their task was difficult, however, because the structure of the hostel obstructed their movement and allowed the strikers to regroup and retaliate. It took the police two days before they had gained control of the hostel. Eleven of the 20 strikers who were injured had to be taken to Middelburg hospital in Witbank.

The mineworkers at the Matla colliery, however, bore the brunt of the violence in Witbank. With a 64 per cent rate of union membership it was less extensively organised than some of the other collieries but this was more than compensated for by the quality of its leadership provided by Crosby Moni and Gwede Mantashe, the chairperson and secretary respectively, who were prominent at the regional and national levels in the union.[56] Once the union had obtained recognition at Matla in April 1985 and was allowed formally to elect shaft stewards from June of that year, it had become involved in a series of disputes with the colliery management, involving illegal strikes, go-slows, sit-down strikes and the sabotage of machinery, which were not concluded when

[56]Crosby Moni, for many years the chairperson of the Witbank region of the union, became a full-time elected Regional Officer of the union in 1997. He was elected as the union's Vice-President at its National Congress in April 2000. Gwede Mantashe's first full-time post in the union was as the first occupant of the position of National Organiser. He then became the Regional Co-ordinator for the Wits region. He was elected as the Assistant General Secretary in 1994 and, in March 1998, was elected as General Secretary in succession to Kgalema Motlanthe who had become the Secretary General of the ANC. He was re-elected as the General Secretary in April 2000.

the national strike was called.[57] The principal long-running issue of contention was about working hours. The colliery was unusual within the South African context in that it was highly mechanised and used continuous miners extensively to cut the coal. In order to repair and maintain the machines without breaks in production it was necessary to make special arrangements which involved altering the starting and finishing times of the shifts. Negotiations had failed to produce an agreement on new shift arrangements so the management imposed its own proposals, which disadvantaged mineworkers by making it difficult for them to get taxis and buses to take them on weekend leave to their rural homes. They expressed their disapproval by engaging in acts of sabotage and by leaving work early on Fridays. The management instituted disciplinary hearings against those who left work early that day and 182 of them were informed they would be dismissed just before the national strike was due to begin. In order to pre-empt the dismissals, and because of the continuing dissatisfaction over working hours, the NUM branch began its participation in the national strike five days early, on 5 August.

During the first week of the strike the mineworkers held meetings, sang and *toyi-toyied* to consolidate their solidarity. They tried to eliminate opposition to the strike by assaulting strike-breakers, in particular, personnel assistants and team leaders. In an attempt to counter this violence, the mine manager called in the SAP on 15 August to assist his mine security force in searching for weapons and for those suspected of attacking strike-breakers. When some of the strikers refused to hand over their weapons the police attacked them. The search developed into a major assault when the police went to the rooms where the strikers were relaxing. The unsuspecting mineworkers were beaten with *sjamboks*, kicked, and hit by rubber bullets. The doors and windows were smashed in 50 of their rooms. The ceilings were damaged in two others. Many private lockers were broken into and personal possessions, including money, were stolen.[58] No-one was arrested but 140 strikers were injured. That evening the shaft stewards were arrested. The following day there was more violence when Matla's mine security and the SAP were reinforced by the security

[57]For a description of industrial relations at Matla between 1985 and 1988, see J K McNamara, *The Underlying Causes and Resolution of Labour Conflict at Matla Colliery, November 1988.* McNamara's investigation was conducted on behalf of the Trans-Natal Coal Corporation, Ltd and the NUM.

[58]The Matla strike committee listed 60 items of damage and theft during the raid on 15 August. Amongst the stolen property were a tape recorder, a carton of cigarettes, a pair of shoes and two shirts. R200 was stolen from one locker.

forces from four neighbouring mines in order to search each room more rigorously. All of the occupants were ordered to gather outside the hostel and, as they did so, the police fired at them. Twenty were injured and were sent to hospital. Many of the others ran for safety to neighbouring farms. When the union's attorneys went to the colliery to try to restrain the manager from ordering further attacks they were refused entry to the mine and then evicted from the area.[59] The mine manager sealed his aggression with an act of inhumanity by refusing to permit a NUM doctor to speak to the wounded strikers.

Three days later, on 19 August, more than 200 Zulus were brought to the colliery to take the place of the strikers and were put in a tent encampment outside the hostel. The hostel manager, and an overseer who spoke Zulu, had gone on a personal recruitment campaign around the KwaZulu homeland, despite advice from Gencor's senior management not to recruit Zulus. The NUM officials claimed that the recruits were told that the mine needed extra labour in order to open a new shaft. Nothing was said to them about strike-breaking.[60] Though the balance of power in Matla had swung back to the management by the third week, the strike, nevertheless, continued to be marked by violence against those who had refused to support it. In the last few days a personnel assistant, who had sided with management, was attacked and then, almost at the last minute, a strike-breaker was murdered.

Emergency Powers and the Courts

The state apparatus was comprehensive in its repression. It imposed physical constraints on action, as the strikers learned each day. But it also worked in non-physical ways through the judiciary and the media. There had been signs from the early part of the year that the judges might be able to use their own understanding of any evidence submitted to them and take a balanced view of a black mineworkers' strike, if one happened to take place. It would, after all, be a legal strike. The first sign that such a balanced approach was possible was in January 1987, after the United Democratic Front, the National Education Crisis Committee and the South African Council of Churches had issued an advertisement commemorating the 75th anniversary of the ANC and calling for the ban on it to be lifted. The advertisement was accepted for

[59] *The Star*, 16 August 1987.
[60] McNamara, 1988, p34.

publication by 16 newspapers but was rejected by *The Citizen* and the Afrikaner media conglomerates. The security police with search warrants then visited 10 of the 16 newspapers around the country and prevented *Business Day*, the *Eastern Province Herald* and the *Weekly Mail*, in particular, from publishing the advertisement. The publishers of the main English language newspapers resented the police intrusion into their affairs and persuaded the Witwatersrand Local Division of the Supreme Court to overturn the regulation that gave the police the power to intervene in the first place. In March, that process was repeated when *The Star* obtained an injunction at the Supreme Court to prevent the police from confiscating copies of the paper that carried an advertisement from the Detainees' Parents' Support Committee. The most dramatic decision, however, came in April when, in response to an appeal by the United Democratic Front and the Release Mandela Campaign, the Natal Provincial Division of the Supreme Court ruled that the state president had exceeded his powers when he issued emergency press regulations in June and December of the previous year. That judgment removed restrictions on campaigning for the release of detainees and on placing advertisements in the press relating to the ANC.[61] But it all turned out to be a temporary judicial aberration for the state president intensified the emergency regulations in June 1987 and, in doing so, considerably reduced the scope for judges to defy the government.

The NUM experienced the full impact of the revised regulations on each of the three occasions it appeared before the courts during the strike. The Anglo-American Corporation was involved each time. The first was on Thursday, 13 August when Anglo-American won the right to evict strikers who were sitting in at its ERGO plant at Springs. The next day, the NUM sought a court interdict at the Bloemfontein Supreme Court to stop the management at Anglo-American's Saaiplaas gold mine in the Orange Free State from interfering with strikers. The union's complaint was that 18 strikers at No 2 shaft had been assaulted by a group of team leaders and were afraid to return to the mine. The strikers alleged that the management had armed the team leaders who had then destroyed the union office. The union claimed that when the team leaders launched their attack they concentrated on the shaft stewards and then separated 200 Mozambicans from the rest to create a potentially volatile ethnic division. In the No 4 shaft the management actually engineered a faction fight which left the

[61]See South African Institute of Race Relations, 1987/88, pp818-830.

mineworkers divided as to whether to strike or not. In case any doubts remained in the minds of the Saaiplaas mineworkers about whether or not they should support the strike, they were subsequently all forced down the shaft at gunpoint. The NUM had further reasons to complain about the behaviour of the Saaiplaas management. It had earlier tried to coerce mineworkers to sign an undertaking that they would accept the wage offer and refrain from striking. It also refused to allow the strikers to withdraw any money from TEBA Cash, the mine bank. The Supreme Court was not convinced by this and refused the union's request because, it stated, it had insufficient evidence on which to base a decision. The union's last court appearance was on 15 August, when a court ruling enabled Anglo-American to restrain 16 strikers at Vaal Reefs from "creating a disturbance, abducting persons, or ... obstructing entry to, and exit from, the mine's premises or inciting others to do so". However, though a victory in one or other of these cases might have raised the morale of the strikers, none of the judicial decisions directly affected the course of the strike.

Restrictions on the Domestic Media

The government's repressive laws also stretched out directly to the media and distorted its coverage of the strike. The government held considerable powers, stemming from the Suppression of Communism Act of 1950 but largely confined to the domestic press, to censure the written word as well as to ban publications and arrest and detain editors and journalists. It had used its powers freely and extensively. Newspapers and journals had been published with blank paragraphs or pages. In some instances, newspapers had been suppressed. The government in 1977 had closed down *The World*, a black daily and weekend newspaper, with the largest daily circulation in South Africa. Its successor, *The Post*, suffered the same fate in 1981. Government pressure had been a key factor, in 1985, in causing the closure of the *Rand Daily Mail*, which, along with the East London-based *Daily Dispatch*, was the most progressive of South Africa's commercial daily press. Editors and journalists had been silenced. Two editors of *The World* had been detained without trial, as had three editors of the *Rand Daily Mail*. Donald Woods, the editor of the *Daily Dispatch*, had been forced into exile in 1978. On the day the strike began, Tony Heard, the liberal editor of the *Cape Times*, was dismissed from his post, allegedly because he had been too critical of the government. The risks, especially to ordinary press reporters, involved more than dismissal.

The New York-based Committee to Protect Journalists stated that in the months leading up to the strike it had "documented dozens of cases of assault, detention, breakage of equipment, and general harassment of reporters …"[62] Zwelakhe Sisulu, the black editor of *The New Nation*, a progressive weekly with a largely white readership and owned by the Catholic Bishops' Publishing Company, was actually in detention at the time of the strike. Black journalists were the most vulnerable police targets amongst media workers but white ones were also increasingly affected by government repressive measures during 1987.[63] It was clear that offending the government at the time of the strike was a high-risk policy for the staffs of newspapers and journals.[64]

The government held no direct controlling interests in the commercial press but its ideology was communicated to the reading public by the Afrikaner language publications and two English language newspapers, namely, the *City Press* and *The Citizen* that were owned respectively by the Afrikaner media conglomerates, Nasionale Pers and the Perskor Group. The Afrikaner language newspapers were insignificant as opinion-formers amongst black people because very few of them spoke Afrikaans. In any event, in the minds of black people they were linked with the government. It was a different matter, however, with the English language *Citizen* daily newspaper, which was increasing its circulation and was an important guide to government thinking about the strike. According to Bobby Godsell, speaking for the Anglo-American Corporation, it went further than that and represented Anglo-American's own views. After a strike press conference on 24 August, he wrote to Peter Gush, describing his satisfaction with *The Citizen's* coverage of the strike. "The attached cuttings from *The Citizen*," he wrote, "accurately and faithfully reflect what I said at yesterday's press conference. The state of reporting on the strike has led me to decide to become a *Citizen* subscriber as it strikes me as the most

[62]John M Phelan, *Apartheid Media: Disinformation and Dissent in South Africa*, 1987, p40.

[63]From 1976 to 1981, out of 250 mainstream black journalists, 50 had been detained for up to 500 days, 10 had been detained more than once, 10 had been banned and one had been imprisoned on Robben Island. By comparison, during the same period, out of 3 500 mainstream white journalists one had been detained, one had been banned and one had been imprisoned. Up until 1987, no Afrikaner journalist had ever been arrested for similar offences. (Phelan, 1987, p65.)

[64]For a discussion of the issues raised in this paragraph, see Chenhamo C Chimutengwende, *South Africa: The Press and Politics of Liberation*, 1978; Davies, O'Meara and Dlamini, volume two, 1985, pp406-418; Phelan, 1987 and Innes, 1984, p205.

reliable newspaper we have."[65]

It was strange that Godsell should have felt it necessary to single out *The Citizen* for playing such a pivotal role in reporting the strike because the Anglo-American Corporation, directly or indirectly, influenced the policies of 17 of the 20 English language newspapers in South Africa. This had come about through Anglo-American's controlling interest in two inter-linked conglomerates, namely the Argus Printing and Publishing Company and the South African Associated Newspapers (SAAN), which, between them, owned the majority of those newspapers. The Argus Company possessed the four largest circulation daily and weekend newspapers, including *The Star*, with the biggest daily circulation, and the *Sowetan*. SAAN, which had owned the closed *Rand Daily Mail*, controlled the *Sunday Times*, the largest circulation weekly newspaper, along with the *Sunday Express, Business Day* and the *Financial Mail*. Between them they processed the presentation of news and views to virtually the whole of English-speaking South Africa. Moreover, the Argus Company, together with the major food group, Premier Milling, controlled the Central News Agency that monopolised the distribution of English language newspapers and magazines within South Africa. Anglo-American's intrusion into the media business had come about when it had taken over JCI in 1960 and had acquired, in the process, a 40 per cent stake in the Argus Company.[66] It perpetuated an intent by the mining houses to control the press that had started in the late 19th century when a consortium of mining capitalists, including Cecil Rhodes, took over the *Cape Argus* newspaper and used it as a base to launch *The Star* in Johannesburg in 1899. The Argus Company they had established, had, by the 1980s, acquired 40 per cent of SAAN's share capital so that Anglo-American was able to play an influential role in both groups. It had additional leverage in SAAN through the 21 per cent of SAAN's share capital held by the nominees of its own holding companies.

The chairperson of Anglo-American, Harry Oppenheimer, often insisted that he did not interfere with the editorial policies of the media under his company's control.[67] It was not necessary, of course, that he should even have tried to interfere because, without any direction, newspapers in general reflected the interests of their owners. This was clearly the case with those owned by the Argus Company and SAAN

[65]Correspondence, "Anglo-American Corporation of South Africa. Note for Mr E P Gush." Johannesburg, 25 August 1987.

[66]Innes, 1984, p205.

[67]*Ibid.*

for they were politically liberal in their approach to the apartheid system in that they were critical of some of its aspects but stopped short of demanding its complete overthrow. Their selective approach to apartheid led them to oppose those aspects of the system that impeded the production of minerals, such as the colour bar in employment, and to expose the atrocities initiated by the security forces in the townships, that adversely affected the flow of investment into South Africa. It did not lead them to support international sanctions or the introduction of unqualified universal suffrage. The bottom line for their opposition to the apartheid system was always the need to perpetuate the interests of capital. They never ceased to be pro-business and, in line with that policy, were consistently strongly anti-communist, critical of trade unions and suspicious of the ANC.

These political preferences were reflected in their reporting of the strike. *Business Day*, owned by SAAN, and *The Star*, belonging to the Argus Company, provided the most extensive reports of the strike. But neither they, nor any of the other newspapers in the two groups, were in any sense supporters of the mineworkers' cause. None of them endorsed the wage demands of the NUM or published in-depth articles about mining conditions that openly criticised the mine owners. They simply reported strike events as they occurred. Amongst all the newspapers, the two independent weeklies, the *Weekly Mail* and the *New Nation*, were marked exceptions. The *Weekly Mail*, in particular, gave the strikers consistent and sympathetic coverage. This newspaper, formed by a dozen journalists after the closure of the *Rand Daily Mail* in 1985 in order to continue its radical tradition, but with an average circulation of only about 16 000, was the only domestic newspaper to provide a detailed analytical coverage of the strike with articles that explained the background to it and, in so doing, exposed the reasons for the mineworkers' unexpected display of solidarity.

The Overseas Media Coverage

When the strike began, the international media was already well represented in South Africa. This was not simply in anticipation of a major disruptive event but because it had been monitoring the South African situation to satisfy expanding international anti-apartheid sentiments ever since television screens around the world had shown the body of 13-year-old Hector Peterson, the first casualty in the Soweto uprising in 1976, being carried in the arms of a distressed youth. The international media had been constantly present, relaying pictures and stories about township violence and repression in its various shades and forms to all of the metropolitan areas of the world. It had had much to report during the major national strikes earlier in the year and there was little doubt that it expected a repetition of violence in the mine compounds. It was also concerned about the possible international implications a mining strike might have through its effect on the world supply of gold and, therefore, on its price to central banks. It raised questions about even the survival of the apartheid system itself.

The overseas media representatives, however, found that it was much more difficult to report on a mining strike than on events in the townships. They were obstructed in their search for first-hand information from the mining compounds and found it almost impossible to get photographic insights. The mine owners, with a mind-set from the inter-war years, still regarded the compounds as their private property and refused to allow journalists and photographers inside them. On the first day of the strike, guards patrolled the fenced perimeters of all of Gold Fields' mines in Carletonville and refused to admit any reporters.[68] When mineworkers from JCI's Randfontein Estates gold mine were being packed into buses to be sent home on the first day, the police sealed off the mine entrances, military vehicles patrolled the roads leading to the mine and journalists standing outside the mine were arrested before being sent away.[69]

The Anglo-American Corporation was also secretive about what it regarded as a private matter. Reuters requested permission to visit its mines but the request was ignored. When one of its journalists and a photographer got into the Western Deep Levels compound, using their own initiatives, they were, as they described below, arrested by the mine security force and turned over to the SAP. "Weapons," the journalist reported, "were much in evidence at Western Deep Levels this week. Mine security armed with shotguns and waving pistols pursued this correspondent and a photographer, forcing our car off the road. Ordered out of the car, we were handed over to the state police who were clearly upset that photographs had been taken of roadblocks at the mine … After being released the Anglo-American security men told us not to set foot in the district again and assigned two cars to escort us out of the area."[70] Anglo-American responded to this allegation by stating that it would investigate the incident. When several other overseas media representatives also complained about its secrecy,

[68] *Irish Times*, 10 August 1987.
[69] *Birmingham Post*, 11 August 1987.
[70] Reuters, 13 August 1987.

NUM members dismissed from Randfontein Estates on their way home to the Transkei.

Godsell, on behalf of Anglo-American, said: "We certainly had some real anxiety about journalists. We've had requests from TV crews to go on to mine properties and whilst we understand that it's the role of the media to report what's going on, we've had some tense and difficult situations and I guess we're putting our concerns about maintaining order and preventing violence above an absolute right of entry to journalists at this point in time."[71]

The exclusion of TV crews and journalists from Anglo-American mines was only one of the overseas media's complaints. It was reported that Gavin Relly had instructed his officials not to disclose which of its mines were on strike even though it was the most affected mining house with at least 100 000 of its black employees involved.[72] When news circulated that there was fighting between security officers and strikers at Vaal Reefs, for instance, Anglo-American officials were reluctant even to admit that there was a strike there at all. The press statements they issued were either intended to counter allegations made by the NUM or to lay blame on the union, as when it claimed that by the second day of the strike there were "disturbing incidents of intimidation" against non-strikers at 10 of its mines.[73] But despite its unwillingness to supply strike data, it complained about mine violence much more than the Chamber of Mines which usually played it down by reporting that it was "gratifyingly low".[74]

Ramaphosa, on the other hand, regarded the media as a window through which to publicise the union's case and both he and Marcel Golding, the assistant general secretary, supplied it with reports of confrontations in the compounds and hostels as they came in from the strike committees. Golding assumed special responsibility during the strike for handling media affairs and established effective channels of communication with journalists. As a result of the attitude of the mining houses, the NUM became the only first-hand source for events within the compounds. This created a difficulty for the media for whenever the NUM presented a report of mine violence it evoked either a denial or a counter-allegation from one or other of the mining houses and there were often no means of verifying which one was accurate. Christopher Wilson, the Reuters correspondent, commenting on this issue, stated: "Without access to the mines, journalists have difficulty sifting through the flurry of accusations flying thick and fast between the union and the

[71] *Weekly Mail*, 14 August 1987.
[72] *Financial Times*, 13 August 1987.
[73] *The Yorkshire Post,* 12 August 1987.
[74] *The Guardian*, 12 August 1987.

mine owners. Whatever the truth behind the exchanges, the mines have built up formidable arsenals to ensure their ability to control black workers who live in single-sex hostels on mine property far from the tranquil villages that house white artisans, engineers and managers."[75] The difficulty of getting at the truth was compounded by the frequency of the union allegations. The NUM recorded incidents of violence between strikers and armed mine security assisted by the police and scabs in all but three of the 26 gold mines on strike and in seven of the 18 strike-bound collieries. In some gold mines and collieries there were repeated instances of violence.

The radio, television and broadsheet newspaper companies in the USA, Britain and a few Western European countries had their own correspondents in South Africa to report on the strike. Other countries relied on the reports of press associations, such as Reuters, that had their own on-the-spot correspondents. The foreign correspondents were often well placed to investigate because they were familiar with the country and were acquainted with the leaders of the NUM and COSATU, as well as those community organisations that supported the liberation struggle.[76] And although they were all formally bound by the government's press restrictions, they generally regarded them more as a hindrance than an obstacle to gathering news. They were also constrained by the political preferences of their respective media organisations. But how they actually reported the strike depended on the extent to which they analysed it within the context of apartheid. Few of them did this. And even those that did were generally ambivalent about their support for the black mineworkers themselves. For instance, in its issue on 13 August, *The Times* warned against treating the strike as a morality play and urged the mineworkers to have faith in the slow process of change of which, it contended, there was already much evidence. The *Independent* from Britain published an editorial on the first day that predicted a dire outcome if a compromise was not reached. The British *Guardian* was one of the few overseas newspapers to put the strike into its apartheid context and provide occasional insights into mining conditions, but even that newspaper ended its

[75]Reuters, 13 August 1987.

[76]A number of British broadsheet newspapers, including the *Independent*, *The Guardian, Observer, Financial Times* and *The Times*, had their own correspondents reporting on the strike. The USA news organisations which maintained news bureaux in South Africa included the *New York Times*, ABC News, *Associated Press,* CBS News, *Los Angeles Times,* NBC News, *Newsweek, The Christian Science Monitor, The Wall Street Journal, The Washington Post, The Washington Times, Time* and *United Press International.*

editorial on 12 August with a cautionary warning to Ramaphosa not to overreach himself. Of the major international newspapers, only the *New York Times* advised its readers and the US government to "Back the Black Miners" unequivocally. It reprimanded the US government for being "oddly insensitive to the rigidities of the white government" and urged it to extend sanctions against South Africa if there was government intervention in the strike in support of the mine owners.[77]

Sources of Information Within the Compounds

Information via the media was virtually the sole source of information about South Africa for people overseas and it undoubtedly played an important part in creating support for the international anti-apartheid movement. In South Africa, however, the situation was entirely different for there the daily experience of apartheid was the prime determinant of the consciousness of both black and white people. In spite of the influence on the media exercised by the government and the mine owners, it played a minimal role in moulding the awareness of black people about the wider implications of the apartheid system, just as it had no influence worth mentioning on the attitude of black mineworkers towards the strike. Some of the newspapers produced black township editions but their reports about the strike were so muted that one commentator in Lesotho believed that a news blackout existed in South Africa.[78] There was no effective alternative progressive press to compensate for the deficiencies in the commercial newspapers for it had been banned. Journals, leaflets and information sheets from the underground ANC and SACP contacts rarely penetrated the hostels. The friends of mineworkers had hardly any public voice. The only domestic publications that were firmly and consistently on the side of the strikers were the *NUM News* and cyclostyled information sheets from both the union and COSATU, but none of them was available to either the black or white public at large. The presentation of the strike through the printed word, therefore, was of significance only to the reading public within the black communities of South Africa. Illiterate people, who comprised the majority of the black population, had to get their strike information from other, primarily oral, sources. The black mineworkers themselves fell largely into this category because as many

[77]Quoted in the *International Herald Tribune*, 12 August 1987.
[78]Transformation Resource Centre, *Work for Justice*, nd, Lesotho.

as from 60 to 70 per cent were illiterate. Thus the majority of strikers received news of the strike beyond their own mines and collieries by word of mouth through meetings in the hostels or at those organised by the NUM outside the compounds, from gossip and from rumour. Occasionally some heard commentaries by managers over the mine public address systems. However, the primary sources of oral information within the compounds were the literate local union activists who, with the authority of the union behind them, passed on their interpretations of what they had read, heard and seen about the strike.

The families of strikers in the homelands had even less access to news and depended on occasional letters from husbands, brothers and sons in the hostels and tales from returned dismissed or sick strikers. There were no publications in the vernacular languages within South Africa that gave them information. In Lesotho the only publication in Sesotho was *Litaba Tsa Lesotho* that was produced by *Work for Justice* and published in Maseru by the Transformation Resource Centre from information gleaned from newspapers, radio broadcasts and personal reports by NUM members. Mostly, then, the families knew little about the strike and received no answers to any of the critical questions they raised about its impact on their welfare.

Two alternative sources of information about the strike were radio and television. Both of these networks were dominated by the state-controlled South African Broadcasting Corporation and were exploited to spread government propaganda. Radio Bantu, which broadcast programmes in the vernacular languages, was the most accessible medium for black people. It was quick to warn the country's black trade unionists that if they followed the mineworkers' example they would put at risk the limited reforms introduced over the past 10 years.[79] Nothing was broadcast that was critical of the government's view of the strike. It was the same with television, though this was a medium that was rarely found in the black communities at that time. However, here, as with the press, the hand of Anglo-American was present. It exercised influence through the privately owned television channel, M-Net, in which Argus and SAAN each held 18 per cent of the shares. The same proportion of shares was held by each of the two Afrikaner media conglomerates. In July 1985, the four major shareholders had reached an agreement with the South African Broadcasting Corporation whereby, in exchange for shares from M-Net, the SABC would provide it with

[79] *The Morning Star*, 12 August 1987.

news and current affairs programmes.[80] Thus, much the greater part of the flow of information about the strike from radio and television was consistent with the views of the government and the Anglo-American Corporation.

NUM's Allies

The NUM had allies within South Africa but the help they could provide was limited. Its principal support base was in the black communities though these had no influence on the government and were unable to provide material support to the strikers. Although COSATU had given its backing to the union, neither it nor its affiliates could influence affairs in the compounds. Other strikes were being held at the time and they put an added strain on the SAP though without seriously impeding its involvement in the mines. There were a few related sympathetic strikes by NUM members in mines and plants that were not a part of the Chamber of Mines negotiations but they were peripheral to the main dispute. Nor could the United Democratic Front assist except by giving moral support. Its leadership structures had been decimated under the state of emergency with the result that its activists were thinly and widely distributed around the country and it had no financial resources. There were some student groups, both black and white, in the major universities that agitated in NUM's favour and raised small amounts of money for the strikers. For instance, they received some material help from students at the University of the Witwatersrand after the union had appealed to them for help with finance, food and accommodation.

The international trade union organisations were the NUM's main source for strike funds. Arthur Scargill, who was the president of the International Miners' Organisation (IMO) as well as the president of the British NUM, launched an international fund through the IMO on the first day of the strike. "We shall channel the money immediately to South Africa," he said, "in a manner that makes it available for use by the black South African mineworkers and their families."[81] The United Mine Workers of America reactivated its South African Miners' Aid Fund, which it had set up to help during earlier strikes in order to donate money directly to the NUM. Money was also

[80]Phelan, 1987, p60.
[81]*Yorkshire Post*, 11 August 1987.

secreted into the country from a variety of overseas trade unions, particularly from the Miners' International Federation (MIF) which had been acting as the conduit for funds from the Scandinavian trade unions since the formation of the NUM.

Ramaphosa had every reason to be circumspect about the NUM strike finances because it was not easy or straightforward to get money from abroad into South Africa. The state president, P W Botha, had warned in his speech after the Nationalists' electoral victory in May 1987 that he would stop certain organisations from receiving foreign funding though he did not specify which ones he had in mind. He said that the foreign funding of extra-parliamentary opposition was intended to "subvert South Africa's constitutional order". There were two pieces of legislation already in existence that he could use. There was the Affected Organisations Act of 1974, which the government used to prevent the financing of the Christian Institute and the National Union of South African Students and which was applied against the United Democratic Front in 1986. And there was the Fund-raising Act of 1978 that had been designed to prohibit subversive organisations or any affected bodies or individuals from collecting money from the public either in South Africa or overseas without the explicit consent of the government.

In the event, neither of the Acts was used. Instead, the South African Reserve Bank issued a secret directive rather belatedly on 24 August 1987 to all authorised bank dealers on how to handle overseas funds destined for the NUM. It stated:

> Authorised dealers are hereby instructed not to accept any transfers from abroad for credit of or intended for the benefit of the above-mentioned organisation (NUM). Should any transfers of a foreign exchange nature come to the notice of authorised dealers such funds should be held in suspense and details of the transaction furnished to the exchange control department of the South African Reserve Bank immediately.

The Standard Bank of South Africa responded immediately by informing all of its branches of the directive. However, the directive did not prevent the NUM from receiving funds from abroad through the normal banking channels. The Reserve Bank simply delayed passing the money on until it had verified its sources so that the NUM received it from about a week to a month late. As Ramaphosa and Motlatsi knew about the directive, they made alternative arrangements for cash to be brought into the country through the 'underground structures' that were being used by other sections of the liberation movement.

The Reassessment of Strike Tactics

The character, purpose and possibilities of the strike began to clarify for both the union leadership and the mine owners by the end of the first week. It was clear, for instance, that the strike was not going to end quickly. On the eighth day, the NUM still claimed that 350 000 mineworkers, from 27 gold mines, all the collieries in Witbank, one colliery in Secunda and the Rand Refinery, were still on strike. An estimate by the Labour Monitoring Group for 19 and 20 August put the total of striking gold mineworkers at 302 000 and of coal mineworkers at 21 000. The participation rate for mineworkers in all of the gold and coal mines controlled by the Chamber of Mines had started at about 65 per cent but during the second week it had risen to more than 80 per cent in 80 per cent of the gold mines.[82]

This increasing degree of solidarity within the union under such adverse conditions gave the NUM leaders a clear signal to continue with the struggle for a living wage. On the other hand, the ferocity of the opposition convinced them that they were not engaged in an ordinary industrial dispute about percentage points on a wage scale that they could possibly win but that they were involved in a power struggle in which the mine owners and the government were prepared to deploy all of their resources to destroy the union. James Motlatsi, who at the beginning had been confident that, because everything was being done within the law, the union could succeed in its claim, significantly revised his opinion. He made this clear immediately after the police attack at Western Deep Levels on Friday, 14 August, when he declared that the defence of the mineworkers had become the overriding issue. From then on, the meaning of the term 'victory' had changed for the NUM. It was going to be a victory for the union if its organisation remained intact. The strike was about survival in a classic example of the class struggle. Until that year, few unions in South Africa had ever survived the impact of bitter and prolonged strikes. The NUM aimed to add its name to the tiny list of survivors.

In an assessment of strike progress, the cost to the strikers in terms of the loss of life, injuries, detentions, arrests and family deprivation due to the lack of income, had to be taken into account. Although there were some mine security and police casualties, the brunt of the violence was felt by the strikers. With this in mind, the specially convened meeting of the National Strike Committee on Sunday, 16 August agreed

[82]Labour Monitoring Group, *Report*, 23 August 1987.

"to examine the escalating violence being committed by the mine security forces and the SAP". It received a report from Ramaphosa stating that during the six days of strike action 240 mineworkers had been detained or arrested and that over 200 had been injured, in some cases seriously, through "barbaric and violent acts committed by mine security and the SAP". The Anglo-American Corporation, however, was making comparable charges against union members. After listening to their report, therefore, the 200 members of the Central Committee resolved to challenge Anglo-American to substantiate the allegations of intimidation and sabotage that it had made against the union by opening its compounds to the media. In consequence, the Central Committee stated:

> Today, we reiterate the challenge issued by our National President that:
>
> 1. To obtain the truth of what is happening on the mines, management should open up their premises to all the media so that the public can be informed about what is really happening.
> 2. A senior member of the Johannesburg Bar Council upon being requested by any of the parties to have access to any mine to assess the actual position.

The increasing and unexpected solidarity within the NUM, however, caused the mining houses affected by the strike to analyse and revise their own approaches in their search for victory. In the first instance, the Anglo-American Corporation clearly needed to debate and resolve the issue of violence because it both tarnished its public image and was costly in terms of resources. But it was not prepared to follow the suggestions of the NUM and allow outsiders to intrude into what it still regarded as its private domain. It wanted to refurbish its liberal image whilst remaining in charge of the flow of information that moulded the public's perception of the strike. It had started the strike by denying that it was using force and blamed the union's tactic of intimidation against non-strikers for any injuries or deaths in the compounds. Both Bobby Godsell and Peter Gush had spoken in those terms. The day before the strike began, Gush, for instance, had denied that management used violence to break strikes. Where there was violence, he stated, it was because it had "been used to force people to strike".[83] But as the strike progressed this was so obviously a misrepresentation of the press reports of events in the compounds that Gush altered his approach by admitting that Anglo-American's security forces possessed arms and used them, but only to protect the lives of its employees. The first

[83] *The Observer*, 9 August 1987.

public expression of this new approach was on the fourth day of the strike at an Anglo-American press conference largely devoted to the issue of violence and presided over by Gush when he admitted that "mine security have guns and armoured vehicles". Another speaker, Theo Pretorius, the chairperson of the Corporation's Transvaal Mines Division, went further and said "it was appalling that we have to use force against our own employees" but, he emphasised, "the mine security forces are not stormtroopers intent on killing people but they have been compelled to use rubber bullets and tear gas to protect lives".[84] Against all the evidence, however, Godsell continued to deny that his organisation was intent on crushing the strike. Just two days after Gush's admission, Godsell declared: "Anglo had not used violence in an effort to break the strike. Force had only been used against miners who had armed themselves with dangerous home-made weapons and who had sabotaged important machinery at plants."[85]

After admitting that its security forces were engaged in violent acts the Anglo-American Corporation argued for conditions where it considered force would not be necessary. The challenge by NUM's National Strike Committee gave it the possibility of doing that but it wanted a different, more widely based agenda. On the day of the meeting of the National Strike Committee, the Manpower Director of Anglo-American's Gold and Uranium Division telexed the following letter to Ramaphosa to explain precisely how the corporation wanted to proceed. It stated:

> In an attempt to prevent further violence on Anglo-administered gold and coal mines we suggest talks with your union on the attached proposals whose purpose is to prevent intimidation, employee violence and the need for mine security personnel to use force. These proposals, if observed, should also remove the need for external security interventions, including those of the SAP. Finally, these proposals seek to guarantee the freedom of both those workers wishing to strike peacefully and those workers wishing to work normally.
>
> The specific proposals are:
>
> 1. All hostels should function normally. Hostel canteens, ablution facilities and other essential services should operate normally under the control of the management personnel normally responsible for these services.
> 2. There should be normal access to hostels and married quarters for workers, management and union officials.
> 3. There should be normal access to shafts and plants, with areas designated for peaceful picketing, should the union request this.

[84] *The Star*, 14 August 1987.
[85] *Sunday Tribune*, 16 August 1987.

> 4. Force or the threat of force should not be used by management, the union, or union members to prevent workers from striking peacefully or working normally.
>
> We suggest that both parties agree to notify each other where they believe that the above conditions have not been observed. Management remains the responsible authority for maintaining peaceful conditions on its property, and will continue to ensure the protection of life and property.
>
> Please indicate your response to this invitation as a matter of urgency, hopefully before the end of today.

Following a meeting of NUM's National Strike Committee, Ramaphosa said that the NUM was prepared to accept the conditions specified in the letter but that he wanted to add eight others. He wanted to include: (a) the Anglo-American Corporation to withdraw the payments it was insisting strikers should make for food and accommodation in the hostels; (b) injured mineworkers to be compensated; (c) the charges to be withdrawn against those who had been arrested; (d) the union head office to be informed whenever a mineworker was held in the mine security barracks; (e) each side to notify the other in the event of a breach of any of the conditions and agree to have it submitted to an independent arbitrator: (f) the mines to be open to the media; (g) the security forces to be demobilised; and (h) no intervention by the SAP in any circumstances.

Godsell replied that they could not accept all of the eight conditions but would "try to work out ground rules for the peaceful conduct of the strike and find some constructive role for the union to play in de-escalating violence".[86]According to Ramaphosa they accepted only one of NUM's conditions, namely to demobilise the mine security, but that, he said, combined with Anglo-American's own proposals, was not "sufficient to restore mines to a non-violent situation".

During the course of the meeting Ramaphosa received news from Anglo-American's President Steyn gold mine in the Orange Free State that 15 unarmed strikers had been injured by the police whilst they were boarding taxis at the entrance to their shaft. Ramaphosa was so angry that he led the NUM delegation out of the meeting. In response, he stated:

> If anything, our meeting with the Anglo-American Corporation had proved that we are dealing with a treacherous, cowardly and ruthless organisation whose negotiations are most deceitful.
>
> Treacherous in that whilst the union proceeds to negotiate in good faith to obtain an

[86]*Financial Mail*, 21 August 1987.

> agreement with the company which will decrease violence on the mines, the same company calls in the police to attack unarmed strikers boarding taxis at the entrance of shaft No 4 at President Steyn.
>
> Cowardly in that they have now firmly come to depend on the state and its forces to assist them in breaking the legal strike and demoralising workers.
>
> Ruthless in that they are going to use everything in their power to destroy our union and the strike which is costing them millions of rands daily.
>
> It confirms the union's suspicions that AAC's utterances are designed to hoodwink the public, whilst underhand and devious activities continue using repression and misinformation to smash our struggle for a living wage.
>
> The blood of members has been spilt on many mines and it is patently clear that the mines have become a war zone in which all the casualties are mineworkers.[87]

Speaking on South Africa's independent radio programme, Radio 702, the following morning, Ramaphosa repeated his accusation that the behaviour of the Anglo-American Corporation was immoral. His anger, however, was not reflected in the reports in the South African newspapers. The breakdown of the talks was described briefly while in a sentence the cause was attributed to police action at the President Steyn gold mine. There was no comment about the hypocrisy of the Anglo-American Corporation in pleading to end violence whilst practising it. Nor were the police criticised for their provocative intervention.

Mass Dismissals and Lockouts

When the mining houses principally involved in the strike, the Anglo-American Corporation, Gencor and JCI, assessed the situation at the end of the first week they recognised that the use of force had not succeeded in bringing the strike to a speedy end. They were confronted, therefore, with the prospect of having to endure a long strike, knowing that the longer it went on, the greater would be their losses. The financial costs to the mining houses were already high due to the loss of production and the destruction of surface property during disturbances. There was also evidence that mining conditions in the deep mines were deteriorating because of rock pressures. Even if the strike were to end quickly, the resumption of production would be delayed.

The Anglo-American Corporation tried to play down the extent of its losses by refusing to provide any figures. Nor would the

[87]NUM press release, 18 August 1987.

Chamber of Mines comment on them. When the Labour Monitoring Group published some calculations at the end of the first week, Godsell described them as "crude … though", he admitted, "Anglo is making significant losses".[88] It was, as Godsell suggested, difficult to make meaningful overall calculations because the mines had different capacities for switching to higher-grade ores to enable them to maintain the value of their output and they possessed different quantities of stockpiled ore on the surface to be used to keep up the volume of their gold output. Nonetheless, sufficient information was available to provide a rough indication of the losses.

The Labour Monitoring Group estimated that the six mining houses, in total, had lost at least R90 million during the five-and-a-half days of strike action. Almost all of this loss had been borne by the Anglo-American Corporation, JCI and Gencor. The first two had lost 100 per cent of their profits for the week while Gencor had lost 79 per cent. Of the others, Anglovaal had lost six per cent of its profits; Gold Fields had lost none. There was no calculation for Rand Mines because it had published no figures for the Harmony gold mine, the only one of its gold mines on strike. The losses incurred by the three worst hit mining houses were equivalent to 55 per cent of the total cost of meeting the union's wage demand for a year over and above what the Chamber of Mines had actually implemented. The whole cost of the wage demand was estimated to be equivalent to the cost of a nine-day strike. These calculations did not make sense to the striking mineworkers. It made less sense to them that the mining houses appeared willing to continue the strike well beyond nine days if necessary.

Ending the strike quickly was, then, a matter of some urgency within the boardrooms of the mining houses. But they did not appear to be willing to settle it through a compromise agreement that would have involved some increase in wages for Gush, their principal negotiator, had consistently refused even to contemplate a further wage increase. Until June of that year he had been the President of the Chamber of Mines and had played a prominent part in formulating the uncompromising stance on wages that had contributed to the strike.[89] The only other considered option was to coerce the strikers back to work by engaging in mass dismissals and lockouts. Johann Liebenberg, on the first day of the strike, had said that there could be mass

[88]*Financial Mail*, 21 August 1987.

[89]Phillip van Niekerk and Jean Leger wrote, in an article headed "Never Say Never" in the *Weekly Mail* on 21 August, that if Gush had "not refused to countenance a further increase in wages, the strike may well have been over by now".

dismissals if the strike was prolonged. Godsell, in support, explained that in South Africa the mine owners had "the legal right reasonably to dismiss strikers". He said that the "moment came when a business had to defend its financial interests for it could not allow workers to continue striking until it was sequestrated. But no employer wanted mass dismissals of workers in whom a lot of money had been invested."[90] The issue was raised again when the South African Manpower Minister, Pietie du Plessis, said, during an interview with the British Broadcasting Corporation on 11 August: "I doubt whether gold production will be disrupted for long because I think the mines will most probably take on other workers; you see, they have a surplus of workers in the mining industry."[91] When the matter was raised with Ramaphosa he thought that mass dismissals were unrealistic from the point of view of the mining houses. "How," he asked, "do you fire 300 000 workers and hope to replace them within the rest of 1987? It is not possible. Even if it were possible it would mean that the mines would remain totally out of production for over six months."[92]

On 14 August 1987, and thereafter on various occasions, lockouts by exclusion and by termination were attempted, with a mixed success.[93] The first mine to issue a threat of dismissal was Anglovaal's Loraine gold mine in the Orange Free State where, by the second day, all but 200 out of a labour force of 2 450 in No 3 shaft had refused to work. It was reported that the decision to dismiss the Loraine strikers had the full backing of the mining industry in general.[94] The threat, however, turned out to be unnecessary. The NUM was not recognised at Loraine so the strike there was illegal and, on the advice of the union, the strikers returned to work. Strikers at the Tavistock, Phoenix and Landau collieries were also given notice of dismissal on 13 August. The union did not have a recognition agreement with either the Tavistock or Phoenix collieries and the strikers there went back to work after being given one day's notice to return. Anglo-American, which owned the Landau colliery, said it intended to close the colliery by the following March but if the 1 500 strikers refused to return to work within four days it would enforce a lockout from then on. Anglo-American carried out its threat on 18 August. The strikers returned to work two days later.

[90] *The Star*, 14 August 1987.
[91] *Twenty-Four Hours* BBC World Service Programme, transcript, 11 August 1987.
[92] *The Scotsman*, 11 August 1987.
[93] Affidavit by Anglo-American during a court hearing in January 1988 to decide whether or not its policy of mass dismissals constituted an unfair labour practice.
[94] *The Citizen*, 13 August 1987.

Mine managers used other threats to persuade mineworkers to end their strike. At the President Brand gold mine the mineworkers were simply reminded that they could be dismissed. The Gencor UNISEL gold mine in the Orange Free State issued an ultimatum to return to work to the night shift on 13 August. Those who ignored it were brought before a mine disciplinary committee. Workers at the state-owned Nuclear Fuels Co-operation in Westonaria began a strike on Monday, 17 August and were immediately locked out. Ultimatums to return to work were given during the first week to strikers at Western Holdings No 1 shaft and at Vaal Reefs No 6 shaft. The following notice was distributed to mineworkers at Vaal Reefs:

VAAL REEFS EXPLORATION AND MINING COMPANY LTD

Incorporated in the Republic of South Africa

Registration Number 05/17354/06

POSSIBLE CLOSURE OF NO 6 SHAFT

Consideration is being given to the closure of the No 6 shaft which is a low-grade marginal operation as prolonged strike action is resulting in an accumulated loss which may make it uneconomic to continue operations. Closure of other shafts may also become necessary later.

Striking employees have been advised of the economic realities of the situation and the National Union of Mineworkers' Branch Committee has been told of the consequences of the strike action.

If the striking employees return to work by Tuesday, 18 August 1987, employment opportunities for approximately 2 000 people will be preserved.

Johannesburg

14 August 1987

AAC

The 2 000 striking mineworkers at No 6 shaft ignored the threat, refused to return to work and then chose to leave their hostels voluntarily, whereupon the manager imposed a lockout and then shut down the shaft. Before the mineworkers left, however, they were put under pressure to sign a document, dated 20 August, which stated that the undersigned person had elected to resign from the employment of Vaal Reefs and that his signature confirmed that he had unconditionally resigned. It added that the lockout would be lifted on the date cited on the document. The publicity in the morning press and the television news bulletins came from an Anglo-American statement that gave a different message, claiming that the strikers had resigned from their jobs and had been told by the NUM to return home. It blamed the NUM for the loss of jobs at No 6 shaft. The NUM, however, sharply contradicted this and declared that the strikers had been dismissed but had preferred to leave their hostel voluntarily to avoid the violence that normally accompanied evictions from hostels. The union emphasised that the mineworkers were still on strike and expected to return to their jobs when the dispute was settled. It added that it was prepared to discuss ways of avoiding the closure of the shaft and even to accept an early return to work. The mine, however, ignored this offer and went ahead with its own plans. The mineworkers went to the mine's stadium where they queued to receive their pay. Only 27 of them changed their minds and returned to work.

It was clear that the NUM did not have a clear strategy to deal with the threats of lockouts and mass dismissals. It had not anticipated that if its members returned home, the mines would be free to fill both their jobs and their places in the hostels with strike-breakers and, in the process, create tensions that would result in violence either immediately or in the aftermath of the strike. This lack of a coherent policy began to assume a greater significance towards the end of the second week when the use of mass dismissals as a strike-breaking tactic ceased to be a sporadic affair and began to be implemented as a matter of policy throughout the industry, thus challenging the authenticity of the legal right to strike. This raised a serious question in the minds of union activists about the usefulness of conforming to the Labour Relations Act. There seemed to be little point in accepting the restrictions that enabled them to strike legally if the employers retained the legal right to dismiss all the strikers because they were acting in breach of contract. The NUM was opposed to the dismissals but it could do nothing to prevent them. It insisted, therefore, that the process should be regulated to give as much protection as possible to the dismissed mineworkers. Ramaphosa wrote about this matter to the general managers of the mining houses on 20 August, stating:[95]

[95]Telex to Anglovaal – Gold and Coal Division (Attention General Managers) from the General Secretary, National Union of Mineworkers, 20 August 1987.

Re Lockouts and NUM's proposals re pay-outs

> We have received several ultimatums to lock out our members by dismissing them in order to induce them to accept the Chamber's offer. Since the mass dismissal of thousands of employees will mean a mass pay-out, a mass evacuation of the premises and mass transportation home, it is of the utmost importance to regulate the above to avoid a potentially explosive situation. We would like to propose an urgent meeting to discuss and hopefully agree upon a mutually acceptable procedure for the paying out of employees, their voluntary evacuation from the premises and their transportation home. We propose in all pay-outs of dismissed workers:
>
> 1. That the pay-out take place in the presence of union officials.
>
> 2. That the pay-out exercise will not be used for any other purpose.
>
> 3. That the employees be given an opportunity to collect their belongings and proceed to the transport provided for them.
>
> 4. That duplicate pay documentation be given to the union so as to allow the union to assist its members in checking their pay packets.
>
> 5. That mine security not be involved in the pay-out, the voluntary evacuation or the boarding of transport, and
>
> 6. That the union be given 48 hours' notice of any pay-out of employees.
>
> The Union is available to meet to finalise procedures to regulate pay-outs. While we believe that the procedures set out above are adequate and appropriate, any meeting should take place prior to any further mass dismissals taking place.
>
> In leaving the mine after dismissal and accepting monies due to them our members are not accepting that their dismissals are either lawful or fair. Our members are not abandoning their lawful and justified strike action and will be willing to resume their employment on the settlement of the dispute between your group or mine and the union or the group or mine acceding to the union's demand tabled to the Chamber.
>
> We await your response.

The union's letter was ignored by all of the mining houses.

The Anglo-American Corporation was the first mining house to apply mass dismissals as a general policy for all of its mines and collieries. When the matter was first raised at a directors' meeting there was strong opposition to it because of its possible international and internal political consequences. Gavin Relly, its chairperson, was concerned about the political fall-out.[96] Peter Gush, however, was dismissive of his

[96]Gavin Relly, the chairman of Anglo-American's board of directors, was in an ambiguous position with the result that he pleased no-one. He had long created a liberal public

argument. He told Relly that if he wanted to play politics he should join a political party. His job, he said, was to produce gold and nothing else. He brushed aside those who disagreed with him with the quip that they knew nothing about the reality of mining. He said that he had direct responsibility for the management of the gold mines and insisted that he was the only director who knew how to control black mineworkers.[97] Gush undoubtedly reflected the attitudes of the mine managers and his view prevailed. Deadlines for a return to work or be dismissed were drawn up and introduced on a mine-by-mine basis. The Gencor and JCI mining houses followed Gush's example on 20 August by setting their deadlines for a return to work. By that time as many as 40 000 mineworkers were under the threat of dismissal.[98]

The number of strikers who had actually lost their jobs rose rapidly as the deadlines were reached. On 24 August, 10 500 strikers had been dismissed but, within three days, this number had risen to 30 000. Strikers at most of the main centres of resistance, namely the gold mines at Vaal Reefs, Western Deep Levels, Western Holdings, Randfontein Estates and Saaiplaas and the Matla, New Denmark and Springfield collieries, had deadlines imposed on them. The mine owners denied that they were issuing ultimatums. "As is the accepted practice," Gencor stated, "workers have been informed that they were absent from work without permission and that failure to return to work today will result in disciplinary hearings."[99] Gencor, however, acted in an uncharacteristically conciliatory way by extending its deadlines, deferring

image for himself by taking a stance against apartheid. In September 1985 he had broken the impasse between business organisations in South Africa and the ANC in exile by leading a delegation of business representatives in discussions with the ANC in Lusaka. But he was also the head of a mining house in a strike situation. Some people expected him to apply his liberal views to the strike and be conciliatory towards the NUM. The United Democratic Front, for instance, had written on 14 August to ask him to settle the strike by meeting the demands of the NUM. It also called for an immediate halt to all of the harassment of striking mineworkers and to attempts to force them back to work. It went on to state: "In the past Mr Relly has claimed to be concerned about a democratic future for all in this country, black and white, and here we refer particularly to his meeting with the African National Congress. However, the reliance of Anglo-American at this time on the help of the South African Police and government in harassing and arresting mineworkers and shaft stewards is wholly against this. We say to Mr Relly, you cannot run with the hare and hunt with the hounds..." (Correspondence to Anglo-American Corporation of South Africa from the UDF, 14 August 1987.)

[97]Conversation with James Motlatsi, October 1998. Motlatsi was told about the meeting shortly after it had taken place.

[98]*Business Day*, 21 August 1987. Bettina von Lieres, March 1988, pp58-60, gives slightly lower estimates.

[99]*The Star*, 20 August 1987.

the hearings and agreeing, the next day, to meet Ramaphosa to discuss the issues arising from the ultimatums. It also decided to delay any disciplinary action where its terms were not met. This marked the beginning of a dramatic change in the attitude of Gencor towards the NUM from competing with Gold Fields in its hostility towards the union to becoming the most conciliatory of the mining houses.

As the number of strikers under the threat of dismissal rose, so did the intensity of the violence. This was illustrated at a number of gold mines such as Bracken, President Steyn, Western Areas, Kinross and Western Deep Levels. At JCI's Randfontein Estates Cooke No 3 shaft, 4 000 were locked out on 20 August and were told that they could only be re-employed if they accepted the wages and conditions imposed by the Chamber of Mines on 1 July. Ramaphosa claimed that management had gone from room to room to put pressure on the strikers, telling them that unless they returned to work they would be fired and evicted from the hostels. At the same time, roadblocks and barricades were erected around the mine to stop strikers from entering or leaving the mine premises. When the strikers broke their way through the barricades the mine manager called the police, who patrolled the streets in armoured vehicles and cordoned the hostels off with razor wire. The NUM went to court over this issue but received an out-of-court assurance from management that the strikers would no longer be restricted in their movements. The union made further claims that on 25 August at least 200 Mozambicans at Randfontain Estates Cooke No 1 shaft were forced underground at gunpoint and that tear gas was used to compel the strikers at Cooke No 2 shaft to go to work underground. When they returned to the surface they were refused food. The management tried to convince the strikers that Ramaphosa had "sold them out" and that they would not get any of the money due to them or be provided with transport to take them home if they chose to leave the mine rather than work.

The problems confronting the strikers during the third week were most clearly illustrated at Western Deep Levels. The manager of its No 3 shaft decided on 26 August to wrest control of the hostel from the strikers and resume production. Before taking any action, however, he broadcast an ultimatum to the strikers in Fanagalo, Zulu, Sotho and Tswana that stated:

> All Number 3 Shaft striking workers irrespective of the shift are therefore informed once again to start work by 8 am tomorrow morning, Thursday, 27 August 1987 or face discharge. Remember, it is still your choice whether you want to go to work or not. Cages will be available up to 8 am. For those of you who choose not to work, pay facilities are available. Take all your belongings at No 3 hostel where you will be paid off and bussed home.

At the same time he sent white miners down the shaft to wait for the return of the strikers. Until that day the strikers had been in full control of the hostel. Early in the morning, however, the mine security force, fully armed in Casspir vehicles, entered the compound after sealing it off with cordons of razor wire and cutting the telephone wires from the union office. The manager left nothing to chance. He had, for instance, called for assistance from Gold Fields in the hope that the presence of its security, with its awesome reputation, would frighten the strikers into submission. The security forces patrolled the compound and, at about 3 pm, they took control of the change rooms and the kitchens. The strikers, however, were still in a defiant mood and they paraded around the compound carrying a NUM flag. A mineworker who was present described the scene: "At about 6 pm people were singing as usual. Hippos came up from behind and in front of the group. I tried to take the banner to defuse the situation but the workers refused to give it to me. At 6.30 pm the shooting started. It lasted 15 minutes. I found one worker who was hit in the face. People were shot in the hostels. I was in room 13 with injured workers. As I wanted to take them to the medical station the mine security came towards me and forced me back into the room."[100] When the mine security moved into the living quarters with their arms, a number of the occupants were injured but no-one was killed.

The security force then escorted more than 3 000 strikers, belonging to both the night and day shifts, to the crush where they were all clocked in before entering the cages to go down the mine.[101] The strikers claimed that they were forced to go underground at gunpoint, regardless as to how they were dressed. Some of them went down with bare feet. Others went down even though they were injured. Though the guns were not necessarily pointed at them, they believed that they would be shot if they refused to go to work. White officials in the mine, however, insisted that the intention of the security force was simply to provide physical protection for those who wished to go to work. When the mineworkers reached levels 66 and 100 where they were expected to work, they began a sit-in protest. Some of them claimed they were protesting at being forced to go down the shaft but those on

[100]The mineworker had been interviewed by Bettina von Lieres in December 1987. (Von Lieres, March 1988, p70.)

[101]Anglo-American agreed that it was unusual for both shifts to go down the mine at the same time but said this was because the mineworkers had been warned that they would all have to report for work by 8 am the next morning or be dismissed. (*Weekly Mail*, 29 April 1988.)

the day shift stated that they were also objecting to having to sit around throughout the night until their shift was due to start. The management accepted neither of these explanations and contended that the return to work had been staged in order to disrupt the mine's operations.

The protesters on level 66, where 2 000 usually worked, fashioned weapons out of metal objects, vandalised mine property and threatened to kill the white miners, who then ran away and hid themselves. The situation on level 100 was entirely different. The safety of the white miners was not threatened and, when a team of white officials eventually reached level 100, they found that the sit-in strikers were quiet and co-operative. The strikers on both levels, however, refused to move until near the end of the day shift when they all voluntarily made their way to the cages and returned to the surface. There they were met by security officials who escorted them, first to their rooms to collect their belongings and then to the liquor outlet where they were formally dismissed and given their pay.[102] Buses were waiting to take them to Carletonville where they boarded other buses or taxis to make the long journey home to Lesotho, the Transkei and Ciskei. In the meantime, the mine manager had announced that if the remainder of the strikers had not returned to work by 12 noon on 27 August they, too, would be dismissed and would have to vacate their hostel premises. He was again assisted by the Gold Fields mine security unit which patrolled the compound to help them make their minds up. A mine captain from Gold Fields, in a hippo-armoured vehicle, explained over the public address system why he was there. He said: "Maybe you are used to mine security at Western Deep Levels. I am not from Western Deep Levels. I am from Gold Fields. If we have not done by tonight what we did yesterday (*that is, shot at strikers*), I will call myself a coward."[103] The shaft, he added, was full of ANC.[104] More than 10 000

[102]Initially, the NUM had claimed that all of the 6 000 mineworkers, who went underground, after the management had regained control of the hostel, had been forced to do so at gunpoint. Von Lieres, 1988, p60, however, put this number at 3 000. But when the counsel for the NUM gave evidence to the arbitrator the following year, in its claim that the dismissals were unfair, he did not say that they had actually been forced underground at gunpoint but that some of them had the impression that such was the case. For differing accounts of this incident, see Von Lieres, 1988, p60; the South African Institute of Race Relations, 1987/88, p683 and the *Arbitrator's Award* in the arbitration between the NUM and the Western Deep Levels Mine, July 1988.

[103]Von Lieres, 1988, p71. The italics are inserted by the author.

[104]The use of the term 'ANC' suggested that there were terrorists amongst them. Some mine managers used the term 'Mandela's children' to imply they were terrorists. It was said that some of Gold Fields' executives laughed at what they described as Anglo-American's naïve pluralist approach to mine discipline. (*Weekly Mail*, 21 August 1987.)

mineworkers had been dismissed at Western Deep Levels by the end of the day.

Strike-breaking

In general, the strikers in all of the mines found that the most difficult period to sustain solidarity was in the early stages. They had never experienced a long strike before and had no idea what effect a period of incarceration in the hostels, without pay and perhaps without sustenance, would have on them. But once the initial apprehension was over and they knew that they could survive it became easier for the union to retain their allegiance. Thus, to the surprise of the mine owners, the strike moved into its second week with no overall sign of weakening. While some strikers returned to work on the advice of the union the strike numbers were maintained by extensions of the strike at other mines. In the main centres of resistance the solidarity of the mineworkers remained so strong that many of them preferred to return home rather than break the strike.

Yet in such difficult and adverse circumstances, unity was a fragile thing that could be punctured by bribery and threats, as was illustrated by the situation at Western Holdings after an ultimatum to return to work had been given to the 4 000 strikers at its No 1 shaft. The ultimatum was due to expire on 21 August but they, like the Vaal Reefs strikers, decided they would rather leave the hostels voluntarily than break the strike. Before the deadline was reached, the strike committee for the whole of the mine organised mass meetings in shafts 2, 3, 5 and 6 to discuss the ultimatum to No 1 shaft. Their common response was that the entire black workforce of 22 000 should leave the mine, shaft by shaft, in solidarity with those threatened with dismissal. The management retaliated by locking out all the strikers at No 2 and No 3 shafts and by telling them that unless they returned to work by Monday, 24 August, they too would be dismissed.

In the midst of this mixture of threats and counter-threats the manager persuaded a senior union branch official at Western Holdings to convince the branch committee, that covered all the shafts, to recommend that mineworkers at the No 2 and No 3 shafts should return to work. He did this without consulting the union's regional committee. He then summoned a meeting of the strikers from the two shafts and told them that if they ended their strike then, after the first shift, they would be paid the wages that had been due to them when the strike had started. Most of them had wages to draw and, as they were all desperately short of money, in a dramatic turn-around, they agreed to end the strike on the management's terms.

The following day, Jan Selepe, the chairperson of the Orange Free State regional strike committee, met Frans Baleni who chaired the shaft stewards' committee in No 2 shaft and was a member of the strike committee.[105] Baleni, however, had not been at the meeting when the decision to end the strike had been taken. Selepe asked him to convene another meeting of all mineworkers at Nos 2 and 3 shafts, but this time to urge them to rejoin the strike. Baleni was apprehensive because he thought he would meet opposition that could turn to violence. But, after discussing the matter with his friends, he convened the meeting. To his surprise, when he addressed the meeting it warmed to him and when he asked them to rejoin the strike he received a rapturous applause. The mineworkers in both shafts had collected their pay and were in the mood to accept his advice. In a display of solidarity with the other shafts they carried Baleni out of the meeting on their shoulders and returned to their hostels. After that, the strike continued to the end at Western Holdings, though the strikers did not implement their resolution to leave the mine in sympathy with the dismissed strikers from No 1 shaft. Baleni, however, became a fugitive from the security force in the mine. He disguised himself and slept in a different place each night until the end of the strike when, with thousands of others, he was formally dismissed.

White workers provided the most constant and reliable supply of strike-breakers and the mine managers, from the outset of the strike, set about using them wherever possible to perform the tasks normally done by black mineworkers. It was claimed that detailed arrangements to use white miners in this way had been worked out well before the strike so that when it began each one of them knew exactly where to go and what to do.[106] The white Mine Workers' Union had advised its members not to do the work of striking black mineworkers during a legal strike but the advice was largely ignored, primarily because the mining houses offered them such generous bonuses to break the

[105]Jan Selepe was the chairperson of the OFS region of the NUM at that time. In 1991 he was defeated when standing for re-election. Following his defeat, he was retrenched at his mine and given a job as a roving trouble-shooter for the union. He moved from that post to become a union education officer in the Secunda region. He died in a car crash in 1993. Frans Baleni became a full-time union organiser in the OFS in 1988. He moved to Johannesburg to be an area organiser in 1993. The following year he was appointed as a national organiser until 1996 when he took the post as a regional co-ordinator in Carletonville.

[106]*The Star*, 11 August 1987.

Scab labour on a shuttle bus to Randfontein Estates.

strike.[107] Those who worked in Anglo-American collieries received an extra R80 a shift while at Gencor's they were given an additional R10 for each hour underground. This meant that the bonus given to a white miner for three days' work during the strike was more than the average wage of R235 that a black mineworker earned in a month.

Early on it was reported that large numbers of unemployed white workers were being flown to Johannesburg to help keep the gold and coal mines in production. However, they had a limited usefulness. They were not able to work where the strike had stopped production altogether so they were concentrated where it was only partially paralysed.[108] Even then, they were able to contribute significantly only where there was relatively small-scale mechanised production, as for example, in opencast coal mines with large drag lines, and in the underground coal mines where coal-cutting machines and shuttle cars were used. An official of the white Mine Workers' Union said, when the strike began, that it would be impossible for white miners to keep up full production in the affected collieries but that they could possibly prevent it falling below 50 per cent.[109] No-one, however, provided any production figures to support this claim.

The position of white miners in the large labour-intensive gold mines was entirely different. In some instances, black team leaders and black personnel assistants who had previously worked as team leaders, who were not supporting the strike, were employed underground to supplement the untrained white strike-breakers but there were never sufficient numbers in total to take the place of the thousands of striking drillers and lashers. They mostly had to do unskilled manual work and that presented particular difficulties for them, as a striker at Vaal Reefs explained. "There were lessons that non-strikers learnt," he said. "The team leaders especially. They went underground to work. When they got there they were forced to '*layisha*' (load). Some of them refused. The white miners then beat them up. Those who worked were forced to work hard. They have learnt, the hard way, that nothing pleases the bosses. If you become their sweet boy, the more they make you work hard for peanuts. One team leader called Tola had his jaws crushed. He refused to '*layisha*', saying he was a team leader. A white miner told him that ...all those who were underground were there to work. Many

[107]The Mine Workers' Union, nonetheless, sympathised with those who worked during the strike because it feared that if the mines were closed by the strike, white miners would lose their homes as well as their jobs.

[108]South African Institute of Race Relations, 1987/88, p683.

[109]*The Star*, 11 August 1987.

workers were injured in this way but the management told them to say they had been beaten up by strikers. Those who were forced to work hard later came to us and begged forgiveness."[110] In the main, in the large gold mines, the white miners were able to do little more than take responsibility for the care and maintenance of underground workings or become volunteer members of the Emergency Protection Units to guard roadblocks and to provide back-up support for the mine security forces in general. The NUM estimated that between 78 and 85 per cent of white mine workers were active in the mine security forces during the strike.

The only way the strike-bound underground mines could maintain any level of production was by recruiting large numbers of black replacement workers through TEBA. The mine managers at Winkelhaak and Stilfontein actually wrote to Ramaphosa saying that they intended to engage "temporary employees" until the situation returned to normal. Other mines simply went ahead and engaged them. But they faced difficulties and opposition. The difficulties were largely logistic ones concerning, for instance, the provision of protected housing accommodation. Where the strikers continued to occupy the hostels there were few places where scabs could be accommodated. The production manager at Western Deep Levels explained the problem when applying to the Supreme Court for the legal right to eject dismissed strikers who were sitting-in underground. He said that it was essential to be able to eject the strikers so that the mine could "employ a fresh work force or move other workers willing to work on to the premises [and] be able to house its employees properly".[111] This problem was illustrated at the Vaal Reefs No 8 shaft in the early days of the strike. On the first day, the mine security told the strikers that if they did not return they would be dismissed, whereupon 500 went back to work. They were not allowed to go back to their rooms in the hostel, however, because the strike committee controlled them. They were housed, instead, in a special camp that was cordoned off from the hostel. The mine manager, in describing their living conditions, said: "The men have no shelter and are exposed to the extremes of the elements. There are only a few toilets and showers available to them and sanitation is extremely poor ... Having lived under these conditions for five nights, these men are understandably tense and tired. The situation is volatile, their work performance has deteriorated and there is a very

[110] *Inqaba*, October 1987.
[111] Von Lieres, 1988, p68.

real risk that … their fear and frustration could lead to further violence."[112] Some of these mineworkers were so dissatisfied that they rejoined the strike. In a similar situation, when a batch of recruits arrived from Natal to break the strike at the Wolwekrans and Vandyksdrift collieries they had to be accommodated in the colliery reception offices.

The NUM itself had neither the resources nor the influence to prevent TEBA from recruiting over the many areas it covered in southern Africa. In many cases potential recruits knew nothing about the strike and even where they did, then TEBA was able to bribe them to work in the mines. Nonetheless, it was not so easy for it to operate in those areas of South Africa, such as the eastern Cape, which were directly associated with the liberation struggle, for the ANC, the UDF and COSATU openly supported the strike. The ANC gave its support in a series of broadcasts from Radio Freedom in Addis Ababa. At the beginning, the ANC urged unemployed workers "not to stab strikers in the back". This was endorsed by the National Unemployed Workers' Co-ordinating Committee's appeal to the jobless not to "scab for the bosses". Towards the end of the strike the ANC returned to the theme and called for the formation of mobile units to guard against "scab labourers".[113] This meant that TEBA had to focus its efforts in those recruitment areas where the influence of the liberation movement was contested, as in KwaZulu, or where it was not present, as in Lesotho.

With the help of UWUSA, recruits from KwaZulu began to turn up in increasing numbers after the second week. Strike-breaking had become UWUSA's principal means of gaining access to mines so the strike provided it with an ideal opportunity to widen its scope.[114] It was reported that 400 Zulus were recruited to work at Anglo-American's Goedehoop colliery and that a further 200 were put to work at Gencor's Matla colliery. Some were also present at Wolwekrans and Vandyksdrift collieries belonging to Rand Mines. They were also said to be working at the President Brand gold mine in the Orange Free State, where the NUM branch chairperson claimed he had "overwhelming evidence … that management was secretly recruiting Inkatha-linked UWUSA members to replace strikers" and, at the same time, to undermine the strike by fomenting violence.[115]

The main source for black scab labour, however, was Lesotho,

[112] *Ibid*, p73.
[113] South African Institute of Race Relations, 1987/88, p683.
[114] *City Press*, 23 August 1987.
[115] *Ibid*.

which provided about 106 000 recruits for the gold mines in 1987, amounting to 22 per cent of their total black labour force. The large numbers of Sotho mineworkers who supported the strike tried to tackle the problem caused by Lesotho strike-breakers by forming strike committees in Lesotho comprised of returned, dismissed strikers. Their intention was both to look after the immediate interests of the dismissed strikers and their families and to try to prevent the recruitment of scabs. The NUM sent a delegation to Lesotho to supply the committees with information about the course of the strike and to assist them to stop the recruitment of fresh labour. The committees had widespread support in Lesotho for there was much public concern there for the strikers and an anxiety about the consequences of strike-breaking. The main body of support came from the churches which responded quickly and issued the following statement in Sesotho to be read by and to Lesotho residents:

STATEMENT BY THE HEADS OF CHURCHES OF LESOTHO

The Heads of Churches in Lesotho are concerned about the situation created by the present legal strike of mineworkers in South Africa, particularly about the hasty and irregular recruitment of Basotho to undermine the just efforts of the mineworkers. To this end we have issued the following statement:

At present men who work in the South African mines are engaging in a legal strike to obtain just wages and working conditions. Unfortunately the Chamber of Mines is recruiting men as strike-breakers to undermine the efforts of the union who represents the interests of the mineworkers. We advise our fellow Basotho:

1. Do not let yourselves be used in this way by mine owners and the recruitment agencies.
2. Do not let yourselves be used against your fellow Basotho who are working in the mines.
3. Be aware that the recruitment agencies are putting you in serious danger by recruiting you for the mines in this way.
4. That the short contracts being offered mean that you may not have adequate training or legal protection and that you may be dismissed at any time without benefits.[116]

But the pressures to work in the mines were too great to be resisted for most young Lesotho men. It was estimated that the rate of

[116]The statement was agreed with representatives from the Christian Council of Lesotho, the Christian Council's Migrant Labour Commission, the Catholic Church's Migrant Labour Project, the Transformation Resource Centre and Basotho representatives of the NUM. (See *Report by the Transformation Resource Centre*, Maseru, Lesotho, nd, but published shortly after the strike.)

unemployment there was about 50 per cent and that many of those who had employment were under-employed, insecure and earned much lower wages than were paid in the mines. At least half of the households throughout Lesotho needed the remittances from migrant workers simply to protect them against abject poverty. In those circumstances, families could ill afford to have scruples about strike-breaking. The government, too, was eager to increase the flow of recruits because more than one-half of Lesotho's gross domestic product came from the remittances of migrant workers while, Lesotho's Central Bank added, the foreign exchange value of the wages brought home by migrant workers paid for 77 per cent of the country's imports.

After an announcement on Radio Lesotho that the mines were hiring men to replace the strikers, thousands of them turned up at the TEBA recruitment offices in Maseru. Most knew nothing about trade unions and strikes and had no understanding of the implications of damaging the solidarity of the mineworkers. They were easy victims for unscrupulous recruitment agencies. The gangster-type Sotho organisation described as the 'Russians', which had acted the year before as a sort of mercenary vigilante group on behalf of mine managers, turned up on this occasion as a supplier of Sotho strike-breakers. And though the mine managers denied any contact with it, there were signs of its involvement. For instance, a rumour spread at Western Deep Levels, just before the mine security and police attacked the chanting strikers on Thursday, 13 August, that "the 'Russians' – Sotho scab labour – had arrived".[117] Nothing could have been more provocative.

A number of Sotho strikers from Vaal Reefs and Western Holdings were amongst those who either left their hostels voluntarily or were dismissed after being told that their shafts were being closed down. But, at the same time as they were receiving their dismissal notices, the TEBA office in Maseru was collecting recruits to replace them. In fact, TEBA had already sent 2 000 recruits there during the second week of the strike.[118] Twenty-six buses filled with new recruits were seen leaving Lesotho for Klerksdorp on 24 August. Two days later a further 600 recruits left Lesotho for President Steyn and Western Holdings No 1 shaft. Dismissed mineworkers who were returning home through Maseru were dismayed when they passed the TEBA recruitment offices and saw queues of young men eager to take their places. "I am

[117]Von Lieres, 1988, p69.

[118]*The Guardian*, 26 August 1987.

very hurt," one of them said, "that management is recruiting my brothers and my friends to go back to the same mine to take our jobs."[119] The Christian Council of Lesotho's Migrant Labour Commission did its best to alleviate their problems. It arranged for assistance to be given to the returned mineworkers and their families. It met strikers at the border post late into the night in order to greet them after their 10-hour trip by bus from Welkom and to prevent them from being exploited by taxi drivers in the final stages of their journey. It provided emergency housing for those who required overnight accommodation during their journeys home and made contingency plans to assist the families of the dismissed strikers.

In general, it was difficult to estimate either the scale of the use of scab labour or its impact on the outcome of the strike. Although there were notorious instances where strike-breakers were used by Anglo-American, it would not disclose how many it had taken on or where they had been employed.[120] At the end of the strike it was estimated that 3 000 were still in employment but that was a gross underestimate of the number actually used during the course of the strike.[121] The only advantage to the mine owners from using them was the help they gave in maintaining production levels. One rough estimate was that where it had been possible to use scab labour, then production levels had been maintained "at tolerable levels" but what was 'tolerable' in those circumstances was highly subjective.[122] The costs of using them were less problematic to measure and were borne by the strikers, their families and the union. They involved violence, injuries, deaths, arrests and a weakening of the union's position. They were high and figured in NUM's final assessment of the strike, but they alone did not determine its outcome.

The Resumption of Negotiations

As the strike entered its third unexpected week the minds of both the union leaders and the mine owners were focusing on specific but

[119]When asked about this the Anglo-American Corporation said that the recruits were being sent to those mines simply in order to be relocated to other mines. The employment contracts that were signed by the recruits, however, specifically stated that they would be employed at Vaal Reefs and Western Holdings. (*Ibid.*)
[120]*Finance Week*, 28 August 1987.
[121]The figure of 3 000 was used by COSATU in its press statement when the strike ended to emphasise the difficulty facing the NUM in getting its dismissed members reinstated.
[122]*Business Day*, 28 August 1987.

different facets of the course it was taking. The omens for the union were worrying. The strikers' families in the rural areas were suffering because they had not received any wage remittances. The strikers themselves had also suffered. The NUM estimated that by 23 August six mineworkers had been killed, 320 had been injured, 300 had been arrested and there was no sign of a respite. There was a rapidly rising rate of dismissals and an increasing number of lockouts. Anglo-American claimed that large numbers of strikers were returning to work, though it did not specify how many or to where they were returning. The NUM reckoned that only a few hundred were involved. In some mines, the NUM declared, mine managers were trying to organise defections through the use of informers within the union. It felt impelled to warn strikers at the Freddies and Free State Geduld gold mines in the Orange Free State that some of their branch officials were acting in that way. The following message in English, Sotho and Xhosa was distributed at the Freddies hostel on 22 August:

THE STRIKE IS STILL ON

> Leaders of NUM advise workers under NUM that there are some leaders against the lawful strike. These leaders are at 5 Hostel Freddies. They are intimidating the workers and they have called an illegal meeting at the arena to mislead the people to go underground.
>
> The chairman is appealing and advising the people not to be misled by such elements. If workers go to work they will not be regarded as members of the union any more. This will be the last time NUM will recognise them. We are in the third week of the strike and it should continue. If workers are not satisfied, they are advised to go to NUM offices at any time.
>
> The branch committee and regional committee advise all the workers that you should all stay on strike in order to succeed like the OK and Railways – they waited five to six weeks and they succeeded.

A slightly different version but with the same message went to the Free State Geduld mine. There, mineworkers were told that management had bribed "a few elements … to cause misunderstanding amongst workers". It stated that they "should forget about NUM if they decide to go to work before they receive a directive from NUM". A similar accusation was made later in the week to account for events at the Free State Saaiplaas gold mine where the mineworkers ended their strike during the third week. It suggested that they had succumbed to "fabrications and threats by Boers" and pleaded with them to return to the strike. The message added that the NUM branch at Saaiplaas was in a state of collapse "due to the wishes of the workers' enemies". These signs of a shift in support for the strike were not sufficient to indicate that it was collapsing but they worried Ramaphosa and Motlatsi. Then there was also the failure to engage the Gold Fields workers in the strike. From their point of view, the morality of the mineworkers' case remained unassailable but they knew that morality did not win power struggles. The issue facing Ramaphosa and Motlatsi was not any more the need to reappraise their tactics in order to win, but how to end the strike with some credit for the union and some rewards for the selfless sacrifices made by the strikers.

The mine owners, too, were facing new questions. It was no longer a matter of forecasting costs but of calculating those that had occurred. The loss in pre-tax working profits suffered by the Anglo-American Corporation, JCI and Gencor had risen to about R190 million which was far in excess of the R150 million it would have cost them to meet the union's wage demand in full for a year. Even where production had continued, it was falling as managers chose to cut down the number of shifts from three to two rather than face additional violence from strikers. There were, moreover, real risks from the use of inexperienced, unskilled labour and actual, not just anticipated, damage to stopes in the deep mines as rock faces collapsed due to insufficient care and attention. Godsell, speaking for Anglo-American, emphasised this point when he stated that "technical reports were coming in from our deep-level stopes that you either had to get people back into those stopes, and it was almost proto-gangs which had to go in to reopen these stopes, or we had to just write off large parts of the mines – with a consequent permanent loss of jobs".[123] The mine owners, however, were in a quandary. After two weeks of violent struggle they could hardly say they now wanted to decide the issue on the merits of the argument because the accumulating costs of continuing it were too great, without suffering a public humiliation. They continued, therefore, with their twin-track policy of mass dismissals and the use of physical force. Its logic, moreover, suggested that it had to be speeded up.

Both sides, however, were under growing pressure to show willingness to settle and this meant resuming negotiations. The NUM had let it be known that it was willing to reopen the negotiations "provided no pre-conditions are set". When this was reported in the press the Chamber of Mines was asked for its reaction. It said that it was willing to negotiate over fringe benefits but not wages because "it is the

[123]*South African Labour Bulletin*, November 1987, p67.

Chamber's view that when the opportunity was there to negotiate wage increases, the NUM did not negotiate in good faith and made a mockery of the negotiation process … Throughout these negotiations the Chamber's negotiating team had the strong impression that your union negotiators were really not interested in arriving at a mutually acceptable settlement, but merely going through the motions to reach a formal deadlock to place your union in a position to threaten, and eventually stage, a lawful strike."[124] It was, of course, in some difficulty over agreeing to resume negotiations over wages because it had said repeatedly that its offer was final. The Anglo-American Corporation, however, had a different viewpoint and, on Sunday, 24 August, offered to start unconditional negotiations to end the 15-day strike. But this, as NUM pointed out, was a political gesture and not a real offer because it was not in a position to negotiate a settlement independently of the Chamber of Mines, which was the collective bargaining agent for all of the mining houses, including Anglo-American.

Despite the restrictive preconditions imposed by the Chamber of Mines, Ramaphosa telexed it, expressing NUM's willingness to reopen negotiations. When the executive committee of the Chamber of Mines met on the morning of Tuesday, 25 August it agreed to meet a NUM delegation early that afternoon in its offices. The negotiations lasted for four hours. Ramaphosa wanted only to discuss the question of wages but the Chamber of Mines' representatives refused. He offered to reduce the union's wage demand from 30 per cent to 27 per cent but that was not sufficient to alter their minds. They were only prepared to improve two fringe benefits. They offered to increase by 10 per cent the pay mineworkers received on their annual holiday though they rejected the NUM demand for an increase in the length of the annual holiday. They also offered to double the death benefit paid to dependants from two to four times an annual salary. The NUM had wanted to increase it by five times. The demands that the anniversary of the Soweto uprising on 16 June should be a paid holiday and that black mineworkers should receive danger money were rejected. Ramaphosa then asked the Chamber of Mines' representatives whether they would convert the cost of the improvements in fringe benefits, estimated to be equivalent to a 1,7 per cent wage increase, into a pay rise, but they refused. His suggestion that arbitration or mediation should be used to break the deadlock was also turned down. The union was given three hours, until 9 pm that evening, to accept or reject

[124]*Financial Mail*, 21 August 1987.

Strikers from Carletonville demonstrating outside the offices of the Chamber of Mines where negotiations were being resumed during the second week of the strike.

the offer. If it were rejected then, the Chamber of Mines declared, it would be withdrawn.

During the negotiations, up to 1 000 mineworkers from the mines nearest to Johannesburg gathered in the rain around the fountain outside the head office of the Chamber of Mines. They carried posters saying "Bullets won't stop us" and "Away with the Chamber of Mines". They danced and sang freedom songs and chanted: "We are not going to work for no money. We want 30 per cent." But very quickly the riot police arrived and gave the demonstrators 25 minutes to disperse. Many remained behind, however, to greet and carry Ramaphosa and Motlatsi on their shoulders when they left the building. At first, Ramaphosa said "progress has been made", but at a press conference later he said that the strike would continue though it could be over by that night. He said that mass meetings of mineworkers in all of the mines and collieries on strike would consider the revised offer. He refused to express his own opinion about the offer to the press but in his hand-written statement to the members' report-back meetings he wrote that "the NUM negotiating team informed the Chamber that this offer would not be acceptable to workers" but that it would nonetheless be submitted to them.

When the union's national executive committee met to discuss the offer they decided not to make any recommendation to the members. It was reported in the press that this was because it was divided over the issue.[125] "There was speculation in industry circles last night," one report stated, "that moderates in the NUM – Mr Ramaphosa among them – had been prepared to accept the benefit offer on the grounds that the strike would have been resolved with both sides' honour intact. But the hawks in the NUM had opted for a struggle to the bitter end."[126] Motlatsi denied that this was the case. During the strike Ramaphosa had made the major union decisions with Motlatsi's endorsement and the NEC's agreement. On this occasion, however, Motlatsi's view carried the most weight for it was his task to explain and justify to the strikers any decisions taken by the leadership. It was his belief that the members had enough information to reach their own conclusions and that the NEC should not try to influence the outcome of the ballot in one way or another.

Voting took place on 26 August at report-back meetings in the

[125]The reports were based on a statement to the press by Kuben Pillay, of NUM's Legal Department, that "some of us believe it would be strategic to accept the offer and others feel we must continue with the strike".
[126]*The Independent*, 27 August 1987.

mines on strike. Only at Saaiplaas, where they had returned to work on that day after receiving an ultimatum, were they refused permission to hold report-back meetings and so could not express a view about the revised offer. Announcing the result, Ramaphosa said: "Our entire membership in the striking mines has decided not to accept the Chamber's offer. The strike continues until our demands are met." The vote was so decisive that it stimulated two shafts at Anglo-American mines and most of the mineworkers at Gencor's Bracken mine to return to the strike. Representatives of the Chamber of Mines heard the result with dismay. Johann Liebenberg said: "Our managements will simply have to do whatever is necessary to bring it to an end as quickly as possible. We think this is a sad miscalculation on the part of the union." He talked of "bad-faith bargaining" by the union and criticised its leadership for the way in which it had conducted the ballot, alleging that the shaft stewards had not explained the terms of the revised offer to the strikers but had simply instructed them to vote "no". The Anglo-American Corporation was more specific about its intentions. Its spokesperson stated that it would proceed with closing shafts and taking 'appropriate action' to resume normal production at others. "We will now concentrate all our efforts to returning our mines to full production," he said.[127]

The Final Stage

Though the composition of the strikers had changed at the margin by the time the ballot was taken there had been little alteration in the estimates of the numbers involved in it. The NUM's figure remained at 340 000 in 45 mines and collieries while that of the Chamber of Mines had fallen slightly to 210 000 at 29 mines and collieries. Whichever figure was taken, however, the strike remained large and remarkably durable. For instance, on 27 August, despite the violence and threats of dismissals at Western Deep Levels, the mine's entire NUM membership, representing at least 63 per cent of the total black labour force there, had been consistently on strike since 9 August.[128] The mine owners' policy of attrition had succeeded in compelling strikers to return to work in a few shafts but in general it did not work out as they had expected. Yet despite this, and irrespective of their decision to resume

[127]*The New York Times*, 27 August 1987.

[128]Affidavit by the production manager. See Von Lieres, 1988, p65.

negotiations, the mine owners continued to use force and mass dismissals to defeat the union. The day during which the negotiations were taking place was a particularly painful one for the strikers.

The Anglo-American Corporation had extended its deadlines to expire after the resumed negotiations had been completed but then set further ones whilst they were taking place. Minutes before the negotiators met, it set deadlines for a further 12 000 strikers to return to work or be dismissed. These were not indiscriminate dismissals for it was using mass dismissals selectively in order to regain control of the hostels in those shafts where support for the strike was greatest. It had done this successfully in shafts at Vaal Reefs and Western Holdings. It had further successes at Western Holding on the day of the ballot. It tried to do the same at Western Deep Levels No 3 shaft when it imposed a deadline to return to work that came into effect on the day of the ballot but, when the deadline arrived, the strikers refused to leave their hostel.

When Ramaphosa spoke at the press conference called to hear the result of the ballot, he drew attention to the violence that had taken place during the balloting process. He said that at Western Deep Levels, "injured mineworkers have been admitted to hospital and we have asked hospital staff and nurses to be on stand-by all night". When the press asked for more information from an Anglo-American spokesperson he said he knew nothing about it, though the next day he admitted there had been violence but refrained from giving details.[129] There were similar events at a number of other mines and collieries on that same day. At the Vaal Reefs shafts, there was a mass dismissal of strikers and the following day, 27 August, the management constructed a compound with razor wire fences and four gates, at the mine's stadium. The strikers were evicted from their hostels early on 29 August and those who were identified as NUM members were dispatched to the razor wire enclosure where they were compelled to queue to sign forms stating they were voluntarily terminating their employment in the mining industry. Armed guards patrolled the queues to prevent any conversation between the mineworkers and to ensure that shaft stewards were unable to listen to complaints from them. About 18 000 strikers were dismissed from Vaal Reefs on that day.

All of the strikers in No 7 shaft at Anglo-American's Freddies gold mine were paid off and transported home in the one day. At precisely the same time the kitchens in the No 5 shaft were closed and the

[129]*The New York Times*, 27 August 1987.

strikers were refused any food, except bread and milk. They were also refused their pay. The mine security at its Elandsrand gold mine in Carletonville went on the rampage on 27 August when it cut off the telephones to the union office and then tear gassed and shot strikers in their determination to get them back to work. Similar events occurred in the coal mines. The mine security had not been allowed into the hostel at Anglo-American's Bank colliery in Witbank during the strike but they forced their way in on 27 August and gave the 1 200 strikers until 9 am that morning to return to work. When they refused they were dismissed, paid the wages due to them and put into waiting buses. The 2 000 strikers at the New Denmark colliery were treated in the same way.

There was a sharp contradiction between Gencor's attitude towards the strikers and its willingness to settle contentious hostel issues by negotiation. There was, for instance, much violence at its Kinross gold mine on 27 August when the mine security tried to force strikers to go to work by tear gassing and shooting at them in their rooms. They caused an explosion in one room after throwing a tear gas canister into it. Fourteen people were sent to hospital and there were many arrests. Yet while this was going on the managers of its Evander group of mines, including Kinross, met NUM officials to talk about "domestic" issues relating to the orderly conduct of the hostels and kitchens during the strike. There were problems at JCI's Randfontein Estates Cooke No 2 shaft. On the day negotiations had resumed, the Acting General Manager complained to Ramaphosa that the catering staff had joined the strike so that he was unable to provide food for the strikers. The NUM said that the mine was at fault because it was refusing to feed the strikers. Two days later the strikers in that shaft were tear gassed and forced to go underground. At mine level, there were no essential differences in the tactics used by the three mining houses in their resolve to end the strike.

The decision to continue the strike, and the subsequent intensified action by the mine owners to end it, renewed discussions in the wider trade union movement about taking sympathetic strike action. The issue had been raised in COSATU two days before the strike began when its Central Executive Committee met NUM officials but at that time the NUM was not asking for solidarity support. It was not illegal under the Labour Relations Act to engage in sympathy strikes though the law had not been tested. In order to support the NUM, the unions would have had to ballot their members under the procedures for legal strike action laid down by the Labour Relations Act and that would have been difficult to complete in time to make any significant impact.

COSATU's support, thereafter, was largely confined to circulating information about the strike to its affiliates, at least one of which, the Paper, Wood and Allied Workers' Union, distributed leaflets about it to its members on the shop floor. In general, industrial workers were aware of the strike and its significance but there were no union-wide attempts to support it. There had been two cases of localised solidarity strike action amongst shaft sinkers and construction workers in the mines and the Transport and General Workers' Union had asked for ballot facilities in the Richard's Bay coal terminal but no action resulted from it. Then, after 200 members of the Chemical Workers' Industrial Union had struck in protest against the relocation of their chemical plant to Bophuthatswana, they declared that their action was also in support of the NUM. The question of sympathetic action was raised again at a COSATU meeting on 22 August when it was agreed that workers should put pressure on their employers to support the mineworkers, but nothing was done about it.[130]

On the day the negotiations between the NUM and the Chamber of Mines were resumed, an organiser of the African Miners and Allied Workers' Union, which organised workers in industries allied to mining, declared that if the negotiations broke down the 9 000 members of his union would "down tools in sympathy with the miners".[131] This gesture of support was followed by a COSATU National Living Wage Conference, on 27 August, at which the spectre of a national sympathy strike was raised when it was declared that the conference would mount a worldwide diplomatic campaign and "national solidarity action" at home in support of the mineworkers' strike. The campaign was intended to influence governments and trade unions in neighbouring countries "to help prevent the recruitment of scab labour"; to enlist the help of rural communities and organisations "to stem the flow to mine recruitment centres" and to press the United Nations and the International Labour Organisation to impose sanctions on South African gold and coal. Jay Naidoo, the general secretary of COSATU, talked about a general strike. He said: "Black workers in this country have few weapons, but one of those weapons is the withdrawal of labour … A general strike is not a last resort. It is one of the weapons we have … The mine owners have made it a test of strength and that is what it is going to be and from our side we are determined to win it."[132]

[130]See Baskin, 1991, pp231-232. Baskin wrote that the action was "too little, too late".
[131]*The Citizen*, 26 August 1987.
[132]*New York Herald Tribune*, 28 August 1987.

But COSATU's resolution and the speech by Naidoo were words of bravado for COSATU was incapable at that time of undertaking the difficult task of organising inter-union sympathetic action.

All the vital strike statistics, except the participation rate, had reached their pinnacle by the evening of 27 August. Nine mineworkers had now been killed, about 500 had been injured, approximately 400 had been arrested and at least 54 000 strikers had been dismissed.[133] There was no sign that the mining houses were weakening in their resolve to use force to defeat the union. It was not a propitious scenario for the NUM. But there was worse to come. That night Ramaphosa was sleeping at the strike centre in Darragh House when, sometime after midnight, he received a telephone call from Robert Reise, an American journalist from *Newsweek*, who told him that the Anglo-American Corporation had decided to dismiss every striker. Thus the policy of mass dismissals, initiated by Peter Gush, was being taken to its logical conclusion. Ramaphosa was devastated by the news. Even if the other two mining houses refused to do likewise it would mean the loss of more than 80 per cent of the union's members. If JCI and Gencor followed Anglo-American's example then the only remaining union members would be those in mines that were not on strike and where, in the main, there were not enough to justify claiming recognition agreements. The union did not have the infrastructure and resources to withstand that intensity of devastation.

Ramaphosa recognised immediately that the strike would have to be called off if the news brought by Robert Reise was true. The union had no answer to legally enforced mass dismissals. He went immediately to the nearby Victoria Hotel where Motlatsi was staying and discussed the crisis with him. They decided, in the early hours of the morning of Friday, 28 August, that Ramaphosa should telephone S W van der Colf, a senior Anglo-American executive, to ask whether it was true that they intended to dismiss every striking employee if the strike continued. When he did so, he was told that it was Anglo-American's intention to do just that. After further discussions with Motlatsi, Ramaphosa rang Van der Colf again to enquire whether Anglo-American would discuss the dismissals question and whether the Chamber of Mines would be prepared to revive its offer the NUM had rejected two days before. As a result of these two telephone calls, Anglo-American both contacted

[133]This figure came from the Chamber of Mines. See M L Crowley and M du Plessis, *Black Employee Mobilisation in the South African Gold Mining Industry, 1987-1988*, Human Resources, Chamber of Mines, April 1990, p15. The media's estimate was usually 50 000.

the Chamber of Mines, which said it would revive its offer and meet with the union, and agreed to hold an early meeting between itself and a NUM delegation at the Carlton Hotel in Johannesburg.

The two NUM leaders decided immediately to recommend that the strike should be called off. They had a round-the-clock arrangement to communicate with the regional strike committees so Ramaphosa rang each one with the request to send delegates to a meeting of the national strike committee later that day. When it met, it reversed its decision to continue the strike and accepted their recommendation to end it. The NUM negotiating team then met with Anglo-American representatives as arranged to discuss the dismissals. The parties left the meeting with a 'mutual understanding' but had different perceptions as to what that meant. The confusion subsequently prompted the NUM to accuse Anglo-American of breaching its undertaking. The NUM's belief at the time was that all of the dismissed mineworkers, including Motlatsi, would be re-employed but that was not what Anglo-American intended. The terms were submitted to the regional committees and then endorsed by the strikers in a series of report-back meetings where decisions were taken by a show of hands. Ramaphosa wrote to Anglo-American on the afternoon of Saturday, 29 August stating that the union was ending its wage dispute on the basis of the offer made four days previously.

After that, Ramaphosa assumed that the ending of the strike would be a formality but that was not the case. He and his delegation met the Chamber of Mines' negotiators, led by Johann Liebenberg, on Sunday, 30 August assuming they would be offered the terms presented earlier in the week. That was not quite what happened for the Chamber of Mines reduced its offer and Ramaphosa had to engage in a salvage operation. He left the meeting with slightly less favourable terms for the offer to increase the death benefit pay-out from two to four times the annual wage was revised down to three times the annual wage. The proposal to increase holiday pay by 10 per cent remained unaltered. Ramaphosa raised the question of the re-employment of the dismissed strikers and was told that there would be no automatic re-employment. The individual mining houses would decide for themselves who amongst the dismissed mineworkers would be re-employed. Nor could he get any guarantee that there would be no discrimination against union activists in the aftermath of the strike. Liebenberg, however, made one slight concession. He said that those who were allowed to return would not lose their benefits. Meetings were then arranged with each the three mining houses to discuss the question of reinstating the dismissed strikers.

The NEC of the union met later on the same day and, without seeking a mandate from the members, instructed Ramaphosa to inform the Chamber of Mines that the strike had officially come to an end. Those strikers who were still in the hostels were told to find out from their mine personnel assistants for which shifts they had to report. The first batch of strikers returned to work on the night shift on Sunday, 30 August, exactly three weeks from the beginning of the strike. The remainder started work the following morning.

CHAPTER TWELVE

AFTER THE STRIKE: MANAGEMENT REPRESSION

Assessments of the Strike

At his first press conference after the strike had been called off, Ramaphosa glossed over the failure of the union to win its main demands and concentrated on its organisational achievement. He said that while the strikers had not won all of their demands, the strike, which surprised the employers by its size and length, was an organisational victory that laid the ground for winning future struggles. "The NUM," he added, "already sees 1988 as the year for winning significant gains."[1] This morale-boosting assessment was repeated, though more aggressively, by COSATU on the same day. "COSATU," it stated, "salutes the NUM and the black mineworkers. In spite of killings, arrests, detentions and security force intimidation the miners have confronted the organised power of the mining conglomerates. We do not see the return to work as a defeat. Though we have not won our wage demand, the mine bosses also have not won. The strike has changed the face of labour relations in South Africa. The bosses know that they will not be able to operate as before. Their prerogative to wield absolute control over our lives has been fundamentally challenged. The attack on the miners must be seen in the context of the apartheid state's attack on the Living Wage Campaign through labelling it a communist plot, banning rallies and confiscating our material." It made reference to the two faces of Anglo-American. "The miners' strike," it added, "has starkly unmasked the real face of the so-called liberal bosses who claim to be opposed to apartheid. Together with the apartheid state they have waged war on the working class. They have used the apartheid security to batter the struggle for human rights and dignity…"[2] Ramaphosa had the same opinion. "There is," he said, "no such thing as a liberal bourgeoisie."

Over the next 18 months, neither reflection nor the impact of management repression had any effect on the union leadership's assessment of the strike. When Ramaphosa came to write his report for the union's Sixth National Congress in 1989, he stated that the NUM had shown that it had possessed the organisational capacity to challenge

[1]*Business Day*, 31 August 1987.
[2]COSATU press statement, 30 August 1987.

one of the most powerful economic institutions in the country. He was to write: "The strike has been viewed by many as a defeat for the union. This is surely a very simplistic assessment of a very complex process. In any complex situation of this nature what is important is the way workers finally assessed their strike. To our members and its leadership, the strike was not a defeat. It was a setback. As miners we know very well that the working class has to fulfil its historic mission and the path to winning that objective is fraught with pitfalls. As miners we know that it is inevitable that we will go from setback to setback and finally total victory. It cannot be disputed that the employers were bent on at least curbing, if not destroying, our union. But they did not achieve this objective even though 50 000 workers were dismissed ... Defeating the bosses' aim of destroying the NUM is clearly a resounding victory..."[3]

The Chamber of Mines, for its part, was officially charitable in its assessment. When Naas Steenkamp, the President of the Chamber of Mines for that year, spoke immediately after the strike, he did not declare a victory but acknowledged that the employers had learned about "the effectiveness of the NUM's muscle, organisational capacity, determination and skill".[4] The Chamber of Mines saw the strike as "the beginning of a new learning phase in the relationship between the union and the Chamber".[5] Steenkamp, however, went beyond generalities when he described the costs of the strike. In addition to the loss of life, the cost had to be measured, he said, "in terms of the strains imposed on the management-worker relationship, the strife and tension between strikers and non-strikers, the disturbing incidence of violence, intimidation and murder, the loss of jobs and the wages sacrificed".[6]

These factors did, indeed, dominate the lives of mineworkers for some time after the strike had ended and compelled them in contradictory ways to reflect on and reassess their allegiance to the NUM. Its failure to exploit their solidarity and counter the mass dismissals had strained its credibility in a way that was difficult to repair in the short run. The mine owners and the government, however, inadvertently aided the process of repair by continuing their historical hostility to black emancipation, thus emphasising the need for a strong trade union movement. The mine owners embarked on a policy of such intensive repression that the NUM was prompted to describe the situation as a 'state of emergency' for black mineworkers. The

[3]NUM, *Sixth National Congress Report*, 1989, pp15-16.
[4]*Business Day*, 31 August 1987.
[5]Chamber of Mines, *Review,* 1987-1988.
[6]*Business Day*, 31 August 1987.

justification for this description was an investigation by the union's Media and Research Department in October and November 1988 into the post-strike experiences of black mineworkers and, through them, into the attitudes of white miners, supervisors and managers.[7] Information was collected through structured questionnaires administered by formal and informal branch committees to union members at the 13 gold mines and 10 collieries belonging to Anglo-American that had been affected by the strike.

The government had revealed a hypocritical attitude towards free trade unionism for whilst its Labour Relations Act was projected as a protection for legal strikers against arbitrary dismissal, it encouraged and allowed the mine owners to dismiss strikers with impunity. A startling statistic was that whereas 30 per cent of all legal strikers were dismissed arbitrarily, only two per cent of those who struck illegally were similarly treated.[8] Although those figures were not available to the strikers they knew the situation to which they referred. Those who had begun to wonder during the strike whether the expense and effort of conducting it legally were worthwhile were convinced afterwards that it was not. An illegal strike seemed to be much the safer and less expensive option for both the union and the ordinary members. The government gave way to the wishes of the mine owners after the strike by amending the Labour Relations Act so that industrial action, whether legal or not, was made more difficult and hazardous. Ramaphosa's claim that 1988 would be "the year for winning significant gains" seemed to be irrationally optimistic.

The Return to Dangerous Work

A positive feature of the 1987 strike, as in all mining strikes, was that for its duration it had removed mineworkers from the dangers of underground work. In consequence, some mineworkers remained alive who would otherwise have been killed and many remained fit who would have been injured or maimed. They also had a break from working in dust-infested environments. There was, however, a negative side to this. For 21 days there had been no daily scrutiny of rock

[7]NUM, Media and Research Department, *Collective Bargaining at Anglo-American Mines: A Model for Reform or Repression?* December 1988. The essential details of the report were compressed into a pamphlet called *State of Emergency on the Mines*: *NUM Under Attack After the 1987 Strike* which was distributed to the membership.
[8]M L Crowley and M du Plessis, April 1990, p15.

conditions and the maintenance of machines that would have come with regular mining operations. Some working places had become unsafe; rocks had moved in unobserved ways and methane gas had consolidated its presence in obscure cracks and crevices in the roofs. In addition there was the possibility that machines had become sluggish, and, perhaps, unreliable through inactivity. Thus it was likely that on the day the mineworkers returned to work they faced greater dangers than usual. But they were ill-prepared for them. The strike had had a disruptive effect on underground working relationships in that, on the one hand, there were white miners, relishing the defeat of the union and eager to teach the black strikers a lesson, and, on the other hand, the union members seeking retribution from those who had engaged in strike-breaking. Safe working required each person to be conscious of his responsibility for the safety of others and that was not present when the strikers returned to work.

The situation was worsened by the cavalier attitude of the mine managers to the participation of black mineworkers in resolving safety problems for during the strike they had set about demolishing the health and safety committees as if they were conducting a vendetta against the very idea of safe working. The members of all union committees were automatically included in the mass dismissal lists, irrespective of the functions they performed, thus ensuring that the structure of safety committees, from branch to national level, was demolished. In the period immediately following the strike, the Anglo-American mines withdrew from the negotiations about establishing safety committees. The only progress made at that time was with the Phalaborwa Mining Company, controlled by Newmont, the American mining multi-national company, which signed the NUM's first mine safety agreement in October 1987, embodying most of the features raised in the draft agreements. It was clear that a vital part of the recovery process after the strike was the re-establishment of the health and safety department on a foundation of branch safety committees.

The urgency of the re-building task was impressed on the union by yet another disaster at a Gencor gold mine on the day following the end of the strike. The frequency of accidents at Gencor's mines had prompted the NUM Safety Officer to describe them as "death traps". This time, the disaster occurred at the St Helena gold mine in the Orange Free State through a methane gas explosion that caused the death of 63 black mineworkers.[9] The mine, which employed 13 000

[9]For a detailed explanation of the disaster, see Michael J Martinson, "A Catastrophic

mineworkers, had been classified as fiery in 1950 after two explosions in which a total of 17 mineworkers were killed. There was a further explosion in 1953, causing the death of eight mineworkers. The fiery mine restrictions in the mine were relaxed in 1956, but because of its history of methane gas explosions, limited conditions were imposed regarding the use of open flame cutting and welding equipment and there was a complete prohibition on smoking and the possession of contraband underground. Those conditions still applied in 1987.

The St Helena mineworkers had participated fully in the strike without any serious incidents of violence or mass dismissals. The night shift had restarted work uneventfully on 30 August. However, shortly after the start of the shift the following morning, when some mineworkers were descending in the cage and others were waiting for permission to work at the intermediate pump station halfway down the shaft, there was a powerful explosion caused by the ignition of a large quantity of methane gas which had accumulated in the pump station. No-one survived to explain how the ignition occurred but it was not difficult to fit together the vital variables to give what was most probably a valid causal explanation. The station was newly constructed so all of the safety precautions should have been in place. However, none of the electrical equipment in it had been flame-proofed. Nor had the methane gas been cleared prior to the start of the shift. The explosion killed the mineworkers in its immediate vicinity but it also entered the shaft "where it shattered numerous reinforced concrete panels separating the shaft into discrete intake and return compartments. Falling debris from the wall and steelwork from the equipped intake compartment piled on top of a loaded cage travelling down the shaft ... The winding rope parted, and thereupon the cage plunged to the shaft bottom, where it was buried under a mass of broken panels and twisted steel supports."[10] The cage had plunged more than a mile and no-one in it survived. It was in such a twisted, compressed state under piles of rubble that it was not recoverable for 15 weeks. The bodies, by then, were decomposed. It was a sad and frightening return to work.

In December 1987, 14 weeks after the explosion, an official inquiry into its causes was opened by the Mines Inspectorate and lasted, with numerous lengthy adjournments, for seven months. From

Methane Explosion in a Main Vertical Shaft Serving an OFS Gold Mine", *Proceedings of the Fourth US Mine Ventilation Symposium*, Chapter 74, University of California, Berkeley, USA, published by the Society for Mining, Metallurgy and Exploration, Inc.

[10]*Ibid.*

the outset, however, the Government Mining Engineer ruled that the counsel for the NUM could not cross-examine witnesses because no black NUM members could be held criminally liable for the accident. On the other hand, he allowed lawyers representing the parent company Gencor and senior management from the mine to cross-examine all the witnesses. It was a repetition of the way the NUM had been treated at the Kinross inquiry. Yet despite this discrimination against the union, sufficient evidence was presented to the inquiry to warrant a charge of negligence against the company.[11] The report was subsequently sent to the Attorney General of the Orange Free State who brought charges of culpable homicide against two of the mine officials.

Reinstatement Problems

The black mineworkers at St Helena had been amongst those who had been able to return to work immediately because they had not been dismissed and had remained in their hostels. For them it had simply been a matter of reporting to a personnel assistant and joining the right gang. Those in other striking mines, who had either returned to their homes voluntarily or who had been dismissed, found it much more difficult to get their jobs back. They all had to reapply through the TEBA offices in their respective home regions rather than at the mines where they had been working. This, however, was not explained to everyone and it caused much confusion and worry. When, for instance, mineworkers who had returned home heard on the radio that the strike had ended they rushed back to their respective mines to reclaim their jobs but were told that they had to go back to their regional TEBA offices and make fresh applications. A mineworker from No 1 shaft in Western Holdings described his experience:

> I stay at Qwa Qwa – we heard on the radio that all workers who were on strike should come, the strike was over. When we arrived we found that we were not allowed into the hostel. We then came to the (union) Regional Office. We stayed here for a week, then we were told that we would have to go home to our respective TEBA offices to apply for employment. We were given money to go home. Two bus loads were re-employed by TEBA but there are still a lot of comrades at home waiting to be re-employed.[12]

[11]This was the opinion of Martinson who was present at the inquiry and acted as a technical advisor to the NUM. (*Ibid.*)
[12]*South African Labour Bulletin*, November 1987.

A striker from Vaal Reefs No 2 shaft described how even some of those who had remained in the hostels had to go through a similar experience. He, like the mineworker from Western Holdings, discovered that not all of them were offered their jobs back. He said:

> We were escorted by armed mine security to the local football ground to collect our pay. We were then paid small amounts of money and told to go home and take up new contracts with our respective TEBA offices before 10 days expired. But when we go to TEBA for re-employment the clerks say there is no job for us. My home is Mafikeng. I have been sent by other workers at home to find out what is happening to our re-employment.[13]

The Anglo-American Corporation had warned that when the strike was settled not everyone who had been on strike would get a job. Godsell had stated that it would re-employ dismissed mineworkers wherever possible but that some jobs had been lost and others would disappear through restructuring. He added, ominously, that those mineworkers who had 'unacceptable' work records would not be taken back.[14] Under this criterion anyone could be justifiably excluded. Matters were not straightforward, moreover, for many of those who were offered jobs were not returned to their previous positions but had to start again from the bottom grade and the lowest pay.

The decision as to whether mineworkers dismissed from the Anglo-American mines would be reinstated was, in practice, primarily dependent on their connection with NUM and their involvement in the strike. Anglo-American introduced a screening process first to identify those who had been active during the strike and then to isolate from amongst them those who had held strike and union positions. Those in this latter group were not allowed to return to their jobs. Moreover, they were victimised further because their names were put on to another list that excluded them from working in any mine at any time. The national office-bearers of the union were not protected from victimisation. The union president, James Motlatsi, was allegedly dismissed on the order of the Anglo-American director, Peter Gush. In consequence, he had no full-time job until his union position was converted into a full-time salaried one at the NUM's 1989 National Congress. Though this removed the need for him to seek reinstatement, he was nonetheless discriminated against by being refused permission to enter Western Deep Levels, even in his capacity as president of the NUM. The screening system, however, did not always work as it was intended.

[13] *Ibid.*

[14] *Business Day*, 31 August 1987.

Frans Baleni, for instance, who had been in hiding from the security force at Western Holdings, applied to be re-engaged there and was accepted. When he presented himself for work at the mine, the personnel assistant dealing with the returned mineworkers looked up from his desk in surprise and said: "You should not be here." But he let him stay.

TEBA officials largely undertook the actual task of sifting through the applicants for reinstatement. The process was not straightforward because, even if there had been no pre-conditions, TEBA could not have given every dismissed mineworker his job back for there were many fewer jobs than had existed before the strike. Six thousand jobs had been lost because of the closure of the two shafts at Western Holdings and Vaal Reefs. Three thousand jobs were occupied by strike-breakers who were not going to be replaced. Many other jobs had disappeared because a number of mine sections had become unworkable. There were other losses because the dislocation caused by the strike had enabled managers to restructure their work forces and reduce their labour requirements. The result was that, when those who were excluded for political reasons were taken into account, as many as 19 000 out of the 42 000 reapplications for work in the Anglo-American mines remained without jobs. This created a major social problem in the rural areas.

The other mining houses responded to the end of the strike in different ways. JCI had dismissed over 7 000 strikers from the Randfontein Estates and Western Areas gold mines but after negotiations with the NUM it agreed to reinstate all of them to their previous jobs with the same conditions of employment. The difficulty there was that JCI had insisted on a deadline for a return to work which was so limited that it was impossible for either the NUM or the mines to tell all of the mineworkers in their homes what to do and give them enough time to process their applications through TEBA. The result was that many workers missed the deadline and were refused work. The union launched an Industrial Court case to get them all reinstated whereupon JCI agreed to its demand in an out-of-court settlement. JCI, nonetheless, weeded out the activists for a number of shaft stewards who returned to the two mines were charged with misconduct and dismissed. The union failed to get them reinstated. Gencor, on the other hand, had dismissed only 312 strikers but it steadfastly refused to negotiate over their reinstatement. Though there were no legal strikes at the Gold Fields' mines, 100 mineworkers were dismissed for being absent for more than six days without permission from management. Only a few of these were reinstated. A number of the dismissed strikers from all of the mines and collieries did not apply for reinstatement because they had decided not to return to mine work.

There was a growing feeling amongst those who had not been reinstated that the NUM had a responsibility for sustaining them until they obtained work. A group of them from Randfontein Estates decided, therefore, to test the union's response by travelling to the union head office by taxis to demand money, food and taxi fares. Motlatsi heard in advance what they intended to do and had visions of a repeat of the 1986 saga when the retrenched mineworkers from Rustenburg had occupied the union office until they were violently evicted.[15] When he was told that the group from Randfontein Estates was actually on its way he instructed the office administrator to send the staff home and to ask them, where possible, to work from home. When the Randfontein mineworkers arrived at the office they found it was locked and it remained so for a week. They were unable to get money even to pay for the taxis that had brought them to Johannesburg. The news of this incident travelled fast around the union and no-one else tried to repeat it.

Settling Scores With the Scabs

Irrespective of the portrayal of the strike through trade union folklore, which regarded a defeat as a step towards victory, and the gloss put on the return to work by Ramaphosa to make it more palatable, the ordinary strikers saw it as a defeat and they looked not only for compensation, but for explanations. Many saw the reasons for defeat in the weaknesses in their own ranks, in breaches in their solidarity. This led them to search for and then punish strike-breakers. It was inevitable that this should happen because punishment for strike-breaking has been an indelible feature of strike action wherever and whenever trade unionism had been practised. Throughout the wider international trade union movement, unity in struggle had been regarded as seriously as nations had viewed patriotism in war. This analogy was expressed in the preface to the rules of the United Society of Cordwainers in Britain as early as 1792. In answer to the question "What is a scab?", it stated: "He is to his trade what a traitor is to his country; though both may be useful to one party in troublesome times, when peace returns they are detested alike by all … In short he is a traitor on a small scale. He first

[15]See Chapter Six.

sells the journeymen and is himself afterwards sold in his turn by the masters."[16] The treatment of scabs as traitors has resulted in their being excluded from work, subjected to abuse at work, scorned in their communities and the object of assaults that resulted in the destruction of tools, physical injuries and sometimes in deaths.

Injury and death through violence were a common characteristic of black South African society at that time. Every form of violation of human rights had become an everyday feature of black lives. Township violence had been particularly intense since September 1984 and political deaths had become commonplace. Most people who died were killed either by the actions of government-sponsored agencies or for collaborating with them. Altogether, between September 1984 and the end of the strike, at least 2 611 people had been killed in political incidents. By far the largest number of deaths had occurred in KwaZulu-Natal, but all the urban areas of the country experienced them. Cases of torture and incidents of severe ill-treatment, perpetrated overwhelmingly by the police, had increased by approximately eight times between 1984 and the end of the strike.[17] In an overview of violations of human rights for the period 1983 to 1989, the Truth and Reconciliation Commission concluded:

> The Commission finds that the period 1983-1989 was characterised by a dramatic increase in the reliance by the SAP, the Ciskei Police and the Ciskei Defence Force (CDF) on: the unjustified use of deadly force in crowd control and protest situations; the use of deliberate ambush and so-called 'Trojan horse' operations in which marchers or protesters were deliberately targeted and killed; the use of assault and torture on suspects and detainees as a systematic pattern of abuse; the fostering of divisions between political and social groupings; the deliberate killing of political activists.[18]

It was unlikely in that climate, therefore, that the strikers would have misgivings about injuring and killing those whom they perceived as traitors in their struggle with the mine owners.

In their desire to settle scores, the black mineworkers looked beyond those who had simply refused to strike on 9 August to the informers who had been planted by managers and the police with the continuing, constant task of undermining the union.

Ever since the NUM had been perceived as a threat to the authority of

[16]V L Allen, *Social Analysis*, 1975, p239.

[17]Truth and Reconciliation Commission of South Africa, *Report*, Volume Three, 1998, pp1-12.

[18]*Ibid*, p122.

the mine owners and seen as a participant in the liberation struggle, some black people had been systematically corrupted to inform on their fellow workers for small rewards. Ramaphosa and Motlatsi were aware of their presence in the union and speculated about possible culprits but, of course, only the incompetent ones were identified. One of the by-products was an atmosphere of suspicion which impeded effective communications in the union's administration. At mine level, however, it was easier to recognise who was and who was not working in the interests of the union. And when there was a strike it became patently clear. Such people, described as '*impimpis*', became immediate targets for retribution.

Some of the NUM branches had anticipated having to deal with strike-breaking before the strike began and had resolved to take action in the event of its happening. The real test for many of the strikers, however, came after the strike had ended when they were compelled to work alongside those who had opposed it. Their reactions were spread over many months and occurred in a piecemeal fashion, depending not only on branch decisions, but also on the initiatives of small groups and even of individuals. Scores were settled in the hostels, outside the compounds and obscurely in the townships. The union leadership had virtually no influence over these events and even their knowledge of them was patchy, based often on uncorroborated tales and rumours.

The mining houses suffered from the same lack of reliable information. Peter Gush, for instance, claimed in his review of Anglo-American gold mines in December 1987 that there had been incidents of violence after the strike but he was not able to quantify or verify them.[19] Indeed, the Anglo-American Corporation listed three, of which only one involved a death, in its memorandum dealing with violent conflict in the mines after the strike.[20] When the chairperson of the gold and uranium division of JCI, Kenneth Maxwell, claimed in September that retribution had been so severe that at least 33 of those who had refused to strike had been murdered, he was forced to retract it after the union had challenged him to substantiate his allegation with details.[21] Following this, the head of the Public Relations Directorate of the SAP went so far as to say that he was unaware of any orchestrated attempt to murder mineworkers who had not participated

[19]South African Institute of Race Relations, 1987/88, p679.

[20]Myburgh, March 1988, pp26-27.

[21]South African Institute of Race Relations, 1987/88, p679. JCI said that the allegation had been made in the "heat of events" and that the number of deaths was lower than 33 and, in any event, had nothing to do with the strike.

in the strike.[22]

There were many alleged cases of punishment where there was no intervention by the police, no outside observers, no evidence sustainable in court and only casual hearsay information that anything had happened at all. Mineworkers who were present when acts of retribution took place were usually unwilling to act as state witnesses even when the victims were their friends. Even a decade after the events, witnesses were still unwilling to discuss the events. There was anecdotal evidence, however, which indicated that as many as between 200 and 300 mineworkers might have been killed after the strike for strike-breaking.[23]

An incident that gave a glimpse of the tensions behind the scenes occurred at the Randfontein Estates gold mine, which was one of the union branches that had decided at the outset of the strike to impose the death penalty for strike-breaking. Shortly after the strike had ended, Motlatsi was invited to attend what he thought was an ordinary meeting of the NUM branch. He was unaware of the branch's earlier resolution. When he arrived, there were about 5 000 mineworkers in the body of the hall and 50 on the platform. He presumed that those on the platform were strike and branch officials. There was some *toyi-toying* before his speech but during it the hall was completely silent. As soon as he stopped speaking, however, a member of the audience stood up and asked: "What are we going to do about the scabs on the platform with you? There are kombis outside waiting to take them away to be killed." Only then did Motlatsi recognise the real purpose of the meeting. He was perplexed because he did not know how to resolve the situation without violence to those on the platform and, maybe, to himself. He argued at length against using violence as a penalty for strike-breaking and then asked the mineworkers in the audience to divide down the middle and create a gangway. He said that he was going to lead everyone off the platform, down the gangway and out of the meeting and that if any one of them were harmed they would have to contend with him first. His request was granted and he led those on the platform out of the meeting without incident. Motlatsi then instructed the drivers of the kombis to take them to a place where they would be safe.

The main substantiated information about post-strike retribution came from court cases and there were few of those. This was partly because the police were not particularly interested in charging black

[22]*Ibid.*

[23]Conversation with James Motlatsi, London, 31 October 1996.

people with killing members of their own communities and partly because it was very difficult to obtain incriminating evidence from witnesses to present in court. Although in total many thousands of mineworkers must have witnessed post-strike violence the vast majority were unwilling to testify before their sworn enemies, the police. Moreover, in many instances they did not question the severity of the sentences that were imposed on scabs and *impimpis* by the mass meetings of strikers. Even where there was sufficient evidence to bring cases before the courts the state prosecutors found it difficult to substantiate them. One such case occurred when nine mineworkers from Blinkpan Colliery, including eight who were members of the safety committee which was responsible for keeping order in the hostel during the strike, were charged with murdering a non-union member who had ignored the strike call on 11 August 1987. The nine admitted that they had entered the murdered person's room to tell him that he either had to join the strike or leave the hostel but they denied using any physical violence against him and there was no other evidence to link them to his death. So, on 3 August 1989, after the trial had lasted for more than a year, they were acquitted.[24]

One of the few cases taken to court that resulted in a conviction occurred after an attack on three team leaders at the President Brand gold mine in the Orange Free State, on 16 September, which left one of them dead and the two others seriously injured. The chairperson of the NUM branch had compiled a 'hit-list' of those who had worked during the strike. He then divided the young members of NUM into Xhosa and Sotho groups and gave them the list with the instruction to punish those on it. He and two others were found guilty of culpable homicide and sentenced to a term of three years in prison. Eight others were found guilty of public violence and given prison sentences of 18 months.[25] There was scant information from other mines about the treatment of *impimpis*. A tribal representative who had worked during the strike at Randfontein Estates gold mine was assaulted and injured on 27 October by about 20 mineworkers who were shouting that they would deal with informers. There was more trouble at the same mine on 20 November when a mineworker was accused of being an informer and almost had his eyes gouged out. There was a threatening but

[24]*Business Day*, 4 August 1989.

[25]These sentences compare with a five-year prison sentence given to Charles Mapetshoawane, a NUM official in Klerksdorp, for urging mineworkers to damage mining machinery if they were forced to work during the strike. No violence or actual damage to machinery resulted from his action.

non-violent demonstration at Western Holdings in the same month, following a union meeting at which the vice-chairperson of No 3 shaft predicted that all *impimpis* would be killed by Christmas. Three mineworkers at this meeting were dismissed for intimidation. Then there was an incident in the liquor outlet at Freddies gold mine which led to an attack on six people who had worked during the strike. But most of the cases of intimidation of scabs were neither recorded by the mine authorities nor reported in the press. They tended, moreover, to be suppressed from the memories of the participants.

Management Revenge

It became apparent from the first day back at work that management triumphalism had resulted in a dramatic change in the working environment. For those with a brief experience of mining the situation was new but for the rest it was as if the days before the formation of the NUM had returned to haunt them. Black mineworkers had always been demeaned and insulted underground by white miners and supervisors but since the formation of the union and the election of shaft stewards such behaviour had been moderated, even if not eliminated. But when they went back to work on 31 August they were treated as though they were part of a defeated, demoralised army. They had lost the back-up support the union had provided and racism resumed its customary predatory poisonous quality. "They had gone back," one mineworker remarked, "to the old system of calling people kaffirs." The white miners underground ridiculed Ramaphosa; they said that the union was dead and asked, rhetorically: "Where is that union of yours?" Taking over an earlier NUM slogan they boasted that 1988 was to be their year. Verbal abuse, however, was only one aspect of racial discrimination in the mines. It became more exaggerated in workplace practices. The slow movement of black mineworkers into whites-only jobs was reversed. In eight gold mines and two collieries, jobs newly acquired by black mineworkers involving the operation of expensive, heavy equipment, such as load haul dumpers and hydraulic drill rigs, were taken from them and given to white miners. It had been the policy of the Anglo-American Corporation to whittle away the privileges that white supervisors and miners possessed but after the strike that process went into reverse, petty-apartheid practices were restored and black mineworkers continued to be prevented from sharing lifts, toilet facilities, change rooms, medical facilities, recreational activities and dining places.

An ominous repercussion was an increase in the incidence of physical assaults by white personnel on black mineworkers underground. The NUM revealed that the number of assaults increased in 10 of the 13 Anglo-American gold mines and that they primarily involved attacks on black mineworkers by white miners, shift overseers, mine overseers and mine security. They used their fists or sticks to inflict injuries in what seemed to be highly personal forms of retribution for the strike but which, in reality, were the simple mechanics of institutionalised violence. At the three Western Deep Levels shafts it seemed that every category of white personnel underground below the rank of mine captain was involved in beating up black mineworkers. It was different in the coal mines where only the Bank and Kriel collieries out of the 10 Anglo-American coal mines reported an increase in assaults. At the Bank colliery they had stopped altogether before the strike.[26]

Black mineworkers were also beaten up more frequently in the hostels, usually by mine security personnel. The violence that was reported in signed affidavits by the victims was of a gratuitous kind, without any apparent reason or justification. For instance, one was beaten for asking why he should be called a 'kaffir'. Another one was hit around the head after repairing a machine. A black mineworker was simply walking past two white miners when he was set upon and assaulted. In another incident, mine security guards assaulted a mineworker from behind while he was standing in a food queue. There were many occasions when black mineworkers were beaten for displaying their allegiance to the NUM by wearing union insignias, such as stickers commemorating the Kinross mine disaster or T-shirts displaying the NUM logo.[27] The victims were able to report such incidents through the mine disciplinary procedures but only in a few cases were white miners disciplined as a result.

Some mine managers tried to push the clock back to the days when the liaison committees and tribal *induna* systems were in place. The management at Vaal Reefs, for instance, appointed *isibondas* to act as room representatives and posted pictures of them on the hostel gates so that the incoming recruits would know that it would be unwise of them to go to the union branch with their grievances. But in order to do that effectively the mine managers knew that they had to do more than demoralise black mineworkers through propaganda and physical

[26]No assaults were reported at all at Western Holdings gold mine and at five collieries.
[27]NUM, Media and Research Department, December 1988, pp46-51.

assaults until they became fearful of acting collectively. They had to dislodge the NUM branches from their positions of influence.

The Impact on the Union Organisation

It became apparent very quickly to those branch officials who survived the strike that the mine managers were intent on separating the NUM branches from their power bases so that they would be incapable of repeating their strike performance or even of obstructing the daily policies of management. All aspects of the post-strike mining life favoured this strategy. The union membership had been severely reduced because at least 40 000 of the strikers who had been originally dismissed had been union members and only a small proportion of them were re-employed. In some regions, but not in others, union officials had been dismissed or arrested. The whole of the regional administrations were eliminated at Klerksdorp and Carletonville but were left largely intact at Secunda, the Orange Free State and Wits. In the collieries at Witbank, the union's structure remained untouched. Throughout the union overall, however, there had been a widespread dismissal of branch officials and shaft stewards. It was estimated that altogether 4 000 shaft stewards had been dismissed. In some instances, as at Western Deep Levels, all members of the branch and strike committees were dismissed. Union education was a particularly important casualty of the strike. It had been decided at the union's National Congress in February 1987 that educational sub-committees should be established at shaft, branch and regional levels, but virtually every one was destroyed or dismembered. In no mine did the NUM branch organisation escape entirely unscathed.

The managers and supervisors knew that their unchallenged superiority relied on the absence of the union from bargaining and disciplinary proceedings. They began, therefore, to cancel their recognition agreements with the NUM. Union recognition was on a mine-by-mine basis and it was withdrawn in five gold mines, including the three shafts of Western Deep Levels. Another two gold mines simply refused to recognise shaft stewards. This meant that almost 84 000 mineworkers were denied any form of representation at disciplinary hearings, in the negotiation of local issues and in the local implementation of national agreements. No coal mines withdrew recognition from the union though restrictions were imposed on the movement and rights of shaft stewards at two of them. But even where shaft stewards were still recognised, they were often obstructed in their union work by being prevented from meeting mineworkers before they were charged with disciplinary offences or by being refused time off from work by their white supervisors to enable them to attend to grievances.

The withdrawal of formal recognition by a mine, however, was often no more than a first step. Because of the remote location of most mines and collieries, the union could not operate without office facilities, including the use of telephones, within the compounds. Before the strike all of the branches with recognition agreements had been provided with them. Immediately after the strike, however, the branches in four gold mines had their office facilities taken away entirely. Only the Elandsrand branch, amongst the gold mines, was left with its own telephone, while all but the President Steyn branch were forbidden to use company telephones for union purposes. The effect was to make it difficult to conduct the day-to-day business of the branches and to isolate them from the NUM head office. Without access to telephones, they were often compelled to spend from one to four hours travelling to Johannesburg by car or taxi in order to communicate with their national officials. They were unable to find alternative office accommodation within the proximity of the mines because the land surrounding each mine was invariably owned by a mining company. Moreover, in order to hold any kind of members' meetings beyond the confines of the compounds they had to seek the permission of a magistrate, under the terms of the emergency regulations, and they were rarely granted that.

There were strict regulations governing all union meetings at Anglo-American's gold mines and collieries. Most of them had existed before the strike but had been enforced in a relatively relaxed manner. This was not the case afterwards when there was a reversion to the attitudes the union had encountered in 1984 and 1985. Managers frequently refused permission for union meetings to be held. And even where it was granted it was rarely possible for branches to hold regular monthly meetings as many had done hitherto. Where permission was given the branches had to submit their agendas and the number and names of their speakers. If there was any deviation then the meetings were cancelled or interrupted and stopped. Permission was refused if there were political issues, such as the Labour Relations Amendment Bill, on the agenda. Outside union speakers, spontaneous speeches and questions from the floor were rarely allowed. Singing was permitted in only three mines but even there the mineworkers were not allowed to sing whilst leaving the meetings.

At the Elandsrand gold mine the union branch officials had to sign a detailed agreement before it was allowed to apply for permission to hold meetings. The agreement contained the following clauses:

> 2.7.1 The Branch Chairman or Vice-chairman must apply in writing for permission to convene the meeting, which application must reach the mine manager at least 10 days before the date of the meeting.
> 2.7.2 The application must indicate:
> 2.7.3 The date, time and venue for the meeting.
> 2.7.4 The points to be discussed at the meeting, together with a short description of each point. Points must be explicit and not vague.
> 2.7.5 The names and places of work of all speakers ...
> 2.7.6 The Branch Committee undertakes to allow any management representative or representatives to be present at such a meeting, including employees from the security department.
> 2.7.7 Any allegation made against management, the company or any employee(s) during a meeting must be able to be substantiated by the Branch Committee after the meeting, if challenged by management.
> 2.7.8 The Branch Committee undertakes to ensure that no derogatory terms or profane language is used during a meeting when referring to the company, members of management or any other employee of the company.
> 2.7.9 The Branch Committee undertakes to ensure that no singing and/or dancing takes place prior to, during and after any meeting.
> 2.7.10 The Branch Committee undertakes to ensure that no political issues or political slogans are used at meetings ...
> 2.7.11 The Branch Committee undertakes to discuss only such business as stated on the agenda. No speaker to address the meeting other than those whose names appear on the letter of application.
> 2.7.12 No publications, pamphlets or printed matter to be distributed prior to, during or after meetings, without management's written permission to do so.[28]

Similar restrictions were imposed on NUM branches in the coal mines. In two collieries union meetings were banned altogether and elsewhere restrictive conditions were imposed, as the following extract from a reply by the mine manager to an application to hold a mass meeting on 1 May 1988 at Kriel colliery indicates:

> I am able to grant you the permission to hold the requested meeting in the recreational hall under the following conditions:
>
> (a) **Only Kriel** employees to attend this meeting;
> (b) **No ex-Kriel** employees to be present;
> (c) You should **rigidly stick** to the agenda as discussed with you on 28 April ...
> (d) The meeting may be held from **8.30-9.30 am** and not from **8.30-10.30 am** as requested;

[28]All the illustrations about the management control of meetings are from NUM, Media and Research Department, December 1988, pp22-34.

(e) Behaviour of the people who attend the meeting is your responsibility and should be acceptable to management;
(f) No cohersion [sic] of people to attend the meeting will be tolerated.

Please note that the granting of permission for future meetings will be based on the results of this meeting with regard to behaviour, conduct and adherence to the aforementioned conditions.

The restrictions on the union were not confined to such regulations. In at least eight gold mines and six coal mines the managers recorded the proceedings of union meetings by tape recorders, video cameras and cameras. One branch reported: "Management puts the tape recorder on the table before the meeting and is not prepared to remove it. If you want to remove it you have to find other facilities." Sometimes, it added, mine security officials attended dressed in private clothes and carrying cameras and tape recorders. It was common for known management informers to be present at union meetings. In general the rules imposed by managers were so strict and the surveillance so apparent that branch officials often believed it was hardly worthwhile holding meetings.[29]

Difficulties in Rebuilding NUM's Membership

The critical task facing the union was that of rebuilding its membership. On the one hand, it had to try to regain the loyalty of lapsed or disaffected members who, through the strike, had become disillusioned with the union. Some turned from the union after counting the cost of the strike to themselves.[30] On the other hand, other members were unhappy with the union because they disagreed with the decision to call it off. The manager of the Randfontein Estates gold mine reported that disaffection there had reached the stage where dissident union members were threatening senior union officials with physical violence.[31] Mineworkers, in some instances, rejected all forms of industrial leadership and returned to their traditional tribal structures and beliefs to resolve their problems. All in all, the NUM had to try and recruit new

[29]See NUM pamphlet, *State of Emergency on the Mines: NUM Under Attack After the 1987 Strike*, 1988, p8.
[30]Naas Steenkamp, the President of the Chamber of Mines, talked of a "reactionary groundswell" by rank-and-file mineworkers (*The Star*, 21 June 1988). The NUM did not deny that there was disaffection but claimed that it was not sufficiently widespread to prevent the union from consolidating its position (*Weekly Mail*, 3 June 1988).
[31]*Weekly Mail,* 3 June 1988.

members from work forces whose understanding of the union and the strike was based on the tales, gossip and misinformation that circulated in the rural areas. This multi-faceted recruitment activity had to be done largely in the hostels and, therefore, under the glare of management and its informers. It was invariably undertaken, moreover, by shaft stewards and branch officials who were known to management and who, therefore, could be easily identified and victimised.

Efforts were made in every gold and coal mine to prevent the NUM from retaining its old members and recruiting new ones but they varied in intensity and complexity according to the whims of mine managers. In no mine, however, was the union accorded the full rights of freedom of association as prescribed by the Conventions of the International Labour Organisation. The anti-union propaganda started in the TEBA offices and continued through every stage of the recruitment process to the underground workings. In the mine recruiting offices in Lesotho, for example, recruits were advised not to join the union because it caused retrenchments. The anti-union pressure was continued in the induction halls where training centre instructors told recruits that management was giving them work whereas the NUM was responsible for dismissals and retrenchments.

Immediately before the strike there were a few cases where mineworkers were prevented from joining the NUM by the destruction of their stop-order application forms, by inducing them to sign union resignation forms and by exerting tribal pressure on them through *indunas* and *isibondas* to reject the very idea of trade unionism. But after the strike, every branch reported incidents of similar anti-union activity. Mineworkers were generally forbidden to form and walk around in small groups. In one mine, it was assumed that wherever there were groups of five or more they must be holding meetings and they were arrested by the mine security. Shaft stewards, former shaft stewards and even long-serving mineworkers were harassed by the mine security whenever they even stopped to talk with anyone in the hostels in the belief that they were trying to recruit new members. Scabs were instructed not to talk to long-service mineworkers who, in turn, were told not to talk with scabs. Shaft stewards were prevented from moving from room to room in the hostels in order to sign up new membership. But when they managed to get new members, the managers frequently refused to implement the stop-order forms and forward the membership applications to the union.

Apart from stopping recruitment activities, the managers spent much time trying to make individual mineworkers fearful of joining the NUM by threats, false promises and propaganda that discredited it.

The intimidation of individual mineworkers was acute. Sometimes personnel assistants and mine security personnel ordered union members to parade and questioned them about their membership. They demanded to know whether they had really signed up to join and whether the signatures on the application forms were actually theirs. They were handed union resignation forms and, if they happened to be illiterate, were sometimes tricked into signing them with their thumbprints. Others were offered incentives, such as promotion to team leader, if they resigned. Those who stood their ground were victimised by personnel assistants who, for instance, did not respond to complaints they made about white miners or altered their charge sheets so that they lost their cases. Victimisation took on many subtle forms.

Anti-union Propaganda

This vendetta against the NUM was backed up by a widely based but disparate propaganda campaign against the ANC, the SACP and COSATU. Many different pro-apartheid pamphlets, typewritten newsletters and leaflets, including publications that were fakes of *bona fide* ones, were distributed throughout the country by obscure political organisations which were either camouflaged as factions within the ANC or COSATU or represented extreme right-wing views within the white establishment.[32] None of these bodies had any acknowledged affiliations in the liberation movement or any recognisable source of funds. The understanding within COSATU and the NUM was that they were financed by the state intelligence services to spread misinformation and to create damaging divisions within the liberation movement. The publications were based on half-truths, information obtained from informers and by telephone bugging devices with the result that, in some instances, their impression of being well-informed made them

[32]See Baskin, 1991, pp271-272. There was, for instance, a glossy newsletter called *VAT (Victims Against Terrorism) Update* which, in its October/November 1988 issue, concentrated on exposing communist influences in the ANC. At that time, Chris Hani was used as a typical example of communist penetration. He had recently been appointed as Umkhonto we Sizwe Chief of Staff and was described as the sort of potential ANC President with whom white South African parents could frighten their children. Organisations such as VAT had their international counterparts. When Oliver Tambo, the President of the ANC, visited the USA in January 1987, the Coalition Against ANC Terrorism and the more broadly based right-wing International Freedom Foundation financed anti-ANC advertisements and mock 'necklace' demonstrations. (South African Institute of Race Relations, 1987/88, p704.)

seem plausible. One of them, directed at trade unions, was the *SA Trade Union Monthly Titbits* which aimed to cause dissension by spreading malicious gossip about union leaders. One issue published after the strike concentrated on the lifestyles of Cyril Ramaphosa and James Motlatsi. It was headed "James and Cyril's Holiday" and described a visit they were making to Sweden on a vacation that included "gambling, theatre, ski and naturally … Nordic blondes". It stated:

> The holiday in Sweden 'can either be linked to informal discussions with Communist bloc comrade agents, which master-minded the NUM strike in SA or – the top officials are spending and gambling away thousands of rands – paid into union funds by workers – and as we all know – the ANC/SACP shall be there to meet the NUM comrades.
>
> MAKES YOU THINK, DOESN'T IT?
>
> The workers are wondering what compensation did Comrade MOTLATSI and Comrade RAMAPHOSA have in mind for the workers who were dismissed during the 1987 mines strike now that they are enjoying the National Union of Mineworkers' funds in Sweden on white women?'[33]

Another issue of that newsletter was also about union funds, but with a different slant. It was written in Xhosa and English and stated:

> IF YOU ARE A WORKER, YOU MUST READ THIS!!!
>
> 1987 – the year the NUM called a strike – what happened? For the good – NOTHING. Everybody suffered. The union controlled by the ANC and COSATU tried to overthrow the workers' source of bread and butter – they failed!!!
>
> 1988 – the year of 40 per cent salary increase. What happened? NOTHING!!! The workers gained nothing. But the union organisers and officials gained. NUM bought 10 new cars. All officials and NUM head office staff workers received 35 per cent increase. They shall deny it, but it is true …
>
> You, the workers, you shout "*AMANDLA*" – for what?
> You pay your subs – for what?
> What is wrong with you? Can't you think?
> Send the union to hell, cancel your membership. Stop your stop orders ...
>
> GO TO THE NUM HEAD OFFICE IN JOHANNESBURG. DEMAND THAT RAMAPHOSA AND MOTLATSI, GOLDING AND OTHER PUPPETS GIVE YOU YOUR MONEY.

[33]*SA Trade Union Monthly Titbits*. Copies were reprinted by NUM, Media and Research Department, December 1988.

DOWN WITH NUM!
DOWN WITH COSATU!
DOWN WITH THE ANC![34]

Many hundreds of copies of the *Monthly Titbits* were distributed by mine managers in the hostels of at least eight gold mines and three collieries. Mine security personnel were seen taking them around. They were left on the tables in the mineworkers' rooms when they were working underground, in the dining halls, at the welfare offices, the concession stores, at the taxi ranks and at the gates to the compounds. The *indunas* and *isibondas*, acting as hostel and room prefects, were instructed to encourage mineworkers to read them. Mine managers also spread their messages through their own media. At the Vaal Reefs gold mine the public address systems were used to blame the union for neglecting the workers' interests, for causing the dismissals during the strike and for the subsequent retrenchments. In some mines the managers used an underground address system called 'Radio Reef' to broadcast defamatory comments about the union to mineworkers waiting at the shaft stations for lifts to the surface. White miners and black team leaders carried this anti-trade union propaganda campaign into the stopes and other working places. In another mine, monthly open-air meetings were held at which the hostel manager and a production manager harangued mineworkers about 'hothead' union officials and the job losses they had caused.

The Intrusion of UWUSA

Not all of the mines and collieries had been able to recruit scabs during the strike but, where they had been introduced, mine managers tried to continue using them as a means of creating an alternative power structure to the NUM which would both challenge its membership basis and adopt policies that were consistent with the aims of the mine owners. The favourite strike-breakers for this purpose were the supporters of Inkatha from KwaZulu-Natal for they fulfilled a double purpose. They helped to break the strike and enabled the managers to establish UWUSA branches in competition with the NUM, even though that union had no recognition or even access agreements with the mine authorities. The mine manager at Western Holdings, for

[34]*Ibid*. The British newspaper, the *Daily Mail*, also blamed the strike on the ANC in its report on 10 August 1987.

example, opened an UWUSA office immediately after he had closed the NUM one. In another mine, NUM had to share an office with UWUSA, with the unions occupying it on different days of the week.

The willingness of the Zulu members of UWUSA to confront NUM overrode the managements' distrust of them because of their generally expressed dislike of mine work. They were not regarded by managers as reliable workers. There were many cases where Zulu mineworkers had broken their contracts and returned home. At the Winkelhaak gold mine, they had departed *en masse* late in 1983.[35] The KwaZulu authorities, however, had pressed TEBA during the mid and late 1980s to overcome its prejudice and take on more Zulu recruits in order to lessen their massive problem of unemployment due to the lack of investment in the region.[36] The strike and post-strike situation, then, came to their aid as mine managers began to give preference to Zulu recruits. The Anglo-American Corporation actually stated on the fourth day of the strike that "all future recruitment for Anglo-American collieries will be done in Natal only, ie only Zulu workers will be recruited in future".[37] The number of recruits from KwaZulu for all the mines increased from 21 834 in 1985 to 39 358 in 1989. Many of these were novices, indicating that TEBA was treating the region as a serious and permanent source for mine labour. At the same time, Lesotho, with its tradition of supplying loyal union members, was downgraded as a labour pool and the number of Sotho recruits began to fall.[38]

There were additional tensions and increased outbreaks of violence in the hostels wherever Zulus who had been recruited during the strike were retained as permanent workers. In the first place, there were antagonisms because they had taken the place of dismissed Xhosa and Sotho mineworkers and had, thereby, weakened the strike resistance. Secondly, by organising themselves in UWUSA they became tools to weaken the NUM. And, thirdly, there were allegations that they acted as management informers.[39] There were a number of instances of bitter fighting between Zulus and NUM members as, for example, at New Denmark colliery in September when seven men were killed. But the methods used by managers to use Zulus as a divisive element irrespective of the cost to both Zulu and NUM mineworkers can best be

[35]Crush, Jeeves and Yudelman, 1991, p166.
[36]Wilmot James, *Our Precious Metal*, 1992, p65.
[37]Von Lieres, March 1988, p79.
[38]The number of mine recruits from Lesotho declined from 105 506 in 1987 to 100 951 the following year. (Crush, Jeeves and Yudelman, 1991, p235.)
[39]M L Crowley and M du Plessis, April 1990, p21.

illustrated by a description of the events at the Randfontein Estates gold mine and Matla colliery.

The failure of the strike, and the retention of Zulu strike-breakers after it had ended, provided the manager at Randfontein Estates with congenial conditions for reviving the tribal *induna* system and fostering tribal animosities to destroy the NUM branch which was mainly comprised of Xhosas. Peter Zazini and Simon Thobala, both members of the NUM who were injured in an attack by Zulus, described how this was done at Cooke 3 shaft.[40] They said that when the strikers returned, two of the managers told them that the NUM no longer existed on the mine. In fact, none of the NUM shaft stewards had been re-employed and a new branch structure had to be created. But that is not what the managers meant. They meant, in effect, that the NUM had no presence in the mine and that they could forget about it. The union had the task, therefore, of re-creating itself amongst the ordinary mineworkers.

Its main difficulty was that UWUSA was present in the mine and was capable, with the help of management, of challenging the NUM's claim to dominance. Management made matters easy for UWUSA. It erected a fence between the two main sections of the hostel to separate the Zulus from the rest and then allowed the Zulus to operate as a distinct tribal group with their own shaft stewards. The hostel manager appointed some of them as 'tribal representatives' or *indunas*. They had permission to carry rubber truncheons and to liaise with the mine security force. Though UWUSA did not have any formally recognised status in the mine it made its presence felt by intimidating the Xhosa mineworkers so that they became fearful of supporting the NUM. The relations between the two groups grew tense and threatening, particularly after the head office of JCI ordered the mine manager to take down the dividing fence. That act was an open invitation to violence which UWUSA accepted at the end of November 1987 when it organised an attack on NUM members at night while they were in their beds. On that occasion there were no serious casualties. A week later, however, a number of NUM members were injured when an UWUSA supporter threw two home-made bombs at them, but he died when the third bomb exploded in his own hands.

There were other incidents at Randfontein Estates. Seventeen NUM shaft stewards protested to the manager about the bombing and asked for the Zulu *indunas* to be transferred to another shaft. They called a meeting to report back to their members but were told that the

[40]See *Weekly Mail*, 29 January 1988. Also see the *Argus*, 22 January 1988.

meeting was illegal and that, as a result, they had to appear before a management disciplinary committee. But before they did so they found that their cards were being rejected by the clock-in machine so they were unable to start work. Whilst they were trying to sort this out they were bundled into a security van and driven to a disused mine building where they were assaulted. This event heightened the tension in the hostel.

On 20 January 1988, members of the NUM heard a rumour that UWUSA was mobilising its members into groups in preparation for a fresh attack so they armed themselves and moved into the UWUSA side of the hostel. That evening from about 7 pm there was fierce fighting between the two groups that resulted in four people being killed and 67 injured. The mine security force fired tear gas at the NUM members and exhorted all of the Zulus in the hostel to come out and join the fight. Despite these efforts, however, the Zulus were driven into the security offices and the fighting ceased for that evening. After it had ended three white mine officials went into the hostel where the NUM members were staying and shone torches in their faces to discover whether or not they had incisions on their faces. Those who did had their cards removed and were told to report to management the following day. On that day, however, there was more fighting though no-one was killed or seriously injured. The incident ended with the dismissal of 450 Xhosa mineworkers.[41]

The situation at Matla colliery, where temporary Zulu recruits became permanent employees in September after the strike, began in a different manner and ended differently for the NUM.[42] Soon after the strike had ended the NUM officials made friendly overtures to the Zulus, who were accommodated in a tent encampment, and invited them into the hostel because they believed that a faction fight would be inevitable unless they took steps to prevent it. They recruited some of the Zulus into the NUM with the result that, contrary to the situation in most other mines, the union membership rose during the last quarter of 1987. At the same time, however, there were signs of a management-

[41]The interviews with the two mineworkers were conducted by Sefako Nyaka for the *Weekly Mail* which contacted JCI, the mine owner, to ask for its response. Some days later, K W Maxwell, the managing director of JCI's Gold Division, denied all the allegations: the two senior managers at the mine had never said that the NUM no longer existed at the mine; UWUSA shaft stewards had not been appointed; there were no Zulu *indunas* and the story about the 17 NUM shaft stewards being assaulted was completely inaccurate. (*Weekly Mail*, 12 February 1988.)

[42]McNamara, November 1988. The account of Matla colliery in this chapter has largely been drawn from McNamara's work.

sponsored anti-NUM group emerging in the hostel. The managers did not hide their preference for Zulu mineworkers. In December, they sent a delegation to Ulundi, the capital of KwaZulu, to study development and education. They then appointed Zulus to prominent positions in the mine, even though they were relatively late arrivals there. The two leading NUM branch officials at the colliery, Gwede Mantashe and Crosby Moni, were convinced that the management was deliberately encouraging the formation of a Zulu faction in order to instigate a fight between it and their members as part of its campaign against NUM.

The NUM branch at Matla had arranged a cultural day in the hostel on 16 December for all residents and invited representatives from COSATU.[43] It seemed that COSATU had not been made aware of the steps the branch had taken to attract the Zulus into its organisation for its main speaker derided Inkatha and its leader, Chief Gatsha Buthelezi. Then freedom songs were sung which referred to Buthelezi as a puppet. The Zulus at the gathering reported this back to the majority who had stayed in the hostel. They were then all mobilised and marched back to the hall to protest. The following weekend there were at least eight assaults in the hostel, some of which arose out of the tensions between the groups. When a NUM delegation met with the management to discuss the deteriorating situation it was told that any group had the right to organise and that, in any case, the union had forced workers to attend its meetings. The management gave the impression of being indifferent to the developing conflict in the hostel.

The differences were politicised during 1988. Inkatha came to represent a non-union group in the mine and though not formally recognised, it was treated as such by the management. Slogans began to appear underground which stated "Inkatha will eat COSATU" and "Kill Inkatha". Rumours circulated about impending violence and faction fights. The management responded by asking representatives of the non-union group and of the union to sign a declaration endorsing the principles of free association and non-intimidation. The non-union group signed but the NUM refused because it said that if it did it would be according the non-unionists the same status as itself in the mine. In order to give it that status, the non-union group asked management on

[43] 16 December was the anniversary of the Battle of Ncome (Blood River) in 1838 when the Boers defeated Dingaan's Zulu army. The Boers laid the foundation stone of the Voortrekker Monument near Pretoria on 16 December 1938 and celebrated that day each year as the Day of the Vow. The ANC, however, also celebrated 16 December as a stage in the liberation struggle because on that day it was announced that Umkhonto we Sizwe had been formed.

15 April for permission to hold a meeting so that it could elect representatives. This prompted the NUM members to form an action committee and to campaign much more explicitly against Inkatha. There was talk about "Zulu spies" and the need to "kill all people who cancel the union".[44]

The climax came five days before the non-union group was due to hold its meeting to elect its representatives. Tuesday, 10 May, was a day of locally organised protests against the Labour Relations Amendment Bill throughout the gold and coal mining industry. In Matla colliery the day started with a demonstration but the members of UWUSA refused to join it. As a penalty, the NUM members chased away the drivers of the personnel carriers that were waiting to transport the Zulu mineworkers to the hostel after they had come up from the morning shift so that they all had to walk seven kilometres from the shaft to the hostel. The demonstration went ahead but a group from it, singing and dancing and carrying COSATU banners and placards, broke through to the general manager's office and threw down their banners and placards in front of it. These events were part of a build-up in tension that erupted into violence between the NUM supporters and UWUSA that evening. The security force used tear gas and rubber bullets to separate the groups and killed one Zulu mineworker.

The news of the fighting spread to those working underground in No 1 shaft. When the Zulus heard it they left work and went back to the hostel, followed by a contingent from the security force, and fought with the NUM members already there. It was a repeat of the earlier clash. The security contingent again separated the fighting groups with tear gas and rubber bullets. Another Zulu was killed.[45] Fighting continued in the rooms throughout the night. The following morning the Zulus refused to return to the hostel so the management took them to a nearby colliery. But after discussion with the KwaZulu authorities the Zulu mineworkers decided to return home because their safety could not be guaranteed. The meeting scheduled by the non-union group to elect representatives on 15 May was never held.

Once the Zulus had left the mine, the NUM members began to take the initiative. They demanded the removal of hostel personnel who had been associated with the non-union group and warned non-union mineworkers that they had to join the union by 20 May. In August they demonstrated over the fact that the security force was externally trained

[44]McNamara, November 1988, p41.

[45]NUM officials in Matla claimed that both of those killed were members of Inkatha. (*Weekly Mail*, 13 May 1988.)

and demanded that it should consist only of personnel from the mine. In the new situation, white supervisors and white miners felt threatened and insecure working underground with large groups of black mineworkers. Some of them resigned. The mine found it difficult to persuade new white recruits to work there. By 1989, the NUM had regained and consolidated its power position in Matla colliery.

Security in the Hostels

The oppression caused by the interaction of the various control mechanisms used by mine managers was sealed by security arrangements within the compounds that were unprecedented in normal circumstances. The security forces were much more visible in day and night patrols in the compounds, in the hostels and even in the dining halls than ever before. The movements of mineworkers in and out of the gold mine compounds were restricted. At all except two of the gold mines they had to carry ID or clock-in cards or wear ID armbands, in most cases for the first time. There were reports of mineworkers being searched going into and out of the compounds. One of them complained that every time he went home his suitcase was opened and its contents were thrown on the ground. He was told to pick them up himself and go away.[46] Restrictions were imposed on visits by wives and friends. Visitors and NUM officials were not allowed to enter the compounds without special permission. It became difficult for mineworkers to visit others from the same mine but in different compounds, as at Vaal Reefs and Western Deep Levels, and even more difficult for mineworkers to enter the compounds of other mines. Within the compounds, roadblocks were frequently set up to restrict the movement of mineworkers in case they happened to be engaged on union business. There was a proliferation of random room searches that were carried out either when the residents were at work underground or at night when the security forces kicked the doors open and entered, without search warrants and without uttering a word of explanation. Mineworkers were never informed that they were to take place. "Hostel rooms," the manager of New Denmark colliery stated in a letter to all employees, "may be searched at random as and when management chooses."[47] They learned about the searches only when they returned

[46]NUM, Media and Research Department, December 1988, p66.
[47]*Ibid*, p73.

from work and found their rooms in a state of utter disorder with their possessions thrown about, damaged and, sometimes, stolen. Union members claimed that this was a means of victimising them and their representatives. Shaft stewards, for instance, could expect their rooms to be searched every five days or so.

The lives of black mineworkers became more restricted than before the union was formed and was more repressive than in the townships under the state of emergency. James Motlatsi, however, refused to make these distinctions. "Repression in South Africa," he said, "is one. It is just that its form and the way it is practised differ in different places." As for the state of emergency, in the townships it was a temporary phenomenon but it "is the order of the day in the mining industry. It has been permanent since the mining industry was started … it started in the mining industry. Recently, though, around the time the NUM was formed, some mining companies started to relax their repression … But immediately after the strike they reversed everything by 30 or 40 years. The company that used to be the worst, Gold Fields – today it is better than Anglo-American."[48]

The Consequences of Repression

The campaign to weaken and, if possible, remove the NUM from the mines had manifold effects for individual mineworkers. Some of them were incalculable, such as their inability to confront management and challenge arbitrary decision-making. The general consequence was a continuing deterioration in their working and living conditions. Without effective union representation the mineworkers were subjected to the arbitrary authority of management. Many more than hitherto were summarily dismissed without any recourse to an appeals procedure. They lost whatever influence they had exercised over their working hours in relation to waiting time for lifts at the end of shifts and overtime. They were made to work unpaid overtime because, managers insisted, they had to recompense for the production losses caused by the strike.[49] They were at a serious disadvantage in matters concerning assaults by white miners and officials because the outcome depended on whether the word of a white or black person was accepted by the

[48]*South African Labour Bulletin*, "Interview: James Motlatsi, President of NUM", June 1989, Volume 14, Number 2, p30.

[49]A mineworker from Vaal Reefs who had been interviewed by Bettina von Lieres (Von Lieres, March 1988, p79).

mine managers. Black mineworkers were rarely believed and only the union could confidently confront managers about their prejudice. To make matters worse, the assaults usually took place without black witnesses being present. But potentially the most serious effect was the deterioration in safety standards. White miners took the opportunity of speeding up the production process, and, thereby, increasing their bonus payments, by neglecting to undertake pre-shift safety inspections and by compelling black mineworkers to work in unsafe conditions. The NUM had only made marginal gains in this regard before the strike but even these were lost in the repression that followed it.

The impact of management tactics on the union was twofold. Its central administrative structure remained intact because it was largely funded by foreign donations. The main casualty was the shaft and branch organisation that had been battered by selective dismissals. There was little difficulty in filling the vacant positions for there were constitutional provisions for providing replacements and an ample supply of candidates. The problem was in obtaining local officials with the requisite education, training and experience to organise and administer a trade union. It was yet another of many occasions when the union had virtually to re-create itself by filling committees with people who had no idea about how to fulfil their constitutional obligations. Fresh batches of mineworkers had to be trained to undertake the simplest of administrative tasks. This was not a matter that could be undertaken quickly or easily in an organisation where more than two-thirds of its members were illiterate. The NUM, as Marcel Golding, the assistant general secretary, remarked in June 1988, had continuing problems in rebuilding its organisation.[50]

The second major difficulty faced by the union was restoring and increasing its membership. The focus of mine managers was to destroy the attraction of the NUM to the mineworkers. They tried to do this by removing its presence from mines in the belief that what was not seen would not be missed. But, more importantly, they attempted to instil a fear of being associated with the union in the minds of mineworkers, particularly the novice recruits. The managers were relatively successful in this respect. The union's paid-up membership fell from 261 901 in July 1987 to 207 941 in May 1988. By January 1989 it had risen by only 545 despite beginning to make inroads into the platinum mines in Rustenburg and the small diamond and asbestos mines in the Eastern Cape, the far Northern Transvaal and

[50]South African Institute of Race Relations, 1987/88, p683.

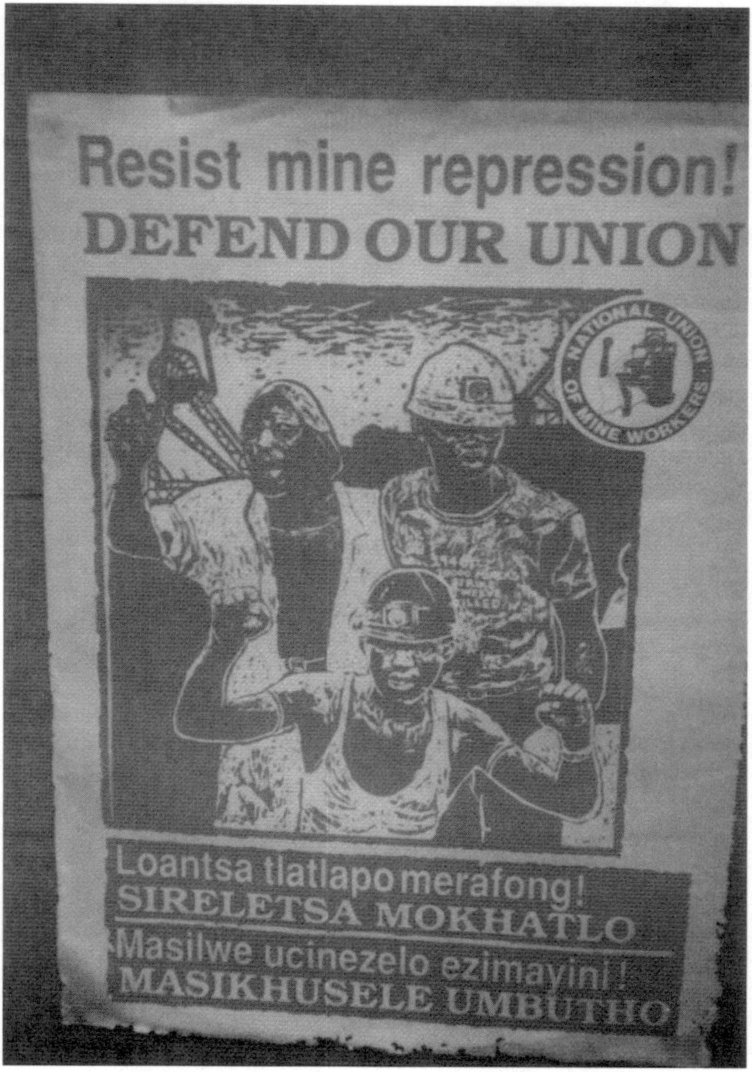

NUM campaign against mine repression.

Pietersburg.[51] Quite clearly the most important task facing the NUM was restoring the confidence of the ordinary mineworkers in it despite the risk of penalties. The length and extent of this operation depended to a large measure on the persistence of managers in maintaining their repressive tactics. There was no relaxation in the situation throughout 1988. The union thus felt it necessary to launch an Anti-repression Campaign at a mass rally in February 1989 and to continue devising ways through which its organisers could gain access to the hostels. A meeting of NUM organisers early in 1989 tried to impress on Ramaphosa the difficulties they faced in gaining access to the hostels. In many instances, they said, it had become an impossible task. He refused to accept their explanation and decided to show them by example that they were wrong. One evening shortly afterwards, he drove by himself to a mine compound in Westonaria, evaded the mine security and, with his bag full of leaflets, made his way to a hostel. The hostel dwellers, however, did not recognise him in the dark and chased him away as an intruder. He returned home somewhat chastened.[52] The union's recovery did not become apparent until the 1990s.

[51]This loss of just less than 21 per cent was important for the NUM but it was not excessive in that period. Eight of the 13 affiliates to COSATU experienced a significant fall in their memberships. Some of them had proportionate falls much greater than that suffered by the NUM. The National Education, Health and Allied Workers' Union, for instance, lost 36 per cent of its members while the membership of the South African Railways and Harbour Workers' Union declined by 31,2 per cent. (Baskin, 1991, p282.)
[52]Ramaphosa told the author about his experience the following morning.

CHAPTER THIRTEEN

AFTER THE STRIKE: THE REBUILDING PROCESS

Conditions for Survival

Throughout the period of repression following the strike, there were counter-tendencies at work which ensured that the union would not simply survive but would recover and develop. Firstly, despite the determination of the mine owners to remove its presence from the mines and the resulting decline in the union's membership, the union's survival was not seriously in doubt for, unlike in earlier periods in the history of black mineworkers, such as after the 1946 strike, the notion of trade unionism was firmly established amongst them. This had been illustrated in the phenomenal growth before the strike and the remarkable degree of solidarity shown during it. The strike defeat had caused some mineworkers to leave the union through disillusionment, but a strong core of militant mineworkers remained that had been stimulated by the aura and the excitement of struggle and who were eager to fight for NUM's recovery. In their eyes the strike defeat had reinforced the need for a strong NUM. Moreover, even in its depleted state, it remained the largest trade union in South Africa with a vital and generally acknowledged role to play in the struggle against the apartheid system.

Secondly, generous financial help from abroad ensured that the fabric of the union's organisation, particularly at national level, could be sustained despite its loss of income from subscriptions. NUM's education, finance, legal and health and safety services were already entirely financed by overseas donations, primarily from trade unions in Scandinavia and the Netherlands. Those donations had been supplemented by financial aid from a number of other foreign trade union bodies during the strike so that in 1987, there had been a massive increase in the amount of money sent in from abroad. Whereas in 1986, about 27 per cent of the union's income had come from foreign sources, a year later overseas funds were more than one-and-a-half times greater than the subscription income.[1] The total amount of money from abroad fell in 1988 but there was sufficient to continue

[1] In that year, despite the costs of the strike, the union was able to invest R2 million in a relief account with an investment bank called Eurotrust but in November 1988 the bank was investigated by the Reserve Bank and placed under liquidation and the union lost its entire investment.

financing the health and safety, education and legal departments and to contribute substantially to the general costs of running the union. This aid enabled the union to maintain its services to its members but, in doing that, it concealed some of the union's internal problems.

The Expansion of the Administration

The central departmental administration, that had been the pivot of the union since its formation, became even more important in the post-strike period for it was the only area of union activity that was insulated from management and, in some ways, government, repression. In this environment, Ramaphosa was able to reorganise and expand the union's central departmental structure and make it even more essential for the union's operations throughout the mining industry. There were nine departments, two of which organised the work of the central office and controlled the union's finances. Of the seven service departments only those dealing with education and legal affairs had staff members and were able to provide rudimentary services to the members. Each of the others, described by Ramaphosa as "under-staffed", was the responsibility of one person.

The legal department accounted for the largest item in NUM's budget account. Its purpose was to provide legal assistance to union members but it was not equipped to deal with the overwhelming workload that resulted from the compensations claims for mass dismissals and retrenchments after the strike. Its main problem was obtaining and retaining staff with legal qualifications because it was not able to pay salaries that were comparable with those given by private legal practices. When it was established, the year before, it had five staff members but four of them quickly left the union and only two were replaced. This was insufficient for the department to provide a satisfactory legal service. Moreover, as none of the staff possessed the legal qualifications that would have enabled them to represent the union in court, it had to pay for the services of lawyers from private practices and this added considerably to the cost of running the department.

The education department faced a similar problem in that it was constrained by its inability to obtain competent instructors to work in the regions. Its tasks were formidable. It had to aim, on the one hand, to overcome the deficiencies of the Bantu educational system by creating a literate membership and, on the other hand, to "politicise, conscientise and enlighten mineworkers as part of the working class" as instructed by the Fifth National Congress in February 1987. The union's educational activities were co-ordinated from the head office by Kgalema Motlanthe who had received his political education whilst in prison on Robben Island. He had only five instructors working for him in the 16 regions of the union. The constraints caused by the shortage of qualified staff were so great that Motlanthe faced the rare problem of not being able to spend the funds allocated to his department.

In addition to providing for the development of these two departments, Ramaphosa initiated significant new administrative developments by establishing departments for organising, collective bargaining, media and research, projects, compensation and health and safety. Organising had always been a basic union task but was regarded as something that union activists did naturally and spontaneously without prompting, guidance or training. It was not until after the Fifth National Congress that it was recognised that organising would not only benefit from centralised co-ordination but actually required it. At that point Ramaphosa formed an organising department with Gwede Mantashe, the regional secretary in the Witbank Region, as the union's first national organiser. Its primary task was to recruit new members despite the constraints imposed by the hostel authorities. It attempted this in two ways. Firstly, in those mines where legitimate trade union activity was forbidden it formed study circles within the hostels that operated as secret, shadowy branch structures in much the same way as the cell system had operated during the strike. Secondly, it expanded its recruitment staff by employing dismissed mineworkers who were already experienced in organising under repressive conditions. By the end of 1988, the organising department was the only one that had staff members in all but one of the union's 16 regions. Fifteen former mineworkers were employed as organisers in all but four of those regions.[2]

The union, however, was acutely short of organisers because, though there was a virtually unlimited supply of dismissed mineworkers who would dearly have liked to work for the union, only a very small number possessed the literacy standards necessary for the job. A small number of dismissed mineworkers whose literacy standards were inadequate for full-time organising work were, however, given jobs as

[2]Dismissed mineworkers were taken on in other departments of the union. For instance, Buyisile Jonas was made an instructor in the education department while four other dismissed mineworkers, including Sazi Jonas from Western Deep Levels, were a part of the union's complement of eight full-time safety officers working in the regions.

cleaners and messengers and were encouraged to educate themselves in preparation for other union work. Some of these were, in turn, employed as recruiters on a recruitment bonus system. They received basic salaries that were less than those given to organisers but were paid a monthly bonus provided they reached a specified recruitment target.[3] They received no bonus if they failed to reach that target. Organisers were subsequently recruited from the pool of recruiters.

The task of organising had been made more difficult when, in October 1988, a Special National Congress decided to double the union contribution from R1 to R2 a month in order to make the union more financially viable. Some members objected to the increase and withdrew from the union. Every member who did not resign had to take out and sign a new membership form authorising management to deduct the higher union contribution from his or her wages and then return it to management. This became a large-scale administrative project because many of the existing members could not see the sense of the exercise. They contended that only new members had to sign stop-order forms. A part of the recruitment campaign launched in 1989 was, therefore, to persuade old members to sign the forms.

The important and highly specialised task of compiling the economic and statistical data for the union's submissions at the annual round of collective bargaining with the Chamber of Mines and for negotiations with the independent mining operators was, until July 1988, the responsibility of two people, Irene Barendilla, the information officer and Manoko Nchwe, the research officer, who both worked in the research, information and publicity department. On that date, a new collective bargaining department was established with Martin Nicol, an economist, as its newly appointed head. Barendilla and Nchwe were transferred to work with him.[4] This new department remained small but was consistently highly productive for the union. About four months

[3]There were five such recruiters. One of them was Moeketsi Sekhokoene who worked in the Westonaria region. He was reputed to earn his full bonus every month. In 1990 he was one of 21 black mineworkers who went to Cuba to study mining engineering. At that time there were only two qualified black mining engineers in South Africa. He became the first foreigner to lead the Cuban student movement and returned with a degree to become a mining engineer in South Africa.

[4]Martin Nicol had been the Labour Research Officer for the Labour Research Service in Cape Town. He and Kate Philip, who was responsible for the projects department, were the first two white officials to be employed by the union. By this time Ramaphosa was firmly non-racialist in his views and no longer held the political objections to employing white officials in the union that had prevented him from giving the post of legal adviser to Paul Benjamin, four years previously.

later, a media and research department was created with five members of staff, including Jerry Majatladi as its co-ordinator and press officer.[5] Majatladi formally took over the press and publicity work previously done by Marcel Golding prior to his election as the assistant general secretary.

Shortly after the strike, a projects department was formed with the aim of developing co-operative projects in the rural areas that would provide work for unemployed dismissed mineworkers and, at the same time, enable them to retain their links with the union. It started with one staff member, Kate Philip, who was responsible for its subsequent development.[6] The decision to establish the department was based on a resolution by the 1987 National Congress which stated that co-operative projects were not only an effective way of organising unemployed members but could be used "as training schools to conscientise workers towards socialism". It added that the Phalaborwa Workers' Co-operative (PAWCO), begun in 1986 as a pilot project, should "be fully integrated into the union's structures".

The Phalaborwa project had been started after the entire 1 800 work force at the state-owned Phosphate Development Corporation (Foskor) opencast mine had been dismissed for supporting the 1986 COSATU May Day stay-away. The NUM had negotiated the reinstatement of all the workers but, when they turned up at the mine, 390 of them were told, without explanation, that they had been dismissed. There was no alternative work in the area for the dismissed mineworkers, primarily because the other employers were suspicious that they might be security risks. Phalaborwa was a highly politically sensitive area that was subjected to close surveillance by the police and military intelligence because of its proximity to the Mozambican border where there were secret Renamo training camps.[7] NUM activists, suspected of having subversive intentions, were particularly closely monitored and were the subject of much police

[5]Majatladi had been a Robben Island prisoner with Kgalema Motlanthe, the head of the Education Department. Both had received much of their education through the ANC on Robben Island.

[6]Kate Philip had been a student activist who joined the staff of the union after completing an Honours degree at the University of Cape Town.

[7]Renamo (Resistencia Nacional Mocambicana) was the rebel group that was then fighting the legitimate Frelimo government in Mozambique with support from the USA and South Africa. The South African government had signed a non-aggression treaty, the Nkomati Accord, with the Frelimo government in March 1984 but had continued to provide undercover military support for the rebels. It was for that reason that the training camps were described as 'secret'.

Jerry Majatladi, NUM's press officer, fresh from eight years on Robben Island.

harassment. The situation had been volatile since the end of 1985 when the police fired on workers from Foskor and the Phalaborwa Mining Company at a union meeting following a Foskor strike. One worker had been killed and others were injured. The incident had caused unrest in the Namakgale Township and, in order to assist them in quelling it, the police had deployed a number of Renamo soldiers who rampaged through the township firing their weapons indiscriminately and killing at least 12 people. Many more were injured.[8] It was in this setting that the May Day stay-away had taken place. Fifty of the dismissed mineworkers had decided to form a union-based workers' co-operative association, employing only part-time labour in the first instance. In September 1986, they had bought a machine to print T-shirts and went to Johannesburg for training in how to use it. It was not until July 1988 that they began to print T-shirts, carrying the NUM logo and the names of the imprisoned and exiled leaders of the liberation movement, even though that was illegal. The police often raided their shop but by 1989 the co-operative was making sufficient money to pay wages to its part-time staff.

New projects were initiated in Lesotho and the Transkei. The Lesotho project started in a difficult situation for the union. The NUM had sent a delegation there during the strike to assist the strike committees set up by Lesotho mineworkers who had returned home. When the strike ended the delegation was instructed to stay there and assist the projects department with the development of workers' co-operatives but it misused NUM funds and was accused of corruption. It also tried to undermine the NUM in Lesotho by setting up a Basotho Mineworkers' Union. As a result, the members of the delegation were expelled from the NUM in June 1988. The projects department then established the Basotho Mineworkers' Labour Co-operative which set about producing low-cost building materials for use in the community. Its first project started at Quthing, in November 1988, with 40 members and began the production of brick blocks the following January.

The project in the Transkei was more ambitious but much less successful. The Transkei Mineworkers' Co-operative Project which was set up after the strike comprised 28 district committees, eight branch committees, planning committees for east and west Transkei and a regional executive committee, but despite this extensive network of committees only two block-producing projects were started. One of

[8] *NUM News*, December 1988.

them was handicapped by its own bureaucracy whilst the other was frustrated by the Transkei's own administrative bureaucracy.[9] The difficulties that were encountered in the Transkei illustrated those faced by the projects department in general. The dismissed mineworkers had no knowledge or understanding of the organisation of production and the department had to devote much time arranging courses for them in acquiring production skills and explaining the role of co-operatives, the meaning of democratic management and simply how committees worked. It was some time before either of the projects was self-financing.

The Long Haul to Mine Safety

One of the most important of Ramaphosa's acts of reorganisation following the strike was the combination of the health and safety department and the compensation department into a single administrative unit, funded by money from the overseas-financed Relief Fund. It was one more step along the trail to mine safety. At the same time, he appointed Mavis Hermanus as the co-ordinator of the new department in place of Hazzy Sibanyoni who became one of eight regional safety officers to work under her direction. Hermanus was an engineering graduate whose competence in mine safety issues was equivalent to that of many mine managers. She raised the status and profile of the department by making health and safety a major union activity in practice and not simply through slogans. The department published safety pamphlets and a newsletter, organised seminars and workshops on safety issues, campaigned for the workplace election of recognised safety stewards and, in August 1989, held the union's Third Safety and Health Conference.[10] An indicator of the extension of the department's work was that the union's expenditure on health and safety rose from almost zero in 1987 to R56 628 a year later.

The situation facing the new department was complex. There was pressure on the government from the NUM and the international media for it to take responsibility for the continuing high injury and death toll in the mines and to initiate a new policy to reduce it.[11] It responded but

[9]NUM, *Biennial Report*, 1989.

[10]In February 1989 Hermanus presented a paper on "The Role of the Safety Steward" to an Institute of Safety Management seminar that clearly delineated the union's view of mine safety procedures.

[11]There were, for instance, 2 227 deaths and 36 713 injuries in the mines in the three years leading up to the end of 1988 and almost 88 per cent of the deaths were in mines controlled by the major mining houses.

within rather strict constraints. Due to the combined effects of the economic depression and the insurrection in the townships it had cut public expenditure throughout the 1980s and mine safety was one of the areas of government responsibility that suffered from the cuts. The Mines Inspectorate, in consequence, was continually short of funds and was unable to function in accordance with its legal responsibilities. The Government Mining Engineer had explained the problem and the proposed solution when he had stated in 1987 that, "with the present clamour for reduced state spending and privatisation there is no prospect of having a sufficient number of inspectors to exercise the necessary direct supervision. With this in mind, the Minister proposed that the safety organisations on mines be strengthened and be given legal status. The relevant regulations have been drafted and provide for the appointment of safety officers and safety representatives. Both appointments will be made by the mine manager."[12] The union played no part in this selection process. The regulations, made law in 1988, compelled each mine employing more than 300 workers to employ a safety officer who had the legal responsibility for ensuring that the mine was safe. The effect of the regulations was to strengthen the authority of the mining houses in safety matters by transferring some of the powers of inspection from the Mines Inspectorate to mine managers and to consolidate the legal exclusion of the union from any involvement in mine safety matters. The government's solution, therefore, was to privatise the responsibility for mine safety.

The new safety officers were required to:

> undertake inspections of the workplaces and machinery, check that safety equipment is properly maintained, ensure that safety and health standards are complied with whether these have the force of law or not, and that workers have the necessary training or qualifications for the safe execution of their work. Safety officers are also required to hold inquiries into reportable accidents, record every accident which results in more than one lost shift and identify places where hazardous work is being performed or dangerous conditions exist.[13]

The black safety 'representatives' selected by managers from the work force were limited to observing and reporting any hazards they saw to the white miners who were their immediate supervisors. They

[12]G P Badenhorst and D Bakker, "Responsibility for Safety and Health in the Production Environment", *Mine Safety and Health Congress Proceedings*, Chamber of Mines, 1988, p9.
[13]Leger, "Key Issues in Safety and Health in South African Mines", p28. Leger described the new regulations as a form of privatisation.

The Safety Campaign at Freddies in the Orange Free State, 1992.

Safety demonstration by paraplegics in the Orange Free State.

were permitted to accompany safety officers on inspections in their own workplaces and were required to meet formally with a safety officer every three months. They had no authority to take any independent action and were entirely accountable to management. When the regulations were promulgated some of the mines broke off their local negotiations with the NUM over the recognition of union safety stewards because they saw the new 'representatives' as a means of bypassing the union. In other instances, however, the NUM increased its influence because some mineworkers refused to be appointed to the new positions unless the union was involved in their selection.[14]

The government's policy concessions were not matched by a comparable willingness by the Government Mining Engineer to take note of suggestions made by the NUM. He continued, for instance, to ignore the international and national evidence about the danger of using polyurethane underground until October 1988 when the fifth polyurethane accident in just over 10 years occurred at Western Deep Levels. The disaster followed shortly after the NUM's annual commemoration of Kinross on 1 October. On this occasion, carbon monoxide fumes killed seven mineworkers. A Joint Inquest and Inquiry was held which exonerated the management, even though the safety regulations had been breached and the danger of polyurethane was more widely known than it had been at the time of the Kinross disaster. The Western Deep Levels deaths, however, convinced the Government Mining Engineer that he had to act, so he notified the mining houses the following month that polyurethane foam could no longer be used underground and that where it was present it had to be removed as soon as possible. But he continued to exclude the NUM from safety policy deliberations.

This question was raised by the NUM's appeal against its exclusion from the Kinross inquiry. It took long to receive an answer, however, for the due process of law was long-winded and the appeal was not heard until 27 November 1989. The issue hinged on the Inspector's interpretation of the Mines and Works Act. He had claimed that the relevant regulation gave him discretion as to whether to allow the union to participate or not. He said that he had weighed up the factors and had decided that it would not assist the inquiry to allow it to ask questions. Kinross Mines, which opposed the appeal along with the Government Mining Engineer, argued that the Inspector had no discretion to allow the union take part but that, in any event, the lengthy criminal trial had

[14]*Ibid*, p29.

made an inquiry unnecessary.

Mr Justice Goldstein heard the appeal and he gave his ruling in favour of the NUM on 2 February 1990. He stated that the regulations issued under the Mines and Works Act gave the Presiding Inspector the responsibility to do everything practicable to find out the cause or causes of an accident. He had to decide what to do in each case and he was allowed the discretion as to whether to permit cross-examinations at an inquiry or not, even though this was not mentioned in the regulations. But in deciding what to do he had to be guided by his obligation to search out the truth. On that matter he had no options. He had to ask, therefore, whether a skilled advocate representing a trade union, which organised the majority of the workers in the mine, could, by cross-examinations, assist him in searching out the truth. The answer had to be 'yes'. It would, Mr Justice Goldstein declared, have been contrary to the Act to do otherwise. He dismissed the argument that an inquiry was not necessary because there had been a criminal trial over the disaster. An inquiry, he stated, had a different purpose from that of a trial. It was possible that an inquiry might uncover causes of an accident that did not constitute an offence whether at common law or otherwise. Moreover, at the trial none of the accused gave evidence so possibly vital data about causes for which they might have been responsible was never put to the magistrate.[15] The NUM welcomed the judgment as a major victory in its struggle for safer mines and hoped that at long last it would convince the Mines Inspectorate that the NUM had a role to play in promoting improved health and safety in the mines. In the outcome, however, the only concession it received was the right to cross-examine at inquiries.

The Chamber of Mines continued to insist that safety was not a negotiable issue. Moreover, it began to adopt a complacent attitude towards it after the government had sanctioned the appointment of safety officers. This was reflected in the presidential address of Colin Fenton, the President of the Chamber of Mines in 1989, when he declared that although gold mining had become more difficult over the last 25 years, it had nonetheless become safer "because of the research being done in safer mining practices and methods, the successful implementation of safety management, and the moves towards improving safety".[16] The individual mining houses, however, were less optimistic and adopted varying positions. Anglo-American, for instance, admitted that

[15]See "National Union of Mineworkers and Others *v* the Government Mining Engineer and Others", Witwatersrand Local Division, 27-30 November 1989; 2 February 1990.
[16]South African Institute of Race Relations, 1989/1990, p671.

mineworkers should have the right to participate in safety issues after Ogilvie Thompson, the chairperson of Anglo-American's Amgold division, had stated that "worker involvement in safety remains a priority".[17] The NUM, however, did not want worker involvement or the right to participate with management. It wanted democratically elected safety stewards in each mine and the right for the union to negotiate industry-wide safety agreements. Neither of those items was on the agendas of the mining houses.

Hermanus discovered, as Sibanyoni had learned before her, that re-establishing the mine health and safety committees destroyed during the strike and negotiating new ones were difficult and slow tasks. Committees had been formed in eight of the largest regions of the union by 1989 but there were union safety stewards in only four of them. The vast majority of the union stewards were in the gold mines in Klerksdorp and the Orange Free State and the Rio Tinto Zinc mine at Phalaborwa, where an agreement had been signed immediately after the national strike providing for the election of 95 safety stewards to represent a work force of just over 5 500. In the Orange Free State, on the other hand, with almost 30 times the number of mineworkers, there were only 140 safety stewards. Klerksdorp was the best-organised gold mining region from the point of view of safety with 187 stewards serving a work force of about 71 000. Alongside this development were attempts to negotiate draft mine health and safety agreements. The only national agreement was with De Beers to cover the diamond industry. Negotiations were resumed at some Anglo-American mines from mid-1988 but the only success by the following year was at the Vaal Reefs complex. An agreement was also signed at Gencor's Stilfontein gold mine in Klerksdorp. The way seemed clear for up to 10 mine level agreements to be signed in 1989 but progress continued to be slow. The agreements gave mineworkers some limited protection on matters neglected by the law, such as the right to negotiate over safety matters, to refuse to work in dangerous places and to elect their own safety stewards but they were voluntary and without the force of law they had a limited impact. Mineworkers were still dismissed, for instance, for refusing to work in dangerous places. They continued to complain about supervisors who disregarded safety measures, refused to carry out inspections, prevented them from carrying out their own investigations and generally ignored protests. There was, for instance, a case at

[17]J P Leger, "From Fatalism to Mass Action – The South African National Union of Mineworkers' Struggle for Safety and Health", *Sociology of Work Programme*, University of the Witwatersrand, Johannesburg, July 1988.

Rustenburg Platinum Mines in 1988 where a stope hand was asked by his white supervisor to check an area for safety. He did so and declared it unsafe. The mine captain was called who confirmed that the area was unsafe. When the mine captain left, however, the supervisor nonetheless instructed the stope hand to work in the area. When he remonstrated again and tried to show the supervisor how loose the rocks were they fell on him, breaking his leg in two places and severing his right arm. After a long spell in hospital he remained severely disabled and was unable even to dress himself. He was put on light duties in the mine at a much-reduced wage.[18]

A Centralised Administration

The main significance for the union of the administrative changes made by Ramaphosa was that they improved the efficiency of the centre when there was no effective administrative infrastructure for the union as a whole. The head office administrative departments, therefore, began to assume a much greater disproportionate influence in running the union. It had always been the case that Ramaphosa had relied on people appointed from outside the industry but once they became heads of departments they began to assume a much greater authority than any of the organising staff. From this period on, appointed staff members, with specialist skills but no background in mining and often with little understanding of class relations, began to mould union policies and supervise their implementation. They were able, moreover, to stand for election to national secretariat positions because the union rules made no distinction between specialist staff appointed to administer the union and union officials, either elected or appointed, with the responsibility for organising it.[19] This anomaly remained in the rules but its defects were slowly, almost imperceptibly, resolved in practice as black mineworkers with the requisite educational standards were recruited to work in both staff and organisational capacities.

The creation of the new departments and the reorganisation of the existing ones entailed a redistribution of the union's income. In theory,

[18]Carey, 1990. His intention had been to save enough capital to acquire a taxi and become independent. The accident made this impossible. He became entirely and permanently dependent on the benevolence of the mine.

[19]The first case of this happening was when Marcel Golding, a qualified research worker and the union's press and publicity officer, stood for election as the assistant general secretary in 1987 and defeated Gwede Mantashe, a former mineworker at Matla colliery.

this redistribution should have indicated the union's priorities and reflected its intention to consolidate its administration and organisation. But, as the members of the secretariat discovered to their dismay, some forms of expenditure took on lives of their own, were difficult to control and actually distorted the union's priorities. Between 1986 and 1988 the wages and salaries bill rose by two-and-a-half times, the expenditure on telephones, telex and postage doubled, the cost of printing and stationery tripled while the expense of rented equipment increased 16 times. In 1986 the largest single item of expenditure, greater than that on salaries and wages or on legal fees, had been on the provision, maintenance and running of motor vehicles. This high level of expenditure continued, almost uncontrollably. The union had bought second-hand cars but its officials used them recklessly. Indeed, fast, reckless driving and a complete disregard for union property seemed to be part of their way of life. Some officials did not have driving licences or, without any difficulty, had obtained them without taking driving tests. Even those who had legitimate licences often had virtually no experience of driving in congested traffic conditions. In consequence, there were occasions when virtually every union vehicle was out of service because they had been crashed or damaged through excessive wear and tear or simply neglected. This situation changed in one respect in 1988 when the union bought a number of relatively new vehicles for its organisers and staff. The book value of its stock of vehicles increased 10-fold during that year. Most officials had the use of these newly acquired cars, except for Ramaphosa who still drove the rather old BMW car he had brought with him to the union. The expenditure on maintaining the vehicles, however, continued to be high as, at any one time, many of them were being repaired at the union's expense because, after so many crash incidents, it was unable to get insurance cover for them.

The figure that explained most about the union's financial difficulties between 1987 and 1989 was the cost of legal fees. The union's legal department still employed no qualified advocates so it had to hire them to defend members' interests in cases of murder, violence, infringements of the emergency regulations, detention without trial, unfair dismissals, mass dismissals and the infringement of safety regulations. It also had to hire attorneys to make representations at inquests and inquiries. The many instances of mass dismissals imposed a particularly heavy burden on the department between September 1987 and the following March when an agreement over them was reached with Anglo-American. Additional legal expenses were incurred by the regions which usually took their own legal initiatives and employed advocates without recourse to the head office and, therefore, without its advice or direction. Legal fees alone rose from R350 808 in 1986 to R1 062 860 in 1988. This was a major drain on the union's resources.

The Broadening Scene

The black public's perception of the union's secretariat, particularly that of Ramaphosa and Motlatsi, was unaffected by either the result of the strike or the problems in their union. Indeed, they were judged by their willingness to confront the mine owners and survive the encounter. For that they received accolades wherever they went.[20] They carried the image of the NUM as a symbol of resistance into the international trade union scene and the national liberation movement. Both of the leaders travelled abroad but it was Motlatsi who mainly carried the banner into the international trade union movement where he was received as a prestigious member rather than as a supplicant seeking aid, as had been the case before. Ramaphosa, on the other hand, became deeply involved in the liberation movement within the ranks of the ANC as the NUM began to assume a pivotal role in the liberation struggle.

James Motlatsi's international role was co-terminus with that of the NUM. He was elected as the president of the Southern African Miners' Federation (SAMF) at its Inaugural Congress in 1988. On a wider front, he became the second vice-president of the African Federation of Miners' Trade Unions, of which the NUM was the largest affiliate. But his main involvement on the international scene was as the representative of NUM in the Miners' International Federation (MIF). He was elected as one of its three vice-presidents, with the highest popular vote, at its 46th Congress in Harare in March 1988 and began to travel the world as its principal Third World representative. The NUM as an organisation, however, continued to maintain a close political relationship with the IMO which was in competition with the MIF. NUM delegates gave it enthusiastic support both in their own conferences and those of the IMO. The contradiction was illustrated at NUM's Sixth National Congress in 1989 when Peter Michalzik, the staunchly anti-communist general secretary of the MIF, had to sit through speeches that lauded the

[20]They were the three original national officials, namely Cyril Ramaphosa as general secretary, James Motlatsi as president, Elijah Barayi as vice-president, along with Marcel Golding, assistant general secretary, and Paul Nkuna, the national treasurer, both of whom had been elected in 1987.

Peter Themba Mthembu, who worked for some years as the Transport Officer in the Finance Department. He had the difficult task of trying to resolve the problems caused by the chaotic use of cars. He had been a long-standing member of MK and had fought during the late 1960s with the MK contingent in its Wankie campaign in northern Zimbabwe, in an alliance with the Zimbabwe African People's Union (ZAPU), to secure a route from Zambia through Zimbabwe to South Africa. The attempt had failed. Mthembu belonged to a small group of ANC activists and former prisoners to whom Ramaphosa gave jobs in the union. He became one of NUM's longest-serving staff members.

values of communism and praised the SACP and the IMO. NUM's official policy favoured a merger, on the basis of equality, between the IMO and the MIF and it submitted a resolution to that effect at the MIF's 46th Congress, for, as Ramaphosa later wrote: "We find the division that exists between the International Miners' Federations regrettable and not serving the interests of the miners worldwide as it weakens miners in many different ways."[21] Motlatsi liaised between them for a time but soon gave up the task for the Cold War barriers they represented at that time were insurmountable. Both he and Ramaphosa began to evade the problem by taking the pragmatic step of conforming to the policy of the MIF as representatives of NUM on its governing bodies, whilst not questioning NUM's official policy of continuing support for the IMO and for a merger between the IMO and the IMF. The effect was to block moves towards a merger because NUM's intentions were neither formally submitted nor advocated in the councils of the MIF. The problem, however, did not go away.

Despite their international obligations, neither Motlatsi nor Ramaphosa could accept invitations to travel abroad in the expectation that they would be free to do so. Motlatsi had a Lesotho travel document to enable him to cross borders though sometimes foreign passport officials, who thought that Lesotho was as fictional as Ruritania, questioned its legality. He had many difficulties entering Britain for that reason and was rarely allowed in without a lengthy explanation. Ramaphosa, on the other hand, required a South African passport and the government was not always willing to issue him with one. He was prevented from travelling to Britain in September 1987 to speak at an Anti-Apartheid Conference sponsored by the British NUM but was allowed to go to Sweden with Motlatsi immediately after the strike to be presented with the Olef Palme Award, worth R36 229. The money was allocated to the union's Education Project Fund. After two days in Sweden they went to Denmark for one day to thank a Danish newspaper for raising R750 000 for the NUM during the strike. They both returned to London but parted there. Motlatsi went on to the USA but Ramaphosa returned to South Africa. He had already been refused a travel document to attend a conference about South African affairs in New York that October.

When in the USA, Motlatsi went on to Washington where he met Senator Edward Kennedy to discuss the post-strike situation of the NUM and the wider issue of sanctions. When Kennedy asked which US

[21]NUM, *Biennial Report*, 1989.

organisations had assisted the NUM during the strike, Motlatsi replied that they had received money only from the United Mine Workers of America. For political reasons, the NUM had refused to accept any help from either the American Federation of Labor/Congress of Industrial Organisations (AFL/CIO) or its adjunct, the African-American Labour Center (AALC) which discharged US anti-communist influences to trade unions throughout Africa with the help of American government money. The AALC, which was suspected of having links with the American CIA, was alleged to have given financial support to UWUSA in its campaign to destabilise the NUM.[22] After talking with Kennedy, Motlatsi gave evidence at a US Senate hearing on sanctions against South Africa.

But the lack of travel documents did not always prevent the two union leaders from travelling abroad. At the end of 1988, they were both secreted out of South Africa to visit Cuba and the Soviet Union. They toured Cuba and were entertained by mining trade union officials and Cuban specialists in African affairs. The highlight of their visit, however, was a four-hour discussion with Fidel Castro on 27 December about apartheid and the implications for the southern African region of the defeat of the South African forces in the Angolan town of Cuito Cuanavale in March 1988. Much of their discussion, however, was concentrated on Castro's scepticism about the policy of *perestroika* in the Soviet Union and on his concerns for the future of socialism throughout the world. He gave them a dismal forecast about the collapse of socialist states in the 1990s. They left Havana a couple of days later to visit Moscow where they had meetings with the miners' trade union leaders and Stepan Shalayev, the chairperson of the All-Union Central Council of Trade Unions of the Soviet Union, primarily to discuss the condition and financial needs of trade unions in South Africa.[23] They left Moscow with the promise of a substantial financial donation that was honoured shortly afterwards.

Ramaphosa's involvement in the liberation movement changed qualitatively after the strike. In an indirect but nonetheless significant way he had played a part in the politics of the liberation movement through his leadership of the NUM and his participation in the establishment and direction of COSATU. He was part of the elite internal leadership of the liberation movement that had emerged

[22]Davies, O'Meara and Dlamini, 1985, p40 and p432.

[23]These meetings were arranged by Arthur Scargill, president of the British NUM, and sponsored by his union. The author and John Hendy, the labour barrister, accompanied Ramaphosa and Motlatsi to Cuba.

through trade unions and community organisations and which was distinctive in that it had not been politicised through imprisonment in Robben Island or in exile. In the government's eyes, the internal organisations not only filled the vacuum in the revolutionary struggle against it but posed a greater threat than the ANC, the PAC and the SACP to the apartheid system. This view was expressed by the head of the government's security police, Lieutenant General Johan van der Merwe, in November 1987, when he said that the government believed that the "legal radical organisations", namely COSATU and the United Democratic Front, were more important than the illegal organisations in promoting "a revolutionary climate".[24]

However, for Ramaphosa to make his personal mark on South African political life he believed he had to do more than lead a legal radical organisation. Since the formation of NUM he had undergone a complete political transformation from black consciousness to non-racialism as the only means to a revolutionary end in South Africa. He had come to believe, therefore, that the political future lay with the illegal ANC. He considered that neither trade unions nor organisations of disparate political alliances, such as the UDF, could give effective political direction in a revolutionary take-over of power. This meant that, from a personal point of view, he had to work with and through the ANC but before he could do that he had to establish his credentials with it and that, in the light of his early association with the Black Consciousness Movement, was not easy. He had redeemed himself to some extent by his role in establishing COSATU with its pro-ANC posture and then by influencing the NUM, first to elect Nelson Mandela as its Honorary President and then to endorse the *Freedom Charter*. His speeches, too, had begun to reflect the ideology of the ANC. All in all, there were many signs by 1988 that Ramaphosa was closely identifying himself with ANC policies.

Given the prominence of Ramaphosa's role within the trade union movement, the external ANC leadership had to treat him as a prominent participant in the liberation struggle. This, however, presented the ANC with a difficulty, partly because it had its own constitutional leadership structure, which it zealously preserved and adhered to, partly because there was an inherent suspicion amongst the exiled ANC leaders of an internal leadership which it neither shaped nor controlled and partly because of Ramaphosa's previous association with the Black Consciousness Movement. Ramaphosa's first physical contact with

[24]South African Institute of Race Relations, 1987/88, p23.

the exiled ANC leaders had been in March 1986 when a COSATU delegation met them in Zambia. There he had had formal meetings with the ANC leadership and informal discussions with some of its members, in particular with Joe Slovo, Mac Maharaj and Chris Hani, and these formed the basis for continuing and close personal and political relationships. Ramaphosa's conduct of the strike helped to consolidate that basis.

The political situation in South Africa after the strike showed signs of an increasing determination by the government and the ANC to fight out their struggle. The state president, P W Botha, made his position clear when he said in April 1987 that he was determined to smash the ANC and then, in July, when he ruled out any negotiation with it. His attitude was reflected in the escalating casualty figures as the state security forces attacked activists in the liberation movement, detained the officials of COSATU and its affiliates, destroyed their union buildings and connived with the white and black vigilante groups that rampaged in black townships, killing and maiming people indiscriminately.

The MK, for its part, reiterated its call in July to take the armed struggle into the white areas of South Africa and declared in November that it was taking "off the white gloves" and was preparing to stage "sensational attacks" before the end of the year. It had developed the capacity to make frequent clandestine movements over South African borders and undertook many more planned acts of insurgency than hitherto. By November 1988, there were 262 reports of 'terror incidents' in South Africa, excluding the homelands, compared with 137 in 1985.[25] There were 'soft targets' amongst them, not for the first time but on an increasing scale. The UDF began to recover from its decimation in 1986 and added to the pressures on the government by resuming its campaigns, though it did so without its leadership which was facing conviction in one of a number of treason trials taking place at the time. This particular one, held in the south-eastern Transvaal town of Delmas, had lasted for three years and came to an end in November 1988. Nineteen people had been charged with various offences including treason and terrorism. Three of the accused, Patrick Lekota, Pope Molefe and Moses Chikane, who were the leading officials of the UDF, were found guilty of treason.[26] In his judgment, Mr Justice K van Dijkhorst stated that the UDF was acting as the internal wing of

[25]South African Institute of Race Relations, 1988/89, p611.

[26]Patrick Lekota, UDF publicity secretary received a 12-year prison sentence, Pope Molefe, UDF's national secretary, and Moses Chikane each received a 10-year sentence.

the ANC and was conspiring to make South Africa ungovernable. The following year the UDF emerged with a reconstituted leadership in which Ramaphosa figured behind the scenes.

Yet whilst this escalating carnage was going on there were increasing signs of attempts to find a non-violent solution. The balance of power shifted slightly towards the ANC when Major-General Bantu Holomisa took power in the Transkei after a bloodless coup on 30 December 1987. Holomisa became an influential ally of the ANC. There were divisions in the white community. An increasing number of white males of military age were either requesting the deferment of their military service or failing to report for it on the grounds that they would not use force against black people. The white End Conscription Campaign, which was affiliated to the UDF, encouraged white men to refuse military service.[27] There was a growing tendency within the white establishment to recognise the ANC in exile as the *de facto* leader of the liberation movement. A number of business, political, media and educational groups in the white establishment had sought meetings with the ANC. Between September 1985 and September 1987 12 of them had gone to Lusaka in Zambia to meet with the ANC, despite government opposition. Some meetings were banned but a number of South Africans, including Afrikaners from different walks of life, learned from first-hand experience that the ANC leaders did not fit the image created by government propaganda. Though the meetings endorsed substantial political differences, some shared ideas about a number of other things emerged.[28] The ANC began to be demystified at the very time that MK was intensifying its insurgency activities.

The government itself was caught up in these contradictory pressures. Rumours spread about discussions between Botha and Mandela and the release of the ageing ANC prisoners from Robben Island. In the third week of the strike the London *Times* had reported: "The South African government is considering the imminent release of an ageing black guerrilla leader from prison, fuelling speculation in government and diplomatic circles that President Botha is exploring ways of freeing the African National Congress leader Nelson Mandela."[29] President Botha had asked the Justice Ministry to

[27]"The Minister of Defence revealed in 1989 that one-quarter of the citizen-force and commando members called up to attend camps applied for deferment, while an additional 15 per cent failed to report. He also said 38 per cent of citizen-force troops requested exemption from township duty." (Price, 1991, p238.)
[28]See *ibid*, p240 for a list of the delegations to Lusaka.
[29]*The Times*, 26 August 1987.

look into the case of 77-year-old Govan Mbeki who, with Mandela, had been jailed for life following the Rivonia trial for sabotage in 1964. Mbeki was eventually released on 5 November. He immediately became a listed person who could not be quoted, except that he was allowed to hold one press conference following his release. He said that he was still a member of the ANC and the Communist Party and would continue to support MK as long as the ANC deemed the armed struggle to be necessary. He emphasised that there could be no solution to South Africa's political problems without the ANC. These comments ensured that his freedom from restrictions would be short-lived. His first rally was banned and, in December, a government order was made which prohibited him from being interviewed by the media, from contributing to any publication and from speaking in public. No rallies were allowed to protest at the restrictions imposed on him. He was placed under house arrest from 6 pm to 6 am each day and his home was under constant surveillance. He was not allowed to move out of the magisterial district of Port Elizabeth. Not long after he was released, the ANC decided that it was time to re-establish its structures inside South Africa and Mbeki was given the task of creating a national underground internal executive. Despite the restrictions on his movements, he found no difficulty in doing this from his home in Port Elizabeth.[30] He worked closely with Kgalema Motlanthe and Mac Maharaj, an exiled ANC official who had secretly returned to South Africa to help Mbeki. The three formed the nucleus of an internal ANC executive.

It was at that point that Ramaphosa was recruited into the ranks of the ANC. Maharaj had a positive view of Ramaphosa after the COSATU delegation meeting with the ANC in Lusaka in 1986 but it was Kgalema Motlanthe, the recently appointed union co-ordinator for education, who passed on Ramaphosa's credentials to Mbeki. Motlanthe had been Mbeki's protégé in Robben Island and had developed a close and trusting relationship with him that continued after Mbeki's release. From the time he left prison earlier in the year, Motlanthe had been a critical observer of the political scene for the ANC and he saw Ramaphosa's potential for the organisation, especially after working with him during the strike. He recommended to Mbeki that Ramaphosa should be asked to join the new executive. Thereafter, Ramaphosa made regular friendly visits to see Mbeki in Port Elizabeth

[30]Conversation between the author and Govan Mbeki at his home in Port Elizabeth, 2 January 1994.

and became an integral part of the internal ANC organisation. He became, therefore, politically active on four different fronts: as an underground but nonetheless official ANC activist, as a member of the UDF, as one of the leaders of COSATU and as the general secretary of the NUM. On many occasions these fronts overlapped and were expressed through the pursuit of union policies that depended very much on the interventions and initiatives of Ramaphosa himself.

Taking Anglo-American to the Industrial Court

The union's primary aim immediately after the strike was to obtain the reinstatement by Anglo-American of the 19 000 dismissed mineworkers who had not been taken back. Its first step was to try and resolve the matter through negotiations. These started with a meeting with Anglo-American on 1 September but it was seven months, after several unsuccessful rounds of negotiations and referrals to arbitration, before an agreement was reached. The union, at the same time, was put under pressure by the Chamber of Mines and the Anglo-American Corporation to agree to a code of conduct in relation to mine violence. Their primary concern was to get Ramaphosa's signature on an agreement protecting the right of mineworkers to refuse either to join the union or go on strike. The Chamber of Mines resumed its pressure at a meeting on 8 October but nothing was agreed.

The NUM had been legally advised that if any of the mining houses engaged in selective dismissals after the strike it would be able to raise the question of reinstating those who remained dismissed before the Industrial Court under the terms of the Labour Relations Act. It, therefore, initiated legal action against three mining companies, namely, Vaal Reefs Exploration and Mining Company, Western Deep Levels and Free State Consolidated Gold Mines (Operations), all of which belonged to the Anglo-American group. Free State Consolidated Gold Mines, normally described as Freegold, had been formed in 1986 from the consolidation of the mining operations at President Steyn, Free State Geduld, President Brand and Western Holdings. By the time of the strike, Freegold consisted of all Anglo-American mines in the Orange Free State, controlling 25 shafts and employing 112 000 mineworkers. When negotiations failed to produce a solution both sides agreed to refer their disputes to a private adjudicator on 5 January 1988 as provided for in the recognition agreements between them. Private adjudication was quicker and cheaper than the Industrial Court where usually each side had to employ counsel. The NUM claimed before the adjudicator that the dismissal of the mineworkers constituted an unfair labour practice under the Labour Relations Act.

The dismissals at Vaal Reefs and Freegold were adjudicated separately from those at Western Deep Levels and different adjudicators were appointed with separately timed hearings. In the Vaal Reefs and Freegold hearings, Anglo-American contended that it had not been able to re-employ the full complement of dismissed mineworkers because the labour forces in those mines had been restructured. There were, it claimed, no jobs for them. As the union attitude was that everyone should be re-employed there was no serious debate during what seemed to be interminable sessions, interspersed with adjournments, to try to reach an out-of-court settlement. But, under pressure from the adjudicator, an agreement was eventually reached on 17 March 1988 that provided some form of compensation for each of the 19 000 dismissed mineworkers. Of these, 9 500 were offered 'non-transferable' jobs and were to be taken back in equal three-monthly batches. They were also to get compensation of up to 10 weeks' pay. The job offers were to be made to union and non-union mineworkers in the same proportion as that which had existed at the beginning of the strike.

The remaining 9 500 dismissed mineworkers were to receive at least 18 weeks' notice and severance and compensation pay depending on their length of service with Anglo-American. The settlement, however, raised its own problems, because both the NUM and Anglo-American, not for the first time, went away with different perceptions of its meaning. The NUM wanted the position of union officials who had been dismissed addressed specifically. It also wanted to know whether 'service' meant with the Anglo-American group or with a specific mine when determining retrenchment pay. The most troublesome difference was over the meaning of pay in determining the amount of compensation. Anglo-American insisted that pay meant cash earnings while the NUM declared that it should include non-monetary benefits, such as the cost of board and lodging and other benefits. The NUM's calculation pushed the total cost of the settlement up by R10 650 000.[31] The issues were referred back to the adjudicator for clarification.

The dismissals from Western Deep Levels were covered by the terms of the same agreement but it was left to a different arbitrator to decide who was and who was not unfairly dismissed and, therefore, which of them should be reinstated. His findings, published on 18 July

[31]*Business Day*, 21 March 1988.

1988, were that those 2 000 mineworkers on level 66 who had behaved violently and provocatively had been justifiably dismissed but that was not the case with the 400 on level 100 who, though some may have shouted and uttered a few threats, generally protested peacefully. He recommended that only the 400 mineworkers on level 100 should be reinstated.

The agreement overall was regarded by the NUM as a recognition by Anglo-American that dismissals were not an appropriate response during a legal strike but that was not Anglo-American's view for it expressly stated that the settlement did not set a precedent. In other situations it considered that it remained free to act as it wished.[32] The settlement, nonetheless, removed one part of the bitter legacy of the strike at an estimated cost to Anglo-American of about R60 million. The legal costs to the NUM were R300 000. It was not, however, the end of the matter for each of the mineworkers entitled to compensation had to be contacted and, because the mine records of home addresses were incomplete, that was difficult. Many mineworkers, therefore, were never notified and received no financial payments. Indeed, as much as R1,8 million of the total settlement was never distributed.

Changes in the Labour Law

While the process of protecting the dismissed mineworkers was taking place, the unions were confronted by the government's proposals to change the labour law. The Minister of Manpower had announced in Parliament just before the strike that he was preparing legislation to "bring order to the trade union movement" but he gave no details and aroused no concerns amongst trade union leaders.[33] When, therefore, he published a draft Labour Relations Amendment Bill in September 1987, "it hit union leadership like a bolt from the blue" according to Ramaphosa and, as a consequence, it was some time before COSATU and its affiliates were able to organise opposition to it.[34] Surprise, however, should have been one of the least expected reactions for it was clear that the government, mine owners and employers in general were determined to prevent a repeat of the strike experiences of 1987. This was not simply about avoiding the experience of the mineworkers' strike but also of preventing black workers from striking successfully

[32] *Weekly Mail*, 8 April 1988.
[33] *The Observer*, 9 August 1987.
[34] *Weekly Mail*, 15 April 1988.

in other areas as they had done at OK Bazaars, on the railways and in the Post Office.

The amendments to the Act were designed to prevent such a reoccurrence. They proposed putting severe limits on the freedom of workers either to strike or to take other forms of industrial action. It was to be more difficult to engage in a legal strike. That would not have been an especial penance for mineworkers because many of them, with recent memories of mass dismissals, already thought that legality was an empty gesture. But the opportunities for employers to end legal strikes were to be extended by enabling them to obtain injunctions whilst they were in progress. Other suggested measures were more oppressive. It was to be a criminal offence to call sympathy strikes. Workers could only strike if they had a grievance against their own employer. They could not strike over the same issue more than once in a 15-month period nor engage in intermittent strikes over the same issue. The power of employers to dismiss workers arbitrarily and selectively after a strike was to be increased. The Bill proposed that the definition of an unfair labour practice should be extended to cover consumer boycotts such as those frequently organised in the hostels. But, most importantly, it provided for unions to be sued by employers for damages caused by industrial action. If the mine owners had possessed that option during the 1987 strike they, possibly, could have damaged the NUM irreparably and certainly much more effectively than by mass dismissals. There was, then, a serious possibility that unions could be forced into bankruptcy. The Bill also contained a clause to enable employers to negotiate with unions that did not have a membership majority in a particular category of workers in a mine or shop or plant, such as was the case with UWUSA in some gold and coal mines. The NUM had every reason to oppose the proposals.

COSATU was the central organising body for the campaign against the Bill. In February 1988, it endorsed a programme of weekly factory demonstrations but, in a bid to blunt their impact and to disarm COSATU and its allies, the government made it illegal for it to continue with its political activities from 24 February.[35] At the same time it

[35]The restrictions imposed on COSATU included bannings on: "(a) publicity campaigns for unlawful organisations, repeal of orders against organisations, release of detainees, sentences on people, abolition of local authorities; (b) opposition to detention, system of local government, constitutional negotiations; (c) calls to commemorate unlawful organisations, or ones for which an order has been issued, incidents of riot, death of a person; (d) setting up alternative structures; (e) interfering with local authorities; (f) calling for sanctions, disinvestment or severing ties with the Republic; (g) organising meetings to discuss matters in this order". (COSATU *Biennial Report, Third National*

banned 17 other organisations, including the UDF and many of its affiliates. COSATU decided to ignore the restrictions but to act cautiously in doing so. It delegated some of its political campaigning to its affiliated unions who were not constrained by the order. This did not, however, lessen the repression by the security forces. There was increased police activity in March and, in a manner familiar to mineworkers, fake pamphlets began to be circulated, purporting to come from a legitimate trade union but condemning COSATU and urging the acceptance of the Bill.

The situation was clearly critical for the government and the trade union movement as a whole. The government treated the attack on the Labour Relations Amendment Bill as an assault on itself and refused to make concessions in the belief that if it retreated its credibility amongst its own supporters would be damaged. Already it was suffering a loss of support to the right-wing Conservative Party that had inflicted decisive defeats on it in the by-elections in the southern and western Transvaal in March. However, on the question of labour legislation the country's big employers backed it. The Association of Chambers of Commerce regarded the Bill as an advance in industrial relations and the South African Federated Chamber of Industries added its endorsement. Ramaphosa commented on the close similarity between the attitude of the mine owners and that of the government. There was, he said, something "strangely familiar" about some of the clauses of the Bill. It was that they "had almost all been demanded by various managements during recognition agreement negotiations: an end to sympathy strikes, no-strike clauses for the duration of an agreement, unions to be responsible for management losses during industrial action".[36]

COSATU responded to the crisis by calling a Special Congress on 14 and 15 May 1988 in an attempt to generate unity around the issue throughout the trade union movement. It was a delicate issue for COSATU. There were some within it who believed that if unity failed to materialise, or mistaken approaches were made in other ways, then the outcome could be the destruction of COSATU itself.[37] The underlying source of such a possibly damaging division were the different ideological approaches to mass action that had marked the debate about the *Freedom Charter*, namely whether unity and action

Congress, 12-16 July, 1989, Johannesburg.)
[36]*Weekly Mail*, 15 April 1988.
[37]Baskin, 1991, p277.

should be based on non-racialism or class. The differences were reflected in the debate at the Special Congress and in the resolutions submitted to it for adoption. The NUM, the principal advocate of non-racial action, tabled one of the three resolutions that aimed to determine the future strategy against the Bill.[38] It proposed a united front to bring together COSATU with "tried and tested mass-based organisations" which would constitute a multi-class non-racial alliance but which was led by organisations that subscribed to the *Freedom Charter*, namely COSATU and the UDF. The main alternative resolution was submitted jointly by the National Union of Metalworkers of South Africa (NUMSA) and the Chemical Workers' Industrial Union (CWIU), calling for a broad front of organisations working for the end of apartheid which included the rival trade union federation to COSATU, the National Council of Trade Unions (NACTU) and black consciousness organisations.[39]

Whereas in the debate about the *Freedom Charter*, the political divisions had been about perceptions and policies for the long-run transformation of the apartheid society, this time they were about ideology as a guide to immediate protests against actual state repressive policies and, in this respect, the NUM resolution had a distinct advantage because the bodies that backed it, the UDF, the ANC and the SACP, were the only ones with the capability for putting into practice what they had resolved. In the end a compromise was reached in which the essence of the NUM resolution was retained in that a programme of action was to be formulated and organised by COSATU's Central Executive Committee and its close allies, the UDF and the South African Youth Congress, but without excluding from participation the black consciousness bodies outside the tradition of the *Freedom Charter*. The Congress also decided to stage a three-day national peaceful protest against the Bill on 6, 7 and 8 June 1988 and, later in the year, to call a conference of a "broad range of anti-apartheid organisations" to take the protests further.

The NUM engaged in COSATU's protest campaign but its ordinary members were not always free to participate as they might have wished. They generally lacked the freedom in the hostels to join the poster and pamphlet campaigns or to stage any kind of protest demonstrations similar to those organised by other unions. Nonetheless, a few days before the Special COSATU Congress, mineworkers in various mines

[38]*Ibid*, pp279-281 for a discussion of the Special Congress resolutions and debate.
[39]Seekings, 2000, pp231-235.

expressed their opposition to the Bill. Ramaphosa claimed that up to 18 000 mineworkers demonstrated spontaneously in collieries and gold mines owned by each of the mining houses in the eastern Transvaal and the Orange Free State. "It is quite clear," he said, "that workers are not prepared to accept this draconian piece of legislation." The demonstration at Matla colliery resulted in two deaths. There were underground demonstrations at Vaal Reefs during which pamphlets opposing the Bill were distributed.

This decision to call a stay-away evoked strong reactions from most employers. The Chamber of Mines claimed that COSATU "was either unwilling or unable to back up its allegations that the Bill was a repressive measure aimed at curbing trade union activity". It added that this, "coupled with COSATU's refusal to call off the protest, suggests that its campaign against the Bill is politically inspired". In other words, it aimed to "destabilise workplace relations and to engineer a confrontation with the state".[40] The South African Consultative Committee on Labour Affairs (SACCOLA), an umbrella employers' body chaired by Godsell of Anglo-American, tried to persuade COSATU on 2 June, in the first formal meeting between the two bodies, to stage a protest action that did not include a stay-away. Godsell contended that the Bill was not a union-bashing measure in that it did not curtail union rights in any respect.[41]

SACCOLA and COSATU engaged in a press propaganda campaign against each other over the merits of the Bill. On 3 June, SACCOLA inserted an advertisement in the press putting its case in favour of the Bill. The advertisement repeated the view, amongst other things, that it was incorrect to claim that the Bill was repressive and an attack on trade unions. It went on to compare the Bill's provisions favourably with those enacted in Western democratic countries but failed to state that many of their labour laws were also essentially hostile to trade unions. COSATU, in its own press advertisement, contradicted SACCOLA's claims and stated that the Bill was part of a broader strategy to suppress democratic opposition which consisted of the systematic undermining of its "right to meet, speak, publish and organise … by banning all our rallies, constant police raids on our offices, numerous anonymous bomb attacks on our property, regular confiscation of our media, ongoing detention of our members…" It recalled that the Minister of Manpower had described the Bill as "a

[40]Baskin, 1991, p286.

[41]South African Institute of Race Relations, 1987/88, p652.

sword over the union's heads".[42]

The stay-away went ahead as planned and was fully supported by NACTU. It was the biggest, strongest and longest protest action in the history of the liberation struggle.[43] On the first day, approximately three million people stayed away from work. This was approximately double the combined membership of both federations. There were wide regional variations in support with the strongest in the Witwatersrand and Natal, which was surprising given the hostility of Inkatha to the action. Overall support fell during the second and third days due mainly to the propaganda, broadcast on radio and television, which threatened the protesting workers with dismissal but, even so, it remained an historically massive protest. The black mineworkers, however, gave it minimal support. The NUM claimed that 35 000 had stayed away from work on 7 June in mines throughout the country, but the Chamber of Mines contradicted this with the report that only 9 000 out of 550 000 mineworkers had supported the action on the first day and that less than that had stayed away from work on the two remaining days.

Outside of the mining industry sporadic protests continued but in general the mood of the workers was mooted. The police had assaulted some of them; many others had been penalised by employers by being dismissed or disciplined; recognition agreements had been cancelled and union facilities had been withdrawn. Restriction orders were placed on a number of union officials while police raids on the offices of COSATU and its affiliates continued. Immediately after the stay-away, SACCOLA asked to meet with COSATU again to agree on amendments to the Bill. The Minister of Manpower had given the impression on 8 June that the Bill was not the final word on labour legislation and that it might be possible to amend it the following year. The various parties opposed to the Bill met subsequently and agreed to ask for the postponement of some of the clauses, but it became law on 1 September 1988 without any alterations being made to it.

Following the resolution of the Special Congress in May 1988, the central executive committee of COSATU began to plan its Anti-Apartheid Conference. It met planning difficulties, however, due

[42]*Ibid*, p653.

[43]*NUM News*, December 1988. The Labour Monitoring Group estimated that on the first day there was a 79 per cent absenteeism rate in the Pretoria/Witwatersrand/Vereeniging area; 67 per cent in the Durban region and 64 per cent in Pietermaritzburg. There were wide variations between industries as well as regions. For example, 90 per cent stayed away from work in the metal industries in the PWV area. (South African Institute of Race Relations, 1987/88, p652.)

to the political divisions that had marked the congress but by September there was sufficient accord around a set of proposals to make it possible to name a date and a venue. The conference was scheduled to be held late in September at the University of Cape Town. At the last minute, whilst delegates were actually travelling to Cape Town from various parts of the country, a number of delegations, including those from NACTU and the black consciousness organisations, withdrew. Then, on the day the conference was due to convene, the commissioner of police banned it because, he said, it was "obviously" intended to fill the vacuum caused by the restrictions imposed on COSATU and the anti-apartheid organisations by the banning orders in February. He also imposed restriction orders on a number of COSATU officials who had been responsible for organising the conference.[44] The failure to present a broad and united front against the government was a set-back in the struggle to build alliances because, as Ramaphosa was to state, unity only came from action and organisation and not from slogans.[45]

Mine Wage and Poverty Levels

During the 1987 strike the Chamber of Mines had tried to show how ordinary mineworkers had benefited from dramatic and unprecedented wage increases since the wages explosion in the mid-1970s. It recounted how the average money wages for underground workers had risen by 430 per cent since 1975 and how this represented a 32 per cent increase in real wages.[46] These calculations were roughly confirmed by the Labour Research Service of the University of Cape Town which stated that the average real wage of black mineworkers in the gold mines had risen by one-third since 1975.[47] NUM did not dispute them.

These increases, however, had to be viewed in the light of three

[44]Three of them, Chris Dlamini, the first vice-president of COSATU, Sidney Mafumadi, its assistant general secretary and Donsie Khumalo, a regional secretary, were not allowed to leave their respective magisterial districts for 10 days from 6 pm on 21 September until 6 am on 3 October. Nor could they leave their homes between 6 pm and 6 am each day.

[45]NUM, *Biennial Report*, 1989.

[46]*Finance Week*, 17 March 1988. On the first day of the strike, Bobby Godsell had claimed in a radio interview with the BBC that mineworkers' wages had risen by 85 per cent in real terms over the previous 10 years but his figure was never confirmed and later was revised down by Johann Liebenberg to 32 per cent.

[47]*Weekly Mail*, 14 August 1987.

factors. First, the basic minimum wage rate in the early 1970s had been so low that to concentrate on percentage changes alone was grossly misleading. Second, they had occurred during a period of high inflation when rapid wage rises were common throughout industries so that the relative position of mine wages had remained largely unchanged. Lastly, the real wage improvements were short-lived. They were concentrated in the years between 1975 and 1983 and then began to decline. It was, therefore, a misrepresentation to use a single set of percentage figures, as both the Chamber of Mines and Anglo-American had done, in such a way as to give the impression that there had been a constant and significant improvement in the standard of living of black mineworkers since 1975. The evidence contradicted that contention. After the strike, black mineworkers and their families were still living under conditions of unrelieved poverty.

Black mineworkers remained amongst the worst-paid workers in the country. It was estimated in 1987 that the lowest minimum wages in industry in general ranged between R416 and R427 a month whereas the minimum monthly wage in mining was R246. The minimum monthly wage for all labourers, set by the industrial council for the steel and engineering industries, was already R509 while even that for novices in civil engineering was R410 a month. The highest starting wage for black workers set by any industrial council was R838 a month for labourers in the automobile manufacturing industry in the eastern Cape. When the NUM, therefore, embarked on its strike for a basic minimum monthly wage of R309 its aim was modest. A comparison between base rates, however, did not reveal the full extent of the disparities because the wages of 90 per cent of all black mineworkers were so low that they earned between 40 per cent and 46 per cent less than the national average wage for black workers in industry.[48]

In international terms, it was exceptional for mineworkers to be so badly paid. In the other major mining countries they were amongst the highest-paid workers. It had been estimated in 1986, for instance, that "in Poland miners earn twice as much as manufacturing workers, in

[48]For a comparison of minimum rates see *Labour Research Service*, "Bargaining Report for the National Union of Mineworkers", 9 May 1988. It was not possible to compare average wages for black mineworkers with wages in the manufacturing industry because the employers, including the Chamber of Mines, had stopped publishing a racial breakdown of wages. Secret wage surveys were undertaken by various management consultancy firms on behalf of employers which gave a variety of estimates. None of them, however, disputed the general contention that mine wages were much lower than manufacturing wages.

Australia they earn 64 per cent more, in the Soviet Union 55 per cent, in Canada 43 per cent, in the United States 32 per cent and in the United Kingdom 20 per cent more".[49] These differentials, however, were meagre compared with that between the wages of white miners and black mineworkers within South Africa. It had narrowed considerably since the early 1970s but white miners still earned four times more than black mineworkers and were the best-paid workers in the country, earning 113 per cent of the average white wage in the manufacturing industry. So, no matter how and with whom comparisons were made, the black mineworkers in South Africa were highly disadvantaged.

Percentage figures and differentials, moreover, gave no indication of the subsistence problems faced by black mineworkers for the vast majority of those in the unskilled grades earned less than that required to avoid poverty. The minimum wage for black mineworkers fell below all of the 'poverty datum line' estimates made in South Africa at that time.[50] At least one-third of all mineworkers hardly reached the Minimum Living Level on which it was stated no family could survive in the long term. Even if the 1988 demand for a 40 per cent wage rise had been granted, one-half of all mineworkers would still have been living below the Supplemented Living Level, which, it was estimated, represented only a "modest low-level standard of living".[51]

The response of the Chamber of Mines to the allegation that it was paying exploitative wages was to insist that as mineworkers received free food and accommodation, its value, calculated at about one-half of the minimum wage rate, should be added to the money wage to show the full value of the wages received by mineworkers. If that were done, it contended, it would be shown that black mineworkers were relatively well paid amongst black workers. Most black mineworkers, however, had to maintain a home in the rural areas, which involved supporting their extended families, as well as covering their living expenses in the mine compounds. Living away from home invariably involved additional, often unanticipated, expenses. It was, however, misleading

[49] *Weekly Mail*, 14 August 1987.

[50] There was some controversy about the meaning of the various poverty datum line estimates. The Director of the Industrial Relations Unit at the University of Port Elizabeth believed that the Household Subsistence Level and the Minimum Living Level were much more approximations of a poverty or bare subsistence income than one which could provide a decent and healthy standard of living. Many employers used the levels as a basis for wage negotiations and in that way kept wages at a poverty level. If this criticism was valid then mineworkers' wages were even more inadequate than claimed by the NUM. (See South African Institute of Race Relations, 1988/89, p425.)

[51] Labour Research Service, 9 May 1988.

of the Chamber of Mines to advertise the provision of single-sex accommodation and subsistence as fringe benefits because, in reality, they were an integral part of the functioning of hostels within the migrant labour system that enabled the mine owners to keep the level of wages invariably and excessively low.

When the NUM leaders next sat down, early in 1988, to decide on their annual wages claims they had to prepare for approximately 30 sets of negotiations for different mining sectors, mining houses and relatively small individual companies. The union's primary negotiations were with the Chamber of Mines over its 525 000 black employees in the gold and coal mines. The NUM had still not yet obtained negotiating rights for mineworkers in the expanding platinum mining sector even though those companies that had recognised NUM in the gold and coal sectors owned it. Its other principal negotiations were with Eskom, employing 57 000 black workers in the electricity supply industry, and De Beers, with 12 000 black employees in the diamond mines. In this set of negotiations, Eskom paid the highest minimum wage rates. De Beers came next with rates that were 55 per cent more than the minimum rates in the gold mines. The coal mines paid 10 per cent below the gold mine rates. The lowest reported minimum rate in mining overall was that paid to asbestos mineworkers employed by the African Chrysotile Asbestos Company, owned by Gencor, in the eastern Transvaal. Though there was no uniformity in the wages structure in mining, by the end of the 1980s, the rates negotiated with the Chamber of Mines were pivotal for the whole industry but more as reference points rather than as levels to be copied and applied.

The leaders of the NUM knew that the annual round of negotiations was not going to rescue their members from the poverty trap but they had no alternative but to decide on figures and put them into the negotiation process. If they refused or were not serious then the Chamber of Mines would set its own rates and impose them on 1 July without any reference to the NUM, as it had done on previous occasions. So they decided to press for a 40 per cent increase in wages for all wage categories and a minimum wage of R350 a month for gold and coal mineworkers. They added improvements in various fringe benefits to their package, including a proposal for an innovative black mineworkers' provident fund. The union anticipated that the wages struggle would be even more difficult than it had been in 1987 because of the deteriorating financial situation in the gold and coal mines. Profit levels were falling during 1988 and the mines were paying out smaller dividends.

The wage negotiations commenced at the end of May when the Chamber of Mines offered an across-the-board increase of 10,5 per cent in response to the union's demand. It rejected all the other demands, including that for a minimum wage, except for the provident fund, which it accepted in principle, provided it was negotiated separately from wages. After a number of adjournments the NUM had reduced its wage demand to 22 per cent while the offer from the Chamber of Mines was raised to between 13 and 17 per cent. The Chamber of Mines put pressure on the union to accept by threatening to give wage increases to non-members only, if its offer was rejected.[52] The impasse was broken when the Chamber of Mines agreed to a phased introduction of a provident fund over three years. The NUM accepted this package, though none of its other demands was included. The agreement was the first since 1983 to be concluded in time for the Chamber of Mines' customary 1 July introduction and the mining houses received it enthusiastically. The chief executive of Gencor, for instance, claimed that it signalled the end of "the era of wage awards to workers above the consumer price index level without corresponding increases in productivity".[53]

There were many hitches in the negotiations over the provident fund and several issues had to be referred to an arbitrator. On most of them, the union failed to get its way but an agreement was eventually reached that provided black mineworkers with benefits on retirement, death, for funeral expenses and on dismissal and retrenchment for the first time in the history of mining. Disability benefits and housing loans were to be brought in at a later stage. The scheme was to be run by trustees appointed by the NUM and the Chamber of Mines and was timed to start on 1 May 1989. For the first two months each mineworker paid 1,5 per cent of his salary into the fund with an equivalent percentage paid by the employer. From 1 July 1989 the percentage rose in stages up to 5 per cent by July 1990. The NUM regarded the fund as a compulsory savings scheme for times of need and asked its members to attend branch meetings to learn about its operation. Other unions had successfully negotiated provident funds for their members but, given the insecurity in mining and the tendency of the mines to return workers they regarded as undesirable, unusable or sick back into the rural areas where they had to rely on sustenance from their poverty-stricken families, a provident fund for migrant mineworkers was an important new welfare provision.

[52] *Weekly Mail*, 24 June 1988.
[53] *New Nation*, 14 July 1988.

It was important for the union in another respect. There was a small but increasing number of black mineworkers who were taking jobs previously done by white miners and who should have been able to participate in the Mine Employees' Pension which had been established in 1949 and which held substantial funds. However, its rules restricted membership to "Europeans only" and the Council of Mining Unions, representing all grades of white workers in the mines, would not relax them to allow black mineworkers to join. It insisted that black employees should have their own fund and was supported by the Minister of Manpower who said that he had sympathy with the white unions that had fought for their own rights. The Chamber of Mines contended, however, that all employees in the same occupational group should make the same contributions to the same pension fund and receive the same benefits. It alleged that the Council of Mining Unions was guilty of an unfair labour practice and took the matter to the Industrial Court which ruled in its favour in October 1989. But that did not resolve the issue because membership of the fund was confined to members of the Mine Workers' Union and that union refused to allow black mineworkers holding blasting certificates to join in any circumstances. The new provident fund, however, compensated for the discriminatory features of the Mine Employees' Pension by catering for all black monthly-paid staff, as well as underground and surface job categories from 1 to 8.

The union pursued more aggressive negotiating tactics against Eskom and De Beers, which had not been involved in the strike and where, therefore, there was no post-strike management repression. Eskom had already restructured its wages policy by eliminating all differentials based on race, gender and region and, in the process, had given many workers substantial increases. In 1988, however, it offered an annual increase less than the rate of inflation and was successfully taken to arbitration by the NUM. In the case of De Beers, the NUM demanded a 20 per cent wage increase and four paid holidays on 21 March, 1 May, 16 June and 1 October that had been designated as Kinross Day. While the negotiations were in process there was a strike at three of De Beers' five diamond mines over the paid holidays issue. Then the wage negotiations broke down when De Beers offered a wage increase for unskilled mineworkers that was less than that obtained in gold mining. There was talk of a strike but a settlement was reached when De Beers raised its offer to 15 per cent and agreed to regard 1 May as a paid holiday and 16 June as an unpaid one. Gold Fields maintained its implacable opposition to the NUM and refused to give more than a 12 per cent wage increase to 600 of its black employees at Zincor, its zinc processing smelter at Springs, which was still its only mining subsidiary where the NUM had a recognition agreement. At the end of September 1988, the dispute was taken to the Conciliation Board but that failed to resolve it. Gold Fields closed the matter, so far as it was concerned, by locking out all the workers. There was a similar lack of progress at the Impala Platinum and Rustenburg Platinum Mines where the union wanted to speed up the negotiations for recognition and get a wages settlement. There was talk of strike action in May and June but nothing materialised. The main difficulty was still that Impala Platinum's mines extended into Bophuthatswana where the NUM could not legally organise.

The Share Ownership Scheme

The Anglo-American Corporation had its own agenda in the post-strike situation. It did its utmost to weaken the union by intensifying its repression of union activities at mine level and, at the same time, by trying to tie it down to a set of jointly agreed rules that would make the union responsible for the behaviour of every shaft steward and branch official. It also embarked on a policy of subversion by launching a share ownership scheme as a means to divert the loyalty of individual mineworkers from the NUM to the company.

Share ownership had already been established in a number of retail businesses in South Africa but during 1987 it became popular amongst some of the large corporations. The largest of the schemes was that launched in November 1987 by the Anglo-American Corporation and De Beers. Through it, everyone who had worked for the companies for two years or more was offered five free shares with the option of having a similar offer each year for another four years. The scheme was to be reviewed after five years. No trade unions, including the NUM, were consulted about it because Anglo-American believed it was a matter for individual workers to decide whether or not to participate. In this particular case, the long-term intention was to subvert NUM's policy of public ownership and, in the process, counter the influence of the SACP, backed by the NUM, over a possible future ANC government. In the words of Gavin Relly, the chairperson of Anglo-American, share ownership was regarded as a way of giving black mineworkers a stake in the free enterprise system as South Africa moved towards a difficult era in management-employee relations.[54]

[54]US intelligence had reported early in 1987 that "the ANC was beholden to the SACP,

It was a policy that involved few risks to Anglo-American for, it stated, it did not intend to allow mineworkers to accrue sufficient shares to alter the pattern of ownership.[55]

The NUM rejected the scheme. At the time the share offer was made, Ramaphosa described it "as political and economic blackmail … to seduce workers away from socialism". After seeing the scheme operate for a year he wrote: "These shares are cheap bribes to divert workers away from our struggle for a living wage. This is clearly the most degenerate tokenism by a company."[56] About 120 000 mineworkers participated in the scheme during 1988 and received R31 each in dividends. The shares, of course, were not free to the company. It had to buy them. Ten shares in 1988 cost R590 so Ramaphosa suggested that Anglo-American should give that amount to each of its employees and let them decide whether to spend it on shares or on other items that would contribute to their well-being.

The Code of Conduct

The NUM's negotiations with the Chamber of Mines about mine violence had begun on 6 November 1987 when each side described what it regarded as its principal causes. The view of the Chamber of Mines, according to Johann Liebenberg, was that the responsibility for the violence lay with the mineworkers in general and the NUM in particular. He declared that NUM members or local office-bearers had committed many acts of violence and that, therefore, its resolution lay with the union. He added that, whether or not the NUM was complicit or had prior knowledge that violence was about to be committed, it had a responsibility to use its authority to persuade its members not to engage in it. Ramaphosa, on the other hand, asked for the 'root causes' of mine violence, namely the migrant labour and hostel systems, to be included in the discussions. He did not deny that mine violence could be influenced by joint action but, he said, the mine owners wanted to treat violence in isolation from the migratory labour system that caused

the Union of Soviet Socialist Republics and Angola for its arms and military training". Quoted from *Communist Influence in South Africa*, US Intelligence Report. This report had been written because the US Comprehensive Anti-Apartheid Act of 1986 had provided for an investigation into communist influence in the ANC. (South African Institute of Race Relations, 1987/88, p703.)

[55]NUM, *Biennial Report*, 1989.

[56]*Ibid*.

it. The gap in understanding between the two sides was so wide, he believed, that he could see little use in continuing discussions and refused to respond to requests for subsequent meetings.

The Chamber of Mines persisted, however, and, in a letter to Ramaphosa on 10 December 1987, pleaded with him to respond for the reason it had given at the November meeting, namely that as much of the violence was caused by union members and officials, the union head office was in a position prevent it, adding, rather ingratiatingly, that "your head office has intervened successfully in the past to put an end to acts of violence". In order to remind Ramaphosa of the union's involvement, the letter contained a list of 10 acts of violence by union activists from February 1986 to October 1987.[57] The NUM replied on 22 December 1987 with a detailed memorandum explaining its position and emphasising again that the migrant labour system with its attendant institutions generated violence. Ramaphosa wanted the Chamber of Mines to set a timetable to destroy these structures but he got no response. He regarded its attitude as being "hopelessly parochial" and indicative of the fact that the mine owners wanted to keep the mine compounds in order to maintain authoritarian control.

At that point the Chamber of Mines decided that it was going to be impossible to negotiate a code of conduct at industry level and left the individual mining groups to decide for themselves whether or not they wanted to negotiate with the NUM. Ramaphosa questioned the sincerity of the Chamber of Mines, calling it a "cowardly turn-about" and accusing it of having "a sinister double agenda … of reaping maximum profits from poorly paid workers and of setting conducive conditions in motion to eliminate the union".[58] He was to repeat his arguments in his report to the Sixth National Congress in February 1989 and added that he questioned the sincerity of the Chamber of Mines in wanting to eliminate mine violence. "For mineworkers," he wrote, "this cowardly about-turn … means that the Chamber can never be trusted, and that as workers we must never let our guard down."[59] Liebenberg described the statement by Ramaphosa as "arrant nonsense" and accused the NUM of using the negotiations "to introduce an entirely different agenda – the migrant labour system". He said the Chamber was willing to analyse how its members might reduce their dependence on migrant labour but that would be a long-term project.[60]

[57]Correspondence: "Violence on Mines": Chamber of Mines to National Union of Mineworkers, 10 December 1987.
[58]*Weekly Mail*, 24 June 1988.
[59]NUM, *Biennial Report*, 1989.
[60]South African Institute of Race Relations, 1988/89, p492.

Without the option of an industry-wide code the issue was left in the hands of the union and each of the mining houses. The question of drafting a code of conduct had already been raised with Anglo-American. The March 1988 settlement between NUM and Anglo-American concerning compensation for the dismissed mineworkers contained a clause that "the parties will immediately negotiate in good faith a code of conduct and an agreement providing for disputes arising from individual dismissals". Anglo-American promptly asked the NUM for a meeting but there was no contact between the parties until December. In the meantime, each side had begun to prepare its own version of a code of conduct.

The primary reason for having a code was to tackle the issue of mine violence. Ramaphosa, however, found it as difficult to discuss it with Anglo-American officials as it had been with the Chamber of Mines and for the same reason. Since the publication of their joint report, *Reaping the Whirlwind?* in 1986, Anglo-American had persistently blamed union members for every outbreak and had charged the union leaders with complicity. Moreover, the proposals it had made for resolving the problem implied blame for they were designed to place constraints on NUM in its relations with competitive unions and non-unionists during disputes. This had been the case when mine violence was discussed during the strike. Anglo-American's proposals on that occasion suggested having areas in the compounds designated for peaceful picketing and an agreement not to prevent mineworkers from working normally during strikes.

Peter Gush, the chairperson of Freegold, was explicit about the union's responsibility for violence in his annual review in December 1987. He wrote that, although the NUM had been effective in calling out its members during the strike, there was evidence of widespread coercion and intimidation. He had earlier wondered whether or not intimidation was a part of a deliberate union strategy.[61] He claimed: "The 18 months preceding the wage strike saw an escalation of acts of violence and brutality by certain union members and leaders towards fellow workers … Similarly, during the strike when some workers wished to return to work, violence increased … This programme of violence highlights a disturbing trend whereby workers are increasingly being denied free choice and freedom of association." This view was repeated in Anglo-American's submission to the adjudicator in the case of mass dismissals in March 1988 when it

[61] *Business Day*, 4 November 1987.

stated: "Anglo's perception of what occurred during the 18-month period preceding the strike was that there was an increasing level of illegal action involving union members, including murders, intimidation, illegal strikes, boycotts, kidnapping, petrol bombing of premises ..." Gush put these statements into context when he said that "initially Anglo management put too much faith in the NUM's ability to handle its own affairs. It turned out the NUM leaders were either unwilling or unable to control some of their members. We wound up by allowing too much intimidation and coercion on our mines..."[62] The advent of the NUM had done little, it seemed, to alter the mine management attitude of paternalism that had prevailed before 1982.

Anglo-American clearly needed the co-operation of NUM's national officials but by prefacing its proposals in such a way it seriously misread them. Both Ramaphosa and Motlatsi resented the hectoring and patronising tone of white miners, managers and executives in their relations with them. They had each matured within their respective leadership positions in the liberation movement to the point where they resented being talked down to by anyone, whether white or black, disliked being contradicted and took offence when they were treated in a racist manner. Ramaphosa carried himself with complete composure in the company of white people in general, irrespective of their formal status. There were many tales about his behaviour towards white wage negotiators. In one set of national negotiations, for instance, he ridiculed one of them by mimicking the stereotypical cringing black person, beginning with "please, *baas*". His need to take firm, disciplined action within his own organisation had shaped his manner outside of it. Whether amongst his colleagues or friends he insisted on loyalty. Those who unfairly criticised or opposed him or in any sense treated him unequally could not remain friends with him. White people were the main casualties of this characteristic for they were the main offenders.

It was not surprising, therefore, that when Anglo-American sent Ramaphosa a copy of its code of conduct on 8 December 1988, along with a litany of violent events in its mines since 1985 and a request for a meeting to discuss it, he responded by sending back by return of post a copy of the NUM's own report, *Collective Bargaining at Anglo-American Mines: A Model for Reform or Repression?*, with its bitter and incisive criticisms of mine managers. The distance between the two was as wide as it had been before the strike. Indeed, the proposals were

[62]*Financial Mail*, Supplement, *Anglo-American: A Corporate Report*, 18 May 1990, p45.

remarkably similar to those they had presented to each other during the strike. They agreed not to publish their respective codes but when Anglo-American did so early in January 1989 the NUM distributed its report to the press the following day. The NUM had, however, pre-empted Anglo-American by publishing extracts from its report in the December 1988 issue of *NUM News*. Under the headline *Workers Charge Anglo* it stated: "Widespread complaints against Anglo-American mines have been made by black mineworkers. Every day NUM offices country-wide hear chilling reports of dismissals, arbitrary disciplinary charges, physical assaults, random police-styled room searches and general harassment and intimidation … Everywhere the story was the same. Anglo claims to be a model of 'civilised labour practices' have been unmasked … No other conclusion can be drawn than that profits count most … At Anglo-American the very survival of NUM is at stake…" It gave 22 examples of "How Anglo Rules Mines" and devoted the two centre pages to describing in more detail how that rule was exercised and within what kind of compounds. It stated:

> The hostel compounds, housing up to 13 000 males, are fenced with three-metre walls topped with barbed wire. The outside mine premises are security-fenced with main entrance gates manned by mine security … On entering mine premises mineworkers and non-mineworkers are subjected to security checks. Mine compounds have their own gates manned by mine security who do a recheck sometimes with workers having to show armbands … with identification numbers like prisoners … With small rooms with welded windows, housing up to 22 mineworkers, the conditions are horrible and unhealthy.

The NUM's statement added that Anglo's report ended without a mention of "the open unholy alliance of the state and mine bosses supported by white miners … The apartheid state, mine bosses and industry in general are clubbed together through the so-called Defence Manpower Liaison Committees (DELACOMS). It is not surprising that the administrative manager of the Johannesburg Chamber of Mines, Col Chris du Toit, is chairman of the Johannesburg DELACOMS … Through the DELACOMS, the unpopular Joint Management Centres structures set up by the state throughout the country to counter anti-apartheid activities, are given information about workers' activities."[63]

Anglo-American did not rebut these claims against its mine managers but simply described them as "presently unsubstantiated allegations" and, in its code of conduct, proposed that mine based joint AAC/NUM committees should be established to investigate them.

[63]*NUM News*, December 1988.

In the same week as its code was issued to the press, however, it acknowledged that some mines had restricted union facilities, banned union meetings, videotaped or recorded others and had searched hostels for illegal firearms. Such acts, it added, were taken because mine managers "had reason to believe emotive language would be used and issues raised leading to possible violence".[64] This illustrated the distance between the two parties for the NUM wanted those rights the managers exercised to be vested in democratically elected hostel committees.

The NUM published its own code of conduct in full before the end of January 1989. It used the same format as Anglo-American but the similarities did not obscure profound contradictory differences in means and aims. The AAC code sought to achieve a balance between collective freedom and the right of individuals to do as they wished, namely not to join a union or not to strike, and to protect the right of managers to manage. Thus, in its preamble it declared that "employees … are expected to … respect the right of shareholders to have their interests protected by the profitable and efficient management of companies in which they invest". This stood in marked contradiction to the NUM's emphasis on the collective rights of mineworkers. The NUM stated that its code was intended to do more than discourage violence. Its purpose was to eliminate "harassment, victimisation, racial discrimination, union-bashing, mass dismissals and unnecessary violence and conflict by promoting and maintaining acceptable norms of behaviour and restoring fundamental human rights to mineworkers". The NUM wanted all the rights of association, namely to strike, to picket, democratic representation, assembly and freedom of movement to be guaranteed. Moreover, it wanted an assurance that there would be no mass dismissals and that the SAP and the security forces would not be used in industrial disputes. Indeed, it wanted Anglo-American to safeguard the very rights mineworkers were being denied in its mines. The union stated that its demands were negotiable at company and mine level but that it retained its belief that the problems of violence would remain until the migratory labour system for mining in general was abolished. The obstacles to a negotiated settlement with Anglo-American were different from those that led to the end of negotiations with the Chamber of Mines but they were as great.

Following the publication of the NUM's report on mine repression in *NUM News* and of Anglo-American's version of a code of conduct

[64]*Business Day*, 12 January 1989.

in the South African press, the two sides argued in public. Indeed, Marcel Golding for the NUM condemned Anglo-American for attempting to negotiate through the press by releasing the contents of its code before a planned meeting with the union in February.[65] When eventually they met they agreed to keep their discussions secret until they were satisfactorily concluded. They continued to negotiate but there was no satisfactory conclusion. Mine violence not only continued to be a possibility but subsequently actually erupted on a scale as awesome as at any other time in the history of the union.

The character and inconclusiveness of the negotiations over mine violence heightened the suspicions each side had of the other's motives and they began to exchange accusations. Towards the end of 1988, the NUM accused the Chamber of Mines of trying to get sensitive information about its planned activities by means of security police-type operations and by using TEBA recruitment offices as regional spy centres.[66] In support of its allegations, the union's collective bargaining department published extracts from documents that had been secreted out of the office of the General Manager of the Chamber of Mines during the past year, which indicated that he had been keeping the union under surveillance with records of union meetings and of the movements of its officials. The Chamber of Mines responded to the exposures by searching for NUM sympathisers amongst its head office staff who might have provided the NUM with sensitive information about its concerns. This behaviour, on the part of both sides, was not an illustration of paranoia but a reflection of the heightening of real tensions in South African society as the crisis in the apartheid system deepened.

[65] *The Star*, 11 January 1989.
[66] *Cape Times*, 23 November 1988.

CHAPTER FOURTEEN

AN UNCERTAIN FUTURE
April 1989-July 1991

Mixed Messages

South African society by 1989 was showing signs of acute instability. The state was clearly in crisis and there was no obvious indication of how it would be resolved. The system abounded with contradictions. The government had concocted a desperate cocktail of political reformism, legislative controls, police violence and military repression that eliminated dissent either by outright suppression or through subversion. It aimed to undermine trade union opposition through its Labour Relations Amendment Bill and the Disclosure of Foreign Funding Bill that denied trade unions access to foreign funds.[1] Behind the protective shield provided by legislation and the continuing state of emergency, the government tried to present an image of strength and determination. Its counter-revolutionary efforts on all counts, however, had been seriously undermined by its ignominious defeat by the Cuban forces in Angola during March and April 1988 and by its forced acceptance of the United Nations Resolution 435 which heralded the transformation of South-West Africa into the independent state of Namibia. The South African government had agreed in December 1988 to withdraw from Angola and cease its support for UNITA and to end its occupation of South-West Africa. By conceding defeat, the government thus curtailed the military drain on its resources at the very time that the international distaste for apartheid was reducing the flow of investment into South Africa which, in turn, was increasing the disenchantment of domestic entrepreneurs with the government's policies. No coherent, viable, generally acceptable vision of South African society was discernible through this confusion.

It was also a confusing picture for black people. Despite the fact that the state suppression of dissent was intense, some apartheid structures

[1]The government had introduced the Promotion of Orderly Internal Politics Bill in Parliament in March 1987 as a means of preventing South African organisations from obtaining foreign funds. After widespread criticisms of the Bill, it was dropped and replaced by a Disclosure of Foreign Funding Bill that became law in March 1989. The new Act required all organisations and people to reveal all sources of external funding and to have their books audited. Both the South African Council of Churches and the Southern African Catholic Bishops' Conference declared that they would defy the Act.

were being demolished following the abandonment of the pass laws and formal job reservation regulations in commerce and industry in 1987. Rumours, moreover, were circulating about the possibility of negotiations between the ANC and the government and about the release of the remaining leading ANC prisoners. In November 1988, a year after Govan Mbeki's release from prison, Zephania Mothopeng, the president of the Pan-Africanist Congress, and Harry Gwala, the Natal union leader and communist, had also been freed from Robben Island, allegedly on humanitarian grounds. Gwala was suffering from motor neuron disease. The rumours were then centred on the release of the senior ANC leaders, including Nelson Mandela. During that time, however, there was no sign of the armed struggle being relaxed. The Minister of Law and Order, Adriaan Vlok, stated that during 1988 there had been more terrorist incidents than in any other year in South Africa's history. Nor had the banning of organisations and the arrest of their officials ceased. Thirty-two organisations had been banned and 3 707 people were detained. There had been 99 trials through which 294 people had been condemned to death. One hundred and seventeen of those people had been executed in 1988.[2]

The contradictions in the system were reflected in events at all levels of society. Where local authorities were concerned they were clear and incisive. There were movements in some local authorities to relax or remove petty apartheid restrictions. In Johannesburg, for instance, the City Council was considering desegregating the city's buses, swimming pools and recreation centres. The city council in Durban was evenly split over a proposal to desegregate the city's five remaining whites-only beaches. Following the elections on 26 October 1988, however, when the Conservative Party, which advocated an inflexible apartheid system, had gained political control of more than 90 out of 529 white local authorities, there had been an extensive reintroduction or intensification of petty apartheid measures. The support for the Conservative Party was strongest in the mining towns of Boksburg, Carletonville, Springs, Welkom, Phalaborwa, Brakpan and Nigel. There they decreed that black people could no longer share in the use of public amenities and they restored 'whites only' notices. In those towns, the pressures to strengthen apartheid came mainly from the white mine employees who constituted the core of support for the CP and the extremist paramilitary organisation, the Afrikaner

[2]South African Institute of Race Relations, 1988/89, ppxxx. In 1988, there were, on average, 14,3 incidents of unrest and violence every day.

Weerstandsbeweging (AWB). This resurgence of political support for apartheid in the mining towns drew black mineworkers into conflicts about segregation within mining communities for the first time when they backed black consumer boycotts in Boksburg and Carletonville. More importantly, their retaliation against white racism began to disrupt work relations underground.

The divisions amongst the supporters of apartheid compounded the confusion about what was actually happening in and to the apartheid system. The governing National Party, already split by the electoral successes of the right-wing Conservative Party, was further divided by attempts early in 1989 to force the State President, P W Botha, out of the party leadership and state office after he had suffered a stroke. He remained in office but, on 2 February 1989, F W de Klerk, who became the principal power-broker, took his place as leader of the National Party. The two men represented rival factions but there was little public evidence that they were pursuing significantly different political agendas. On 5 July, Mandela and Botha had a clandestine meeting at the state President's Cape Town office. Before anything could materialise from their discussions, Botha was forced to resign in August and F W de Klerk took his place in an acting capacity until after the whites-only election in September when he became the constitutional head of state. However, there was still no indication in public of a radical change in government policy towards the liberation movement. The government continued to insist in public that there could be no reform so long as the ANC continued with the armed struggle.

Nonetheless, negotiations about negotiations between representatives of the ANC and the government were being conducted in private, in one form or another, as they had been for the previous three to four years. The central ones involved Mandela. During that time, it was reported, "before the rest of the world knew anything of it, the future of South Africa was being explored in secret conversations in hospitals, prisons, and a cabinet minister's home between government officials and their principal political prisoner".[3] The rumours were fed in December 1988 by the transfer of Mandela from Pollsmoor Prison to Victor Verster Prison near Paarl where he was accommodated in his own cottage and freely allowed to have visitors. There had also been clandestine meetings in Britain between ANC officials, led by Thabo Mbeki, and prominent Afrikaner church leaders and business people. In a public

[3]Allister Sparks, *Tomorrow is Another Country: The Inside Story of South Africa's Negotiated Revolution*, 1995, p36.

sign that negotiations were possible, the ANC issued the Harare Declaration on 21 August 1989 by which it made explicit the conditions under which it was prepared to negotiate with the apartheid government.

'Charterism' *v* 'Workerism'

The government's political gestures and military repression, the rumours and black disbelief led to much debate in COSATU and its affiliates about how to take advantage of the perceived weaknesses of the regime. The liberation movement was a forum characterised by debate. Problems were generally posed as questions in prepared papers that were examined in workshops and debated in conferences until a consensus was reached. In both the NUM and COSATU issues were rarely determined by votes. The primary subject of debate at the beginning of 1989 was that concerning alliances. It was recognised that the widest possible unity should be achieved in order to confront the government. But how much unity was necessary to achieve its objective? How far should it extend? Who should be excluded and for what reason? These questions raised a fundamental theoretical matter concerning political orientation that divided the liberation movement. They had plagued the trade union movement since the debates over the *Freedom Charter* and had dominated the discussions at the Special COSATU Congress the previous May when attempts were being made to mobilise opposition to the Labour Relations Amendment Bill.

In colloquial South African terms the issue was between 'Charterism' and 'Workerism'. 'Charterism' was the term used to describe the anti-apartheid strategy enunciated by the SACP and adopted by the ANC. It involved starting with the limited objectives of the *Freedom Charter* on the premise that the achievement of a liberal democratic state was a prerequisite condition for socialism. The SACP, however, did not see this as two distinct stages but as a process. It argued: "Freedom and national liberation are the **immediate** goal, not the end of the road. They are a way-station on the road to a socialist goal, worthwhile and valuable and worth fighting for in themselves; but yet only a way-station on the road ahead."[4] In terms of practical politics, however, the strategy did involve two stages in that

[4]*The African Communist*, Number 114, Third Quarter, 1988, p32. The issue was much debated in the pages of this journal in 1987, 1988 and 1989 and was explained in some detail in The Path to Power, the programme of the SACP adopted at its Seventh Congress in 1989.

the anti-apartheid forces had to be mobilised to achieve national liberation and then, and only then, and with a different coalition of forces, could there be a socialist revolution.

'Workerism', on the other hand, insisted on the single-minded advocacy of socialist ideals through which the achievement of national liberation would be an integral part of the process. It advocated exclusive working class leadership, gave priority to struggles in the workplace and was, therefore, trade union-centred. It rejected the notion of nationalism and posed the issue of liberation exclusively in class terms that would be attained simultaneously with that of socialism. Workerists rejected the ANC because it was a national rather than a class organisation and, therefore, implied working with organisations they defined as their class enemy. This raised the critical difference between the two strategies. There was no distinction between their ultimate goals expressed in abstract terms of freedom and liberty. The differences emerged, in the first instance, when meanings were given to the terms. These were present, moreover, within each of the strategies. Workerists consisted of syndicalists who envisaged the replacement of the capitalist state by a socialist one run exclusively by trade unions after its overthrow by revolutionary means, such as a general strike. They believed that 'freedom' and 'democracy' were bourgeois notions.

Opinions amongst the Charterists were more diverse. They ranged from groups that believed abolishing apartheid would be enough because there could be freedom within the prevailing economic system, to those who were convinced that the fulfilment of the *Freedom Charter* would be sufficient and then to members of the Communist Party for whom freedom ultimately depended on the achievement of economic emancipation that could only be reached through the public ownership of the means of production. The real differences between these groups, however, were practical ones that perpetually dominated relationships within the broad liberation movement and which arose from the consequences for organisation that followed from adopting one strategy compared with the other. It was an issue that dogged the UDF during the whole of its brief life.[5] For the Charterists there was virtually no limit to the range of possible alliances. The only precondition was a belief in the need to end the apartheid system. Any group that did that was a welcome ally without having to spell out what should follow from liberation. Thus they welcomed the support of bantustan-based

[5]See Jeremy Seekings, *The UDF: A History of the United Democratic Front in South Africa, 1983-1991*, 2000.

groups of students and civil servants and of white liberals such as the Five Freedoms Forum and Black Sash.[6] The Workerists, on the other hand, rejected such alliances and aimed only to bring established anti-capitalist organisations together.

After the UDF had been banned in February 1988, the mantle of leadership in the liberation struggle fell to COSATU. The argument about alliances, therefore, was primarily conducted between its affiliates. Ramaphosa on behalf of the NUM argued the mainstream Charterist position while Moses Mayekiso, the general secretary of NUMSA, put the Workerists' case.[7] Ramaphosa continued to press for the broadest possible alliance. It was necessary, he said, "to strengthen our structures through action on the ground in the unions, civics, students, youth, women, church, sport and the cultural formations of the MDM".[8] He urged bantustan anti-apartheid groups, organisations of white liberals and relatively conservative interest groups that stood in the 'middle ground' to join with the affiliates that had comprised the UDF, including Indian and black community organisations.

The Search for Unity

The NUM's concern about alliances went further than getting a broad, common anti-apartheid platform. It wanted the whole trade union movement to be united within a single federation and that, in the first instance, meant trying to unite COSATU and NACTU. The ground had been prepared for collaborative action by the ANC when it met NACTU leaders in Harare in May 1988 and explained that it was not

[6]The Five Freedoms Forum was formed in 1987 by a number of small white liberal and left-wing organisations to campaign against apartheid. Black Sash comprised white middle-class women who participated in public demonstrations against various aspects of apartheid wearing black sashes. The organisation was originally formed in 1955 as the Women's Defence of the Constitutional League. It retained its character over the years but in 1981 it associated itself with the *Freedom Charter*.

[7]Moses Mayekiso had been arrested in 1986 and charged with treason for allegedly running 'people's courts' and street, area and block committees in Alexandra. He remained in detention until December 1988 when he, and his four co-accused, were released on bail but under stringent, restrictive conditions. In April 1989 the five of them were acquitted of all charges. When Mayekiso was first detained he was the general secretary of the Metal and Allied Workers' Union (MAWU). In 1987, whilst still in detention, his union amalgamated with several small motor unions to form the National Union of Metalworkers in South Africa (NUMSA). During the whole of that period he retained his union position and emerged from prison as the general secretary of NUMSA.

[8]NUM, *Overview*, Sixth National Congress, 1989.

necessary for NACTU to endorse the *Freedom Charter* in order to join a united front. The obverse of this was that COSATU was free to engage in campaigns with organisations that opposed the *Freedom Charter*. A consequence of the meeting was that COSATU and NACTU joined forces for the first time to call a widely supported three-day protest against the amendment to the Labour Relations Act from 6-8 June.

The collaboration continued when, at the suggestion of NACTU, a Workers' Summit was called for 4-5 March 1989 to explore the possibility of further united trade union action against the Act. The two organisations made preparations for the summit and agreed on an agenda comprising the "LRA Act, Building Unity and State Repression". However, less than two weeks before the summit was due to start, NACTU withdrew from it because COSATU refused to delete the second item on the agenda and replace it with one that confined them to discussing the differences between the two federations rather than the possibilities of eliminating them. The summit was held, nonetheless, with COSATU, 16 non-affiliated unions and 11 NACTU affiliates representing one-third of its membership, in attendance.[9] Thus the purpose of building unity was partially served. Seven hundred delegates from 43 trade unions were present and agreed on a campaign to alter the Labour Relations Act. The summit, however, was so preoccupied with the Act that it left no time to debate the question of unity. Nonetheless, Ramaphosa believed that the first step had been taken towards "building unity in action which will lead to one union in one industry and one united federation in South Africa".[10] The division within NACTU led to the resignation of its general secretary, Phiroshaw Camay, who had first given the task of creating a black mineworkers' union to Ramaphosa.

The NUM was a disciplined affiliate of COSATU and it expressed its primary political concerns through its organisation. It was only where the mining industry provided a particularly bad example of an apartheid abuse, such as with racial and job discrimination, that the NUM's own liberation profile was highlighted. This happened over the issue of management repression and, on 5 February 1989, it launched an 'anti-repression campaign' at a mass rally of 4 000 mineworkers at the University of the Witwatersrand where it was announced that special committees were being set up to collect data about repressive activities.

[9]See Ray Alexander, "An Historic Workers' Summit", *The African Communist*, No 118, Third Quarter, 1989.
[10]NUM, Sixth National Congress, 1989, p20.

The union also raised with the Chamber of Mines the question of the Labour Relations Amendment Act which, in combination with the Riotous Assemblies Act, the Internal Security Act and the Emergency Regulations, seriously curtailed its ability to protect its members. It was difficult to strike legally and impossible to picket and hold outdoor meetings. The Emergency Regulations were also used by the security police to raid NUM offices in Namaqualand, Nelspruit, Carletonville and Rustenburg during the first six months of 1989. They also gave the police the powers to stop and search NUM members whenever they pleased. By April 1989, there were at least 20 union members in prison. Two others remained on death row.

NUM's Sixth National Congress

This congress was held in April 1989 under the slogan 'Defend and Consolidate the NUM'. Its tone was set by Ramaphosa when he said: "What is clear is that the days of the apartheid system are numbered and the working class must lead ... as we collectively chart our way to a society free from oppression and exploitation." However, he cautioned against over-optimism. "The scenario," he declared, "reflects cracks within the ruling class, where no coherent orientation exists on the way forward and where the potential for disintegration exists. But we must not confuse the cracks and divisions within the ruling class as a sign of the imminent collapse of the apartheid system."[11] Nonetheless, the slight scent of success encouraged the congress to go on to adopt a defiantly anti-capitalist and anti-apartheid stance. It demanded that Nelson Mandela and all political prisoners should be released, that the state of emergency should be lifted, that all the apartheid laws should be scrapped and that there should be universal suffrage. It went on to declare that such were the problems in the South African economy that "capitalism and 'free enterprise' can never provide justice for all of South Africa's people". The solution, it proposed, lay in following the guidelines set out in the *Freedom Charter*.

The reality at branch level, however, provided no grounds for optimism. None of the individual mines showed any desire to extend or consolidate their institutional relationships with the NUM. On the contrary, they seemed intent on sidelining it as an organisation of little or no significance. This applied as much to Anglo-American's mines as

[11]NUM Sixth National Congress, 1989, p6.

The Sixth National Congress, 1989.

to any others even though, at national level, it was involved in devising conciliation and share-ownership schemes. From their superior power positions they acted as if the NUM barely existed and were quite prepared to settle for uninhibited and undisguised hostile relations whenever they were forced to recognise that it did exist. In consequence, virtually every move made by the NUM to defend and consolidate its position involved confrontation and struggle.

Regaining and retaining recognition agreements were problematic for many branches. Some in Anglo-American mines were still struggling to recruit more than 33 per cent of the work categories one to eight in order to press for recognition but they were making little progress. Where memberships fell below that figure, as at Gencor's Leslie gold mine in Secunda, their recognition agreements were scrapped. The union had made no progress towards recognition at any of Gold Fields' gold mines irrespective of how many members it had recruited. In general, Ramaphosa told the Sixth Congress, relations with mine managers had reached "an all-time low" with the result that very little conventional industrial relations activity was taking place.[12] The situation in the multifarious non-Chamber mines was even worse for there the union had virtually no rights and, in consequence, no influence over wages or conditions of work. In addition, as Motlatsi emphasised in his presidential address, there were the endemic problems in the mining industry, "the hated migrant labour system, undemocratic structures in the hostels, racist practices, unsafe working conditions..."[13]

The debates reflected the seriousness of those issues yet the delegates were neither defeatist nor sombre. The congress fulfilled various union purposes. In the first place, it was an assembly where the delegates mixed socially, enjoyed themselves and made new friendships. They continually interrupted the debates with their singing and *toyi-toying*, which was an expression of their political intent rather than a form of relaxation. By bringing delegates together from all parts of the country the congress generated a spirit of national solidarity which was absent for much of the time as the ordinary members grappled with their respective disparate shaft and branch problems. It made statements of policy but on this occasion, after the important policy-making congress of 1987, that was more a matter of re-affirmation than formulation.

The congress did, however, allow for reassessments and it conveyed messages to those members who were not present as well as warnings

[12]*Ibid*, p23.
[13]NUM, *Overview*, Sixth National Congress, 1989.

to the mine owners and the government. Its pronouncements ranged widely over wages policy, job discrimination, working hours, overtime, retrenchments and repression at work. It dealt with internal organisational issues that arose, for instance, from the management's failure to process stop-order forms through which mine managers deducted union subscriptions at source and forwarded them on to the union. One resolution accused mine managers of deliberately denying the union finance by being slow in processing stop-order forms and then refusing to submit monthly subscription cheques to the union head office when they had done so. There was particular concern about branch finances. When a mine withdrew recognition from a union branch, the stop-order facilities were cancelled and the branch officials had the task of collecting subscriptions by hand from each individual member. The delegates were concerned that hand-collected subscription fees were being retained by the branches instead of being forwarded to the regional offices for submission to the head office. This practice did not affect the operation of the head office because it was largely funded from foreign sources but it had a direct disruptive impact on the work of the regional offices because they were dependent on subscriptions. There were cases, as there had been before the national strike, where some branches were more prosperous than their respective regional offices.

The Emergence of the Mass Democratic Movement

The main impact of the NUM Congress was to reaffirm it as an emphatic ally of the banned ANC and to make proposals to unite the anti-apartheid forces behind it. The ANC had distributed a document called *Constitutional Guidelines for a Democratic South Africa* in July 1988 which proposed that the future South Africa should comprise a multi-party democracy in a unitary state, practise universal suffrage based on the principle of one person one vote, have a bill of rights that was enforceable by an independent judiciary, protect religious, cultural and language rights and the freedom of speech and association, redistribute land but recognise private property rights, ban all forms of racism and tribalism, endorse a mixed economy and declare the country to be a non-aligned state within the United Nations. The NUM Congress decided that those guidelines should form the basis for discussions within the mass democratic movement. It underlined its identification with the ANC by re-emphasising its commitment to the *Freedom Charter* and insisting that it should become a permanent feature of NUM policy by being included as an annexure to the union constitution. It went on from there to propose the creation of an anti-apartheid coalition of all organisations between which there was principled agreement, but with COSATU and the UDF at its core.

This proposal was submitted to COSATU's Third National Congress in July 1989 and formally gave rise to an alliance that would be called the Mass Democratic Movement (MDM).[14] The conduct of the delegates at the congress and the tone of its debates were defiant. The NUM delegates lived up to their reputation as spontaneous singers of protest songs that revered the ANC, Umkhonto we Sizwe and the SACP. Illegal ANC and SACP flags were unfurled. Delegates from the rostrum taunted the security services by prefacing their remarks with such statements as "If I were a member of the ANC, but, of course, I am not saying I am a member of the ANC..." The same inference was made in relation to membership of the Communist Party. These were gestures that would, in the past, have led to arrests. They were made despite a strong police presence. The police were seen in and around the hotels where delegates stayed, and they kept a watchful eye on the proceedings from outside the congress hall.[15] Fake pamphlets were circulated that were clearly intended to create disruptive political divisions between the delegates. Moreover, in what could have been an attempt to sabotage the proceedings, a large number of the delegates who ate communally in the large dining hall suffered from painful stomach disorders. The congress, nonetheless, went ahead with a greater degree of unanimity than at its previous meetings. The political

[14]There had, however, been references to the MDM before the congress. A four-page leaflet distributed by COSATU at the congress mentioned the cohesion between COSATU and the UDF which expressed itself through the mass democratic movement. It then added: "The challenges that face the MDM at this juncture are the strengthening and building of grassroots structures in all sectors. The building of structured links between the different sectors of the MDM at local and regional level." This was written prior to the debate on the NUM resolution and anticipated its contents. The COSATU congress agenda, moreover, listed the presence of an MDM speaker who happened to be Murphy Morobe, the National Publicity Secretary of the UDF. Cyril Ramaphosa had also anticipated the emergence of the MDM when he spoke about it at the NUM congress in April. The MDM first appeared on the political scene as an entity when it launched a National Defiance Campaign on 26 July 1989. After that it was an acknowledged organiser of liberation activities. The term 'mass democratic movement' was commonly used at the time to describe the unity of people and organisations in actions and campaigns. It seemed appropriate, therefore, to use it as a label for a coalition of anti-apartheid forces. It was, however, just one part of the liberation movement and did not include the ANC or the SACP. (See South African Institute of Race Relations, 1988/89, pp681-682.)

[15]Baskin, 1991, p345.

hostility between the NUM and NUMSA, that had characterised the earlier congresses, was largely absent as they confronted the question of negotiations with the government.

It became clear at the COSATU congress that those directly involved in the internal struggle, and who knew nothing about the clandestine meetings except through rumours, were either opposed to negotiations or sceptical about them in one way or another. Harry Gwala, addressing the congress as a visiting speaker, equated the "so-called negotiations" with "imperialist manoeuvres". He concluded: "Our path lies in the determined united action of the people, not to reform apartheid, not to fight for structural reforms … but to destroy and bury apartheid so that on its ashes they may build a new, non-racial, democratic South Africa."[16] Others were less dismissive but still sceptical. In his presidential address to the COSATU congress, Elijah Barayi declared: "F W de Klerk is not a new man at all … His troops continue to emasculate the will of the people through brute force. He is attempting to win legitimacy and respectability that is denied him by the overwhelming majority of our people. Please do not sell him to us for any price. He is not a good buy."[17] Jay Naidoo, the general secretary of COSATU, wrote in his report that the South African government was not serious. "To date there is not a single signal from the regime that it is prepared for genuine negotiations. Our cadres and organisations are still banned and restricted, political prisoners and detainees still languish in apartheid dungeons, virtually all extra-parliamentary opposition activity is illegal, the SADF is still occupying our townships and the ANC remains banned with thousands of our people forced into exile by apartheid."[18] In the end, after an intense debate and without a vote, an approach to negotiations was agreed. The congress accepted a NUMSA resolution, amended by the NUM, which favoured a negotiated settlement but under strict conditions that involved "the participation of the people". There was strong opposition to secrecy and to the notion that negotiations should displace other forms of struggle. But, in the end, negotiations remained the central issue. In the words of the NUM, the MDM had to agree on a common approach to "the South African road to power through negotiation".[19]

[16]*Ibid*, p346.
[17]*Ibid*, p348.
[18]COSATU Third National Congress, 1989, p14.
[19]Baskin, 1991, p349.

The Defiance Campaign

Within a few weeks, the MDM did just that. In July 1989, following discussions between the ANC, the UDF and COSATU in Lusaka, the MDM announced that it would embark on a nation-wide defiance campaign. It was to be the first such action since the defiance campaign in 1952. The announcement coincided with continuing action against the amendment to the Labour Relations Act. A number of regional protest rallies had been organised but they did not all materialise because of the intervention of the security forces. At Vereeniging, Klerksdorp and Secunda, the police simply intervened just as the rallies were about to begin and ordered everyone to disperse. In other cases, banning orders were issued. The one for the western Transvaal was issued too late to stop the meeting so its closure, too, was left to the police at the last minute.[20] Some rallies, however, went ahead. About 15 000 people, mostly mineworkers, attended a rally at Thabong stadium in the Orange Free State to hear James Motlatsi describe the Labour Relations Amendment Act by its code-name "*taqama keosetjhe*", meaning "lie still while I pick your pocket". He told them that the torch of liberation was in the hands of the mineworkers and spoke of the role being played by Nelson Mandela. "It is clear," he stated, "that the government realises that in order to have the country running smoothly, Mandela has to be released."[21]

The MDM explained that its campaign was starting on Wednesday, 2 August with peaceful, mass action against segregated public facilities, beginning with the hospitals. It began with a warning from the Minister for Law and Order, Adriaan Vlok, that the MDM was under police scrutiny and would have to face the consequences if there was violence or intimidation. Sick black people were asked to present themselves for treatment at designated whites-only hospitals in Johannesburg, Krugersdorp, Vereeniging, Durban and Pretoria from the first day of the campaign. This was to be followed later in the month by banned individuals and banned organisations defying the government and 'unbanning' themselves by doing those things that their banning orders specifically proscribed, such as organising and attending meetings, issuing statements and demonstrating in public.

Some doctors in 'blacks-only' hospitals sent their patients for treatment to 'whites-only' hospitals whilst a number of sick black people also went

[20] *Weekly Mail*, 28 July 1989.
[21] *New Nation*, 28 July 1989.

there on their own volition. It was estimated that about 200 black people went for treatment to the 'whites-only' hospitals on the first day. There were no incidents of confrontation or violence because, after some reflection, the hospital authorities gave them treatment. This campaign lasted for only a short period and did not amount to the desegregation of the hospitals. Black mineworkers participated where they could. They faced no difficulties from the Chamber of Mines because in June it had announced its intention to merge its own whites-only Cottesloe Hospital with the Rand Mutual Hospital that catered only for black mineworkers. The only reported hospital action at mine level was at Matla colliery in Witbank where 10 black mineworkers who were injured in separate incidents reported to the whites-only mine hospital for treatment. Meanwhile, other protests were organised. In Johannesburg and Cape Town an 'All Schools for all People' campaign was launched to desegregate the educational system. The Pretoria Council of Churches launched the 'Standing for the Truth Campaign' to obtain open facilities, at first on the buses. The organisers negotiated with the Pretoria City Council, which agreed that if black people possessed valid tickets they should be allowed on the buses. The campaigners bought tickets but were arrested by the police nonetheless.[22] This type of police behaviour was commonplace throughout the defiance campaign.

The day after the MDM campaign began, the NUM embarked on its own attack on racially segregated facilities in the mines. There were many examples where preferential treatment was given to white miners. No-one was allowed to take cameras, radios or firearms into the mines but only black mineworkers were searched for them at the compound gates. In the segregated tea-rooms the white miners drank from mugs and were provided with seats while black mineworkers had to stand in their section and drink from plastic containers. The black mineworkers had to use toilets with no doors. They were not allowed to mix with white miners in the change rooms or the cages. Both groups had time-sheets which they had to sign at the beginning of their respective shifts but whereas the white miners could sign out at the scheduled end of a shift, the black mineworkers were not allowed to do so until they had completed the tasks that had been given to them. There was not a single black underground supervisor.

Representatives of the NUM and the Chamber of Mines had met on 28 July 1989, shortly after the MDM had announced the launch of the

[22]Learn and Teach Publications, *Learn and Teach*, No 5, p34, Johannesburg, 1989.

campaign. The mine owners wanted to know whether the media reports that the NUM supported the MDM campaign were correct and, if so, to what extent it would get involved. They warned that union action could "lead to polarisation, confrontation and even violence" and suggested that instead of protesting they should establish a working party "to address specific practices which [the] union regarded as racially discriminatory".[23] Ramaphosa did not respond on the spot but when he did so it was at length and in detail. By then, however, the union's campaign was in full swing and the reaction of the mining houses was known. There was concerted action at a number of Vaal Reefs shafts. On 7 August, black mineworkers washed in the white changing rooms, forced white miners to stand in queues with them for the cages and used the senior canteens. Anglo-American had warned that it would take disciplinary action against employees at Vaal Reefs who used facilities "to which they were not entitled in terms of their seniority". It added that it was its policy "to provide facilities, wherever possible, to all employees according to skill and seniority, regardless of race".[24] The NUM believed that Anglo-American was playing with words because there were no black persons with the qualifications to enter and participate in the facilities hitherto provided for white miners. Anglo-American, however, was true to its word. When 11 black mineworkers at the Vaal Reefs South shaft forced white miners to queue in line for the cage with everyone else, they were arrested, charged with intimidation and incitement and then dismissed from the mine.

Other NUM branches initiated their own courses of action and received mixed responses. As the campaign progressed, it became clear that it was proceeding without overall direction from either the NUM or the Chamber of Mines. In effect, it was being played out at mine level by union activists on the one hand and white racist miners and managers on the other. No action was taken against a black mineworker who washed in a whites-only change room at Vaal Reefs No 9 shaft but

[23]Correspondence from the Chamber of Mines to the NUM, 4 August 1989.

[24]*The Citizen*, 31 July 1989. The chairman of the Anglo-American Corporation, Gavin Relly, published a statement at about this time called *Racial Discrimination*: *This is Where we Stand*. He emphasised the need for a "society freed from racial barriers" and added that "anything that drives the races apart, like racial discrimination of any sort, is terribly damaging" to the process of achieving prosperity for all. He ended by stating: "In the workplace they [employees] behave in a way that supports our non-racial position. Firm disciplinary action is taken against employees who allow their political convictions to prevent them from performing in our spirit of non-racialism, the job they were employed to do." (nd.)

when others on No 7 and 8 shafts in the same mine attempted to do likewise the mine security force detained them. At the Impala Platinum Refinery in the Witwatersrand, white employees went to the canteen and toilets armed with pick handles and sticks and assaulted black workers who were using their facilities. The police subsequently charged none of the white miners. But it was the mine managers who most overtly expressed their prejudices. There was much controversy at Anglo-American's Springfield Colliery in Witbank. It employed 2 000 black mineworkers and housed them in overcrowded hostels without any recreational facilities. The union members decided to use all of the colliery's segregated facilities. They entered buses reserved for whites and then went to the communal lounge in the white suburb and ordered drinks. The manager responded by insisting that they should be charged double for anything they wished to purchase. When the black waiters in the lounge refused to do as he had demanded, steps were taken to clear all of the black mineworkers out of the suburb altogether. Members of the SADF, who were brought in to assist the mine security, harassed the mineworkers by stopping, searching and interrogating them. The news of the use of the SADF appeared in *The Citizen* and the *New Nation* but Godsell, speaking for Anglo-American, denied that it was present on colliery property and said that the manager had not summoned it.[25] The manager at Arnot colliery, also in Witbank, locked the whites-only toilets rather than let black mineworkers use them. The director at Mintek, the state research establishment, posted guards at the doors of its canteen in order to prevent black workers from entering it. Black mineworkers were sometimes not aware of the extent of the discrimination against them until the campaign began. For instance, when they entered the whites-only stores at the Premium diamond mine, near Cullinan, they discovered that the prices of goods there were significantly lower than those they were being charged in the concessionary stores. This prompted them to boycott the purchase of all food at the mine until they were provided with the same food as white customers at similar prices. They also supported a black community boycott of the local Spar shop because of its racial practices. The manager at the President Steyn gold mine in the Orange Free State banned all meetings to discuss the defiance campaign and disciplined every black mineworker who entered the senior mess and bar, reserved for members wearing ties. The union claimed that the manager of the mine was trying to instigate a faction fight by spreading rumours that

[25] *The Citizen*, 18 August 1989; *New Nation*, 22 August 1989.

the Sothos were dissatisfied with the campaign. The manager at Anglo-American's Ergo gold mine tried to dissuade black mineworkers from using the whites-only change rooms by saying that the facilities were divided according to seniority and not colour. The union members retorted that if that were the case then seniority should be based on age and not occupation.

Whilst the struggle in the mines was taking place there were public arguments between the officials of the mining houses and the NUM about the extent and meaning of segregation. Ramaphosa responded to the meeting that had been held late in July with a four-page letter to the Chamber of Mines on 10 August 1989 in which he explained the union's position and countered charges, made particularly by Anglo-American, against it. He rejected the suggestion that they should set up a joint working party to examine the incidence of racial discrimination. "We do not need a research project," he wrote, "to uncover practices which are part of the daily reality of the lives of our members. We need only honesty and a commitment to social justice." He compared the mine owners' public opposition to racial discrimination with their practice. The South African mining industry, he wrote, "has initiated and institutionalised many of the worst aspects of racism. We have in mind the migrant labour system, racial segregation in the mines and the attitude of mining management to free trade unionism … There is no economic or social justification for a system of labour utilisation which operates like a penal institution. There is a mountain of evidence … to show that forced, oppressed, subjugated people contribute inefficiently and ineffectually to the labour process. If the Chamber, therefore, is really interested in generating wealth it would be doing itself a good turn by taking prompt, immediate steps to abolish the migrant labour system with its attendant evils." Ramaphosa went on to question the Chamber's claim that it had abolished racial segregation at work. "Although," he stated, "there is an enormous pool of talent among black workers there are virtually none who are in positions higher than team leader and artisan's aide … The Chamber has been devious in substituting educational standards for racial segregation for it knows that the standards it demands for skilled positions are unobtainable for the vast majority of black miners because of the inadequacies of the system of black education." It was the union's view, he wrote, that it was not its actions which had led to polarisation but the system of labour controls that had divided black workers from white ones and which had been set up and nurtured by the mining houses. As for violence, that arose from the fact that managers controlled the physical means of repression and used them with painful regularity. "It seems petty in the extreme," he stated, "for your Chamber and its member mining houses which dominate South Africa's mining industry to blame its migrant labour force for the ills of the industry simply because they have decided to stand up for their rights." He reiterated that the NUM had no other option but "to use every means at its disposal to protest against the discourteous, harsh, repressive and inconsiderate treatment which its members receive from white supervisors, managers and officials of the mines. Such practices are obnoxious and are irreconcilable with humane, democratic and civilised standards ... We hope now that you recognise our feelings and understand why we fully support the protest campaign of the Mass Democratic Movement."[26]

NUM's letter was released to the press and was followed by public denials and counter-allegations. The Chamber of Mines replied on 23 August and reiterated its commitment to abolish job segregation. It regretted the union's refusal to join a working party and deplored actions by NUM members that "have gone beyond actions against racially discriminatory practices" such as the disruption of underground queues and the invasion of senior dining room facilities.[27] Anglo-American issued its own denials and counter-allegations because, as its mines employed the majority of NUM members, they were subjected to most of the disruption. Godsell insisted that since the legal colour bar had been removed from the Mines and Works Act two years before, segregation in the Anglo-American mines was based on rank and not race. White miners, he insisted, enjoyed privileges because of their occupational seniority and not their racial classification. The Chamber of Mines endorsed this view.[28] It had improved its liberal image towards the end of August 1989 by obtaining a Supreme Court decision that set aside two potentially discriminatory regulations that had been inserted in the Mines and Works Act early in the year. One of the regulations required the manager of a mine to provide adequate latrine and change room facilities "after the negotiation with individual employees' organisations as to the needs and preferences of their members and having regard to their physical, moral and social welfare". In effect this allowed the white trade unions to demand segregated facilities. The second regulation made it necessary for the mine authorities to have regard for the "proven labour needs of mines and works before permitting

[26]Correspondence from the NUM to the Chamber of Mines, 10 August 1989.

[27]Correspondence from the Chamber of Mines to the NUM, 23 August 1989. See also *The Star*, 28 July 1989.

[28]*The Star*, 28 August 1989.

a person to be admitted as a candidate for certain of the certificates of competency". This could be and was used to prevent black mineworkers from qualifying for jobs hitherto reserved for whites. At the time of the judgment the Chamber of Mines was unable to train black mineworkers as winding engine drivers or upgrade sufficient of them to supervisory jobs to fill all the vacancies. The Chamber of Mines had reported two months earlier that there were about 100 black mineworkers with blasting certificates, comprising roughly one per cent of the total, but that there were 760 vacancies.[29] Of course, none of this satisfied Ramaphosa's criteria for non-racial mining.

This correspondence occurred against a background of nation-wide protest and police retaliation as the defiance campaign gained momentum. Throughout the country, meetings were banned, attacked and broken up and scores of people were detained. The police set up roadblocks to prevent black people from attending meetings and demonstrations and stopped and searched those who were nonetheless determined to go ahead. They sealed off the campus at the University of the Witwatersrand to prevent a rally taking place in support of freeing banned organisations. In the eastern Transvaal the police were given the authority to be present at all campaign meetings, to video their proceedings and to confiscate documents. All meetings had been prohibited in the western Transvaal by the end of August. In Johannesburg, the police harassed black people who were waiting at bus stops and tried to prevent them from entering union offices. Police sensitivities were already high in Phalaborwa because of its proximity to the Mozambique border but there was even greater violence during the campaign. In Kimberley, the police were everywhere, arresting activists and preventing black people from meeting in groups. When a meeting was banned in Grahamstown, in the eastern Cape, the MDM transformed it into a church service.

The protesters targeted whites-only beaches in Cape Town and Durban. Sixty-three buses were organised to take black protesters to the whites-only Strand Beach in Cape Town but the police stopped them by setting up roadblocks. This did not prevent Archbishop Desmond Tutu from leading about 800 people on to the beach but the police retaliated when they left by firing at them with tear gas and guns. Thousands of protesters in Durban marched to the whites-only Addington Beach under the slogan 'The People Shall Swim'. When they arrived there they were met by armed police and a gang of white

[29]South African Institute of Race Relations, 1988/89, pxliii.

men who were shouting abuse and chanting, "*Kom, kaffirs ... kom, coolie ... julle was nie*" which meant, "come *kaffirs*, come coolie, you people do not wash". The police ignored the abusive white racists and arrested 10 of the protesters for holding up an ANC banner and a further 48 for attending an unlawful gathering.[30] About 500 people were arrested in Cape Town during a march to the Parliament buildings, under the banner 'The People Shall Govern'. The police dispersed the marchers with a water cannon, using purple water, and by swinging *sjamboks* at them. Graffiti on buildings in Cape Town the next day repeated the slogan.[31] There were occasions when teachers marched alongside pupils who were demonstrating against segregated education. The South African Youth Congress was the first organisation to unban itself on 20 August, the anniversary of the formation of the UDF. Many other organisations, including the Congress of South African Students, the End Conscription Campaign, the Soweto Civic Association and the UDF itself, did likewise. In Johannesburg, a group of more than 100 travelled to Soweto, in violation of their restriction orders, so that they could meet activists. Overall, the campaign brought together youth, women, students, church leaders, workers and union officials. As the number of protesters grew, so did the number of arrests. Altogether, at least 2 000 people were arrested during the protests, of whom about 240 were detained without trial.[32]

The defiance campaign officially culminated on 20 August but more demonstrations, including factory protests involving bans on overtime, were planned in the run-up to the whites-only election on 6 September. These plans were laid at the second Workers' Summit on 26 and 27 August. The summit's intention was to discuss further action against the Labour Relations Act and the possibility of a stay-away before the election. On this occasion, NACTU was fully involved along with COSATU and the independent unions. The deliberations, however, were complicated by the presence of heavily armed police. The police, wearing riot gear and armed with rifles, gas, grenades and sten guns, were already in the hall when the 750 delegates arrived. The meeting, the police declared, was restricted according to the terms of the state of emergency and could continue only "on condition there were no flags, banners or stickers; no speeches other than those dealing directly with the LRA; and no singing or talking after 5 pm".[33]

[30]Learn and Teach Publications, No 5, 1989, pp34-35.

[31]*Ibid.*

[32]*Ibid.*

[33]Baskin, 1991, p388.

No invited guests could attend. Police video units were already installed in the hall so that it was impossible for anyone to disregard the restrictions without being detected. "The union leadership came up with an imaginative plan. Delegates were asked to break into smaller groups. Each would meet separately and appoint a scribe to take notes. Other delegates would act as messengers shuttling ideas and proposals back and forth between groups. It was an effective move. There were simply not enough video cameras to film every group and there were not enough police to monitor the proceedings. This made it impossible for the police to hold any particular individual accountable for decisions taken. Using these tactics the summit adopted a plan of action under the very noses of the police."[34] COSATU and NACTU also tried to get the Rand Supreme Court to declare the police restrictions illegal but their application was dismissed on the grounds that "the police had acted within the scope of the law and with restraint", although one particularly clumsy intervention had been to arrest a young man for reciting a poem about Nelson Mandela.[35] The summit decided to hold brief stoppages or demonstrations in all factories on 1 September and a two-day stoppage on 5 and 6 September 1989. There would be a month-long consumer boycott of white-owned shops from 13 September and an indefinite overtime ban. The unions, however, also agreed to continue negotiating with the employers as a means of amending the LRA.

There were many demonstrations in factories and shops on 1 September, following the advice of the Second Workers' Summit but the action in the mines mostly took place during the following day, which was a Saturday. There was a peaceful protest march at the De Beers Kimberley mine and thousands of black mineworkers, mostly from the coal mines in the Witbank region, demonstrated by singing, *toyi-toying* and carrying placards against the Labour Relations Act and the elections. The police raided the union regional offices at Empangeni, Namaqualand and at Welkom where they detained three organisers. At Grootgeluk in the Rustenburg area a union member was arrested for wearing a union T-shirt bearing the words '34 Years of the *Freedom Charter*'. There were no demonstrations in Phalaborwa where there were roadblocks on all the main roads and where the NUM regional secretary and regional organiser were in detention while the rest of the officials were afraid to sleep in their own beds at night.

[34]*Ibid*, p389.
[35]*The Star*, 28 August 1989.

The Killing of Jeffery Njuza

There had been protests against segregation at the purification plant at Rustenburg Refineries, owned by JCI, for some time. Every three months for the past year the NUM branch there had approached the management to dismantle the segregated facilities and each time it had been told that the matter would be investigated. Nothing was done on any occasion so when the MDM defiance campaign began the branch was quick to join in. There was much to protest about at the plant but there were signs, during the last week of August, that the white workers were preparing to fight to retain their privileges. One of them actually threatened a black protester with a gun. There was talk of white miners wearing chains with inverted swastikas on them and talking about themselves as members of the 'Wit Wolwe', an ultra-right-wing white vigilante organisation that had been issuing death threats to MDM activists.[36]

As a part of the NUM branch protest, on Saturday, 2 September, Jeffery Njuza, a shaft steward and another union member went to the kitchen and drank from two whites-only mugs. Njuza was an active supporter of the union's anti-segregation campaign and had already been charged by the manager for sitting on a whites-only chair. A white supervisor took the mugs from them and smashed them. Another made notes of their names and the reasons why they were there. Later in the day, Ockert Vermeulen, a process supervisor in another section, entered the purification plant, walked over to Njuza's supervisor and asked where he worked. The supervisor was seen to point to Njuza whereupon Vermeulen approached him, drew a gun and shot him in the head. Njuza died instantly. Vermeulen ran away and could not be found for four hours. He was eventually discovered dead in a basement. He had apparently committed suicide. Two days later, JCI issued a brief even-handed press statement, expressing its deep regret to both families for "the unfortunate incident".

All of the black workers at the refineries reacted to the killing by taking immediate strike action but the management ignored their feelings and threatened them with dismissal. A NUM delegation, led by Ramaphosa, eventually had a meeting with Basil Davidson, the

[36]Wit Wolwe (White Wolves) was active in other parts of South Africa. It had issued death threats to members of the defiance campaign in Pretoria and had attacked NUM members during the consumer boycott in Carletonville. It had claimed responsibility for bombing the headquarters of the South African Council of Churches the year before. Its alleged leader, Barend Strydom, was sentenced to death in May 1989 for opening fire on lunchtime crowds in central Pretoria in November 1988 and killing seven black people.

managing director of the refineries. Ramaphosa put four demands to him. His first was that the segregated facilities should be abolished. Secondly, he asked that the white supervisor who had pointed out Njuza to his killer should be suspended pending an inquiry. His third request was that, as it was known that white miners carried guns at work, the black workers should be guaranteed their safety at work. Lastly, as the strike had arisen directly out of the killing, he believed that it was right that the strikers should be paid for the working time they had lost. The manager said he would investigate the first demand and agree to the third so that in future all workers, black and white, would be searched on entering the refineries. But he rejected the second demand, because he saw no justification for suspending the supervisor, and dismissed the fourth outright. Ramaphosa was angered and astounded by the arrogance of the response. He leant across the table towards the managing director and taunted him. "Do you," he asked, "consider yourself a racist, *Baas* Davidson?" But he got no reply, for Davidson did not have the power to penalise Ramaphosa for what a white person would normally regard as intolerable insolence.[37]

Njuza's death was only one of many. Indeed, there were 239 politically related deaths in August and September of that year.[38] However, the public circumstances of the death and the political significance of its cause to the NUM created a martyr of Njuza and he entered into the annals of the folklore of black mineworkers in their struggle against apartheid. The union's Central Committee made a special tribute to him at its next meeting.[39]

The Election Protests

The pace of events was quickened by the imminence of the election on 6 September. The police searched the home of Jay Naidoo, the COSATU general secretary and spent eight hours raiding the COSATU offices on the sixth floor of National Acceptances House in Rissik Street, Johannesburg. The NUM had its offices on two floors of the same building but the police did not enter them. Nor did they interrogate NUM's secretariat.[40] While the police were carrying

[37]From the notes of the author who was in the NUM national office during that period.

[38]South African Institute of Race Relations, 1989/90, p236.

[39]There is a 'Jeffery Njuza Room' in the NUM's Elijah Barayi College in Yeoville, Johannesburg.

[40]There was an occasion at about the same time when Ramaphosa was arrested but it was

documents out of the COSATU office, staff members of the NUM and COSATU demonstrated in the street. When the demonstrators refused to disperse, 10 of them, including Gwede Mantashe, the national organiser of NUM, were arrested and detained overnight in Johannesburg's John Vorster Square police station. Led by Mantashe, they sang to such an extent throughout the night that the police were relieved to release them the next morning. One incident followed another. When F W de Klerk gave his last election speech at the Johannesburg City Hall on 4 September, the staff of the NUM and COSATU, led by Naidoo and Ramaphosa, marched up Rissik Street to the City Hall, without police permission, in order to deliver a letter to him deploring the exclusion of black voters from the election. It was the first such demonstration in the centre of the city since the 1950s. The marchers were stopped before they reached their destination and were surrounded by armed police with their fearsome dogs, whereupon everyone sat down on the road. The letter was delivered but none of them was allowed into the hall. Immediately afterwards the leaders instructed the protesters to stand up and march the way they had come. Police vans followed a short distance behind. After a few minutes, a woman started singing *Nkosi Sikelel i'Afrika* and others joined in. At that point, the vans stopped and the police, threateningly, scrambled out. The response was instantaneous. Someone gave the word to disperse and everyone disappeared into the darkness, as if the march had been a mirage. The sum of these experiences, leading up to the election, convinced Ramaphosa that De Klerk's government "while on the surface calling for talks with black leaders, is in fact conducting a secret war of attrition aimed at silencing and removing from the public arena mass democratic movement activists and spokesmen for the trade union movement".[41]

The election stay-away, which resulted in more than three million workers and students staying at home over the two days, was accompanied by continuing protests against segregated facilities in the mines. One such protest by NUM members in Rustenburg revealed how the law was weighted against them. Thirty-four of them played

an illustration of a somewhat comic element in the administration of repression. He had returned to his home in Soweto to find a thief running away with his television set. He gave chase but failed to catch him so he called the police. When the police arrived they had an ideal opportunity to search his home. To their delight, they found an ANC sticker on the door of his refrigerator and promptly arrested him for possessing an illegal document.

[41]Baskin, 1991, p405.

soccer on the day of the election inside a stadium at Rustenburg where the Conservative Party, which was in office, had imposed petty apartheid restrictions. They were assaulted by a group of white men and then charged by the police with either trespassing or loitering. No police action was taken over the assaults. Indeed, though the police had been present to arrest the NUM members they contended that they knew nothing about the assaults until after the NUM had prepared a civil action against the town council on behalf of its members.[42] The NUM participated in the stay-away and estimated that up to 85 000 black mineworkers supported it on the first day and about 100 000 on the second day, despite serious opposition from the mine managers. "In some mines, including Vaal Reefs," Ramaphosa later reported to the union's 1990 Central Committee, "the management forced our members to work at gunpoint and, in others, security police were sent underground to arrest miners for refusing to work." Ramaphosa believed that the extensive use of the police during the stay-away was a calculated ploy to convince the white voters that it was not only necessary to retain the apartheid system for their own well-being but that more control by the security forces was needed to ensure their safety. After the election, the union was still trying to free its officials from detention in Welkom and Phalaborwa.

Further Protests and Secret Negotiations

The election result returned the National Party to office but not to full power. It obtained 93 seats out of 166 but, for the first time since 1963, with less votes than the combined opposition. It lost seats on both of its wings. The right-wing Conservative Party increased its representation from 22 seats to 39 but the newly-formed Democratic Party haemorrhaged the white left of the NP by taking 21 per cent of the votes and gaining 14 seats. The white electorate was polarised by the prospect of negotiations, though within a limited political spectrum. Seventy-nine per cent of the white electorate voted for the retention of some form of white supremacy. The NP had entered the election with a commitment to devise a five-year plan where there would be equality before the law and every South African would have the right to participate in decision-making at all levels of government, but where "no individual or group would be allowed to dominate or be dominated. The new

[42]*Sowetan*, 2 November 1989.

South Africa would be based on the principle of self-determination regarding its own affairs, with joint decision-making on general affairs".[43] Under the NP, there would still be a 'white' South Africa that controlled the economy. The Democratic Party that favoured a federal constitution similar to those of the USA, Canada and Australia and which was not, therefore, based on race rejected this route out of the impasse for whites. It distinguished itself from the ANC by supporting the continuation of a capitalist economy as well as by its preference for a federal government.

The election confirmed F W de Klerk as the State President. Apart from that, it was an irrelevancy. Indeed, the domestic politics of South Africa were being driven primarily by international events in that period. The position of the ANC had been strengthened by the endorsement of its guidelines for a negotiated settlement by the Organisation of African Unity.[44] It had, however, been weakened in its capacity to wage the armed struggle by the impending collapse of the Soviet Union and the German Democratic Republic, its major suppliers of arms and military expertise. On the other hand, as the apartheid state had been nurtured by the USA and Britain primarily because of its strategic value as an ally in the struggle against international communism, the same international factors were removing the Western need for South Africa's allegiance, and thereby isolating it even further as a pariah state. Moreover, in the new international circumstances, the Western nations had little need to fear the emergence of a black communist-led government in South Africa for the context in which it might have emerged was disintegrating. The questions, therefore, as to whether or not there should be negotiations, and the conditions under which they should be held, were being answered by forces outside of South Africa though the answers still had to be articulated by the government and the ANC. The situation in this respect was no different after the election from what it had been beforehand.

In the period immediately following the election, the confidence in collective action generated by the defiance campaign was extended to traditional industrial relations issues and reflected in a wave of strikes, primarily about wages. Black mineworkers were at the centre of this with strikes in different parts of the country. Eight thousand black mineworkers at Sasol's coal mines in Secunda, organised by the Chemical Workers' Industrial Union, went on strike for a food

[43]South African Institute of Race Relations, 1988/89, pxxiv-xxv.

[44]The guidelines were published in full in *The Star*, 25 August 1989.

allowance for those living outside the hostels. There were also strikes at an antimony mine in Phalaborwa and at the copper mines in Messina. The biggest strike, however, was a rare national one at five De Beers' diamond mines over a wages claim that had first been submitted in April 1989. The grievance among diamond mineworkers was that, though they were paid substantially more than gold mineworkers, their wages lagged behind those in the manufacturing sector. The issue was essentially about the wage distinction between workers in labour-intensive industries compared with those in capital-intensive ones. The negotiations started with a NUM demand for a 60 per cent wage increase and a De Beers offer of 10 per cent. The dispute ended, after two three-day strikes involving about 10 000 workers, with the acceptance by the union of an 18 per cent wage increase.[45] A long-running wage dispute at the Iscor mines at Sishen, Thabazimbi and Grootgeluk was also settled to the distinct advantage of NUM members compared with that received by members of the rival Black Allied Mining and Construction Workers' Union. There was no settlement, however, in the case of the three-month-long food boycott by NUM members at the Rietspruit Opencast mine. The boycott had been organised after the manager had unilaterally increased the prices of breakfast and lunch by two-thirds. The manager refused to talk with the shaft stewards because, he said, the issue did not relate to work.[46]

Ramaphosa was not directly involved in these negotiations, which were primarily the responsibility of Marcel Golding, the union's assistant general secretary, and Martin Nicol, head of the union's collective bargaining department. Ramaphosa handled the negotiations with the Chamber of Mines and they had not presented a problem in 1989. In this political interregnum, he continued with his MDM work, assisting, for instance, in the organisation of the Conference for a Democratic Future in Johannesburg on 7 October when all organisations, including the churches, concerned with changing the course of South African history, were due to come together. He also had two important Soweto community responsibilities. He was part of the Soweto People's Delegation that had been formed late in 1988 by civic organisations in Soweto to negotiate a settlement of its financial problems with the Soweto Council caused by the long-running rent boycotts. He was also a member of the Crisis Committee that had been established in Soweto, in effect, to control Winnie Mandela and her

[45]*The Citizen*, 14 October 1989 and *Financial Mail*, 10 November 1989.
[46]*New Nation*, 17 November 1989.

Mandela United Football Club that was allegedly terrorising local residents in Soweto.[47] Then, on 10 October 1989, he was a member of a delegation of four that had an historic, confidential first meeting with Nelson Mandela in the Victor Verster Prison near Paarl.[48] Events were moving him fairly rapidly into a position of such national political prominence that he was being separated, though not necessarily distracted, from his union work.

The new State President, in his search for a solution to the political crisis, had begun tentatively to dismantle petty apartheid restrictions. By mid-September he had eased the restrictions on gatherings, protests and marches. As a result, there was a succession of mass marches against repression, racial segregation and the Labour Relations Act in all of the main urban centres. Between 30 000 and 50 000 people marched through Cape Town on 13 September. Over the next few days there were marches of 25 000 in Johannesburg in protest about police brutality, 40 000 in East London and 20 000 in Durban. These were unprecedented numbers but the largest, possibly the biggest protest march in South African history, took place early in October in Uitenhage in the eastern Cape when approximately 70 000 people responded to a call from COSATU and NACTU to protest against the Labour Relations Act.[49] Altogether, 150 000 workers in 17 centres throughout the country marched in support of COSATU's protest against the Act on 14 October.[50] An estimated crowd of 50 000 demonstrated in Port Elizabeth, a town with a long history of black protest activity.

It is not known whether this public manifestation of dissatisfaction influenced De Klerk in his deliberations about the future but by then it was clear that the principal source of the momentum had moved from the streets to the ANC, operating behind the scenes. This situation bred

[47]Winnie Mandela had formed the Mandela United Football Club which operated from her home in Soweto and came to act as a vigilante group, allegedly responsible for criminal acts within the community, including killing, torture, assaults and arson. The Crisis Committee had attempted but failed to persuade her to disband the club. The Truth and Reconciliation Commission in 1998 concluded that in part the Crisis Committee was responsible for not bringing her under control. (See Anthony Sampson, *Mandela*, 2000, chapter 26 and the Truth and Reconciliation Commission, Volume Two, 1998, pp577-582.)

[48]Also present were Albertina Sisulu, the co-president of the UDF and Walter Sisulu's wife; Elijah Barayi, president of COSATU; and Murphy Morobe, the acting publicity secretary of the UDF. Albertina Sisulu first heard the official news of her husband's release whilst watching television with Mandela on that occasion.

[49]South African Institute of Race Relations, 1989/90, pp226-227.

[50]*Ibid*, p358.

rumours and speculation. One such speculation, early in October, concerned the imminent release from Robben Island of the trade unionist Oscar Mpetha and Walter Sisulu, the former Secretary-General of the ANC. Aged 80 and 77 years respectively, they were the oldest Robben Island prisoners. Then, in a surprise announcement on 10 October, De Klerk declared that not two but eight ANC leaders would be released from prison, though no dates were given.[51] Only Nelson Mandela, amongst those sentenced to life imprisonment at the Rivonia trial in 1964, was to remain in prison. This was the news the MDM delegation saw announced on television during their visit to Mandela. Events then moved quickly. On the following day, three church leaders, Desmond Tutu, Frank Chikane and Allan Boesak, met De Klerk and spelt out six conditions for negotiations with the ANC.[52] Almost simultaneously, the US government published what it considered to be the appropriate terms for negotiations to start. By then there was serious talk of a negotiated settlement though, following his meeting with Mandela, Ramaphosa warned against undue optimism. It was believed that the announcement of the releases had been timed to help Margaret Thatcher, the British Prime Minister and steadfast supporter of the apartheid system, to resist pressure for increased sanctions against South Africa at a meeting of the leaders of Commonwealth countries in Malaysia on 18 October. As if to confirm this caustic interpretation, all of the remaining Rivonia trialists, except Mandela, gained their freedom on 15 October. Ramaphosa remained cautious about the course of events. Thatcher, he said, "should not delude herself that De Klerk's releasing of our leaders indicates that he is going to abolish apartheid and give in to the demands of the people".[53]

The NUM's New Role

In different ways the NUM became directly involved in the subsequent political events. The union, with its steady flow of income from abroad, acted as a service provider for the liberation movement as the events

[51]The eight, including Sisulu and Mpetha, were Ahmed Kathrada (60), Elias Motsoaledi (65), Jeff Masemola (61), Andrew Mlangeni (63), Raymond Mhlaba (68) and Wilton Mkwayi (65).

[52]The Reverend Frank Chikane was the general secretary of the South African Council of Churches; Dr Allan Boesak was the president of the World Alliance of Reformed Churches while the Most Reverend Desmond Tutu was the Anglican Archbishop of Cape Town.

[53]*Business Day*, 11 October 1989.

unravelled during 1989 and 1990. It began by providing material support for the ANC leaders as they were released from prison. When Govan Mbeki was released from Robben Island, the NUM executive instructed its secretariat and Kgalema Motlanthe, the co-ordinator of the Education Department, "to look after his well-being". The union did not pass money to Mbeki but provided him with a car and paid his bills. It had done the same for Harry Gwala after his release in 1988. When Walter Sisulu and the other senior ANC leaders were unexpectedly released, NUM provided them with a similar service. It was concerned that none of the released ANC leaders should experience any form of material deprivation on their return to freedom.

The release of the political prisoners also gave Ramaphosa a new public role in the liberation movement, projecting him into the limelight as a key figure in the ANC. He represented one of a new generation of activists who had been entirely politicised by the internal experiences of apartheid as distinct from the political prisoners and the exiles. His new role was as the chair of the National Reception Committee that had been established to help reintegrate the released ANC leaders into contemporary South African society. His job, in the first instance, entailed arranging mass meetings for them in various parts of the country so that they could be introduced to the black public. For most black people they existed only in folklore. Their faces had been hidden from public view and their exploits had never been recorded as a part of South African history. Their transition from historical obscurity to national eminence, however, was instantaneous. Seventy-five thousand people gathered to greet them at Soccer City outside Johannesburg on 29 October. Ramaphosa, with a large banner displaying the ANC logo and the SACP flag behind him, chaired the proceedings and introduced the unknown faces to the crowd, which cheered, clapped their hands, sang, raised clenched fists and *toyi-toyied*.[54] There was a similar situation when they visited Umtata in the Transkei. There, 80 000 dancing, singing, brightly dressed people turned out to greet them. Ramaphosa shared in this transformation for, as he chaired some of the meetings and introduced figures from the past, he became, in the public mind, the link between two heroic periods. He also, therefore, shared in the glory.

The next stage in this process of unravelling the apartheid state came when De Klerk announced on 16 November that all South African

[54] *The Citizen*, 30 October 1989. Govan Mbeki, already out of prison but a banned person, was allowed to attend Soccer City and stand on the platform with the others but only after he had received special dispensation from the government.

Ramaphosa at the mass rally in Umtata to welcome the ANC leaders who had been released from prison in Robben Island.

beaches were to be immediately open for people of all races and that the Separate Amenities Act was to be repealed. A week later he stated that the Group Areas Act, which prevented the different racial groups from living in close proximity to each other, would not be applied in four 'free settlement areas' that were to be pronounced later. There, members of all racial groups would be able to live legally alongside each other. These steps were followed by more significant ones early in 1990. On 15 January, the former ANC prisoners were permitted to travel to Lusaka in Zambia to meet the ANC exiles for the first time in over 25 years. Ten days later, a detailed political statement by Nelson Mandela was legally published in South Africa in which he rebutted the reasons given by the government for not talking with the ANC. He said that the ANC would not renounce the armed struggle before negotiations began; that it would not break its alliance with the South African Communist Party and that it would not compromise on the issue of black majority rule. In a separate statement he gave public support to the ANC policy of nationalising key sectors of the South African economy, including the banks and the mines. He wrote: "A change or modification of our views in this regard is inconceivable."[55] De Klerk's next step was unexpected and dramatic. On 2 February 1990, he announced to the opening session of Parliament that Mandela was soon to be freed; that the 30-year ban on the ANC and PAC was to be lifted; that the SACP was to be legalised and restrictions on 33 other opposition groups and 334 individual activists were to be removed. He added that all non-violent political prisoners were to be released, news censorship was to be ended and that there was to be a suspension of executions. "The season of violence is over," he declared, "the time for reconstruction and reconciliation has arrived."[56] Mandela was unconditionally released from prison on 11 February 1990.

Mandela's imminent release was unexpected. Ramaphosa was in hospital in Johannesburg, recovering from pneumonia, when the announcement was made. Everything happened so quickly that he discharged himself from hospital to fly to Cape Town but arrived late for the scheduled time of release. When he arrived there his job as the chair of the National Reception Committee was temporarily changed to that of a crowd controller as he cleared the way for Nelson and Winnie Mandela as they walked along the road from the prison. The hospital identification tag could still be seen on his wrist as he waved his arm to

[55]Riley, 1991, pp214-215.
[56]*Ibid*, p215.

persuade people to move away. Later in the afternoon he held the microphone for Mandela when he spoke to a crowd of thousands of people on the Grand Parade in Cape Town. All of Ramaphosa's skills as an organiser, however, were required to cope with Mandela's re-entry into public life because it had many manifestations and coincided with other political demands arising from the new situation.

De Klerk's announcements created a situation for the exiled ANC and SACP organisations for which they were unprepared. With few resources, except for some external funding, they had to construct new, extensive administrative structures within the country. For that they required skilled and committed personnel more than they needed finance. At that time the UDF and the trade unions were the only organisations where black people had been able to obtain organising and administrative experience and skills. The unions, alone, possessed resources they could devote to political activities. The ANC and the SACP were compelled, therefore, to seek help from them. However, not all of the unions were able to respond because many of them had minimal administrative structures and were not financially viable. The NUM was one of the exceptions. When its executive met, therefore, it was able to offer material assistance. It extended its service for released ANC prisoners to include the ANC and SACP organisations. It set up an ANC office in central Johannesburg and provided transport for its senior figures. It acted similarly for the SACP to which Ramaphosa made clear NUM's attachment when, at the union's Central Committee meeting in April, he said that it was the only political party qualified to articulate the interests of NUM members "through a national democratic revolution to a society free from exploitation".[57] The union provided an office for the SACP on the third floor of the National Acceptances House, where its own offices were located, and obtained finance for it by collecting donations from people who wanted to support the SACP but did not want to be publicly identified with it.

Consequential Strains in the Union Administration

All of this occurred, however, just as the strains from three different sources began to compound and undermine the NUM's ability to function effectively. Indeed, the unbanning of the ANC and SACP was one of the contributing factors for, in order to provide them with skilled

[57]South African Institute of Race Relations, 1989/90, p350.

organisers and administrators, it denuded itself. Ramaphosa was already a significant figure in the incipient ANC organisation. Kgalema Motlanthe, a pivotal member of the internal executive of the ANC, became the chair of its Witwatersrand Region though he was still nominally the union's Education Co-ordinator. Ramaphosa's secretary, Nomsa Nkwanca, moved over to run Mandela's office while Manne Dipiko, the union organiser in Kimberley who was recently released from prison, became responsible for establishing the ANC in the northern Cape.[58] The biggest strain on the union, however, arose from the fact that many officials, staff and members at regional and branch levels moved over voluntarily to establish local ANC and SACP structures. In doing so, they tried to do two jobs at once but their priorities lay with their new ventures. This process intensified as the two organisations moved into mainstream political life as mass political parties. For the NUM members and officials, it was as if a political pressure valve had been released. For the first time in their lives they were able to participate legally and openly in party political activities. They clamoured for party positions and gained status in NUM for doing so. The effect on the NUM was to haemorrhage talent away from it to such an extent as to impair its ability to operate effectively as a unified union administration.

The second source of internal stress arose from the legacy of the strike. The legacy had different dimensions but an especially important one was the legal task of obtaining either reinstatement or financial compensation for the large number of dismissed strikers. The union had always lacked the legal skills that would have enabled it to challenge the mine owners in the courts, without recourse to expensive outside legal services. The consequence had been that its legal costs had been consistently high. But, in the aftermath of the strike, those costs were crippling. As a means of reducing them and, at the same time, improving the efficiency of the legal department, Ramaphosa hired Fiona Labalestier, a young lawyer who had been trained in the London Chambers of John Hendy, QC, who was the legal adviser to the British NUM. She arrived in the middle of 1989 but without a formal work permit so that her position in the country was tenuous. Nonetheless, she applied herself as an advocate throughout the various regions of the

[58]Manne Dipico, a member of the underground ANC since 1982, was sentenced to five years in prison in 1987 for terrorism, subversion and promoting the aims of a banned organisation. He was released after three years under the terms of the *Groote Schuur-minute*, signed by the government and the ANC in May 1990, that provided for the release of political prisoners who had been charged with organising illegal acts.

union and, for the price of a modest union salary, reduced both the backlog of cases and the union's legal costs. Her successes, however, attracted the attention of the government and created an official animosity towards her. On one occasion, whilst walking in the street she was accosted from behind by an unknown person who threatened her with violence unless she left the country. In order to protect herself, she lodged clandestinely in the home of a colleague. Later, when she returned from a visit to London, the customs officials at Jan Smuts Airport in Johannesburg submitted her to an intense, lengthy and aggressive interrogation. Then, after she had been in South Africa for not much longer than six months, the government ordered her out of the country.[59] This was a blow to her desire to assist the NUM and a setback for the union, which was left with a legal department that was as deficient as before and still having to rely on expensive, hired legal help.

The union's internal problems, however, were exacerbated by a third source of stress that was fundamental for its proper long-term operation. Ramaphosa, aware of this source, compiled a Strategic Plan in 1989 to identify what he described as the "weaknesses that could cripple our union". He wanted the means to assess "from time to time whether its activities programmes, campaigns, officials, leadership and membership are geared in a way to enable it to achieve its aims and objectives".[60] At that time, the union employed 85 full-time officials, some of whom were inclined, the memorandum stated, "to regard working for the NUM as another job and not as a calling for our revolution where commitment and loyalty are the most important qualities of any trade union official".[61] It added that "several officials in the union have become accustomed to bad and unacceptable practices which manifest themselves in absenteeism, latecoming, neglect of duties, inefficiency and unprofessionalism..." It was especially critical of the lax behaviour of the staff in the regions and made suggestions as to how that could be rectified. It expressed concern about the state of reporting by regional officials to the national office. "Some reports," it stated, "are drafted by regional chairmen without the knowledge of the regional committees ... (others) ... are drafted on the way to NEC meetings."

The plan was distributed to the regions with the recommendation

[59]The second half of 1989 was a period of contradiction and great uncertainty in the apartheid establishment. The security services were particularly nervous, unsure of the future and, in consequence, illogically repressive. Most activists expected, at some point, to be harassed by them.

[60]NUM Memorandum, *Strategic Plan*, 1989.

[61]*Ibid.*

that it should be "thoroughly discussed and implemented". However, nothing was done about it. It seemed that the plan itself was a casualty of the deficiencies it was intended to eradicate. More than a year later, the issue of union competence was raised again when a second Strategic Plan was presented to the NEC. A month later, in June 1990, a Strategic Planning Workshop, comprising all of the main officials and administrators, made an in-depth assessment of each union department. It noted, firstly, that staff shortages contributed to the problems of every department. No department escaped criticism but the education, media and research, and administration departments were said to have defects that were almost beyond repair. It stated that they were deficient in leadership and co-ordination, had no clear perspectives and comprised people who lacked the requisite training and, therefore, skills and political commitment to do their assigned tasks. The National Executive Committee, consisting of the chairperson of each regional committee and a representative from each region, was also assessed and criticised for a range of defects. It was described as lacking discipline, failing to carry out congress resolutions, neglecting the implementation of Strategic Plan I, and not communicating with the regions. Finally, it was criticised for considering itself to be above criticism.

A characteristic common to all of the criticisms, including that of the NEC, was the low level of political consciousness of the people concerned. This put the Education Department at the centre of the solution yet it was described as being in such a deep crisis that it was a part of the problem. The scope of its responsibilities, however, was daunting. In 1990, it was supposed to service the union's 228 branches with their respective lay officials and executive members. The branches were located within 16 regions, each with its own complement of full-time and lay officials. There were, in addition, 2 213 shaft stewards drawn from the ordinary membership and requiring information and guidance. The Education Department, however, was suffering most from the transformation in the political situation. Its head, Kgalema Motlanthe, was described as being so "overloaded with ANC work" that he needed to be provided with an assistant. He had only five instructors working in the regions. Other politically conscious officials, at regional and branch level, were also overburdened with tasks as they took on the responsibility of establishing ANC and SACP branches throughout the mining areas. The union's political educational task should have been made easier by the new freedom of discussion in South African society but the change was slow to penetrate the mines. There were virtually no mines where managers allowed black mineworkers to engage in the self-educating process of free and

NUM students at Whitehall College.

uninhibited debate. Without union facilities and guidance, therefore, young black mineworkers could not be seriously politicised to understand the full significance of the changes that were taking place.

Ramaphosa and Motlatsi appealed to the international trade union movement for assistance in overcoming its education crisis. The response was the provision of a two-month educational course in Britain from 10 September 1990, for 15 NUM officials, through the co-operation of the British National Union of Mineworkers, the Manufacturing, Science and Finance Union (MSF) and the International Miners' Organisation (IMO). The IMO gave a grant to cover the cost of fares to Britain and incidental expenses whilst there. The MSF union provided subsistence, accommodation and teaching facilities at its residential Whitehall College in Bishops Stortford, England for five weeks of the course.[62] The subjects covered different areas of social analysis and included classes in trade union organisation and administration. The lecturers were drawn from British universities and from the community of South African ANC exiles in Britain.[63] The remainder of the course was subsidised by the British NUM and was spent in either the Yorkshire or Durham areas of that union where the students learned about the practice of mining trade unionism in Britain.[64] It had been intended that this course should be held yearly but, even though the students returned to fill important positions in their union,

[62]None of the students had travelled outside South Africa before and had never been served by a white person. When they arrived at Heathrow Airport, a white coach driver took their baggage from them and loaded it on to the coach. He unloaded the cases at Whitehall College where they were welcomed by the white catering staff with smiles, cups of tea and biscuits. The experience was traumatic – a mixture of amazement and pleasure that they found difficult to comprehend.

[63]The lecturers were: Ben Fine, Professor of Economics at Birkbeck College, London; Norman Levy, a South African exile and former president of SACTU; John Williams, a solicitor for a number of trade unions; John Hendy, legal adviser to the British NUM; Mzala Nxumalo, ANC and author of a book about Gatsha Buthelezi; Vic Allen, Professor of the Sociology of Industrial Society at Leeds University; Frene Ginwala, author and specialist in the social structure of South Africa; Harold Wolpe, ANC, sociologist at the University of Essex and formerly a co-accused with Mandela in the 1964 Rivonia trial; Elaine Unterhalter, South African exile and specialist in South African education; Derrick Swartz, research student at the University of Essex and specialist in community struggles in South Africa; Lawrence Harris, Professor of Finance Economics at London University; Brian Bunting, South African exile, SACP official and author of several books on aspects of the liberation struggle in South Africa and Sarah Howard, Education Officer for MSF and Director of Whitehall College.

[64]The course was financed by a grant of £20 000 from the IMO and approximately £10 000 from the area offices of the British NUM and the Fire Brigades Union. MSF provided free accommodation, subsistence and some tuition at Whitehall College for a total of five weeks which was roughly equivalent to a grant of about £30 000.

the events in South Africa turned the opinion of NUM members against educational provisions that could appear to be a relic of colonialism. The union decided, therefore, to establish its own residential college.

The shortcomings of its own officials, however, were not the only educational problem identified by the NUM. It was concerned at the lack of black people with the professional qualifications that would enable them to play a part in the management of the mines in what might be a different future for South Africa. There were major obstacles in the way of black students obtaining engineering qualifications in South Africa. Few of those who matriculated had had a science-based education with the requisite grade in mathematics. As a result, there were only two black people in the country with mining engineering degrees and neither one worked in the mines. In April 1990, therefore, Ramaphosa announced that the union was initiating a major educational scheme to train approximately 100 black mineworkers overseas in disciplines such as mining engineering, geology and metallurgy. In order to make this possible, the union needed to have group scholarship provisions and the only countries that were prepared to provide them were the socialist ones in Eastern Europe, the Soviet Union and Cuba.[65] Unfortunately, the Soviet Union and the socialist countries of Eastern Europe were in a state of political crisis at the time and were not in a condition to negotiate such a mundane matter as group scholarship provisions. The only country that was both able and willing to respond was Cuba.

Through their embassy in London, the Cuban education authorities agreed to provide 22 scholarships that included subsistence and tuition. The only cost to the union was to be the airfares from Johannesburg to Havana. Twelve of the scholarships were to be based at the university in Havana and 10 at a technical college, also in Havana. All were in mining-related subjects and the courses ranged from four years at the technical college to six years at the university. The first year, in each case, was to be devoted to learning Spanish. The potential students had to have matriculated with at least the equivalent of an O Level in mathematics.[66] All of the regions of the union were circulated with

[65]There were, of course, scholarship provisions for individuals to study at universities in various parts of the Western world. Indeed, the United Mineworkers' Federation of Australia had offered to finance one or two black mineworkers to study at an Australian university at that time.

[66]They were also required to have no family members dependent on them because the union was unable to provide subsistence for them. Two of the students gave misleading information with the result that their wives and children later turned up at NUM's head office to claim financial support.

Ramaphosa with Rodney Bickerstaffe, general secretary of UNISON, the British public service union, in London, 1990. Ramaphosa was on his way to visit the NUM students at Whitehall College.

this information and 40 mineworkers applied for the courses. After interviews, 21 with the requisite qualifications were chosen and travelled to Havana, via London and Madrid, on 5 September 1990. Their fares were paid for out of the grant given by the IMO. The students began their courses just as Cuba entered a severe and prolonged economic crisis due to the collapse of the Soviet Union. They suffered from the consequent deprivation along with the citizens of Cuba but there was no interruption in the educational provisions arranged by the Cuban authorities. The NUM, however, took a special interest in their welfare and sent officials on systematic visits to the country to ensure that they received assistance.[67]

The Campaign for Justice, Peace and Democracy

The mine owners had made no concessions over segregated facilities as a result of the union campaign late in 1989. To black mineworkers this was a paradox when, in the same period, the government had begun to dismantle segregated facilities at large. Early in 1990, therefore, the NUM resumed its desegregation campaign but widened its scope from simply demanding the removal of petty apartheid practices to include the demand for "Justice, Peace and Democracy". In effect, this meant that, in addition to the removal of segregation it wanted the recognition of the union, an end to union-bashing and the introduction of democratic hostel conditions. By March, various union branches had joined the campaign and were taking direct action through marches, underground sit-ins, demonstrations and petitions. The pattern of past protests, however, was repeated. In his report to the union's Central Committee the following month, Ramaphosa listed the difficulties they faced. "There are," he declared, "still too many mines where we are not represented; too many where we have no recognition agreements; too many where we have no access. Management repression has continued to handicap our work. We are constantly denied access to our members and are prevented from recruiting non-members..."

About 20 000 members of the NUM, primarily in Anglo-American

[67]One of the students died whilst on the course. The majority of those who remained completed their courses. When they returned to South Africa, however, they found it difficult to obtain appropriate work because the government body responsible for validating academic qualifications refused to recognise Cuban diplomas and degrees. One student who left South Africa with a matriculation at Grade 12 was told that the diploma he had obtained was equivalent to a school Grade 8 in South Africa.

mines, took part in the protests. Their demands ranged from the abolition of the hostel system itself to the removal of barbed wire from the hostel perimeter fences. They experienced varying degrees of success. At Vaal Reefs West, on 16 February, black mineworkers forced white miners to queue in line with them for the underground cages. As many as 8 000 staged an underground sit-in at Vaal Reefs East on 8 March because of delays in being hoisted to the surface. There had been a similar reaction to racial queuing at Western Holdings in the Orange Free State and at Bracken gold mine in Secunda, where they succeeded in forcing white miners to queue underground and in abolishing the 'piccanin' system. Union members had staged a protest sit-in at President Steyn but when they returned to the surface the police attacked them with tear gas and rubber bullets. The NUM claimed that 19 of its members were injured when the mine security at Freddies mine broke up an illegal march. The racial tension at the Harmony mine was heightened when white miners began arming themselves after the mine security had fired rubber bullets to disperse a protest march. When the occupants of the mine hostel at Genmin's Unisel mine in the Orange Free State began their protests they were tear gassed and fired on by the mine security force but that did not deter them from expressing their support for four union members, who had been charged with resisting the racial queuing system, by refusing to return to the surface on 14 March. At Western Deep Levels, a gathering of 10 000 mineworkers were prevented by force from even starting their protest march. The police also injured 40 black mineworkers when they fired rubber bullets to disperse a mass meeting at the Saaiplaas mine in mid-April. In a different kind of protest, mineworkers at the No 4 shaft of President Steyn ignored the regulations and took their wives and girlfriends into their hostel. Overall, during this episode, there was one fatality when white vigilantes belonging to the AWB killed Mnikelo Ndamase, a union member from St Helena, on 21 March. In addition, 500 were injured and 300 were selectively dismissed.[68]

The union's Central Committee met and debated the protest strategy while many of the protests were underway. Three hundred and twenty delegates were present and each one was asked to describe how apartheid was practised through his own experiences. They listed 59 different forms of discrimination that were then categorised in the following manner:

[68]For details, see South African Institute of Race Relations, 1989/90, pp372-373 and NUM leaflet, *NUM Leads Mineworkers Against Racism: Campaign Update,* nd.

1. WORKING CONDITIONS

1.1 The wage structure is racially applied.
1.2 Bonuses are paid on a racial basis.
1.3 There are separate cages for black and white workers.
1.4 Black workers are compelled to queue for cages and white workers are not.
1.5 Acclimatisation of workers for underground work is applied in a discriminatory manner.
1.6 The allocation of working clothes is discriminatory.
1.7 There are separate pay offices and administration points for black and white workers.
1.8 Black workers are compelled to act as the personal servants of white workers in the work situation through the 'piccanin system'.
1.9 Black workers are subject to abuse, humiliation, insults and assaults by white supervisors and management personnel, mainly because of the colour of their skin.
1.10 White workers are allowed to rest and eat underground while this is denied to black workers.

2. CONTROL MEASURES

2.1 Black workers are made to wear armbands as though they are cattle, whereas white workers are not.
2.2 The searching of workers is racially applied.
2.3 Only black workers are fingerprinted.

3. FACILITIES

3.1 Facilities such as toilets, change houses and sports are racially segregated.
3.2 Medical facilities such as clinics, ambulances, hospitals and medical schemes are racially segregated.

4. TRAINING AND JOB MOBILITY

4.1 Job grading is racially applied.
4.2 Opportunities to upgrade skills are more available to white workers than to black workers.
4.3 The fact that all workers are not recruited through a single agency is racial.
4.4 Promotion opportunities are less available to black workers.

5. LIVING CONDITIONS

5.1 Only black workers are subjected to the migrant labour system.
5.2 Mines provide better houses for white workers than for black workers.
5.3 The transportation of workers to and from work is racial.
5.4 There are racially segregated living areas and schools in living areas administered by the mines.
5.5 White workers accommodated in single quarters share better facilities than black workers.

The Central Committee instructed the secretariat to send its catalogue of discriminatory practices to the Chamber of Mines with a demand for a timetable for their removal. Ramaphosa subsequently wrote to the Chamber of Mines with the comment that the only change he could detect since the 'colour bar' had been legally imposed many decades before was in the way in which it was justified. He dealt separately with some mine managers who wrote to complain about the breaches of mine regulations by NUM members. For instance, in response to a complaint by the manager of Douglas Colliery that NUM members were disrupting the bussing system, he wrote: "From your perspective, what is happening … is an infringement of rules and regulations whereas from the point of view of the NUM it involves a matter of principle. There could hardly be a wider degree of divergence."[69] Many of the individual mine complaints were about the "unacceptable behaviour" of NUM members. At the Winkelhaak gold mine in the east Rand, for instance, the manager had complained that, on 29 April, a "group of NUM members returning from a meeting in Standerton was singing, dancing and chanting and some members had sticks and *sjamboks* in their possession. They regrouped inside the hostel where they continued the singing, dancing and chanting. The group dispersed only after the security had to intervene." There had been a repeat performance on 1 May but on that occasion they also displayed a flag.[70] In general, mine managers either enforced their existing rules concerning NUM activity in the hostels or issued more stringent ones, as in the case of the Free State Geduld Mine, which is illustrated below.

[69]Correspondence, 23 May 1990.
[70]Correspondence, 8 May 1990.

Violence at Welkom

The NUM's campaign was directed against the mining houses and the government but in practice it was a matter of black mineworkers within different mines, acting in the name of the union but going outside of its procedures for the resolution of grievances, in their confrontations with white miners who were uncompromisingly opposed to desegregation because segregation was an essential condition for the privileges they received. There had frequently been tense relationships between them underground because, though white employees possessed the authority of management, they were invariably physically outnumbered. This tension was implicit in the relationships between black and white communities in the mining towns and their adjoining black townships. For most of the time an uneasy peace prevailed in and outside the mine compounds but it was threatened as black mineworkers began to engage in open political activity under the banners of the ANC and the SACP. They wore ANC T-shirts, waved ANC and SACP flags, carried Mandela's picture underground and wrote pro-ANC slogans on the rock faces. In retaliation, white miners joined vigilante groups, armed themselves when they went underground, patrolled the streets as if they were the police and threatened to take the law into their own hands. Within this context, there were an unprecedented number of incidents of intimidation and assaults by black mineworkers against white miners but many more acts of reprisal by white miners.

Welkom in 1990, with its cluster of gold mines, possessed all the ingredients for racial violence. It was a modern white conservative town with a population of 65 000 of whom 80 per cent were directly or indirectly dependent on mining. The mine compounds and hostels were spread around the white town and accommodated about 150 000 black mineworkers. Nearby was the township of Thabong with an estimated black population of up to 450 000. The town and the township had entirely separate administrations. Welkom was controlled by the right-wing Conservative Party while an official black local authority administered Thabong. Welkom was a prosperous white town. It was known as the 'white by night' town because no black people were allowed on its streets after dark. That had been a legal requirement for most of its history but remained the practice in 1990. In some of its suburbs, the building regulations did not allow even for the construction of servants' quarters because servants were expected to get out of the town at night and live in Thabong.[71]

[71]Witwatersrand Business School, *Thabong-Welkom: Is This the New South Africa? A Case Study*, nd.

Thabong, on the other hand, was a deprived black town. Anglo-American had built it and for most of its history it had no electricity or running water. It had insufficient educational facilities to meet the needs of those pupils who tried to respond to the ANC's 'back-to-school' campaign. When the schools had opened for enrolment in mid-January, many young people were refused entry. By the end of the month, the pupils had begun a non-violent school boycott that fairly quickly became violent. There were increasing incidents of violence in Welkom on both sides of the racial divide. For instance, on one occasion, several black men attacked a white man with pangas and, on another occasion, a white woman was murdered by her black gardener. Young black men drove the streets in taxis waving replicas of AK-47s and ANC flags. There was a rush by the white population, women in particular, to buy firearms. The white vigilante groups then began to patrol the streets.

The process of escalation could clearly be seen. As the black school-children became more violent, the police began to detain them. The residents of Thabong, in consequence, organised a consumer boycott of white businesses and posted activists outside their premises. White vigilantes attacked the activists and killed one of them. Black people who were found in Welkom after dark were invariably assaulted and sometimes killed. A second black consumer boycott was launched on 7 May to protest against the activities of the white vigilante groups, one of which, Blanke Veiligheid (BV), declared that it was going to clear Welkom of its black population. A NUM spokesperson described the situation as "very angry, very tense".[72] In an attempt to defuse it, a meeting was called by the local Methodist minister, involving the mayor of Welkom, the police chief and representatives of business, COSATU, the NUM, the UDF and of Bronville, a nearby coloured township. The group became known as The Forum and played an important role in trying to reduce racial tensions in the area.

Events in the nearby mines paralleled those in Thabong and Welkom. Already, 31 black mineworkers had been dismissed for organising sit-ins, refusing to attend disciplinary hearings and neglecting to use grievance procedures. Then, on the day the second boycott was launched, members of the NUM at the President Steyn mine were assaulted by white miners for going underground wearing ANC badges and colours. The mine manager arranged a disciplinary hearing into the assaults but pre-empted the findings by dismissing three of the black

[72] *Ibid*, p10.

mineworkers involved before it was held. This prompted about 50 black mineworkers to demonstrate by *toyi-toying* through the compound, carrying ANC flags and placards and allegedly shouting slogans such as "Kill management! Mandela is our new manager".[73] They refused to stop when they came across a group of 20 white mining officials who were attending a presentation on mining equipment, so the mine security force intervened and, without warning, fired on them. They all fired rubber bullets, except one who used live ammunition and he injured 14 of the demonstrators. In the ensuing mêlée one of the security guards was killed and a white manager died after being stoned in his car. Two other white officials were injured.

The NUM issued a statement regretting the deaths and extending condolences to the families but the two main right-wing groups, the AWB and the BV, vowed revenge. Their members, armed with shotguns, rifles, whips and knives, gathered outside the police station and then, in a mass meeting, decided to organise a white boycott of wholesalers to prevent supplies of food reaching Thabong. The BV leader expressed the views of the meeting when he said: "We will starve the *kaffirs* into submission."[74] White people again bought up supplies of ammunition from the gun shops while the government deployed extra police and troops in the town and appointed a special team of investigators to track down the killers of the white men. Fourteen South African Police vehicles with scores of armed police and dogs sealed off the mineshaft. They put barricades across the mine entrances and searched the hostels for weapons.

In a dramatic move to defuse the situation, Anglo-American, which controlled the mine, called a meeting in Welkom between the white Council of Mining Unions (CMU), the NUM, the Chamber of Mines, the SADF, the SAP and Adriaan Vlok, the Minister of Law and Order. Ramaphosa and Motlatsi represented the NUM.[75] The meeting agreed "to set up working structures – involving the CMU, the NUM and the Chamber of Mines – to resolve differing problems". The trade unions were to appeal to their members to refrain from acts of 'intimidation,

[73]*The Star*, 18 May 1990.

[74]*Ibid.*

[75]Ramaphosa drove from Johannesburg to the meeting in Welkom with three passengers, including Motlatsi and Jay Naidoo. It had been reported that armed white vigilantes were patrolling the streets of Welkom but as they approached the town Ramaphosa suggested that, instead of going straight to the meeting place, they should drive slowly through the centre so as to see the situation for themselves. Naidoo said: "OK, but slow down so that I can have my last cigarette." There were, however, no incidents and they reached the meeting safely. (Notes by the author who was the fourth person in the car.)

violence and persecution' while the mine owners were to ensure there was security on their properties. The rest was to be left to the police and defence force.[76] The only dissenting voices during the meeting were those of Piet Ungerer, the general secretary of the MWU, who said his members would not return to work unless their safety was guaranteed, and Jay Naidoo from COSATU who called for the immediate disbanding of vigilante groups and criticised the mine security for provoking the violence in the first place. Both Motlatsi and Ramaphosa regarded the meeting as 'historic', but because of its composition rather than its achievements. There had never before been a meeting at which representatives of the NUM, the white miners' unions and the mine owners had been present. And, although they both deplored the racist policies of the white unions, they came away from the meeting deeply impressed by the uncompromisingly class approach of the white union officials in their attitude towards the Chamber of Mines and the government minister. It was new in their experience to see reactionary racism combined with progressive class politics and they wanted to explore it further so as to see more clearly the possibility of a collaborative relationship between the NUM and the white miners' unions.[77]

After the meeting, a local newspaper described Welkom as a "besieged town" swarming with policemen and soldiers.[78] Within a few days, there was more violence in Thabong township as black youths attacked people and property linked to collaboration with the state. Local black councillors had to go into hiding. An assistant manager at President Steyn was injured when his car was stoned on the road between the mine and Thabong. A police car was also stoned when the police arrested hawkers and confiscated their goods near to the township. Burnt-out cars and tyres blocked the streets that were patrolled by the youths. The police moved into the township in force on the day after the agreement was reached and shot and killed seven youths and injured 40 others. The death toll rose to 10 within three days. In Welkom, members of BV demanded an immediate curfew and reinstated their own street patrols.

The level of violence, however, was reduced as a stalemate developed

[76]*Saturday Star*, 19 May 1990.

[77]NUM, *Twenty Points on the Relationship Between the NUM and the CMU*, June 1990. A report prepared for Ramaphosa that described how the racial divide worked for the advantage of the mine owners and explained that there could be no genuine collaboration so long as racial discrimination was practised in the wider society. Despite their comparable class origins the NUM and the CMU did not subsequently collaborate in their approach to the mine owners.

[78]Witwatersrand Business School, p13.

between the contesting sides. They were weary after months of stress and trauma and sought a way out. The Forum met representatives of the meeting attended by Ramaphosa and Motlatsi and set up mechanisms to investigate sources of tension in the mines. They established a fact-finding commission of enquiry into hoist queuing, underground protest meetings and disciplinary procedures. The measures, however, did not eliminate organised incidents of violence. Three weeks after the agreement between the three parties in the mining industry had been reached, the NUM office was bombed. NUM members continued to organise political meetings that were consistently broken up by the mine security forces. Twelve people were injured on one such occasion. The AWB also returned to the streets and on 16 June organised a march through Welkom to protest about the detention of some of its members. These events, however, were accompanied by other factors in Welkom that focused the minds of both black mineworkers and white miners. The economic condition of mining was deteriorating and the mines were forced to reduce their labour forces. Social welfare organisations moved in to help feed the families of unemployed white miners while black mineworkers, as usual, were returned to the rural areas with little or no assistance. The impact of retrenchment, however, spread well beyond Welkom to encompass the whole mining industry. Coping with it became a major factor in the work of the NUM.

The Retrenchment Crisis

From the point of view of the NUM, variations in the size of the labour force were critical indicators of the success or failure of the mining industry and, in the early 1990s, there were pronounced signs of failure. Between 1987 and 1992, the size of the labour force in the gold and coal mines, the main areas of union activity, declined by 33 per cent. Altogether, 173 957 jobs were lost in the mines controlled by the Chamber of Mines.[79] There was expansion elsewhere, as in the platinum mines where the number of mineworkers employed increased by 26 per cent, but that did not absorb the retrenched mineworkers from the widely dispersed recruitment areas because the majority of employees in the platinum mines were local Tswanas. The loss of mine jobs was highest within South Africa and the 'homelands' and was by far the lowest in Mozambique. The numbers recruited from the Transkei and

[79]Chamber of Mines, *Statistical Tables, 1992*, p28.

the Ciskei fell by 41 per cent and 52 per cent respectively between 1987 and 1992 whereas the comparable figure for Mozambique was 7 per cent. Even in Lesotho, that had always been a favoured recruitment area, the number recruited fell by 23 per cent.[80]

The manner in which the job losses were distributed was a matter of particular concern for the NUM because the areas most affected were the ones from which many of its members came. The Mozambicans, on the other hand, who were least affected, tended to be compliant towards management and distrustful of the NUM, largely because of its intense involvement in the affairs of South Africa. Moreover, because they usually worked for long contracts they were frequently the most experienced and, therefore, relatively highly skilled mineworkers. Retrenchments, in effect, enabled the mine managers beneficially to shed their mines of their most troublesome employees. Anglo-American, Gencor, Gold Fields and JCI, four of the six major mining houses, all experienced significant job losses in the period after 1987 but employed a higher proportion of foreign mineworkers at the end than they had done at the beginning. In the Anglo-American mines, four South Africans were laid off for every foreign mineworker while in the Gencor and JCI mines the proportion was three to one.[81]

The NUM had initially expressed the concerns of its members at the level of retrenchments by marching and demonstrating under the banner "Fight Retrenchments – No Job Losses" and by pressing the Chamber of Mines to negotiate a nation-wide retrenchment agreement, with the intention of setting minimum standards relating to compensation, retraining and finding alternative work. The Chamber of Mines rejected that proposal outright. However, even while the union was arguing for a protective agreement against retrenchments in 1989, there was a sense amongst its active members of an imminent general crisis in mining for at its congress in April of that year it adopted a resolution to investigate how the mining industry could be shaped in the future to benefit mineworkers. It specified a timetable for the production of a blueprint that would deal with all aspects of mine work and life of significance to them.

The government was also alerted to the possibility of difficult times, though in part they would result from its own doing. It had begun the process of reshaping its form of capitalism to resemble more closely the current western prototype. In essence, this involved ending state

[80]*Ibid.*

[81]Crush and James, 1995, p21.

support for private enterprise and transferring publicly owned enterprises to the private sector.[82] A key factor in this process was reducing state subsidies to the gold mining industry and allowing it to be shaped by market forces. In January 1988, this resulted in the repeal of the 1968 Assistance Act that had enabled state assistance to be given to gold mines either in the form of tax relief or as direct financial aid. Thereafter, mines were at liberty to apply for assistance but the process was designed to reduce state subsidies rather than expedite them and that left a number of mines in financial difficulties.

The gold mining industry had visibly begun to move into a state of permanent decline after 1988, however and, therefore, the implications of having a market-driven gold mining industry were so dramatic that the government was compelled to examine the consequences of its policy. It appointed the Marais Committee of Inquiry to investigate the "position of marginal-profit mines" in the light of the financial problems a number of mines were facing.[83] Neither the NUM nor any other union was represented on the committee. Nor was the NUM asked for its comments on the draft report distributed to "interested parties". It made its views known, however, in a press statement late in August 1990. The Marais Committee was motivated by macro-economic concerns and paid virtually no regard to the social consequences of closures and downscaling. Indeed, it made only one, rather brief, reference to black mineworkers when it stated: "It is generally accepted that the closing of marginal-profit mines will have a negative effect on unskilled labour, which it will be difficult to employ elsewhere in the present circumstances." Most of its comments related to the prospects for white skilled and managerial personnel who might be made redundant.[84]

Acting on the assumption that the government's adherence to the free market would and should continue, the committee recommended that gold mines should not be singled out for special assistance to avoid perpetuating an over-dependence on a single primary commodity. It further recommended that the policy of limiting state assistance to them

[82]"The government had committed itself to the market mechanism and effective competition measures in this regard (eg privatisation, corporatisation, contracting out and deregulation)." Marais Committee Report, 1990, p26.

[83]The Committee of Inquiry Into the Position of Marginal-Profit Mines was chaired by Dr G Marais, MP, Deputy Minister of Finance. Members of the committee were drawn from the Inter-Departmental Committee on State Assistance to the Minerals Industry, the Chamber of Mines and the South African Reserve Bank. A draft of the committee's report was distributed "to interested parties for prior examination and comments" before being submitted to the Cabinet.

[84]Marais Committee Report, 1990, p43.

should continue on an *ad hoc* basis, provided it was in the national interest. It based its calculations of retrenchments on the assumption that the price of gold would rise and this led to a gross underestimate of their extent. It predicted the loss of 77 600 jobs over a five-year period from the end of 1989 but that had been exceeded in little over a year. The NUM responded by demanding the establishment of a central fund for retraining retrenched mineworkers with the provision that each unskilled worker should receive six months' paid retraining to improve his chances of getting work elsewhere or of obtaining skilled work in mining.[85]

The scale of the crisis increased from June 1990 when the dollar price of gold fell to its lowest point for four years. Large mines were affected. ERPM at Boksburg, with almost 10 000 employees and a history of liquidity problems, began to operate at a loss. When the price of gold fell, therefore, the government established a commission of inquiry to investigate whether it ought to receive state assistance. Then, in November of the same year, Harmony in the Orange Free State with a work force of more than 25 000 sent off alarm signals when it began to contract. It was estimated that only 11 of the 31 mines that were members of the Chamber of Mines would be out of danger if there were a further gold price fall.[86] Thirteen gold mines, including Harmony and others with up to 10 000 employees, were categorised as marginal ones that were unlikely to survive unless they were given governmental assistance.[87] Stilfontein, a gold mine in Klerksdorp that had reduced its black work force from 9 000 to less than 3 000 over a three-year period, was the first of the mines to close. It announced in April 1991 that it would finish its underground operations by the end of the year and would retain only just over 500 workers for short-term maintenance purposes.

The precarious financial position of so many mines brought the perennial problem of marginal mines back into the limelight. The meaning of 'marginal', however, had changed. A marginal mine was no longer one suffering from the exhaustion of its gold-bearing seams but was one that was failing to make sufficient profit for its shareholders

[85] *Business Day*, 24 August 1990.

[86] *Saturday Star*, 9 June 1990.

[87] The 13 marginal mines, each with its 1989 workforce in brackets, were: West Rands Consolidated (1 889), Bracken (2 640), Durban Roodepoort Deep (3 363), Barberton (1 860), Grootvlei (2 528), Marievale (386), ERPM (9 258), Blyvooruitzicht (9 125), Doornfontein (9 213), Libanon (7 580), Venterspost (8 205), Western Areas (8 552) and Harmony (25 551). Source: The Employment Bureau of Africa Limited, *Report*, 1989.

for a variety of reasons. In this respect, it was no longer a matter that affected a few mines. The difficulty for many mines was that during the period when the gold price had been relatively high they had maximised their profits by mining and exhausting their high-grade ores. Thus they were left with low-grade ores at the very time that the gold price made it uneconomic to mine them. Given this situation, it became clear to the NUM that a simple single campaign against retrenchments was not feasible. It recognised that a number of factors, some of which were beyond the control of the mine owners, were leading to a permanent downscaling of the industry. It concluded that this could create a crisis so complex that it could have long-term effects on the level of unemployment, the standard of living of the unemployed, the viability of mining towns, the profitability of industries that served the mining industry, the tax revenue of the government, the country's foreign exchange earnings and the survival of local economies that supplied workers for the mines. The NUM believed that such an array of related issues could only be tackled by restructuring the industry in a deliberate and pre-planned manner and that before that could happen there needed to be national discussions between all the major stakeholders.

C G Knobbs, the President of the Chamber of Mines, indicated the scale of the crisis from the point of view of the mine owners when he wrote to Ramaphosa in February 1991 about the forthcoming wage negotiations. He stated that the gold crisis was the most severe for at least three decades and cited the declining ore grades, rising working costs and the declining gold price as the cause. "At today's gold price," he stated, " … 16 mines employing 108 000 people are operating at a loss. Against this background the single most important challenge facing all of us is that of survival."[88] He went on to state about wages that "the industry will be unable to agree to an increase this year at anything but a very low level unless more mining operations are to be put at risk … A case could in fact be made out that we are entering a phase in which 'concession bargaining' would be the most responsible course to adopt." This call for wage restraint, however, was made to mineworkers whose productivity, measured in terms meaningful to them, was rising rapidly and for that they believed they should have been rewarded. Unfortunately, there was no uniformly agreed measure of productivity in gold mining. The gold content of ore had fallen

[88]Correspondence from C G Knobbs, President of the Chamber of Mines, to the General Secretary, NUM, 28 February 1991.

continuously from 13,34 grams per ton of ore to 4,99 grams in 1989 so that the actual production of gold per mineworker had also fallen. This measure, however, was meaningless as an indicator of productivity because the industry had no control over either the price it received for gold or the quantity of gold it produced. The decline in the gold content of the ore did not create a problem for the mine owners so long as the gold price was rising but once it stabilised and then fell, their only means of maintaining their profit levels was to mine more ore, if possible with less labour. Thus the burden was placed squarely on the shoulders of the working mineworkers. How they responded was shown by the increase in the monthly output of ore per man from 30,75 tons in 1980 to 35,5 tons in 1989, without receiving any commensurate rewards.[89]

Grand Gestures and Practical Politics

Ramaphosa wrote an angry reply to Knobbs. He accused him of trying to 'soften' up the union prior to the wage negotiations, of being hypocritical, weak and unimaginative, of treating the union with contempt over the issue of retrenchments and of failing to rectify any of the incidents of racial discrimination the union had related in a memorandum to the Chamber of Mines.[90] On one item, however, there was agreement. In the conclusion of his letter to Ramaphosa, Knobbs had stated that the mine owners "would welcome a discussion on the longer-term challenges facing the industry". The union responded positively and they decided jointly to convene a Gold Summit to which the mine owners, mining trade unions and the relevant government departments would be invited. The NUM's 1991 Congress endorsed the idea and invitations were sent out in May for a one-day summit meeting on 3 June. Its purpose was intended "to explore a common and co-ordinated approach to promote the long-term viability of the mining industry during a period of change in the operational environment". Its themes were: "(a) the co-ordination of mine closures and the downscaling of operations; (b) prolonging the lives of mines and mining operations as long as possible; (c) protection of agreed conditions of employment and the maintenance of acceptable standards of health, welfare and safety of

[89]See the Marais Committee Report, 1990, p7 for a brief discussion of the measurement of productivity.

[90]Correspondence from Ramaphosa to the President of the Chamber of Mines, 10 April 1991.

employees, and (d) the training and retraining of retrenched miners and assistance to communities affected by mine closures and the downscaling of mining operations". This was an ambitious programme for a single day's meeting.

In April 1991, before the summit met, the union received a copy of the Blueprint for the Mining Industry it had commissioned at its 1989 Congress. Since then the situation had altered in important respects in that changes in the political climate indicated that it might be possible for black mineworkers to exercise some control over their industry but under such adverse conditions that the industry could only survive if it were radically restructured. The arrival of the blueprint, therefore, seemed opportune. It was framed within parameters set by the *Freedom Charter* and stated, in consequence, that nationalisation, in which the right of mineworkers to participate in mining decision-making was embodied, was essential for the survival and rational growth of the mining industry.

The possibility of political power had released grand visions of the future amongst officials in the NUM and this was reflected in its blueprint, its perception of a Gold Summit as a means of resolving mining problems and its proposal to establish a state-funded Permanent Mining Commission, with trade union participation, to oversee a planned downscaling and restructuring of the industry.[91] The union's optimistic view of political possibilities, however, had a shaky basis. The ANC was finding it difficult to make progress in its talks with the government. Thirty-eight ANC exiles had been granted a temporary immunity from prosecution to allow them to return to South Africa at the end of April 1990 and engage in the reconstruction of the ANC. On 2 May 1990, a small number of them, including Joe Slovo, the general secretary of the SACP, had joined with Nelson Mandela and Walter Sisulu in an ANC delegation to meet with the government at Groote Schuur (Cape Town) in order to discuss the obstacles to negotiations. They agreed on a document known as the *Groote Schuur-minute*, which stated in its preamble: "The government and the ANC agree on a common commitment towards the resolution of the existing climate of violence and intimidation from whatever quarter, as well as a commitment to stability and to a peaceful process of negotiations."[92] The *Minute* went on to stipulate the manner in which this would be achieved. One

[91]The last proposal was made in a special resolution at the Seventh National Congress in April 1991.

[92]See South African Institute of Race Relations, 1989/90, pp731-732 for a copy of the *Minute*.

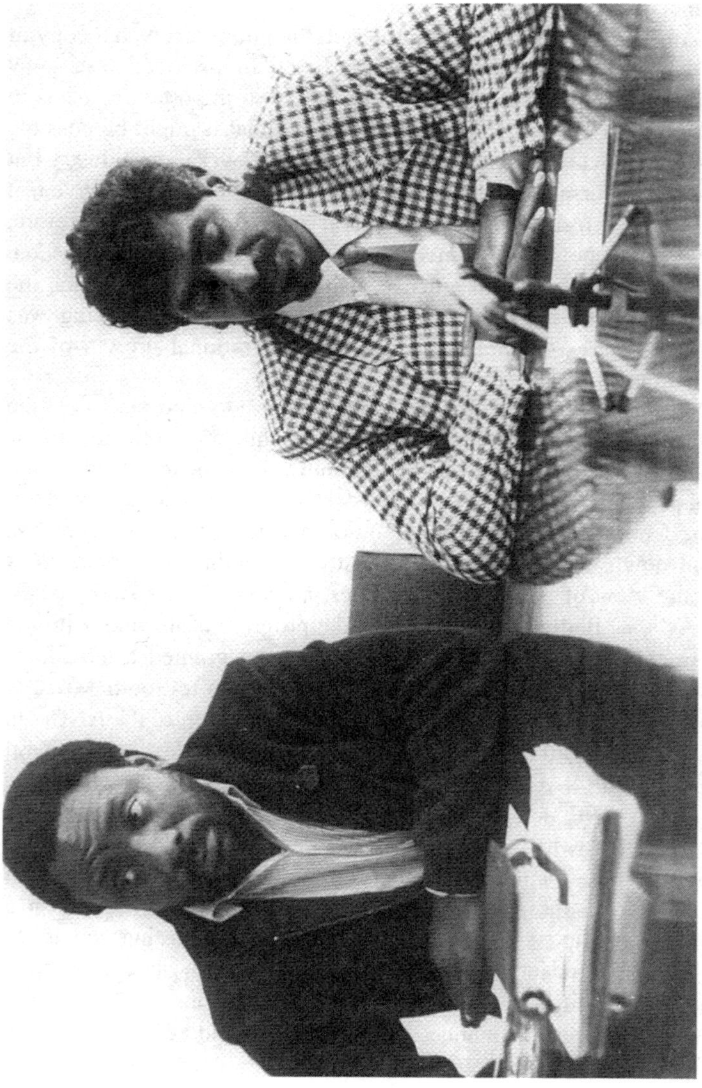

Ramaphosa with Jay Naidoo, general secretary of COSATU, at a press conference on violence in KwaZulu-Natal.

of the clauses related to the four-year-long state of emergency, which was lifted a month later, except in KwaZulu-Natal. A working group was formed to discuss removing the obstacles to further negotiations, such as the ANC's commitment to the armed struggle and the continuing high level of violence in the townships.

Grand visions about a 'theoretical' future had little attraction for ordinary black people who were being exposed to increasingly sustained, cruel and murderous attacks in their homes and neighbourhoods. Such was its intensity that ordinary ANC supporters accused the ANC organisation of abandoning them to the police, the security forces, black and white gangs of vigilantes and the IFP. Many asked that the ANC should provide them with arms so that they could defend themselves. The death toll through political violence during the seven months following August 1990 was more than 1 200 people.[93] Massacres involving multiple deaths of men, women and children were a feature of the time. In 1990 and 1991, there were 65 massacres resulting in 621 deaths in the Transvaal alone.[94] Overall, they occurred in the streets, on the trains and in the buses. Where there were political antagonisms there were no safe havens for black people. For instance, there was carnage in the streets of Sebokeng, south of Johannesburg, in March 1990, when the police fired on a crowd of 60 000 UDF supporters, killing nine people and injuring 447. Following a faction fight in the same area in September, the army intercepted an unarmed crowd of 5 000, killing 42 people and injuring 124. There was slaughter on the railways in the same month when an armed gang rampaged through a commuter train leaving 26 people dead and more than 100 injured. Crowded buses were also targeted. In July 1990, armed vigilantes had attacked a bus in Durban and left 26 dead and 60 injured in it. A sole white person fired on a bus at KwaMashu, north of Durban, in October and killed seven black people. The residents of Sebokeng suffered again when 13 people were killed during an AK-47 attack on a funeral vigil on 12 January 1991. Two months later, 15 people were killed and 18 injured at a prayer vigil in Alexandra.[95]

The intensity of the violence and the government's inability, or refusal, to counter it, had persuaded Nelson Mandela to advise the ANC national executive committee to withdraw from the pre-negotiation talks in September 1990 but it was not until April 1991 that a decision

[93] *The Star*, 8 April 1991.

[94] Truth and Reconciliation Commission, Volume 3, 1998, p675. A list of the more serious massacres is included in the *Report*, p676.

[95] *The Star*, 8 April 1991.

ANC supporters in Cape Town demanding a constituent assembly, in January 1992.

NUM delegates at the COSATU congress in July 1991, where the crisis in the negotiating process was discussed.

The delegates, from left, are:
Second row, Kate Philip.
Front row: Frans Baleni, Manne Dipico, George Nkadimeng, (not known), Xhola Williams.

to follow his advice was actually taken. On that occasion, early in the month, the ANC issued a seven-point ultimatum to the government, the central feature of which was the need for action to prevent violence and the arrest of people who instigated it. If the government had not complied with the demands by 9 May the ANC intended, it declared, to withdraw from all discussions on the future constitution of South Africa.[96] The ANC did not enforce its ultimatum because the risk it faced was that the violence might have been orchestrated to compel it to withdraw from the talks and, thereby, end the possibility of peaceful negotiations.

The ANC's ultimatum exposed the relationship between its discussions with the government and the violation of people's rights in the townships. Both were clearly a part of a dialectical totality. Although formal negotiations about a future constitution had not begun it was generally known what each of the main parties in the frame was aiming for. The context was the ANC's demand for a constituent assembly to draw up a constitution for a unitary state with a universal franchise. The National Party government, however, was opposed to simple majority rule, implicit in the ANC's proposals, and wanted power-sharing arrangements that would guarantee the rights of racial minorities. Other white parties, such as the Conservative Party and the AWB, simply wanted to reinforce apartheid. Inkatha, the other main participant in the power struggle, wanted a democratic federal system in which minorities were protected.

However, within the groups opposed to the ANC were segments that were determined to fight to retain their old-established privileges and, in order to do that, they used a combination of death squads, security forces and vigilante groups, going under the rubric of a 'Third Force', to commit violence on such a scale that it would destabilise the political process. Their acts invited retaliation from ANC supporters. Thus an ever-widening circle of violence was created. In many instances, Inkatha gangs attacked black people in their own communities but it could not be described as ethnic violence because many of the ANC supporters who were attacked by the Inkatha Zulus were themselves Zulus. White vigilante groups and their supporters used what was described as 'black-on-black' violence as propaganda to illustrate their contention that black people were incapable of managing their own affairs.

The circle of violence encompassed the mining industry because all

[96]*New Nation*, 12-18 April 1991.

of the antagonistic groups were present there. Inkatha had for long campaigned to destabilise the NUM. On this occasion, the NUM was viewed with even more hostility because it was an open advocate of the election of a constituent assembly to draw up a democratic unitary constitution that specifically protected workers' rights. Ramaphosa made this clear when he said that all governments abused their powers so that those rights had to be very clearly prescribed and protected.[97] It further antagonised both Inkatha and the white vigilante groups by moving closer to the SACP. It was reported that, by April 1991, black mineworkers formed the bulk of its membership with most of them concentrated in Klerksdorp and the Orange Free State. In the light of its strong presence in the mining areas the Internal Leadership Group of the SACP decided to assist the NUM in its campaign against retrenchments and representatives of the two organisations met to discuss how they could collaborate.[98] The close proximity of their offices encouraged fraternisation and friendships between their respective leaders.

However, neither the involvement of the NUM with the SACP and its repeatedly emphasised commitment to the nationalisation of the mines, nor the rebuffs it had received from the Chamber of Mines, detracted from the union's desire to discuss the future of the industry with the mine owners and the government. It was not an anomaly that seriously concerned Ramaphosa. He wanted to explore all options. He was eager to attend a group-level conference when invited to do so by Gengold, the owners of Stilfontein, which was being closed down.[99] He also met the executives of Anglo-American to discuss its retrenchment plans. And he pressed ahead with arrangements to host the Gold Summit jointly with the Chamber of Mines on 3 June.

The union's proposals at the summit to restructure the industry did not mention ownership. They contained three time categories. For the short-term, the union wanted immediate relief that included state assistance for marginal mines and retrenched mineworkers, the establishment of working groups to tackle, for instance, racial discrimination and the creation by the government of a Permanent Mining Commission. For the medium-term, the proposals included training, a review of health and safety provisions, abolition of the migrant labour system and the provision of mine houses. The long-term proposals concerned job creation schemes in mining towns and the

[97] *Sunday Times*, 7 April 1991.
[98] *New Nation*, 12-18 April 1991.
[99] *Weekly Mail*, 12-18 April 1991.

rural areas and planned downscaling over five-year periods by each of the mining houses. A Summit Steering Committee was formed to oversee the work of four task groups that were instructed to submit proposals by November 1991. The groups met and exchanged ideas but did not meet the deadline, which was extended indefinitely. The proposal to establish a Permanent Mining Commission was 'put on hold' by the government and forgotten. The summit, in effect, was simply a 'grand gesture' that allowed the mine owners and the government, with the complicity of the NUM, to engage publicly in discussing solutions to deep-seated problems without having to commit themselves to apply them. Nothing came of other restructuring schemes devised by the union. The Blueprint for the Mining Industry remained a blueprint. In an equally cosmetic gesture, a NUM delegation was dispatched to Germany to discover how retrenchments were handled when mines in the Ruhr coalfield were closed with union collaboration between 1970 and 1987.

Ramaphosa Leaves the NUM

The pace of political change in the union's environment, however, did not slacken, though its impact was varied. The return of the leading ANC and SACP exiles and the re-establishment of their organisations had inevitably altered the political balance in the liberation movement. The ANC in its new form began to make policy decisions that hitherto had been formulated within the MDM in general or in COSATU in particular. In practice, decision-making about the form and timing of protests had been highly fragmented and, in consequence, had reflected the intensity of local feelings. This displacement caused some resentment. For instance, the South African Clothing and Textile Workers' Union criticised the ANC for announcing a consumer boycott in April 1991, as a means to press for a consultative assembly, without consultation with it or any other union. It printed a country-wide advertisement objecting to its exclusion.[100] There was some simmering resentment among the internal leaders of the liberation movement over their exclusion from the secret talks Mandela was having with the government. But none of this counted. Without formal positions in the ANC, the internal leaders were displaced from their positions of influence for, though the protests continued, the future of the apartheid

[100] *Weekly Mail*, 5-11 April 1991.

system had moved into the hands of the national representatives of the ANC and government.

Ramaphosa was directly affected by this situation. From the time of Nelson Mandela's release, he had appeared as a potential successor to him. He was constantly at his side, seemingly advising, helping with speeches and regulating access to him. When Sisulu and the others came out of prison he continued with his managing role, always in the public eye, organising and addressing mass meetings, the popular internal leader of the liberation movement who had been entirely politicised by the day-to-day experience of apartheid. In that sense, he had a common link with ordinary black people. He had, moreover, a powerful base in the NUM. Never once had his position been challenged in the union. Nor had anyone seriously questioned his right to speak for COSATU whenever he thought he should do so. He was described, on more than one occasion, as an 'undeclared' communist who had the backing of the nascent Communist Party.[101] None of this, however, mattered within the highly structured ANC organisation with its hierarchy of officials guided by a national executive committee elected in exile in 1985. When the exiles returned he was sidelined and pushed into relative obscurity. When the future leaders of South Africa were discussed in September 1990, Ramaphosa was described as having "mysteriously faded from prominence" and of occupying a "faded role".[102]

The change in Ramaphosa's fortunes altered his perception of his role in the NUM. He began to talk privately of taking time out to study abroad. He favoured a post-graduate law course at Harvard for a year or so but was unsure what he would do later.[103] He wanted to escape from the unrelenting pressures from nine years of public exposure as a union leader. He emotionally distanced himself from his union work as he prepared to leave it. This attitude grew and persisted throughout the second half of 1990 and was uninfluenced by the proceedings of the ANC's Consultative Conference in December 1990, which he attended as an ordinary delegate. Unknown to him, however, divisions were crystallising in the ANC that would affect his future. The conference was the first occasion at which the internal leaders were able to air their

[101] *The Guardian*, 14 September 1990 and 6 July 1991.

[102] *The Guardian,* 14 September 1990.

[103] As an indication of the changed circumstances of the NUM, when James Motlatsi was offered a scholarship at the Northern College in Britain in 1984, through the intervention of the British NUM, Ramaphosa was aghast at the thought that he should even contemplate accepting it and accused him of running away from the union.

resentment at being excluded from ANC policy-making as if they were not a legitimate part of the liberation movement. The ANC began to be seen as a combination of three factions, namely the former Robben Island prisoners, the returned exiles and the internal leadership from the MDM. They were not equally placed within the ANC. Formal authority lay with the first two factions that were composed of people with strange faces who, in effect, sat at the high table. The third faction comprised a small number of relatively young representatives of the ordinary members of the MDM who had loyalties and allegiances to their own organisations and whose commitment to liberation was not derived from the historic role of the ANC but from their own experiences in the townships during the previous decade. There was not yet a commitment to the ANC as such amongst the ordinary members.

This was the situation when the first full conference of the ANC in South Africa for over 30 years was held in Durban from 2-7 July 1991. Its primary purpose was to elect its national officials and a new and entirely reconstructed national executive committee, and to determine policy for the negotiated liberation of the country from apartheid. Ramaphosa had received many nominations for the position of Secretary-General of the ANC but he was undecided about what he should do, even on the opening morning of the conference. The pressure on him to stand for election from the ordinary members was so great, however, that he had no real alternative but to allow his name to go forward. He stood in competition with the incumbent, Alfred Nzo, who had held the office since the Morogoro Conference in Tanzania in 1969, and Jacob Zuma, a returned exile. On 5 July 1991, at the relatively young age of 39 years, Ramaphosa gained an overwhelming victory with 58 per cent of the total number of votes cast.[104]

Thus, on 24 July 1991, the NEC of the union announced that it had agreed to second Ramaphosa to become the Secretary-General of the ANC. It had been possible that Motlatsi, his collaborator in the union for the whole of its existence, would also soon leave the union. Before the ANC conference, Motlatsi had stated in a press interview that he had no desire to serve as NUM's president for more than 10 years and that period was due to expire the following year.[105] However, the prospect of denuding the union of its two principal officials in one sweep convinced him to stay on. The loss of Ramaphosa's guiding hand

[104]He received 1 156 votes compared with 450 votes for Jacob Zuma and 371 votes for Alfred Nzo.
[105]*New Nation*, 12-18 April 1991.

NUM's secretariat before Ramaphosa's move to the ANC.
From left:
Back row: Paul Nkuna, national treasurer; Marcel Golding, assistant general secretary and Cyril Ramaphosa.
Front row: Elijah Barayi and James Motlatsi.

alone was sufficient to alter the way in which the union would cope with the worsening conditions in the mining industry. His departure, however, signalled the beginning of a new political environment for the union in which it would be able to act with greater confidence, though not necessarily with more success, in its struggle against the mine owners.

CHAPTER FIFTEEN

UNDER NEW LEADERSHIP 1991-1992

Threats to NUM's Unity

The departure of Cyril Ramaphosa from the union had an immediate impact on its administration. He had exercised a moral authority that had extended throughout the union but which had been especially marked in its central office and in a way that had been largely of his own making. Its personnel had been his choice. It had been unknown for any official or staff member to question his organisational decisions or the political direction he had given. His personality had determined the functional relationships between the various members of the secretariat. Because he came from the Venda tribe, that had virtually no presence in the mining industry, his actions had never been associated with tribal dominance. He had not been able to prevent tribal faction fights at branch level but through his neutrality he had prevented tribalism from being institutionalised in the organisation. In other words, he had epitomised the unity of the NUM. It was quite clearly going to be difficult to replace him.

The union, however, did not have to search immediately for a replacement. It was bound by its constitution and, at the September 1991 meeting of its National Executive Committee, it announced that Ramaphosa had been seconded to the ANC until his term of office had officially expired and that during the period of his secondment, the assistant general secretary, Marcel Golding, would act as the general secretary. Golding had been responsible for specific areas of work and had closely watched Ramaphosa's methods of control. He knew the difficulty of the tasks he was to confront. But he had to work in a situation that was essentially different from the one that had been faced by Ramaphosa. Ramaphosa had been aided by the generally agreed need to organise against apartheid. Now, not only were his status and authority absent but the apartheid system appeared to be facing imminent collapse. For the first time in the history of the union, its central purpose began to be unclear to many ordinary members and this encouraged both tribal and political divisions that threatened the union's cohesion.

James Motlatsi, too, had to carve out a new role for himself. His responsibilities had always been complementary to those of Ramaphosa. He had, in consequence, neither administered nor negotiated.

His role had been as the titular head of the union, addressing regional, branch and mass membership meetings and acting as the union's principal ambassador. For much of the time he had not even possessed his own office. All of that altered with Ramaphosa's departure. He had to engage in a multiplicity of tasks, many of them new to him. His office now became the place where members, officials and visitors, both national and international, congregated for discussions and advice. No major organisational or administrative decisions could be made without his participation or advice. He endeavoured to lessen his burden by relying on Marcel Golding to handle staff and administrative matters but, to counter that, he acquired increasing obligations as the union's principal spokesperson in South Africa and abroad as the number of public platforms on which the union's views were sought multiplied. The members perceived him as the principal link with the past and expected him to carry on where Ramaphosa had left off but they did not give him the same unanimous, unswerving backing. Although he was accorded some of the reverence that had been given to Ramaphosa, it was not sufficient to protect him from political, personal and tribal attacks as the organisation's unity began to fracture.

The political freedom that had followed the release of the political prisoners and the unbanning of the liberation organisations enabled disparate views to be expressed and acted on. Without the common unifying goal of defeating apartheid it began to matter whether the future of South Africa lay in the resolution of class conflict and, therefore, in the rejection of capitalism or whether it lay in carving out a role for itself as an integral part of the global capitalist system. Within each of these broad categories, there were different perspectives and emphases which, given the new-found freedom to express hitherto subversive views, to agitate and organise, were represented by different groupings. The benchmark that had distinguished one section from the other was how the *Freedom Charter*, the ANC and the SACP were perceived but this began to change in the new circumstances. The importance of the *Freedom Charter* began to recede as the ANC contemplated how to reconcile opposing class divisions in the interest of national unity. In the same process, some people set aside their former opposition to the ANC as it became clear that it was a leading player and would be dominant in the final reckoning for apartheid. People with different, hitherto irreconcilable political views, in the white as well as the black communities, began joining or supporting the ANC for simple personal, opportunistic reasons. Apart from the violent opposition of Inkatha to the ANC and the hostility of white right-wing groups, perceptions of the ANC as a puppet of the Communist Party, nurtured by the Soviet Union, began to mellow and fade. This left the SACP as the prime issue of contention and, therefore, the new benchmark. In political terms, the critical division between South Africans, after 1991, was between those who supported the Communist Party and those who did not. This did not necessarily reflect support for different, contrasting policies since many black people backed the Communist Party because of its historic role in the liberation struggle and because of the popularity of Chris Hani, its new general secretary, who was viewed by many as a future president of the country.[1]

The situation in the NUM reflected that in the country as a whole but was more starkly defined. Here the benchmark was not simply support for the SACP but membership of it. This was emphasised by Motlatsi in an interview with *NUM News* in December 1991. When asked about the role mineworkers could play in effecting political change, he replied: "Mineworkers would have to join the ANC and the South African Communist Party, not only as card-carrying members but to play an active role in realising their aims and objectives."[2] Small groups of mineworkers had pre-empted his advice and had used their new-found political freedom to establish SACP branches in the mines, particularly in Klerksdorp, Carletonville and the OFS. Mineworkers joined them in relatively large numbers and came to constitute a substantial and influential segment of the party. A significant element of the NUM's head office staff also joined the SACP.

Motlatsi was not himself a member of the SACP but his call for mineworkers to join it and his support for its policies reflected the presence in the union of an alliance that was united around the *Freedom Charter*. It was, however, a rather fragile alliance that was threatened by the possibilities for political fragmentation. It was disrupted, for instance, by the zeal of the SACP activists who, even before Ramaphosa had left the union, mounted political challenges for positions in the union and, in some cases, succeeded.[3] In their eagerness to gain positions they neglected to convert mineworkers to the policies they advocated and, in consequence, left them without a basic belief in the need for change. Such people were relatively easily turned around

[1]Chris Hani, who had been the Chief of Staff of the ANC's military wing, Umkhonto we Sizwe (MK), was elected as the general secretary of the SACP at its Eighth Congress in December 1991 in succession to Joe Slovo, who became its chairman.
[2]*NUM News*, April 1992.
[3]Their campaign had started early in 1991. In March 1991, for instance, it had led to the defeat of Jan Selepe, the long-standing chairperson of the OFS region, who had been a member of the initial group that had formed the union in 1982. Thomas Qhina replaced him.

and became fodder for other factions that were emerging. Mostly, these factions were a reaction to the successes of the SACP and were united by their anti-communism. Here, Motlatsi was in an ambivalent position. Though he advocated policies that were broadly in accord with those of the SACP and had a friendly and co-operative relationship with the SACP through its general secretary, Chris Hani, he became increasingly critical of its activists in the union. He blamed them for the emergence of disruptive factions that made his work more difficult. It was, moreover, a complex situation that had overtones of tribalism. Many of the SACP members were Xhosas whom the Sothos, who had their own political agenda, suspected of seeking positions of domination in the union. Even the rivalry between union officials came to have the same tribal overtones.

The Struggle With Impala Platinum

This mixture of trade union, political and tribal loyalties was present in the material conditions of South African society as a whole and was reflected in the violent disruption of everyday life in many areas of the country. There was a prevailing mood of uncertainty that led to internecine struggles to retain or gain advantages. Violence was a common feature and there were many deaths, most of which were in Natal and the Pretoria/Witwatersrand/Vereeniging area (PWV) where the Inkatha Freedom Party (IFP) and the ANC were competing for political dominance. But elsewhere there were various disaffected groups of white vigilantes and members of the security forces, allegedly operating covertly as a 'third force', wreaking havoc in black communities.[4] From the middle of 1991, it became potentially life-threatening to take a train journey in the Witwatersrand area for there were at least three attacks on train commuters every week.[5] Some of the

[4]The Truth and Reconciliation Commission reported: "Many came to believe that a 'hidden hand' or 'third force' lay behind the random violence, which included military-style attacks on trains, drive-by shootings and a series of massacres and assassinations." It added, however, that the Commission of Inquiry Regarding the Prevention of Public Violence and Intimidation, headed by Judge Goldstone, could find no evidence of a 'third force'. Nonetheless, non-government organisations, violence monitoring groups and some national newspapers remained convinced that a 'third force', comprising members trained and supplied with arms by government security forces, did exist and were responsible for much of the violence. (*Truth and Reconciliation Commission of South Africa Report,* Volume Two, 1998, pp585-586.)

[5]South African Institute of Race Relations, *Race Relations Survey*, 1992/93, pp245-246.

violence originated from within the townships as blacks were attacked for collaborating with the government authorities. There was conflict, too, within the liberation movement, between members of the ANC and the PAC. The overall intensity, the indiscriminate brutality and the apparent mindlessness of much of the violence led the government, the ANC and the IFP to initiate a Peace Convention, which, in September 1991, compiled a National Peace Accord containing a code of conduct for political parties and organisations and also for the police and the security forces.[6] The accord, however, was an expression of intent that left the causes of the violence untouched. The number of deaths continued to rise at an increasing rate. In the nine-and-a-half months after it was signed there were as many as 2 328 deaths through politically motivated violence.[7]

One of the sources of the violence was the continuing refusal by the leaders of some of the 'homelands' to identify with the liberation struggle and agree to their reintegration into South Africa. Bophuthatswana was one such 'homeland'. Its president, Chief Lucas Mangope, in alliance with right-wing forces in South Africa, was determined to retain its limited independence with its own defence forces, police and internal security regulations. He was opposed to the intrusion of the ANC with its trade union allies into Bophuthatswana's affairs. This was an issue that directly concerned the NUM because of the presence of platinum mines there.[8] The union had been defeated in 1986 in its attempt to organise the complex of four mines belonging to Impala Platinum but had succeeded in getting a foothold just over the border in Rustenburg.[9] Afterwards, Impala Platinum remained effectively non-unionised except for the presence of a small local registered 'sweetheart' union called the Bophuthatswana National Union of Mining Employees (BONUME) that had been formed in 1988 and had about 3 000 members. Since then, platinum mining had expanded considerably and, with a labour force of more than 104 000, was second only to gold mining as an employer of black labour. In the new political climate, and in the belief that Bophuthatswana would

[6]See National Peace Convention, *National Peace Accord*, 14 September 1991. There were 29 signatories to the accord.

[7]South African Institute of Race Relations, 1992/93, p124.

[8]Much of the following narrative about the Impala mines is drawn from the *Report of the Commission of Enquiry Into Events at the Bafokeng North Mine During March 1992*, December 1992. The commission had been established jointly by the NUM and the mine management and consisted of a panel of three that was chaired by Advocate Paul Pretorius. It is referred to below as the Pretorius Commission, 1992.

[9]See chapter 6 for details of the defeat.

soon return to South Africa, the NUM resumed its attempts to organise the Impala Platinum mines and, quite quickly, recruited 43 000 members.[10] It was not able to represent them directly in the mine, however, because it had no formal relations with the company. Its only contact with the management was indirectly through its members on the Workers' Committee with which the management occasionally met. A catalyst in its attempts to gain recognition occurred when Genmin, the owner of Impala, imposed a wages award on 1 July 1991 that was significantly lower than that negotiated by the NUM at the Rustenburg Platinum mine across the border in South Africa.[11] The mineworkers at the Impala mines were angry when they heard of the wage discrepancy and started a series of protests involving strikes and violence that lasted for some months and drew them closer to the NUM.

The Impala mineworkers lived in a situation in which it was relatively easy to agitate, spread rumours and generate animosities. Ten thousand of them were accommodated in a barrack-like hostel complex that was surrounded by a fence tipped with razor wire. They were segregated along ethnic lines in rooms that contained about a dozen bunks and were supervised by 38 tribal representatives who had been appointed by management after consultation with tribal elders. Though it was possible, in some of the older rooms, for them to do their own cooking, the majority ate in large communal dining complexes. In this completely institutionalised existence, their lives were distracted only by the plentiful supply of liquor, dagga that was on sale outside of the hostel, and sex that, if necessary, they could purchase in the informal trading sector. It was an environment, however, in which attitudes easily became infectious so that once, for instance, the idea of joining NUM started to circulate, it spread rapidly. And with the spread came a new phenomenon. Isaac Mayoyo, who had left BONUME to join the NUM, began to recruit militant young black mineworkers to act in a new role of marshals to assist with crowd control at mass meetings and suchlike events but with no specific union functions. He had got the idea, he stated, from watching COSATU organise its meetings. However, because NUM officials had no legitimate access to the mines,

[10]It even uniquely recruited three Impala white shift bosses who had been members of the Mine Workers' Union. They joined in NUM's recruitment campaign. The three men with their families were shunned and threatened by members of Rustenburg's conservative community. NUM, believing that they were all in fear of their lives, provided them with physical protection and material sustenance.

[11]Genmin's wage was between 7 and 12 per cent compared with between 12 and 15 per cent negotiated by the NUM.

the marshals developed a degree of autonomy that enabled them to operate broadly in parallel with the official organisation of the formally unrecognised NUM, though not always in harmony with it. They began to organise the ways in which work and leisure activities were conducted. The union acknowledged that some of them lacked control and discipline. Later they were described as being primarily "the storm troopers of the ANC, perhaps the SACP, in the mine".[12] Their initial tasks, however, were to persuade mineworkers to join the NUM and then to encourage them to engage in protest actions. They were uncompromising in their recruitment tactics and allegedly used force to back up their powers of persuasion. Behaving like the police, they commandeered mine vehicles and searched persons and vehicles entering the hostel area.[13]

The Workers' Committee, set up by the company as a substitute for the NUM, compiled a catalogue of grievances for submission to management that was remarkably similar to one that any NUM branch in the gold mines would have drawn up during any of the previous five years. It included their wages grievance, dissatisfaction with incentive payments, the removal of privileges for team leaders and the abolition of the *induna* system. It met with management at the end of July 1991 but no agreement was reached. At that stage, with encouragement from the marshals, the Impala mineworkers were prepared to protest against a range of issues. They were presented with just one such issue when the manager failed to take action against a white worker who had assaulted two black mineworkers and another white worker who had shot and killed a black mineworker. Feelings about the manager's behaviour came to a head on 15 August, when, in their first truly collective act since 1986, 18 000 of the Impala mineworkers staged a sit-in underground. On the following day, NUM members were attacked during a meeting and nine of them were killed. The union alleged that the attackers were vigilantes called in by the manager. This caused the dispute to escalate. The Workers' Committee then began to assert its authority amongst the mineworkers and imposed its own working conditions. It reduced the working day from eight hours 40 minutes to four-and-a-half hours and instituted a series of 'go-slow' campaigns.

The situation became chaotic. The Workers' Committee had a further meeting with management on 19 August when it demanded an

[12]Pretorius Commission, 1992, p17.

[13]*Ibid*, p19. The union's views were presented in evidence to the commission.

access agreement, a recognition agreement and a code of conduct as a prelude to serious negotiations with the NUM, but all that the manager would concede was that the union should apply for recognition under the law of Bophuthatswana and provide proof that it had members at Impala. However, applying for legal recognition in a nominally independent homeland involved a major switch of policy by the NUM because hitherto it had refused to do so on the grounds that it would have implied recognition of its 'stooge' administrations. Nonetheless, Marcel Golding, only recently appointed to act in Ramaphosa's place, made a formal application for government registration on 28 August and submitted almost 25 000 stop-order forms to the company as evidence of its recruitment successes. Nothing happened, however, as a consequence of Golding's initiative. The NUM did not get recognition but, to make matters worse, the management failed to process at least two-thirds of the stop-order forms, thus denying the union of income.

By the beginning of September, managerial discipline had collapsed under pressure from the Workers' Committee so the manager used the only sanction he had left. On 3 September 1991, he locked out the entire black work force of 8 000 at the Bafokeng North mine. The lock-out was short-lived but the violence continued when the police shot and killed one of the marshals on 10 September and a bomb exploded in the Sotho section of the hostel complex. Although, towards the end of September, an agreement was reached with management over the July wages issue, it did not lessen the tension for on 28 September intensive fighting broke out between supporters and opponents of the NUM in the hostel complex and 14 people were killed. This outburst prompted a meeting between the union and the management when they discussed a draft agreement but, as with the previous agreement, this was not an indication of the real state of affairs. Nothing changed in the conditions of work and the 'go-slows' continued throughout October, as did hostility towards team leaders over their privileged treatment. Moreover, when the manager of the mine responded to a request to discuss contentious issues with the mineworkers in a general meeting on 28 October, he was severely assaulted by them and spent three weeks in hospital in intensive care.[14] From this incident onwards there was no doubt about the company's policy. Union officials who were called to the mine after the assault had taken place were arrested and detained by the police. Senior managers of the mine and officials of BONUME, the local minority union, then

[14] *Ibid*, p23.

met with Chief Lucas Mangope to elicit his official support in countering the NUM. Following that meeting, a lawyer representing the NUM was ordered off the company premises and the members of the Workers' Committee were arrested and allegedly tortured by the police. Isaac Mayoyo was actually detained by the police in the office of the manpower manager whilst he was having discussions with him. These acts temporarily terminated any contact the NUM had with Impala Platinum.

Deaths at President Steyn

The events at Impala Platinum overlapped with the call by COSATU and NACTU for a two-day stay-away starting on 4 November 1991 over the introduction of a value-added tax (VAT) by the government. Three-and-a-half million workers participated in it but it did not receive the NUM's full support for only about 100 000 of its members joined it. The stay-away was conducted peacefully both in general and in the mining industry. The NUM branches took their decisions over whether to participate or not at mass meetings and accepted whatever was decided. That was how it seemed to be at the President Steyn gold mine in the OFS when a mass meeting of members in No 2 and No 4 shafts decided to support the stay-away, though agreeing that any worker would be allowed to work if he wished to do so. In those two shafts, however, there was a potentially explosive mixture of union, political and ethnic loyalties.[15] The two shafts, between them, had exactly equal complements of Xhosa and Sotho mineworkers. Most of both groups had been members of the NUM until, not long before the stay-away, a number of the Sothos left the union because of their dissatisfaction with the rise of one per cent in the union subscriptions that had been implemented since the previous August. Moreover, as they were from Lesotho, they were unhappy about being compelled to strike, and possibly lose pay, over a specifically South African issue. For these reasons, many of them were unhappy about participating in the stay-away.

[15]For details of the violence, see the *Report of the Inquiry Conducted by the Committee of Inquiry Into the Violence at the President Steyn Gold Mine in Welkom During November 1991* (Goldstone Commission, April 1992). This committee was set up following the government's decision to establish a general 'Commission of Inquiry Regarding the Prevention of Violence and Intimidation' with a continuing brief to initiate investigations into incidents of violence as they occurred. Mr Justice Richard Goldstone was appointed to chair it. (Goldstone Commission, 1992, p13.)

A sour note entered when the union branch officials approached the manager of President Steyn for an agreement to allow the stay-away to go ahead and were rebuffed by him. Hitherto their relationship had been relatively harmonious. At first, the manager suggested that those who wanted to strike should take leave but added that he would follow the principle of 'no work, no pay and normal disciplinary action'. Then he distributed a 'brief' to all mineworkers, outlining the action he would take against anyone taking part in the stay-away and pointing out that it was both illegal and futile. Those Sothos who were already doubtful about striking over an issue that did not concern them became even more hesitant when they learned that they would not only lose pay, but could be disciplined and lose their jobs as well. In consequence, a number of them in No 4 shaft decided to defy the strike call and gathered at the tunnel that led to the mine shaft at the normal starting time of 8.30 pm on 3 November, with the intention of going to work. They were met by a crowd of *toyi-toying* Xhosas, who, contrary to the decision of the mass meeting not to intimidate non-strikers, had decided to stop anyone going to work. There was a commotion in the tunnel, a Sotho and a Shangaan were killed, and the Sothos were chased back to their rooms. Both sides immediately regrouped, confronted each other and fought throughout the night until the SAP dispersed them by firing rubber bullets. During that night 15 people were killed and 55 were injured. The Xhosas bore the brunt of the violence for nine of them were killed and 30 were injured compared with three Sotho deaths and 13 with injuries. The NUM office was burned down.

The confrontation developed into faction fights as the Sotho and Xhosa mineworkers rallied to support their respective sides. This became clear when the Xhosas set about rectifying the ethnic imbalance amongst the casualties the following weekend. Then, in an orgy of violence, 25 were killed and 263 were injured. A number of ethnic groups were caught up in the violence but the Sothos suffered most and had twice as many casualties as the Xhosas. Moreover, the attacks were sadistic. The Goldstone Commission reported that "they had been savagely killed, having been beaten, stabbed, their throats cut and their heads smashed to pulp".[16] These revenge killings were replicated at No 2 shaft during the same weekend when, without any obvious provocation other than to show solidarity with their respective ethnic groups in No 4 shaft, Sotho and Xhosa mineworkers clashed, leaving 38 dead and 59 injured. On this occasion, the ethnic balance among the casualties shifted

[16] *Ibid*, p13.

FREE STATE GEDULD MINE M/28/90

TO ALL EMPLOYEES

JG McCALLUM
MINE MANAGER
17 MAY 1990

POLITICAL ACTIVITIES

MANAGEMENT WISHES TO INFORM EMPLOYEES THAT, IN THE INTEREST OF MAINTAINING ORDER AND GOOD RELATIONS, THE FOLLOWING WILL NOT BE ALLOWED ON FREE STATE GEDULD MINE:

- DISCUSSIONS OR INDICATIONS OF POLITICAL AFFILIATION DURING NORMAL WORK OPERATIONS AND PROCEEDINGS.
- DISORDERLY BEHAVIOUR.
- DISPLAY OF FLAGS OR BANNERS OF A POLITICAL NATURE.
- POLITICAL MEETINGS.
- MARCHING.
- SINGING AND CHANTING OF A POLITICAL NATURE.

EMPLOYEES ARE REQUESTED TO ADHERE TO THESE RULES, AS IT IS IN THEIR OWN INTEREST TO AVOID ANY POTENTIAL CONFLICT SITUATIONS WHICH CAN ARISE, DUE TO THE DIVERSE INTERESTS AMONGST OUR WORKFORCE.

POLITIEKE BEDRYWIGHEDE

IN DIE BELANG OM GOEIE VERHOUDINGE EN ORDE TE HANDHAAF, WIL BESTUUR GRAAG ALLE WERKNEMERS INLIG DAT DIE VOLGENDE NIE OP FREE STATE GEDULD MINE TOEGELAAT SAL WORD NIE:

- BESPREKINGS OF AANDUIDINGS VAN ENIGE POLITIEKE AARD GEDURENDE FORMELE WERKSAAMHEDE EN PROSEDURES.
- WANORDERLIKE GEDRAG.
- VERTONING VAN VLAE OF BANIERE VAN 'n POLITIEKE AARD.
- POLITIEKE VERGADERINGS.
- MARSJERING.
- SINGERY OF DREUNSANG VAN 'n POLITIEKE AARD.

WERKNEMERS WORD VERSOEK OM BY DIE BOGENOEMDE REëLS TE HOU AANGESIEN DIT IN HUL EIE BELANG IS OM POTENSIELE KONFLIK SITUASIES, WAT DEUR VERSKILLENDE BELANGE IN DIE WERKSMAG VEROORSAAK WORD, TE VOORKOM.

---

again for only six of the dead and nine of the injured were Sothos. Most of the rest were Xhosas.

The final act in this killing spree occurred on 24 November when armed members of the 'Russians', a criminal gang from the nearby township of Thabong, were collected in two kombis and taken to a place just out of sight of a roadblock near the hostel at No 4 shaft. They surreptitiously entered the hostel and attacked the Xhosas in their rooms. Eight of the Xhosas were murdered and 26 were injured. This clandestine incident had the appearance of being a carefully planned operation with the participation of the mine security force. One of the taxi drivers recalled seeing a white person standing near a mine security vehicle, waving the 'Russians' over to the mine, though the other taxi driver could not confirm this. In any event, there was evidence that the killings had been planned for there had been no confrontation between the Sothos and the Xhosas in the hostel that night, only Xhosas had been attacked and the dead were found lying near their rooms. All the Xhosas in No 4 shaft were subsequently evacuated for their own safety to a hostel at the Saaiplaas mine near Welkom.

The Goldstone Commission blamed a militant element among the members of the NUM at President Steyn for initiating the violence on 3 November. But, although the violence was localised, the commission made comments and recommendations about mine violence in general. It contended that the genesis of the problem at the President Steyn mine lay in the relaxation of restrictions on black political activity in general. Hitherto, it stated, political activity by black mineworkers had taken place, but covertly, and because of its illegality there were no union/management agreements to regulate it. The commission suggested, therefore, that collective agreements should be drawn up to decide on the legitimacy of political activity in the hostels that would cover the nature and control of meetings, *toyi-toying*, the carrying of weapons and the formation of 'peace committees'. Mine managers had already made their own attempts to regulate such activities, as the leaflet from the manager of the Free State Geduld mine illustrates.

The difficulty in the way of drawing up collective agreements at mine level was that mine managers were generally opposed to the NUM's belief in the freedom of association and the right to strike. Indeed, from their point of view, the intention of such agreements was to regulate and constrain trade union activities. The NUM wanted, in particular, to reduce the role of the security forces, especially that of the Emergency Protection Units, and to make the carrying of weapons illegal. But neither the commission nor mine managers regarded these proposals as being feasible so no progress was made at that level. The experience of the NUM and the Anglo-American Corporation illustrated the difficulties for they had been trying to negotiate a code of conduct for the previous four years. When the violence at the President Steyn mine broke out there was a draft agreement in circulation, waiting for a formal signing.[17] Ramaphosa, who had negotiated the code, was not satisfied with it because it contained so many restrictions relating to strike action that it had little relevance for most of the protest activity at branch level. This had placed Ramaphosa in a dilemma for he knew that it was as far as Anglo-American was prepared to sign up to.[18] It was either that draft or nothing. He chose the draft.

Another Change in the Leadership

By the end of 1991, there were a number of issues about which the union needed to formulate its attitude. There were mine violence, wages and retrenchments, continuing mass dismissals and, according to the Kimberley Regional Committee, the post of the general secretary. The secretariat decided, therefore, to convene a meeting of the Central Committee, whose purpose was to consider major policy issues in between National Congresses, on 24-25 January 1992 in Johannesburg.

The question of mine violence was at the top of the agenda. As an indication of its priority, the NEC had established a special Crisis Committee to handle incidents of violence immediately they occurred. Marcel Golding, in his report to the NEC, contended that the violence was primarily directed against the union and was a denial of the members' rights. He warned the employers and the government that, unless it was redressed, there would be greater conflict because mineworkers would be forced to protect themselves. "For peace to exist," he stated, "there must be democracy on the mines ... Employers cannot continue to hide behind apartheid laws and collaborate with the security forces in undermining the union, detaining its members and arbitrarily dismissing workers."[19] The mine owners, he claimed, were paying lip service to workers' rights. They had dismissed mineworkers for taking part in the stay-away in the previous November and were still using mass

[17]The code was included in the NUM National Executive Committee minutes for 23 March 1991 but it was not formally adopted by both sides until June 1992.

[18]NUM NEC, 23 March 1991. The minute stated: "This draft contains the provisions that the Anglo-American Corporation is willing to take insofar as a settlement in regard to this matter is concerned."

[19]NUM Central Committee (Documentation), 24-25 January 1992, p4.

dismissals as a means of resolving disputes. Gold Fields had applied the policy of mass dismissals to end disputes at its Doornfontein gold mine in Carletonville and at its Black Mountain copper mine in Namaqualand. In addition, in March 1992 after Golding had submitted his report, the Iron and Steel Corporation (ISCOR) dismissed almost the whole of its 2 000 black labour force at its Durnacol Colliery in northern Natal, in order to destroy opposition to its management there. The dispute took a particularly unpleasant turn when the Minister of the Interior at KwaZulu tried to get the dismissed mineworkers reinstated by persuading them to sign a document, in the presence of a commissioner of oaths, declaring that they had been intimidated into striking by the NUM. Kgalema Motlanthe urged the minister "to stop meddling in the dispute between mineworkers and ISCOR's Durnacol Colliery".[20]

The year 1991, Golding declared, had been one of the most difficult years for wage negotiations ever experienced by the NUM. The prices of minerals were falling due to the poor state of the South African economy and declining world trade. The majority of mineworkers received an average rise of six per cent that was much less than the rate of inflation. The settlement, however, had contained some positive non-wage items such as the right to negotiate for full-time shaft stewards and for four days' paid leave a year to enable shaft stewards to attend NUM education courses. However, the union had also agreed to accept gold price and performance bonus schemes. This was the second major reversal by Golding of established NUM policies. Hitherto, the union had aimed to achieve a uniform wages structure throughout the industry so that all mineworkers were paid the same wage rate for performing the same work. Now, under the new schemes, a bonus would be paid if the price of gold reached a specific level. But secondly, and more importantly, bonuses were to be paid according to the profitability of each respective mine. This meant that the bonus schemes had to be negotiated at mine level and would vary considerably between mines. An agreement was subsequently negotiated with the Chamber of Mines to establish a framework to govern mine-level discussions about bonuses. Golding insisted that the alternative to the schemes would have been to accept lower wages for all grades in all mines. Under the new schemes, he claimed, at least some of NUM's members would get more money. They were, however, highly contentious and, after an intense debate and despite being supported by

[20]South African Institute of Race Relations, 1992/93, p373.

Delegates at the 1992 NUM Central Committee advertising their political commitments.

memoranda from the union's collective bargaining department, they were rejected. The majority of the delegates believed that they were divisive and unfair and contributed nothing to the struggle to resolve the perennial problem of low wages. It was still the case that the majority of black mineworkers received less than one-third of the average wage of black workers in the motor industry and about one-half of the wage paid to checkers at supermarkets.[21]

The question concerning the position of the general secretary was on the agenda solely because of the Kimberley regional committee's concern, stated in a letter to the general secretary, that because there was no constitutional provision for the appointment of an acting assistant general secretary, Golding was having to do two jobs at once, and the burden of that was overwhelming.[22] The implication of the letter was that the Central Committee should hold an election for the position of general secretary and then, depending on its result, proceed to choose an assistant for him. The Klerksdorp and Wits regions supported Kimberley. The constitutional position was that all national office-bearers were elected at each triennial National Congress. Since Golding had been elected in 1987, this had been a ritual with unopposed elections. The only diversion was when each person was confirmed in his position and then carried uncomfortably on the shoulders of delegates from the rear of the hall to the platform. As Ramaphosa had been re-elected at the 1991 Congress to hold office until the next one in 1994, Golding's appointment had been for the remainder of Ramaphosa's tenure. There should have been no reason to interfere with that process but there was a strong feeling amongst some of the delegates that an election should be held. By January 1992, the political animosities in the union had coalesced around the personality of Golding who was perceived as a leading member amongst those opposed to the SACP. As soon as he was appointed to act for Ramaphosa, there were moves in the regions to replace him. The Central Committee meeting seemed to provide them with that opportunity but the issue was not clear-cut.

When Motlatsi delivered his address to the Central Committee he referred to the "leadership question" and made proposals that would have made it possible to remove it from the agenda. He emphasised the

[21]In 1991, the highest minimum wage for black workers in South Africa was R1 500 a month at Volkswagen's factory. Most motorcar workers received within R100 of that figure. The minimum monthly wage for supermarket checkers was R900. The majority of black mineworkers in the gold mines received between R380 and R520 a month.

[22]NUM NEC, September 1991, p51.

growing burden of the general secretary's job through the increase in the size and complexity of the union. Constitutional provisions, he stated, that had been devised in 1982 for a membership of about 14 000 gold mineworkers were not appropriate for a membership of about 300 000 spread around the mining of six minerals. Like the Kimberley region he believed that the general secretary was overburdened but for a different reason. His solution was also different. He suggested two options. "I have reached the conclusion," he stated, "that the job of general secretary is indeed many jobs. In fact, in our union it is four jobs and I am going to propose to you that instead of electing a new general secretary in 1994 we should abolish that post and that of an assistant general secretary and create instead four equally rated positions called National Secretaries."[23] His other option was to retain the position of general secretary but replace that of assistant general secretary by four national secretaries. Neither of the options interested the Central Committee. They were referred to the NEC and forgotten.

The issue, therefore, remained on the agenda. Its outcome hinged on whether or not the Central Committee had the constitutional right to hold an election for the post of general secretary. Motlatsi, in his speech, hinted that it had no such right. The Kimberley, Klerksdorp and Wits regions, however, insisted that the election should go ahead according to clause 8.2 of the constitution that gave the Central Committee the authority "to consider and decide upon policy issues of major importance in between National Congresses". In order to break the impasse, Motlatsi called on Ramaphosa, who had originally drawn up the constitution and who was present at the meeting, to give his advice. Briefly and simply, Ramaphosa said that, strictly according to the constitution, the Central Committee had been called in response to Kimberley's demand to look at the position of the general secretary and that was what it should do. The election should go ahead. His interpretation was accepted and arrangements were made to hold a ballot. Unfortunately, as Golding had not anticipated this situation, there were no nomination forms, ballot papers or ballot boxes available and the process was delayed. Everything was in place before the end of the meeting, however, and the ballot went ahead. Eight of the regions nominated Golding but, angry at the turn of events, he refused to stand. Some of the regions responded by saying that unless he did stand they would walk out of the meeting. In order to prevent such a strategically important union occasion from being disrupted, Motlatsi intervened

[23]NUM, James Motlatsi, *Presidential Address*, 24 January 1992.

Kgalema Motlanthe listening to the election results that made him the Acting General Secretary of the NUM.

and convinced Golding that he had a democratic duty to stand for election. Golding gave way and the election went ahead with just two names on the ballot paper, that of Golding and Kgalema Motlanthe, who was the head of the Education Department but also a key figure in the recreation of the ANC. Ramaphosa was asked to take charge to ensure that the ballot was secret while he and two guests acted as scrutineers. Motlanthe won the election, with 185 votes compared with 106 for Golding. Motlanthe, then, became the acting general secretary until the constitutionally scheduled election was held in 1994. Golding was returned to his previous position as assistant general secretary.

The advent of a new, relatively inexperienced, general secretary altered Motlatsi's task yet again in that he had to assume a much greater practical responsibility for administrative matters. Motlanthe did not, at that time, possess Golding's experience in administration. He was a modest, self-effacing person who was highly intellectual in his approach to the NUM and its environment. In his new job, however, he was not given much scope for analysis for he was confronted immediately with the same network of structural problems that had bothered his predecessor. Golding had stated the previous September that the union was in an unhealthy financial state, spending approximately 30 per cent more than its income whilst carrying a large debt. The union, he had concluded, had insufficient checks and balances, especially in the regions.[24] The cohesion of the union was affected by the changing balance of influence between the regions as the old established gold and coal regions lost members due to the year-by-year decline of their respective industries, while Rustenburg, with its platinum mines, suddenly rose to become the second-largest region with a membership that was almost entirely new to the NUM and inexperienced in both trade unionism and politics. The effect on the total membership was that it rose slightly over the previous year to just less than 300 000.

The Resurgence of Violence at Impala Platinum

One of the issues that Motlanthe had to confront immediately on taking office was the long-running dispute with Impala Platinum. The arrest of the Workers' Committee in November 1991 had merely ensured that the dispute would continue its violent course and become an integral part of the ANC's struggle for the reincorporation of Bophuthatswana into

[24]NUM NEC, 7 September 1991.

South Africa. This became explicit when, on 23 November, at a mass meeting of mineworkers at the Catholic Church in Phokeng, not far from Rustenburg, a call was made to use industrial action at Impala specifically to undermine the Bophuthatswana government. Less than a week later there was rioting in the hostel that caused much damage to mine property. When the security forces intervened, they injured many people. The whole of December was punctured by stay-aways, go-slow campaigns, the intimidation of non-unionists by the marshals and repression by the security forces. This culminated in a sit-in on 6 January 1992 at one of the shafts. Six hundred and eighty mineworkers were subsequently dismissed but they remained around Impala, participating in agitations and trying to get reinstatement. There were further incidents of strike action and some cases of arson. The NUM claimed that the basic cause of the trouble was an ongoing struggle between Sotho and local Tswana sellers of dagga outside the hostel. The bitterest disputes, however, were over NUM's support for internal South African liberation causes that pitched Tswanas, who supported the NUM, against Sothos and brought in the security forces.

However, the cause of the violence and the identities of the perpetrators became further confused when it was discovered that a group of 20 Koevoet mercenaries, armed with machine guns, were accommodated in the security barracks of the Swartklip shaft at the Rustenburg Platinum Mines, owned by JCI. The Koevoet were the notorious counter-insurgency group that had been deployed in Namibia by the SAP to track and kill cadres from the South-West African People's Organisation (SWAPO). It was reported that they were performing a similar task in Rustenburg but this time their targets were ANC and NUM members.[25] Strangers, who did not speak the local

[25]*The New Nation*, 31 January 1992. The Koevoet, meaning 'crowbar', was the name given to the Special Operations K Unit of the South African Security Police. It had been set up in 1979 in Natal but was moved to South-West Africa to act as a counter-intelligence insurgency unit because of the failure of the South African Defence Force to mobilise local opposition to SWAPO. It had been the equivalent of RENAMO in Mozambique. It was a mobile unit with about 1 000 members, led by white officers but with recruits from the local population. It had practised a systematic pattern of torture and killing that was institutionalised as an operational military and policing technique. The unit kept no records of the identities, number or whereabouts of the people it killed but, as it operated a bounty system whereby its members were paid for every corpse, it kept a scorecard. In the mid-1980s its scorecard showed that it had killed between 300 and 500 SWAPO supporters every year. It was supposedly disbanded in 1990 and its members were said to have been absorbed in other police units. Its re-emergence in South Africa in 1992 was not accompanied by an official explanation. See *Truth and Reconciliation Commission of South Africa Report*, 1995, Volume Two, pp61-84.

languages and who were seen in the vicinity of other mines in the district, were suspected of being members of the Koevoet. The effect was to create an atmosphere of fear in the district. JCI confirmed that the Koevoet were staying at the Swartklip shaft but insisted that they were there to search for criminals on the farms in the area and not to attack union activists.[26]

The presence of the Koevoet, however, did not alter the behaviour of the marshals who continued to insist that all of the hostel inmates, including the Sothos, should support the NUM. They generally had the support of the local Tswanas who made up about 70 per cent of the labour force but they faced much resistance from the Sothos. This came to a head when they attempted to coerce the Sotho hostel inmates to attend a political rally on 21 March 1992, the anniversary of Sharpeville, at Phokeng, where the ANC had much local support. The Sothos resisted and fighting broke out between them and the Tswanas, during which 13 mineworkers, mostly Sothos, were killed. Immediately afterwards, the police escorted the majority of the Sotho mineworkers to a TEBA depot from where they were repatriated to their homes in Lesotho.

Both Impala Platinum and the Bophuthatswana government consolidated their control and repression of the mineworkers after the violence in March. Impala built a second razor wire-tipped perimeter fence 30 metres outside of the first security fence, with access points controlled by armed security guards. The remaining members of the Workers' Committee were dismissed. This was an act the manager said he regretted not having taken at the outset.[27] The government built a defence force camp in the middle of the four mines and installed a 24-hour rota of patrols at the mine entrances. These, however, were merely adjuncts to its system of repression for it had arrested more people for political offences in the past year than had been similarly arrested in the whole of South Africa.[28] It also responded to pressures from the ANC by amending its Internal Security Act to prevent any political organisation that was not a registered political party from meeting in Bophuthatswana and to forbid any individual who was not a citizen from speaking publicly. In its amendment, in case there should be any misunderstanding, it included a 33-line definition of acts that were banned as public protests.[29]

[26]*Ibid.*

[27]Pretorius Commission, 1992, p33.

[28]*The Weekly Mail*, 8-14 May 1992.

[29]*Ibid.*

Despite these obstacles, however, the union secured a breakthrough in August 1992 when Impala Platinum agreed to a form of recognition that would circumvent the government's legal restrictions. The reasons for Impala's change of mind were never made clear though some of them were obvious. The struggle for recognition had cost the company more than R100m in lost production while 30 people had been killed in clashes between union and non-union members and NUM members and the mine security force.[30] It is possible, moreover, that Impala's parent company, JCI, had reasoned that the ANC was going to be in the ascendant within South Africa and over the 'homelands'. It could ill afford to be seen as an ally of a discredited and defeated president Lucas Mangope. Under the new arrangements, the union was to be seen as a central council, consisting of shaft representatives who would, in effect, be NUM shaft stewards. It would have the right to call on two NUM officials for advice, as and when it was required. The council would be able to bargain over wages, conditions of employment and new issues as they arose, as well as negotiating procedures and other agreements.[31] The NUM, then, obtained a presence in Bophuthatswana even though, for the time being, it was a rather subdued one.

[30]South African Institute of Race Relations, 1992/93, p360.
[31]*Ibid*, pp360-361.

CHAPTER SIXTEEN

THE WIDER NATIONAL AND INTERNATIONAL CONTEXT

Short-term Strategies *versus* Long-term Aims

The leaders of the NUM did not anticipate the retention of the existing class structure, based on the private ownership of the means of production and a free market mechanism, after the apartheid system had been dismantled. The union's adoption of the *Freedom Charter* as an addendum to its constitution bound it to its contents, which included the statement: "The mineral wealth beneath the soil, the banks and monopoly industry, shall be transferred to the ownership of the people as a whole." Motlatsi emphasised this commitment at the end of 1991 when he declared that "the *Freedom Charter* is still our guiding document – more so where it addresses the question of monopolies. Nationalisation of the mines is still the best option to address that question."[1] On a number of occasions during 1992, plans for the mining industry were discussed that did not go so far as to suggest nationalisation but they did anticipate a restructuring of the industry that entailed an extensive involvement of black mineworkers in running its affairs. Ramaphosa who, in 1992, was the ANC's leading negotiator, had been the architect of the NUM's policy relating to the running of the mines and it was presumed that he had carried those ideas with him to his new job.

Demands for a different kind of society in which the balance of class relations would be altered were also made in other sections of the liberation movement. Joe Slovo, speaking on behalf of the SACP at its Eighth Congress in December 1991, said: "The transformation we seek cannot be brought about without a considerable degree of state intervention and, therefore, the existence of a public sector. And the private sector must be tamed to the needs of the people. This does not imply across-the-board nationalisation. Where state involvement is considered necessary for effective social planning it can take a multiple of forms ..."[2] COSATU was more explicit about nationalisation at its Economic Policy Conference on 27-29 March 1992. There, in a document called *A Path to a Democratic Economy*, it declared: "A future democratic government – based on the support of the majority of

[1] *NUM News*, December 1991.

[2] *The African Communist*, Journal of the South African Communist Party, First Quarter, 1992, p28.

people – will need to play a leading role in making sure that all share in the wealth of the country. The state can make it possible by … nationalising some important parts of the economy …passing laws to prevent conglomerates from controlling important parts of the economy [and] introducing price controls and other types of laws to make sure that wealth is redistributed."[3] In its declaration on the ownership question, COSATU stated: "We believe that in the long term, society as a whole should own the most important parts of the economy. We should strive towards nationalising these 'leading heights of the economy' so that we can produce what our people need – not simply for profit. Such nationalisation should be under the control of the workers." The inference in all such statements was that the mining industry, as one of the largest employers of labour and the country's major earner of foreign currency, was an indisputable target for government control, if not for ownership.

There could have been no doubt in the minds of those who led organisations opposed to the liberation movement, therefore, that the negotiations for constitutional change were perceived by the liberation movement as a means towards a reconstruction of South African society that would reduce the power of private capital within the context of a strong central government. On the other hand, the primary aim of the National Party, separately and through the government it controlled, was to protect the rights of private property that were, at that time, wholly a white preserve, within a free market economy. They saw this being achieved within a political system with strong regional governments. Some white interest groups wanted to go further and press for the establishment of an autonomous regional white capitalist state. Inkatha, too, wanted to preserve capitalism but within a powerful KwaZulu regional government. The bottom line for the groups opposed to the ANC, then, was to protect the *status quo* as far as possible and that was in accord with the interests and the behaviour of the mine owners.

Because these conflicting class aims determined the primary relationship between the liberation movement and the constituent parts of the apartheid state, they might have been an integral part of the negotiating process, determining its course and prescribing its outcome. Yet that was not the case. Very soon the negotiations developed their own identity, momentum and priorities that superseded the structural aims that had been enunciated by Slovo and COSATU. This became clear shortly after the negotiations began. The transition from informal,

[3]*NACTU/COSATU Bulletin*, April/May 1992.

semi-secret discussions to formal negotiations that could lead to a constitutional settlement took place in the second half of 1991. Before then the main obstacle to formal negotiations had been the ANC's insistence that, in order to take the matter out of the hands of the existing Tricameral Parliament, a constitution for a new South Africa should be drawn up by a constituent assembly elected by universal franchise. This had been unacceptable to the government. In mid-1991, however, the ANC had offered a compromise, which was that the procedure for drawing up such a constitution should be worked out by a multi-party conference. The government accepted this and, on 20 December 1991, formal negotiations began with the first meeting of the Convention for a Democratic South Africa (CODESA) at the World Trade Centre, near Johannesburg's Jan Smuts Airport. There were 228 delegates in attendance from 19 political organisations.[4] The main negotiators were Ramaphosa for the ANC and Roelf Meyer, the Deputy Minister for Constitutional Affairs, for the government. Thus Ramaphosa's negotiating skills, developed after many confrontations with the mine owners, were able to come into play.[5]

Ramaphosa wanted a speedy settlement. He was supported by Chris Hani, who stated early in 1992: "We believe that it is absolutely imperative that we now move very rapidly towards a negotiated settlement, with elections for a constituent assembly next year… We need the certainty of an election date…"[6] Joe Slovo, a member of the ANC's Negotiations Commission, entered the debate on the side of Ramaphosa with a controversial proposal to expedite the settlement. This was that the ANC should make major concessions by agreeing with the National Party's demand that the new constitution should provide for a power-sharing arrangement for a limited period and that the continuity of the government administrative should be ensured by honouring the existing contracts of all civil servants. Slovo's proposals,

[4]They included the ANC, the National Party, four other Parliamentary parties, all 10 homeland administrations, the Natal Indian Congress, the SACP, the Transvaal Indian Congress and the SA government. The Azanian People's Organisation (AZAPO), the Conservative Party and the Pan-Africanist Congress (PAC) refused to attend.

[5]Not long after the negotiations had started, a section of the ANC leadership, that was opposed to Ramaphosa and critical of the leading role he was playing in the negotiating process, persuaded Mandela to remove him from the negotiating team and confine his activities to internal ANC affairs. However, representatives from other organisations involved in the negotiations, who had witnessed the quality of his work, informed Mandela that without Ramaphosa the negotiations had come to a halt. He then restored Ramaphosa to lead the ANC team.

[6]*The African Communist*, Fourth Quarter, 1992, p2.

described as the 'sunset clauses', provoked much controversy within the ANC and the SACP. His critics argued that the strategic goals of the liberation movement would be watered down for the sake of negotiating tactics.[7] Nonetheless, Slovo's views were largely incorporated in the ANC Negotiations Commission document and the negotiations continued on that basis.[8]

Gold Fields, Koevoet and the Boipatong Massacre

Whilst the debate about Slovo's proposals was taking place, the relationship between the ANC and the government was deteriorating. There was escalating indiscriminate violence on the Witwatersrand railway network and in the Vaal townships. For instance, a random shooting at the Daveyton railway station, in the east Rand, on 15 June, killed five people and injured 18. Gunmen cruising in a minibus at nearby Evaton casually shot down and killed nine people. There were many cases of individual and group atrocities for which the ANC held the government ultimately responsible. There were some, however, that were clearly the work of ANC supporters. For instance, a young woman was necklaced in Boipatong, a black township in the eastern Transvaal, for having a romantic relationship with a Zulu hostel-dweller. Later the same day, a former schoolteacher, who was a supporter of Inkatha, was murdered. The following afternoon another Inkatha supporter in Boipatong was killed.

The negotiations, at that stage, were deadlocked over the refusal of the government to consult with the ANC about new security legislation as well as over a disagreement over whether it should be necessary to have a two-thirds majority in the constituent assembly to adopt a constitution, as favoured by the ANC, or a three-quarters majority as demanded by the National Party. Frustration led the ANC to try to force the government to retreat by mass action involving 70 rallies throughout the country in celebration of the Soweto revolt on 16 June, a general strike and then mass demonstrations. The government countered with its own publicity campaign and by giving the police

[7]See *ibid* for criticisms of Slovo's 'sunset clauses'.

[8]Some critics of Slovo accused him of deserting his principles. When this point was put to Walter Sisulu, then the deputy president of the ANC, he said that the very opposite was the case and that Slovo had made a very painful, personal and courageous decision in the interests of national unity. (Interview with Kate Carey, 7 January 1994, Shell House, Johannesburg.)

extra powers.[9]

Then, in Boipatong on the night of 17 June 1992, in what at first sight appeared to be a revenge attack for the deaths of the young woman and the Inkatha supporters three days before, 45 people were slaughtered in their homes while a further 22 were seriously injured. The victims included nine children, two babies and 17 women. The residents were raped, hacked, stabbed, shot, beaten and disembowelled. Hundreds of homes were attacked and looted. The massacre gave rise to a massive campaign of investigations, accusations and counter-allegations but with everyone expressing grief and culminating in a national day of mourning on 29 June.

It was contended that a group of up to 300 men armed with pangas, knives and guns from amongst the 600 Zulu residents at the KwaMadala hostel, in the compound of the Iron and Steel Corporation of South Africa (ISCOR), less than two kilometres outside of Boipatong, had committed the massacre. One of the hostel residents described how everyone there had to belong to Inkatha and join a unit that trained them in the use of weapons.[10] The police, however, claimed to have found no direct link between the hostel-dwellers and the massacre. When about 250 detectives entered the hostel to interrogate the inmates about it, they discovered a cache of weapons but contended that they had insufficient evidence on which to base arrests. The ANC believed that the police made this statement because they had assisted the KwaMadala hostel-dwellers in carrying out the massacre. Evidence from survivors of the massacre endorsed that view.[11]

They said that before the attack took place, a large contingent of police had arrived in the town and had cleared the streets of barricades and of patrolling youth members of the ANC-endorsed self-defence units who were reacting to rumours that an attack was imminent. Police vehicles were seen in the vicinity of the attacks before they took place. An attendant at a petrol station about 300 metres from Boipatong said that during the evening of 17 June he had seen a large group of armed men move into Boipatong so he had activated an alarm connected to the police station. Two white men had arrived in response about a quarter of an hour later but they had ignored his request to investigate and returned to their station in

[9]*The Guardian*, 16 and 30 June 1992.

[10]*The Star*, 26 June 1992.

[11]Truth and Reconciliation Commission, 1998, Volume Three, pp683-690. This commission's *Report* contains an account of the massacre that was largely based on evidence from the survivors.

Vanderbijlpark.[12] Shortly afterwards, a security guard at the filling station radioed his employer about the presence of the armed men who then telephoned the police but their response was the same. Workers on the late shift at two ISCOR factories near to the township reported seeing two groups of police at different sides of the township, dropping men off from their Casspir vehicles. On no occasion did the police take any kind of precautionary action to protect the residents of Boipatong. A police Casspir was seen following the attackers to their hostel after the massacre. Then, when survivors in Boipatong rang the police for help, they got no response. There was little doubt that the police colluded with the attackers.[13]

There was much speculation about the identity of the attackers but the evidence from the survivors indicated that there were three groups. The first two groups consisted of black men wearing red and white headbands so that they would not mistakenly attack each other. The first group moved ahead, shouting, breaking windows and causing confusion. The second group went into the houses and attacked the residents. The third group, allegedly consisting of armed white men, some of whom had blackened faces, surrounded the houses and shot anyone who tried to escape. The black participants who spoke Zulu were clearly from the KwaMadala hostel. The white ones, who spoke Afrikaans, were not recognisable members of the police. At first, nothing more than the colour of their skin was known about them.

The first clue about the possible identity of the white participants came when a security guard at the eastern Transvaal Greenside colliery, owned by Gold Fields, informed the ANC that a member of a former Koevoet unit, based at the colliery, had told him that the Koevoet had been responsible for the Boipatong assault. The Koevoet

[12] *The Star*, 26 June 1992.

[13] The Truth and Reconciliation Commission reached an unequivocal conclusion, blaming the hostel-dwellers, in collaboration with the SAP, for the violence. Its findings stated: "The commission finds that KwaMadala hostel residents, together with the police, planned and carried out an attack on the community of Boipatong and the surrounding informal squatter settlement, Slovo Park, on 17 June 1992. The commission finds that the police colluded with the attackers and dropped them off at Slovo Park." Subsequently, in July 1993, 16 KwaMadala hostel residents were charged with complicity in the attack. The police participated in the trial but vital tape-recorded evidence in their possession was mysteriously erased from a tape. The trial judge, who dismissed the charges, said that the erasure of the tapes was the result of incompetence rather than a deliberate attempt to hide evidence. The commission, however, concluded: "The police were responsible for destroying crucial evidence in that they erased the tape of transactions in the control room of the ISU." (Truth and Reconciliation Commission, 1998, Volume Three, p689.) The ISU was the Internal Stability Unit of the Security Services, notorious for using violence against witnesses.

had white as well as black members. The ANC passed this information to the Goldstone Commission of Inquiry which, the very next day, raided the colliery in the company of the police and members of the ANC Intelligence Department. There, in a hostel normally used by Gold Field's security force, they found 40 members of the Koevoet with a cache of arms. This aroused their worst suspicions. The commission held an emergency sitting and began to question all of the relevant parties but they were presented with contradictory and inconclusive evidence. The chief of the security police denied that the unit had been involved in the Boipatong massacre. On the contrary, he said that he had evidence that residents from the KwaMadala hostel had been responsible for it. The members of the Koevoet, he alleged, were expert trackers who were used to combat crime, such as the theft of farm stock and copper telephone wire.[14] The police, moreover, insisted that they had rented the colliery premises from Gold Fields through a normal commercial transaction. Gold Fields made the same point to the commission.

Mineworkers from Greenside colliery said that their curiosity had been aroused when they noticed a group of strangers in the hostel who were armed, carried radios, spoke a strange language, were fed separately and were generally kept apart from the mineworkers.[15] They noticed that they did not work during the day but left at night in kombis carrying objects wrapped in blankets. A local NUM official said that on the night of the massacre he had seen two police personnel carriers collect a number of the men from the hostel. They were carrying two large canvas bags. Survivors of the massacre declared that they had seen and heard white Afrikaans-speaking men, in tracksuits and carrying torches, arrive and depart in police vehicles. The white men had kicked open the doors of the houses but had remained outside while the black attackers did the killing.[16]

The Goldstone Commission, which had been unable to reach a clear conclusion about the complicity of the Koevoet unit in the massacre, had been handicapped by the fact that the investigating officer it had sent to Greenside colliery had himself been a member of the Koevoet and had refused to interrogate the Koevoet members. Judge Goldstone did not rule out the possibility that they had been implicated but, in July, he stated that whether or not the Koevoet unit of the SADF was

[14]*The Guardian*, 27 June 1992.

[15]*The Star*, 26 June 1992.

[16]Allister Sparks, *The Observer*, 21 June 1992. Sparks had interviewed the survivors himself.

involved in political violence, "the very existence of such a group in South Africa in 1992 is calculated to cause yet further distrust of the security forces".[17] The ANC, however, continued to believe the evidence of its witnesses from Greenside colliery and, after accusing the police, and thereby the state, with organising the massacre, Ramaphosa announced on 23 June 1992 that the ANC was withdrawing from all bilateral and CODESA negotiations. The collapse of the negotiations gave encouragement to those who had come to believe that they were diluting the primary aims of the liberation movement. Chris Hani was one such person. In July, at a rally in Natal, he said that the withdrawal of the alliance from CODESA was his "happiest day" since 1990 because CODESA had been destroying the liberation movement. Anticipating the revival of the armed struggle, he called on MK to teach black people how to defend themselves because the government and its allies were counteracting mass action with violence.[18] The negotiations, however, had only broken down at a formal level. They had not ended, because Ramaphosa and Meyer had continued to meet in secret to prepare for their resumption.

The NUM's part in this episode had centred on the presence of the Koevoet unit at a Gold Fields mine. The union, however, had other reasons for being concerned about the behaviour of that mining house. It had negotiating rights at Greenside colliery but this was virtually the only Gold Fields mine where that was the case. Irrespective of how many of its employees NUM organised in any of its large, rich gold mines, Gold Fields refused to grant the union recognition. It obstructed the union by being lethargic in processing stop order forms and, wherever possible, by dismissing union activists. NUM also had a current dispute with Gold Fields because it was one of only two mining houses that had recently rejected a profit-sharing agreement during the 1992 wage negotiations with the Chamber of Mines.

Motlatsi reacted quickly, therefore, when he heard about the presence of the Koevoet unit at Greenside colliery. He called for an urgent meeting with the Chamber of Mines and asked it to make a clear statement about the massacre. Bobby Godsell from Anglo-American, who led the Chamber's team, said that it was committed to the peace process and dissociated itself from the violence. The Gold Fields representative went no further than endorsing Godsell's statement. Motlatsi thought that the statement was neither sufficiently clear nor

[17]South African Institute of Race Relations, 1992/93, p28.
[18]*Ibid*, p446.

precise in view of the fact that Gold Fields had not given a satisfactory explanation about the presence of the Koevoet unit at its Greenside colliery. The government and employers' organisations had already agreed to support a day of mourning for the Boipatong victims on Monday, 29 June by allowing their respective employees to attend services. The Chamber had agreed to arrange a two-hour memorial service but at a time when it would not disrupt production. Gold Fields, however, issued its own statement to the effect that it would not allow a memorial service to be held on its mines. At that point, the NUM delegation staged a walk-out and held an *impromptu* demonstration, led by Crosby Moni, the chairperson of the Witbank region that included Greenside colliery. Simultaneously, mineworkers demonstrated outside of Gold Fields' offices in Johannesburg, waving placards saying, "Gold Fields means Killing Fields". Gold Fields reacted by stating that it knew nothing about the presence of the Koevoet unit at Greenside colliery because it was entirely the responsibility of the colliery manager.[19]

The NUM's Role in the International Labour Movement

The contradiction between pursuing either a pragmatic or a politically principled course in its international trade union affiliations, that had arisen when the International Miners' Organisation (IMO) had been formed as a breakaway from the Miners' International Federation (MIF), in the mid-1980s, had, by 1992, been influenced by the actions of Ramaphosa and Motlatsi and the dramatic change in the international political scene caused by the collapse of the Soviet Union and the communist governments in east and central Europe. Although the NUM still regarded the IMO as a political ally and, for the reasons given by Ramaphosa in 1989, still favoured a merger between the two organisations, its representatives on the executive board of the MIF, concerned about maintaining the flow of funds that came through the MIF, had taken no steps to press for one.[20] Their actions had been reinforced by the international political events that had put the capitalist world in the ascendancy and favoured the MIF that served its mineworkers. The NUM's decision, in 1989, to press for a merger on the basis of equality was a reflection of political realities at the time but

[19] *The Guardian*, 27 June 1992.
[20] See chapter 13, p436-438.

that was not how the leaders of the MIF saw it by 1992. Believing that the IMO's influence would decline with that of communism, they lost interest in making any kind of accommodation with it. This became apparent as soon as the cracks began to appear in the communist system.

The change in the balance of power was reflected in the attitude of Peter Michalzik, the general secretary of the MIF when, on 8 January 1991, he was approached by Arthur Scargill, the president of the IMO, for a meeting between the officials of the two organisations to discuss ways in which they could co-operate as a prelude to working out the details of a merger.[21] Michalzik did not reply for more than three months, because, he said, it was his "duty to consult with the MIF's statutory bodies".[22] In his reply, after committing the MIF to the principle of international unity, he wrote: "For numerous reasons and in particular recent developments in central and eastern Europe, we reckon ... that your proposal would create more difficulties and obstacles than it would assist us in achieving our prime objectives."[23] Ramaphosa was sent a copy of the correspondence by Scargill but did not respond to it. At the level of the individual unions, however, there was co-operation for, six months later, the mining unions from South Africa, Britain, Australia, Germany and the USA met in Geneva, whilst at an ILO conference, and agreed on joint action in relation to support for coal mining generally through the maintenance of international coal prices.[24] The leaders of the same five unions, who were in Johannesburg as guests of the NUM's Central Committee, met again on 20 January 1992, at the instigation of John Banovic, the Secretary-Treasurer of the United Mine Workers of America (UMWA), who wanted to discuss the question of international unity. Michalzik was also present at the meeting though he was not a representative of any of the five unions. The outcome was an 'agreed statement' that they would "examine a proposal to establish one single international organisation for mineworkers and energy workers" at a meeting in Haltern, Germany, in April and that they would co-operate to prepare another statement, this time for the United Nations Earth Summit on Development and Environment, at Rio de Janeiro, in June.

From then onwards, however, events moved in reverse as the MIF

[21]Correspondence, Scargill to Michalzik, 8 January 1991.

[22]Correspondence, Michalzik to Scargill, 23 April 1991.

[23]*Ibid.*

[24]The unions were: the South African NUM, the British NUM, the United Mineworkers' Federation of Australia, the German Industriegewerkschaft Bergbau und Energie and the United Mine Workers of America.

British trade union leaders, Arthur Scargill, president of the British NUM, and Ken Gill, general secretary of the Manufacturing, Science and Finance Union (MSF), with James Motlatsi and Marcel Golding during the meeting of the Central Committee in January 1992.

disentangled itself from the commitment made on its behalf. Within two months of the January meeting the situation had been transformed. Though the president of the German miners' union had endorsed the 'agreed statement' at the Johannesburg meeting, his union was unwilling to negotiate any merger that contained communists. Its objection was focused on the inclusion of Alain Simon, the general secretary of the IMO, in the British NUM delegation for the April meeting. Simon was a member of the French Communist Party. The executive of the British NUM reacted by declaring that unless Simon was included, then it would not attend the meeting. John Maitland, the president of the United Mineworkers' Federation of Australia and vice-president of the IMO, acting as an intermediary, informed Scargill on 2 March 1992 that the South African NUM and the UMWA had joined the German union in objecting to the presence of Simon at the April meeting.[25] Without giving a reason, Motlatsi said that he had no option but to fall in line with the MIF, which by then had rejected outright a merger with the IMO.[26] But no-one had mentioned this as a possible difficulty to anyone in the IMO. Scargill had first heard about it in a letter from a third party after the MIF had made its decision not to merge with the IMO.[27]

Motlatsi's support for the MIF decision soured his relationship with Scargill to the point where it became difficult for their respective unions to collaborate over issues of common concern in mining. This was a difficult situation for the NUM for the British NUM had always been admired and emulated by the South African black mineworkers. They had turned out in large, enthusiastic numbers to greet Scargill wherever he went on his visits to the South African mining areas. Moreover, some of the leading officials of the union were unhappy about the turn of events. Motlatsi was the leading player in the international affairs of the NUM and this made his stance particularly important. But his views about international affiliations were not shared by either Kgalema Motlanthe, the NUM's general secretary, or Gwede Mantashe, the national industrial organiser, for both of them wanted closer ties with the British NUM and, by implication, with the IMO. In a meeting with Scargill in August 1992, Motlanthe said that he felt uncomfortable sitting with the MIF because he had far more in common with the IMO.[28] Mantashe was so concerned at the breakdown

[25]Correspondence, Maitland to Scargill, 2 March 1992.

[26]Author's notes of meeting between Scargill and Motlanthe, London, 28 August 1992.

[27]Correspondence, Scargill to Maitland, 9 March 1992.

[28]Notes of meeting between Scargill and Motlanthe, 28 August 1992.

of the relationship that he suggested the national executive committees of the two unions should meet in South Africa as soon as possible in order to reaffirm the bond between them.[29] Such a meeting, however, could not take place without the support of Motlatsi and nothing more was heard about it.

The attitude taken by Motlatsi was doubly important because of the stature of the NUM in the world of mining trade unionism. It was recognised that if it affiliated with the IMO it would expedite a merger and that, even if it remained in the MIF, it could have a similar effect if it argued for a merger within the executive bodies of the MIF. With those possibilities in mind, Michalzik sought to tie the NUM even closer to the MIF by suggesting that it could have a permanent position of influence in the new international body following the amalgamation between the MIF and the International Federation of Chemical, Energy and General Workers (ICEF). The MIF Education Officer was due to retire and the idea was to replace him with an Assistant General Secretary who, ideally, would be Marcel Golding, the existing Assistant General Secretary in the NUM. Golding, it was said, had the international experience and commitment to the MIF that other NUM officials lacked.[30] The amalgamation process, however, took time and, before it was completed and the new post was agreed upon, Golding had become a Member of Parliament in the 1994 democratic elections and was no longer a NUM official.

The last time the question of a merger was seriously discussed was at the Third Congress of the IMO in Algiers from 9-12 May 1993. On that occasion, Thomas Qhina, the chairperson of the OFS region of the NUM and Manne Dipico, the NUM's area organiser in the northern Cape, were in attendance as observers. Qhina gave a rousing speech in support of the IMO and they both submitted a resolution calling on the IMO and the MIF to dissolve and establish a single international organisation for all mine and energy workers throughout the world. The South Africans emphasised that the new international body would have to accept into affiliation all mine and energy unions irrespective of colour, creed, religion or political beliefs. This meant that not only the French CGT but also the Cuban miners' union would have to be allowed to affiliate to it. The resolution called for immediate discussions and asked the IMO vice-president, John Maitland, to convey its terms to the MIF

[29]Correspondence, Scargill to Motlatsi, 14 February 1993.

[30]According to Motlanthe, who had been present in Stockholm when the appointment was discussed. (Notes of meeting between Scargill and Motlanthe, 28 August 1992.)

Congress in Budapest that was being held immediately following the IMO one. It was unanimously endorsed by the IMO, then taken by Maitland to the MIF, but it was not discussed there, even though it originated from one of its own leading affiliates. The final blow to the merger discussions, however, came, when, eight days after being re-elected as a vice-president of the IMO, Maitland announced that his union, the United Mineworkers' Federation of Australia, a founder member of the IMO, was withdrawing from the IMO and joining the MIF.[31] The question as to which federation the NUM should be affiliated ceased, thereafter, to be a contentious issue in the union.

Domestic Mining Issues

Retrenchments had become commonplace since 1987 through a general rationalisation of the mine labour force and the closure of specific shafts. Between then and 1992, almost 160 000 jobs had been lost from all mines in South Africa. Eighty-five per cent of those, amounting to about 142 000, had been lost in the gold mining industry alone.[32] In 1992, however, a retrenchment crisis of a different dimension loomed when Harmony, near Virginia in the Orange Free State, the Rand Mines gold mine with about 20 000 employees, appeared to be on the brink of closure. The Marais Committee had listed it as a marginal mine in 1990 but had not anticipated its closure until 2008. The company informed the NUM in May 1992, however, that it intended to shed 8 000 jobs and that unless costs were cut drastically there would be further job losses, perhaps leading to complete closure. The following month it actually retrenched 6 200 mineworkers. It then asked the NUM for advice. They jointly approached the government and asked for improvements in state policy towards marginal mines. Then the NUM, Rand Mines and the white mining union produced a rescue plan that entailed a degree of consultation and co-operation that the NUM had never previously experienced with any mine.[33] In a compromising mood, the union signed a wages deal in August but the price of gold fell again and a new set of measures was proposed in October in order to produce an extra

[31]Simon in an address to the IMO Presidium, 14 and 15 December 1993.

[32]Department of Minerals and Energy, *Minerals – South Africa, Statistical Tables, 1980-1999*.

[33]Martin Nicol, head of the NUM's Collective Bargaining Department, compiled a number of memoranda dealing with different stages of the situation in Harmony between May and September 1992.

300kg of gold a month at no extra cost. To reach that figure they had to introduce new working conditions, including Sunday work. In return for NUM concessions, the management agreed not to impose wage cuts or increase the monthly hours of work. Some aspects of the measures were unacceptable to the white mining unions but they were applied nonetheless.

The only other positive features for the union during 1992 were the signing of the Code of Conduct Agreement with Anglo-American in June and the compromise recognition agreement with Impala in Bophuthatswana shortly afterwards. The economy was in recession for the third year without a break. The South African Reserve Bank had stated that "the economic downturn had not only been abnormally long but that it also had more serious repercussions for living standards and employment than most previous downswings in the economy, except for the depression of the 1930s".[34] Real production in the agricultural sector fell by almost one-quarter in 1992, thus contributing to a food price inflation rate of about 30 per cent. The overall rate of inflation in 1992 was 14 per cent. The resulting decline in living standards was made worse for people in the rural areas by the drought that was destroying subsistence agriculture. The dollar price of gold fell for the second year running, causing a further contraction in gold production and a continuing decline in South Africa's foreign exchange reserves. The value of the rand fell in the foreign exchange markets, thus pushing up the rate of inflation by making imports more expensive. The prices of exports that had a significant import content were also increased, thus making them less competitive in the world markets. A higher import bill and lower returns from exports resulted in a balance-of-payments crisis. All of the principal economic and production indices reinforced each other in this recession process with desperate consequences for South African society.

When the NUM began its annual round of wage negotiations in 1992, this situation set its parameters. Its members were determined to get a substantial wage increase but they were in an extremely weak bargaining position and had little chance of getting one. For its members in the gold mines, therefore, the union opted to continue the policy it had adopted in 1991, namely to seek to augment whatever small basic wage increase its members were given with bonuses from the gold price and performance bonus schemes. Most of the mining houses had entered the schemes and agreed to pay a five per cent

[34]South African Institute of Race Relations, 1993/1994, p372.

increase in basic wages. The two outside of it, Gold Fields and Anglovaal, agreed to increase wages by six per cent. Whether or not the schemes operated, however, depended on particular conditions. Firstly, the gold price scheme only came into operation if the price of gold exceeded a given level. Secondly, only those mines that had a ratio of profit to total revenue of more than five per cent were obliged to pay out under the performance bonus scheme. Under this scheme, the Chamber of Mines insisted that mines were entitled to count the royalties paid to land owners as a cost before calculating their profits. This last condition was challenged by the NUM because it would push many mines below the five per cent level. It was taken to the arbitration court but the ruling went in favour of the Chamber of Mines with the result that not only were many mines relieved from paying any bonus but the profit levels of those obliged to pay were considerably reduced. The ruling virtually destroyed the usefulness of the scheme for the union members.[35] When the NUM's Central Committee examined performance bonus schemes in February 1993, it found that the payouts were surprisingly low given the huge profits of some of the mines. "Three of the top five payouts," it stated, "were given at marginal mines. If these mines can afford to share large profit amounts, the formulas for sharing at the rich mines that paid small profit shares needs to be critically reviewed ..."[36] The committee, however, did not, on that occasion, reject profit-sharing because it regarded the defects simply as 'side-effects'. It decided to continue with it "as a strategy on gold mines during the present phase of mine downscaling".[37]

The mine owners used the fall in the profitability of mining as a reason both to worsen working conditions, for instance by insisting on Sunday working, and to undermine the union. When lists of retrenched mineworkers were announced they usually included the names of local union officials and activists. It was as if there had been no change in their attitude to black trade unionism over the last decade. For instance, as a reprisal for supporting the COSATU stay-away on 3 and 4 August 1992, the De Beers Finsch diamond mine, in the Kimberley region, dismissed the entire branch committee and 88 of the 90 shaft stewards.[38] The Central Committee decided to initiate an industry-wide campaign against those mine owners who used retrenchment as a means of

[35]See *ibid*, 1992/93, pp368-369, for details of the ruling. Also see *New Nation*, May 8-14, 1992, p27.

[36]NUM *Central Committee Report*, 11-13 February 1993.

[37]*Ibid*.

[38]*Ibid*.

bashing the union. It was also dissatisfied with the collective bargaining arrangements. Since its formation, it had not been allowed to represent any members employed above category eight. In recent years, however, some of its members had been promoted to higher category jobs and, although they remained members of the NUM, it was not allowed to represent them. In a situation with slowly broadening opportunities for black mineworkers, the NUM wanted the right to represent its members in whatever job they worked. It also wanted to tackle the problem of non-unionism through an 'agency shop' arrangement whereby those who refused to join the NUM were obliged to pay a reduced subscription in acknowledgment of their receipt of the benefits from collective bargaining by the union. The Chamber of Mines took steps to forestall this move by rescinding the closed-shop agreement that it had signed on 1 June 1937 with the nine established white unions that had members in the mines.[39] It justified its action on the grounds that agency and closed-shop agreements were contrary to the principle of free association and free choice. The Mine Workers' Union, that had been protected by the agreement in the first place, did not object because it believed that without its constraints its militant right-wing policies would attract new members to the extent that it could double its membership.[40]

NUM's Social Plan for the Future

Everything the union did in the strict industrial relations sense, however, was circumscribed by the necessity to break down the apartheid system. There was a clear understanding in the union that the abolition of the punitive restrictions of apartheid and the creation of a democratic socialist society were necessary conditions for alleviating both their working and living conditions. In consequence, the NUM continued its leading role in building up the organisation of the ANC and of the SACP. The membership of the ANC rose phenomenally from an official base of zero, when it opened its first office in Johannesburg in March 1990, to 900 000 in 1 400 branches by April 1993. The membership of the SACP also rose, though less dramatically,

[39]The agreement had been signed in order to prevent a breakaway Afrikaner nationalist union called the Afrikanerbond van Mynwerkers from challenging the dominant Mine Workers' Union, thereby creating dissension amongst the white miners. Dissension was created nonetheless. (Lang, 1986, p359.)

[40]South African Institute of Race Relations, 1992/93, pp373-374.

from 5 000 in January 1991 to 60 000 in October 1993. The NUM officials continued to exploit their skills as organisers by establishing and running branches in both of the organisations. Some exposed themselves to the risks of violent retaliation from Inkatha 'self-protection units', 'third force' groups or disaffected state security forces and were named on their respective death lists. Motlatsi had to have an extension built to his house in Dawn Park, a suburb of Boksburg in the east Rand, to accommodate special security personnel so that he and his family could have armed protection for 24 hours every day.

The disruption to the negotiations, prompted by the Boipatong massacre in April, was prolonged by a ferocious attack on an ANC march by the Ciskei police in Bisho on 7 September. The marchers, led by Ronnie Kasrils and accompanied by Cyril Ramaphosa and Chris Hani, were aiming to reach Bisho, the capital of the Ciskei, to protest against political repression as part of its campaign to re-integrate the homelands into South Africa. Ciskei, at that time, was governed by a military government led by a former ANC supporter, Brigadier Oupa Gqozo, with the support of the South African government. When the crowd of about 70 000 converged on the stadium, just over the South African border, the security forces began firing, killing 29 demonstrators and injuring about 200.

Overall, there were scenes of horrendous violence that resulted in as many as 3 347 deaths during the course of 1992.[41] However, incongruously, the violence seemed to be moving hand in hand with progression towards a negotiated settlement. Despite it all, and particularly the Bisho incident, in which he feared for his life, Ramaphosa continued to have informal meetings with Meyer who represented the government, in an effort to remove the differences between the two sides. The success of the meetings was indicated when, on 26 September 1992, just less than three weeks after Bisho, a Record of Understanding was reached between the ANC and the government as a basis for renewed multi-party negotiations. The following month the NEC of the ANC agreed to return to negotiations but there were many issues to resolve before they could formally start. The ANC, for instance, had a list of 15 preconditions. There were substantial differences over whether there should be a unitary or federal state, whether there should be majority rule or an enforced coalition, and over the transitional process itself. The ANC was particularly suspicious of the government's commitment to the

[41]South African Institute of Race Relations, 1993/1994, pp652-653.

negotiations because of its expressed determination to retain control of all the institutions of white minority authority throughout the transitional period and by the manner in which it was unilaterally endeavouring to restructure the economy.[42]

Nonetheless, the informal meetings continued and were followed by formal bilateral ones, involving the ANC and government, in December 1992 and January 1993 that gradually narrowed the differences. The ANC's preconditions by this time had been reduced to three, namely the release of all political prisoners, including those on death row, the fencing of Inkatha hostels and banning the carrying of traditional weapons. The last two conditions reflected the ANC's concern about the level of political violence caused, it believed, primarily by the Inkatha Freedom Party (IFP). Ramaphosa believed that the government's approach to the negotiations and the level of violence were related. The National Party (NP), he wrote, "hoped that by locking the ANC into protracted and fruitless negotiations, the structure of the ANC on the ground would be rendered ineffective by the ongoing violence. The capacity of the ANC to mobilise any significant resistance to the government would have been thoroughly undermined. The NP achieved a measure of success in this regard. Areas of Natal, the PWV, Ciskei and Bophuthatswana saw the ANC having to operate under the most repressive conditions. The climate of fear bred by the violence also had an impact on many people's attitude to politics. It tended to frighten people away from engaging in any sort of political activity, to the point where they would not join mass action or even take part in an election. For an ANC which relied on a politicised and active population, this proved a great threat."[43]

The NUM delegates who attended the Central Committee meeting in February 1993 reflected the scepticism of Ramaphosa, who was present as a guest speaker, about the intentions of De Klerk and his government. Negotiations were still suspended and there was intense township violence. They declared that when Mandela was released they had "hopes for a rapid transition to democracy, but instead, peace, democracy and the eradication of the inequalities of apartheid continued to elude us in 1992. Instead, the transition to democracy was held hostage to state-instigated political violence; it was held hostage to De Klerk's attempts to protect white minority interests in the new constitution; and to tie the hands of the new government to

[42]African National Congress, *Secretary General's Report*, 49th National Conference, 17-21 December 1994.
[43]*Ibid.*

prevent it from addressing the inequalities that apartheid has so deeply entrenched in our society".[44] This pessimistic view of the future, however, did not prevent the delegates from planning the NUM's contribution to the anticipated democratic elections.

The Central Committee stated:

> Victory for democracy means more than the vote. To address the legacy of apartheid we need to empower our people, to create jobs and provide social services. This can best be achieved through a Reconstruction Programme based on a solidarity pact with the ANC-led alliance.
>
> The NUM Central Committee, therefore, decided to support a Reconstruction Accord between COSATU and the ANC, that includes the following aspects:
>
> - The rights of workers as embodied in the Workers' Charter should be protected and guaranteed by an ANC government.
> - Consultation with the democratic forces and accountability of leadership must be entrenched in the way the new government operates.
> - A programme of action on issues of poverty, job creation, education and training for our people
> - This Reconstruction Accord should be a signed document that binds the future ANC government.
>
> On the basis of such an accord, the NUM decided to give its fullest support to the ANC's election campaign, to make our vision of reconstruction, redistribution and development a resounding success.[45]

The Central Committee then established an Election Campaign Fund and urged COSATU to do likewise. The initiation and co-ordination of the NUM's own campaign activities were put under the control of a National NUM Election Campaign Committee, the primary task of which was to harness the union's 'grass-roots' leadership skills for use by the ANC. In addition, the union was instructed to second "some of its officials, office-bearers, regional leadership and full-time shop stewards" to work with the ANC. In order to make full use of them, the Election Committee was told to begin negotiations with the mine managers for time off with pay for all of its seconded members. The Central Committee also agreed to train and pay "capable retrenched workers in the various regions to do voter education". Finally, it "decided to approach the ANC to ensure that workers from neighbouring countries who have worked in this country for more than five years are entitled to take up South African citizenship, and are included on the voters' roll".[46]

Though, in the months following the Central Committee meeting,

[44]NUM Central Committee, *Declaration*, 15 February 1993.
[45]*Ibid.*
[46]NUM, Central Committee, February 1993.

while the union's mind was focused on the possibility of an election and on the formulation of a programme for a future ANC government, it could not neglect the changes in the material conditions of its members. The mine owners continued to reduce the scale of their mining activities without consulting the NUM either about retrenchments or about the need to soften their impact by encouraging alternative job opportunities, providing job-training facilities or stimulating rural agriculture. Indeed, the contrary had happened. The government had added to the misery of the dismissed mineworkers by taxing their once-for-all retrenchment packages at rates as high as 50 per cent. None of them had ever previously earned enough to reach even the lowest level of taxable income. The union protested to Derek Keys, the Minister of Finance who, as the former chairman of Gencor, knew much about the employment conditions of mineworkers. The union was apprehensive at the time and even thought that he would make matters worse by increasing value-added tax on food. When it came to the budget in March 1993, however, Keys exempted a significant proportion of retrenchment pay from taxation. But it was the loss of jobs that was the real problem. It was, the union's Central Committee declared, "a social crisis in its own right" and it demanded that the companies should negotiate a 'Social Plan' with the union in order to manage the process of change in the size, location or working methods in mines. It wanted, in particular, a fair procedure for selecting those to be retrenched, a reasonable period of notice, a minimum severance package of two weeks' pay for every year of service, provisions for training or retraining in anticipation of retrenchment and retrenchment counselling. The NEC was instructed to start drafting a Social Plan Act and to convene a National Retrenchment Conference in order to try and create a consensus around its provisions.

The union's Collective Bargaining Department began to draft the plan by examining and listing the problems faced by black mineworkers who were retrenched when mine managers closed down or curtailed mining operations, rationalised the work process or introduced new technologies. They found that very few sociological investigations had been made into those issues since the acceleration of downscaling. The most extensive piece of research had been that by Gay Seidman for the Sociology of Work Unit at the University of the Witwatersrand early in 1993.[47] The union, in consequence, relied heavily on her findings. She

[47]Gay W Seidman, "If Harmony Closes, Will the Last One to Leave Turn Out the Lights? The Social Impact of Downscaling on the Gold Fields Area", *Preliminary Report*, 2 April 1993, Sociology of Work Unit, University of the Witwatersrand.

had concentrated on the effects of retrenchments at the Harmony gold mine near Virginia where the work force had been reduced from 32 000 in 1988 to 14 600 in 1993. She had examined the impact on the town of Virginia and on Mandela Park, the informal settlement just across the railway line from Virginia. She had also interviewed a sample of the households of retrenched mineworkers in Lesotho, from where most of Harmony's labour force came. Seidman described the impact on Virginia, where some businesses in the retail sector had reported an 80 per cent drop in trade, as "catastrophic".[48] About 70 per cent of the retrenched mineworkers in Mandela Park had been unable to use their severance pay to create viable, sustainable forms of subsistence. The situation was even bleaker in Lesotho where no families that were interviewed, whether or not they possessed land, could subsist without regular sums from the pay of mineworkers.[49] It was clear from her findings that the welfare of the mineworkers and their families was incidental to the whole process of technological change in the mining industry. Little of substance had changed in this respect since the formation of the union.

Martin Nicol and his colleagues in the collective bargaining department scanned the literature about the mining industry for evidence of the effects noted by Seidman in Virginia. In the preamble to their report on Social Plans they listed the six areas they had discovered where black mineworkers were disadvantaged when retrenchments were contemplated. They were the lack of information, a failure to receive sufficient notice, inadequate recall procedures, an absence of training facilities for non-mining work, no infrastructure to advise them about how to utilise their severance pay most effectively and a lack of concern about the health of those who were actually retrenched.[50] The report proposed, in the first instance, that the union should negotiate the establishment of reserve funds on an industry-wide basis in order to meet the costs of restructuring and retrenchments in the future. It recognised, however, that the full costs of retrenchments could not be met through the normal negotiating procedures with the mine owners. It suggested, therefore, that a Social Plan Act should be passed that

[48]Jonny Steinberg with Gay Seidman, *Gold Mining's Labour Markets: Legacies of the Past, Challenges of the Present*, Labour Studies Research Report 6, Sociology of Work Unit, University of the Witwatersrand, September, 1995, p35-36. Seidman had been primarily responsible for the research.

[49]*Ibid.*

[50]NUM, Collective Bargaining Department, *Background to the NUM Proposal on Social Plans*, 6 June 1993.

would compel the mine owners to negotiate with the NUM over any changes in the production process that would affect the welfare of the mineworkers. Extensive retrenchments in platinum mining in May and June reinforced the union's argument for a Social Plan but the Chamber of Mines did not respond.[51] When Bobby Godsell gave his presidential address to the Chamber of Mines in June 1993, he spoke of "the pain" suffered by all the stakeholders in the gold mining industry and listed retrenchments along with falling profits, declining dividends and reduced tax returns for the government but without making any distinctions between them. He made no mention of a Social Plan or even of the need for one.[52] This was as the mine owners wanted it, because they rejected the proposal to establish a national contributory fund to deal with the rehabilitation of mines faced with recession. The NUM obtained the right to negotiate about the issue with several individual mines in 1993 but nothing materialised because all of those mines were marginal ones that had no reserves to spare for a Social Plan.[53] Nor did the government respond. By the time Motlanthe presented his Secretarial Report to the Eighth National Congress, in February 1994, no progress had been made so he presented the Social Plan simply as a possibility "for effective state intervention in the mining industry".[54] The union, anticipating that there would be no action until after the apartheid system had been abolished, noted in its final report that "the Social Plan should be adopted as alliance policy and marked down as the first law to be passed by a new, democratically elected government".[55]

The 1993 Picture of Black Deprivation

Once an end to apartheid was in sight, the main organs of the liberation movement began to assess the character and intensity of the social and

[51]Western Platinum, in Bophuthatswana, announced in May 1993 that it was laying off 1 500 mineworkers. In the following month Impala Platinum did likewise, making the total number of mineworkers retrenched since 1991 up to 7 000. Four hundred and fifty mineworkers were also laid-off at Trans-Natal Collieries in June (South African Institute of Race Relations, 1993/1994, p600).

[52]R M Godsell, *Presidential Address*, 103rd Annual General Meeting of the Chamber of Mines, 8 June 1993.

[53]The mines were Durban Deep, Harmony, ERPM and Grootvlei. (NUM, *Marginal Mines: The Challenge for the RDP*, 14 July 1994.)

[54]NUM, *Eighth National Congress Report*, February 1994, p9.

[55]NUM, Head Office, *The Social Plan – One Pillar in NUM's Strategy for Worker Rights and Participation*, 18 July 1993.

economic problems with which they were going to be faced and the policies they would need in order to resolve them. There had not, hitherto, been a systematic collection of the evidence of the devastation caused by apartheid on black people's lives. Black people knew about it from their daily experiences and that was sufficient to warrant the struggle for liberation. It was, however, a different matter when it came to the formulation of policies to repair the devastation. At that point it became important to know the precise extent of the damage in different spheres so that a system of priorities could be devised that could be incorporated in planning strategies for a future ANC government.

This understanding had prompted the ANC and COSATU to send a delegation, led by Thabo Mbeki, to the International Bank for Reconstruction and Development (World Bank) in April 1992 to canvass support for a survey of living standards in South Africa. Mbeki framed his request as a need for help in devising strategies to combat poverty. He was heard sympathetically and, as a result, a World Bank task force was sent to South Africa to examine the logistics of conducting such a survey. This was followed by a request from the World Bank to the Southern African Labour and Development Research Unit (SALDRU) at the University of Cape Town to assist in the co-ordination and management of the collection of the data required. The governments of Denmark, Norway and the Netherlands agreed to finance the survey and it became operational from the middle of 1993. The result, published in August 1994, came too late to assist in the formulation of the ANC's policies of reconstruction and development but it provided incontrovertible statistical evidence that they were based in reality.[56]

The society that SALDRU surveyed had distinctive, but confusing, characteristics. South Africa since 1948 had sought international competitiveness that would provide it with the status of a developed capitalist country through industrialisation with the overt protection and support of the state. Its general categorisation was that of a 'Second World' country along with the socialist countries of Europe. It had aspired to 'First World' status but had failed to achieve it, in part because its international competitiveness in the production of minerals, its most valued commodities, was declining. South African production was significant in the international markets only in the field of primary products. Insofar as it had industrialised, its products had failed to

[56]SALDRU, Project for Statistics on Living Standards and Development, *South Africans Rich and Poor: Baseline Household Statistics*, August 1994. The statistics used in the remainder of this section, unless stated otherwise, are drawn from this source.

penetrate the international markets. But the greatest barrier to 'First World' status was that the criterion for entry had changed. Global polarisation was no longer based on the distinction between industrialised and non-industrialised countries. 'First World' countries were those that controlled "the global industrial system through their command of financial power and communications systems, their dominance in decision-making over the use of resources at the global level, and their monopolies of technology and weapons of mass destruction".[57] All other countries, including South Africa, were in the periphery. In that periphery, the 'Second World' was in the process of disintegration. The South African government was frantically privatising and withdrawing from intervening in the running of the economy. It was becoming, in effect, a 'Third World' country, desperately trying to industrialise.

There was a problem, however, with this categorisation because South Africa represented a microcosm of the world capitalist system.[58] Within its territory there existed segments of the 'First', 'Second', 'Third' and 'Fourth' worlds. The mining conglomerates belonged to the industrialised world and participated with those who controlled the global industrial system. This was visible in South African society through the lifestyles of a tiny fraction of the white population. Though the 'Second World' was disintegrating, the majority of the white population subsisted comfortably and securely as state functionaries.The ANC's intention, through Slovo's 'sunset clauses', was to protect their positions of privilege so that they would not be exposed to the tribulations of 'Third World' life that were faced by the black industrial workers who lived in the townships. However, in work and earning wages, they suffered and shared deprivation with the majority of the black population that was either without work in the urban areas or without land in the rural areas and who, along with the vast majority of people in Africa, lived in a 'Fourth World' of sheer absolute poverty. Black mineworkers figured ambiguously in this categorisation for they possessed jobs and regular wages but they were migrants with homes and families firmly located in the 'Fourth World'. As the SALDRU survey clearly showed, the 'worlds' in South Africa in 1993-94 were racially divided with only black people populating the 'Third' and 'Fourth'.

In South Africa, the homes of the different racial groups were spatially separated and displayed glaring, contrasting distinctions. Only

[57]Samir Amin, *Foreword*, in Hein Marais, *South Africa: Limits to Change. The Political Economy of Transition*, Second Edition, 2001, ppvi-x. Amin give an incisive analytical picture of South Africa within the global economy.
[58]*Ibid.*

44 per cent of black South Africans lived in or shared houses compared with 82 per cent of both white and Indian people. The rest of the black population lived in shacks, mainly in shanty towns, traditional huts in the rural areas or hostels in the urban areas. There were different configurations of housing in the black rural and urban areas. Various materials were used to construct the houses, such as mud, wattle and daub, mud and cement, cardboard, wood, corrugated iron sheets, cement blocks and bricks. In all of its forms there was gross overcrowding.

Most black families lived in completely deprived circumstances. They were denied access to the social infrastructure that was taken for granted in the white communities. Piped water was supplied to less than 18 per cent of black households. About 50 per cent obtained water from pipes in yards or public kiosks. The rest got it from boreholes, flowing streams or rivers, or stagnant water sources. The absence of running water meant that there was inadequate sanitation. About 34 per cent of black households had flush toilets. Sixteen per cent had none while the remainder used crude pit latrines. Often, excrement flowed into the rutted mud streets where children played. This was the case, for instance, at New Brighton, the black township on the outskirts of Port Elizabeth. There was a national electricity grid that supplied power to members of the white communities whether they lived in urban or isolated rural areas. However, only about 37 per cent of black homes were connected to it. In the eastern Cape, that percentage was as low as nine. In the rural areas as a whole, just over one-quarter of black homes had access to electricity. It followed that other sources of power had to be used for necessary domestic purposes. For instance, 35 per cent had to use wood for cooking. This was not an easy option because it was estimated that wood collectors, usually women and children, spent between half an hour and four hours each day collecting it. Lighting was also a problem. More than 40 per cent of the black households were dependent on candles for lighting while a further 25 per cent used paraffin.

The level of income was obviously a prime determinant of the quality of life of all ethnic groups. Black households were at the bottom of the list and, on average, received less than one-sixth of the incomes of white households. The average, of course, was a distortion for most black households had either extremely low incomes or none at all while a small number had excessively high ones. Almost 30 per cent of black households were dependent on remittances from absent family members. In the rural areas where the families of black mineworkers lived, this percentage was nearly 37 per cent though, surprisingly, almost 22 per cent of black urban households were in a similar position. Income, of course, was dependent on access to work. When black people in the survey were asked what they thought was the most important thing the government could do for them, 62 per cent said it should create jobs. The reason was obvious. The South African Reserve Bank estimated, in August 1993, that almost half of the economically active population of approximately 12 300 000 was unemployed.[59] This caused many people to join the informal market. About 27 per cent of all black people who were described as self-employed sold goods on the streets while 12 per cent operated illegal shebeens.

Black workers with jobs, however, worked, in the main, in the lowest income categories. Less than one per cent of them, compared with 21 per cent of white workers, worked in managerial and administrative posts. Twenty-eight per cent worked as labourers while only 0,4 per cent of white workers were in that category. Other forms of deprivation followed directly from this. Almost 45 per cent of black people found healthcare too expensive while about 16 per cent could not afford the money for transport in order to get it. Lastly, in this categorisation of deprivation, black people suffered from a lack of education. More than 20 per cent of all children in the age group six to 24 years did not attend school at all. Of those who did enrol, less than half had completed standard six by the time they had reached the 14+ age group. This meant that more than half in that age group were functionally illiterate. This discrimination in educational achievement continued throughout the age groups so that in the 18+ plus group only 11 per cent of black students completed standard 10, which was equivalent to matriculation, compared with 61 per cent of white students.

Discrimination, and the deprivation that went with it, had been normal, unalterable features of life for all black people for as long as any of them could remember. None had played any part in the formulation of policies at any level of political, economic or industrial activity except what might have been achieved through their trade unions. They had been excluded from the political franchise and had, therefore, been denied the basic right to participate equally with others in the choice of their representatives and government. That situation still prevailed in 1993 and informed the thinking of black mineworkers, along with all other black occupational groups. The question before them was not just how to devise policies that would redress the racial

[59]South African Institute of Race Relations, 1993/1994, pp599-600.

New Brighton – serving the opulent white community of Port Elizabeth.

The quality of black urban life. The open sewer down the main street of Red Location in New Brighton.

imbalance but how those policies would be implemented and by whom. This question increasingly concerned the activists in the NUM. In their minds, the exclusion of black people from decision-making was itself a major act of deprivation that had to be rectified.

The Reconstruction and Development Programme (RDP)

By 1993, this situation had altered to the extent that, in addition to trade union membership, black people were involved in the formulation of policies for a future ANC government as members of the ANC and the SACP. But the unions were still the main focal points for involvement in both national and local decision-making. They had their own established democratic practices and carried some power, through their ability to withdraw labour, to influence the decisions of employers and the government. At that time, the trade unions derived additional influence from the fact that the ANC and the SACP were dependent on them for personnel and material support. The unions, however, did not act independently but through their respective federations.[60] Of the four federations, only COSATU, through its alliance with the ANC and the SACP, was able to contribute directly to the formulation of post-apartheid policies. Its affiliates, therefore, were in a privileged political position. This was shown when both the NUM and the National Union of Metalworkers of South Africa (NUMSA), the two biggest affiliates of COSATU, pledged their support in February for a reconstruction accord with the ANC and prompted the leaders of COSATU to do likewise. The accord was intended to be a plan that would commit the ANC to a reconstruction programme that embodied the collectivist values of COSATU.[61]

Between February and July 1993, a draft reconstruction programme was compiled by the NUM's collective bargaining department that attempted to reflect the views expressed by its Central

[60]The federations were the Congress of South African Trade Unions (COSATU) with 1,7 million members in 17 affiliates; the Federation of South African Labour Unions (FEDSAL) with 270 000 mostly white members in 17 affiliates; the National Council of Trade Unions (NACTU) with 327 000 all-black members in 24 affiliates; and the South African Confederation of Labour (SACOL) with 40 000 all-white members in five affiliates. (*Ibid*, pp583-585.)

[61]Some affiliates criticised the leadership of COSATU for committing it to the accord without first consulting its membership. A special congress was called in September 1993, therefore, when the issue was debated and COSATU formally pledged its support for an accord with an ANC-led alliance.

Committee.[62] The NUM, more than any other COSATU affiliate, was influenced by the *Freedom Charter* and this was apparent in its emphasis on the role of government and the inviolate nature of people's basic rights. "The state," it declared, "will have to play an active and developmental role … a key purpose of a reconstruction programme is to ensure that the role of the state is supported and supplemented … by the mass-based formations and institutions of civil society … This has a very important democratising effect because it involves people and institutions in making and implementing policy."[63] The document insisted that the apartheid state would have to be restructured and the civil service reorganised, in order to eliminate its apartheid practices. Likewise, the police and the security forces needed to be converted from being agents of suppression to servants of the community. It provided a list of socio-economic demands. "All citizens of South Africa," it declared, "have a right to decent housing appropriate to family and individual needs and with tenancy or property rights that are transparent and protected by law … The state should play a significant role in the provision of finance for low-income housing … Every household, be it urban or rural, should have access to affordable electricity … to running water from a clean and environmentally viable source … to an effective telecommunications system [and] to cheap, effective and safe transport." The NUM also wanted an "equitable and integrated National Health Service" in which rural services would be given priority and from which no-one would be excluded for financial reasons. The document emphasised the need for education for all and singled out the various facets where improvements were necessary. Reflecting the needs of mineworkers, it wanted training for a variety of skill paths at all skill levels and, in particular, it wanted training in portable skills. There were, however, significant omissions in the draft document. It made no mention of the NUM's attitude to the question of privatisation which the government was currently hurriedly pursuing or to public ownership, particularly of the mining industry, even though that had been the long-standing policy of the NUM. These omissions were possibly intended to reflect a degree of sensitivity to the ANC's

[62]NUMSA was the first COSATU affiliate to submit a draft manifesto. Its proposals, submitted in March 1993, called for a national health service, seven years of compulsory schooling and a housing programme to provide at least 1 million houses within the next five years. It wanted the privatisation programme to be put into reverse. NUMSA supported the alliance for the election period but it was unhappy about its continuation after an interim government had been elected. (South African Institute of Race Relations, 1993/1994, pp590-591.)

[63]NUM Collective Bargaining Department, *Draft Reconstruction Accord*, 18 July 1993.

intention to appeal for support across the class divide in the general election.

The NUM's draft document, along with those from other COSATU affiliates, formed part of the consultation process that resulted in the ANC's final version of the Reconstruction and Development Programme that was published in March 1994. Mandela was proud of this process and contended that it was a continuation of a tradition that had begun with the *Freedom Charter* in 1955. He wrote in the preface to the published programme:

> The process now underway is that of developing the detailed policy and legislative programme necessary to implement the RDP. In preparing the document, and in taking it forward, we are building on the tradition of the *Freedom Charter*. In 1955, we actively involved people and their organisations in articulating their needs and aspirations. Once again we have consulted widely.
>
> However, in 1994 we are about to assume the responsibilities of government and must go beyond the *Charter* to an actual programme of government. This RDP document is a vital step in that process. It represents a framework that is coherent, viable and has widespread support. The RDP was not drawn up by experts – although many, many experts have participated in that process – but by the very people who will be part of its implementation. It is a product of consultation, debate and reflection on what we need and what is possible.[64]

The ANC's published RDP policy repeated many of the aspirations in the NUM document for they were commonly held in the liberation movement. It also contained a section that dealt specifically with mining and minerals and which met the essential demands of the NUM. As with the NUM's own proposals, the RDP did not raise the question of public ownership of the mining industry. Indeed, it presumed a continuing relationship between the union and the mine owners by emphasising negotiating rights and the need to strengthen tripartite structures, such as the Mining Summit. In one respect, however, it went further than the NUM and proposed that private mineral rights should be returned to the democratic government, in line with the practice in the rest of the world. The proposals for mining, in general, involved increased government intervention. They aimed to pass new laws to improve the democratisation of the workplace by compelling employers to negotiate the organisation of work with mineworkers and their union. Mechanisms, they suggested, should be put in place to regulate downscaling in the mines so as to reduce the

[64]African National Congress, *The Reconstruction and Development Programme*, March 1994, Preface.

suffering of retrenched mineworkers and their families. Amongst the mechanisms were provisions for reskilling and training for alternative forms of work that the NUM had specifically requested. The RDP envisaged a new set of minimum standards that would ensure that fair wages were paid and employment conditions were protected. It also wanted a health and safety system that recognised the real hazards of mine work. Finally, the RDP wanted all mineworkers to have the right to live at or near their places of work in decent accommodation and to be provided by the mine owners with educational, training and social facilities for themselves and their families.[65]

The rural dimension of mineworkers' lives was also catered for in sections described as "Land Reform" and "Rural Development". "Land Reform" entailed redistribution and restitution. The case for redistribution was strongly put. "The RDP," it stated, "aims to improve the quality of rural life. This must entail a dramatic land reform programme to transfer land from the inefficient, debt-ridden, ecologically damaging and white-dominated large farm sector to all those who wish to produce incomes through farming in a more sustainable agricultural system. It also entails access to affordable services and the promotion of non-agricultural activities."[66] On the question of restitution, the RDP stated: "To redress the suffering caused by the policy of forced removals, the democratic government must, through the mechanism of a land claims court, restore land to South Africans dispossessed by discriminatory legislation since 1913. This court must be accessible to the poor and illiterate. It must establish processes that enable it to take speedy decisions. In order for this court to function effectively, constitutional rights to restitution must be guaranteed."[67]

Thus, in both the industrial and rural areas, the RDP aimed to make a dramatic improvement in the lives of mineworkers and their families and, in consequence, it was received enthusiastically by the NUM. It held workshops and mass meetings to propagate the merits of the RDP amongst its own members. It had already decided to engage in voter education in preparation for an election. The task was extensive and complex. Prospective voters had, in the first instance, to be persuaded to register and learn about the mechanics of voting. No black South Africans had ever voted, except in trade union elections and the recent internal elections in the ANC and SACP. The vast majority, therefore,

[65] *Ibid*, pp99-102.
[66] *Ibid*, p84.
[67] *Ibid*, p22.

had never seen a ballot paper and, as about two-thirds of the potential electorate of between 22 and 27 million were illiterate, it was likely that they would not understand even a simplified version of its contents. Indeed, it was estimated that about 50 per cent of those living in the rural areas did not know that a cross signified their preference for a particular political party. The high level of illiteracy and the lack of communications with and within the rural areas accounted for the fact that approximately 20 per cent of the potential electorate knew nothing about the negotiating process and the possibility of an election in which they could participate.[68]

[68]James Motlatsi, *Presidential Address*, Eighth National Congress, NUM, February 1994.

CHAPTER SEVENTEEN

ON THE MARGIN OF SUCCESS OR FAILURE

The Continuation of the Negotiating Process

Following the bilateral meetings early in the year, a multi-party planning conference was held on 5 and 6 March 1993. There it was decided to resume negotiations, through the multi-party Negotiating Forum, not later than 1 April. The forum met and agreed to instruct the Negotiating Council to consider and report on six issues. They were:

- The form of state for a future South Africa and its basic constitutional principles.
- A transitional or interim constitution.
  The form of transitional regional government.
- Fundamental human rights during the transition.
- A transitional executive council (TEC) and sub-councils, an independent election committee and an independent media committee.
- The future of the 'independent' homelands.[1]

The ANC and the government, it appeared, had resolved their major differences and formal arrangements for the transition to a democratic government in the foreseeable future seemed possible. But there were still obstacles in their way. The IFP, under the direction of Buthelezi, had withdrawn from participation in the negotiations and had entered into an alliance with the right-wing Conservative Party and the Ciskei and Bophuthatswana governments. They called themselves the Concerned South Africans Group (COSAG) and talked about civil war if their demand for a federal state was not met.[2] The AWB, which refused to participate in the negotiations at any stage and demanded an Afrikaner homeland of up to 25 per cent of South African territory, also threatened civil war if its demand was not met. Both the IFP and the AWB engaged in arson and armed attacks on ANC supporters, particularly in the east Rand townships, that generated retaliatory action. Vigilantes and hit squads based within the state structure supported them in intention. The result was an escalation of

[1]South African Institute of Race Relations, 1993/1994, p544.
[2]Sparks, 1997, p187.

politically related deaths during the course of 1993, many of which were caused by indiscriminate massacres of unknown individuals and families, chosen at random for symbolic reasons.

The Assassination of Chris Hani

The death, however, that carried the greatest symbolism and which resonated throughout all of the black communities of South Africa, to such an extent that it threatened not simply the negotiations but the very idea of reaching a negotiated settlement, was not a random killing. On the morning of Saturday, 10 April 1993, Chris Hani, the general secretary of the Communist Party, a member of the ANC's National Executive Committee and the most popular and respected person in the country next to Mandela, was shot and killed in the drive of his house at Dawn Park, close to where Motlatsi lived. His bodyguard had been released for the day so he was unprotected when he got out of his car. A white man in a parked car across the road fired two shots at Hani and then walked over to fire two further shots at his body. The white man then drove away. A neighbour of Hani, a white Afrikaner woman, witnessed the shooting, memorised the registration number of the car and immediately informed the police. A short time later, the police stopped the car and its occupant, a Polish immigrant called Janusz Waluz, was arrested on suspicion of murder.

Chris Hani was popular amongst young black South Africans because of his already legendary activities as a leader in Umkhonto we Sizwe. But he also had a powerful contemporary presence amongst black people of all ages in the townships and rural areas where he spoke eloquently and with fervour about the need and means to redress their problems.[3] He was unique amongst the returned ANC leaders in that he eschewed the pleasures of life in the white northern suburbs of Johannesburg and moved around the country to all types of areas where he met and listened to ordinary people. He had mass appeal. His murder by a white racist, therefore, gave rise to much contradictory speculation about South Africa's immediate future. There was talk of a complete breakdown in the relations between blacks and whites as young black people vowed revenge for his death. This would have

[3]Allister Sparks, the South African writer, stated: "Next to Nelson Mandela, he was the most heroic figure in the country to the young blacks of the restless townships. He had more charisma than any of the others and was the only real orator in an ANC-alliance line-up." (*The Observer*, 11 April 1993.)

Chris Hani

heralded the end of the negotiation process. Others, on the other hand, including some in the ANC's negotiating team, believed that the murder could be made to hasten a settlement. Ramaphosa was one of those. The alternative, he said, was for the country "to slide into chaos". In order to avert a crisis, he added, a date for elections would have to be agreed upon without further delay, and a Transitional Executive Council would have to be established.[4] Nelson Mandela made two national television appearances to appeal for calm, to allay fears of a racial war and to project the creation of a non-racial South Africa as the only possible outcome.[5] But people in the townships were ambivalent about that message. When, for instance, Thabo Mbeki, speaking before a crowd of thousands outside the Civic Centre in Boksburg, said that the time had come when Nelson Mandela should be president of the country, but "the only way this will happen is if we act in unison in a disciplined fashion", he was heard in silence.[6]

There was ominous violence from the first day after the assassination when two white men were set on fire and killed and a third one was seriously injured in a black township near to Johannesburg. In another incident, gunmen fired automatic rifles at two cars with white drivers just east of Johannesburg.[7] A contingent of 100 AWB supporters taunted, threatened and provoked black people in a memorial march in Vanderbijlpark in the Vaal Triangle.[8] The violence seriously escalated on Wednesday, 14 April, when more than 90 memorial services, combined with demonstrations, were held throughout the country. It was estimated that more than 80 per cent of the black labour force stayed away from work on that day in support of the memorial services, making it the biggest work stoppage in South African history.[9] There were running battles between the police and ANC protesters in Cape Town, Durban, Pietermaritzburg and Port Elizabeth. A number of people were killed, many more were injured, shops were looted and other property was damaged. President de Klerk described that

[4]*New Nation*, 16-22 April 1993.

[5]Phillip van Niekerk, *The Weekly Mail*, 16-22 April 1993. Mandela spoke first on the evening the murder was committed and, therefore, projected the ANC and not the government as the instrument for calming the people. He said: "I appeal to all our people to remain calm and to honour the memory of Chris Hani by remaining a disciplined force for peace." (Nelson Mandela, *The Illustrated Long Walk to Freedom*, 2001, p194.)

[6]*Saturday Star*, 17 April 1993.

[7]*The Guardian*, 12 April 1993.

[8]*Sunday Star*, 18 April 1993.

[9]*The Citizen*, 20 April 1993.

Wednesday as a "dark day" for South Africa.[10]

On that day about 40 000 mourners, who had attended a memorial service for Chris Hani at the Jabulani Stadium in Soweto, marched to the Protea police station to present a memorandum protesting against the lack of police protection for ANC leaders. The police began to act provocatively as soon as the march started by driving their vehicles through the crowd as it went along but without getting a negative reaction. When they reached the police station a delegation of alliance leaders presented their memorandum and, in response, received an assurance from General Braam Strauss, the police chief, that the police provocation would stop. The crowd broke up into groups, turned away from the police station and began to make their way home. As they did so, without giving them time to disperse and without warning, the police fired live ammunition into their backs. Four of them were killed and 250 were injured.

As the leaders of the march were, at that point, at the rear and nearest to the police, they received the full blast of the shooting. One of them who was killed was Sam Tambani,the NUM's National Education Chairperson and the secretary of the Soweto branch of the ANC. He had presented the memorandum to General Strauss. He was widely known and respected as a trade union educationist as well as an ANC leader in Soweto. He had been shot in the back. The police media liaison officer claimed, however, that the police had fired only after the crowd had tried to break into the station. "We were patient," he said, "but they threatened our lives and property. We fired." The government endorsed the police version of events by issuing a statement saying that the police had acted with restraint. All of the eyewitness accounts contradicted the police version. According to them, the demonstration had ended and people were returning home when the firing started. The missions of observers from the United Nations, the European Community and the Commonwealth, when deploring Tambani's death, stated that the police had opened fire on a "well-disciplined commemoration rally". It was an event that was disturbingly reminiscent of the Sharpeville massacre in 1960.[11]

The manner of the deaths was perceived as a reflection of the continuing brutality of the state against black people. It intensified their

[10]*The Star*, 16 April 1993.

[11]*The Weekly Mail*, 16-22 April 1993. This newspaper's account was written by three of its reporters who were present at the memorial service and when the police opened fire. See also *The Star*, 16 April 1993 that gives two alternative versions of the police behaviour.

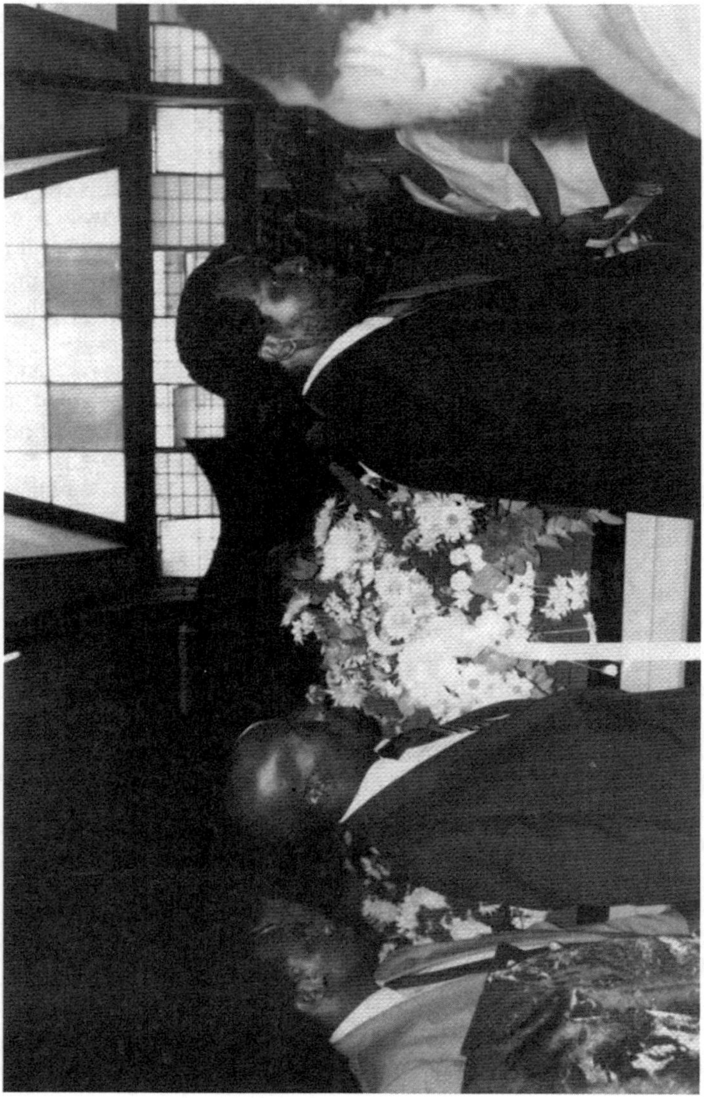

James Motlatsi at Sam Tambani's funeral.

anger about Hani's murder and focused it on the state as well as on racist white organisations. Members of the NUM, whilst mourning the death of their own Tambani, were quick to make that connection. Hani's death, however, meant other things to them. Many mineworkers, particularly in Klerksdorp and Carletonville, were members of the SACP. They had lost their leader. The union had close institutional links with the SACP. It was still helping to finance it, partly by providing it with offices on the floor below those of the NUM. The close proximity of the two head offices encouraged familiarity between Hani and the NUM leadership. In any event, Hani had acted as a respected adviser to the union on national and international political matters. Gwede Mantashe, the NUM's national industrial organiser and a member of the SACP, had accompanied him on a visit to Cuba and then London. Motlatsi, moreover, was personally affected as a neighbour and friend, as well as a supporter, of Hani. He had been one of the first to see Hani's dead body in the driveway of his house.

The leaders of the ANC-led alliance turned to the NUM for assistance in organising Hani's funeral. The NUM agreed and set up a Funeral Committee. Motlatsi was chosen as the committee's chairperson and he played an important role in preventing the funeral from becoming a focal point for widespread social disruption and violence. As neither the ANC nor the SACP had sufficient funds to pay for such funeral costs as the hire of a stadium, transport from the stadium in Soweto to the South Park cemetery in Boksburg, a cortege, food and drink for mourners and portable latrines and tents in the cemetery grounds, the Funeral Committee set up a trust fund.[12] Motlatsi made a personal appeal for financial support to the Chamber of Mines, businesses, embassies and wealthy individuals. The Chamber of Mines responded positively to the appeal, though its president Bobby Godsell said that its contribution "should not be construed as an indication of mining industry support for the views of the late Chris Hani or the policies of any of the organisations of which he was a member. The donation was made as a gesture to mark the tragic death of a political leader in violent circumstances which members of the Chamber abhor."[13] A number of private companies made contributions in kind. Coca-Cola, for instance, made a donation of soft drinks.

There was an ominous sign of what could happen at the funeral when, the night before, 19 people, including three children, were

[12] Cheadle, Haysom and Thompson, the legal firm that served the NUM and the liberation movement, administered the fund.

[13] *Business Day*, 20 April 1993.

Filing past Chris Hani's coffin.

murdered and more than 20 were injured in an indiscriminate attack by unknown gunmen at the black township of Sebokeng. When the crowds gathered on Monday, 19 April 1993, the day of the funeral, therefore, there was apprehension and anger but they were moderated by a universal sadness and disbelief that such a murder could have been committed. The consequence was sufficient calm to enable Motlatsi's funeral plans to work without any serious disruption. Virtually the whole of South Africa's black population mourned that day. It was estimated by the *South African Labour Bulletin* that about 92 per cent of black workers had stayed at home as a mark of respect for Hani, which made it a bigger event than even the previous Wednesday. There were no official figures showing the degree of support from the black mineworkers but the *Labour Bulletin* stated that in general they adhered to the calls to stop work. Anglo-American gave a different estimate. It stated that only 11 per cent of the mineworkers in its gold mines and 10 per cent in its coal mines did not turn up for work on the day.[14] The other mining houses did not publish any stay-away figures.

In the First National Bank Stadium in Soweto, where Hani's body lay, there were between 80 000 and 100 000 people, some of whom had been waiting throughout the night. Many thousands waited outside. Although Ramaphosa appealed to those in the stadium to return home after the ceremony, approximately 600 buses ferried people the 35-40 kilometres to the cemetery in Boksburg, where they had to pass hundreds of armed, uniformed AWB members who patrolled the street leading to the cemetery in order, they said, to protect nearby white homes. There were some incidents of violence and arson outside the stadium and at the nearby NASREC railway station. Three people were killed by gunfire from a workers' hostel as they walked to the cemetery. Five homes in Dawn Park were looted. Altogether, six people were killed and 14 were injured in the PWV region in incidents surrounding the funeral. But the violence did not escalate further, as had been feared.

The investigation into Hani's murder then began to follow the normal legal course. A former Conservative Party MP, Clive Derby-Lewis, was arrested on 17 April for complicity in the murder. On 21 April, five more people, including Gaye Derby-Lewis, the wife of the former MP, were also arrested. They all had extreme right-wing connections. Two of the five were released the same day. Along with Waluz, the main suspects who eventually went to trial were Clive and Gaye Derby-Lewis.[15]

[14]*The Citizen*, 20 April 1993.

[15]Waluz and Clive Derby-Lewis, a Conservative Party politician, were sentenced to death

Negotiations, Mass Struggle and Violence

Almost a week after Hani's murder, the ANC-led alliance had announced a two-phased programme of rolling mass action to expedite the demands made by Ramaphosa, namely the setting of an election date and the formation of a Transitional Executive Council (TEC) before the end of May. John Gomomo, the president of COSATU, stated that the mass action was intended to provide an impetus for those demands and to supplement the negotiations and not displace them.[16] Consumer boycotts, in response to local conditions, were started in various parts of the country. For instance, a boycott at Krugersdorp, where the Derby-Lewises had their home, was about the behaviour of white policemen and white local councillors towards black people. A boycott in the eastern Transvaal was aimed to get the reinstatement of workers dismissed for participating in stay-aways and to force the government to de-electrify the fence along the Mozambican border. The reinstatement of dismissed strikers was a familiar demand made by boycott organisers. In some instances, as in the eastern Transvaal, white farmers and business people retaliated with their own boycotts.[17] In June, the AWB attempted to leave an impression on the political situation when some of its members drove an armoured vehicle through a police cordon and smashed into the World Trade Centre in Kempton Park where the multi-party negotiations were taking place. The negotiators were forced to flee. COSATU called a stay-away in protest at the action of the AWB but the responses were patchy, ranging from more than 85 per cent in East London to between 10 and 20 per cent in Durban, 10 per cent in Johannesburg and five per cent in Cape Town. Rolling mass action was proving to be difficult to maintain.

As was intended, the negotiations continued alongside the mass protests. Slovo described their relationship in the following terms. "Negotiations," he wrote, "are a terrain of struggle which … depend upon the balance of forces outside the process. It was the link between the negotiations and our mass struggle that played an absolutely key role."[18] In any event, the ANC and government chief negotiators met in secret

for the murder at a Johannesburg court on 15 October 1993. Gaye Derby-Lewis had been acquitted of the murder the previous day. Neither of the two was executed because an indefinite moratorium on hanging had been applied for the last three years.

[16]*Business Day*, 20 April 1993.

[17]South African Institute of Race Relations, 1993-1994, p630.

[18]Joe Slovo, "The Negotiations Victory: A Political Overview", *The African Communist*, Fourth Quarter, 1993, p7.

the day following the funeral. Two days later there was a formal meeting of their planning committee. There were no disagreements between the two sides over the need for fixing an election date and establishing a TEC as quickly as possible. Whatever problems there were arose primarily from COSAG, that was opposed to setting a date for the election before the character of the constitution had been decided. COSAG quit the negotiations whilst the issue was being debated but there was still a 'sufficient consensus' in the Negotiating Forum to endorse a proposal from Ramaphosa that the election should be held on 27 April 1994.[19] The verdict was made public at its meeting on 2 July 1993. This first universal election in South Africa was to be for a transitional parliament that would also be a constitution-making body. The forum also decided that a transitional constitution would have to be drawn up and adopted before the elections could be held.

There was much scope for the negotiations to be sabotaged in one form or another. In the first place, the NP still exercised state power through the government and, if it wished, could hinder, or even stop, the movement towards democracy. Also, there could be a rear-guard action from right-wing forces both within and outside the state apparatus. The most significant threat to a negotiated settlement, however, came from the IFP immediately after the decisions to set up the TEC and fix the election date had been made. Indeed, there seemed to be a direct correlation between those decisions and the level of violence in both the PWV region and KwaZulu-Natal, though it was not the only causative factor.[20] The number of politically related deaths in those two areas rose from 236 in June 1993 to 549 the following month. The number of deaths in the PWV region alone rose almost fourfold during the same period. The new levels maintained their momentum until April the following year. It was the worst period for political violence in the history of South Africa. The situation was worsened by the availability of AK-47 rifles to supplement traditional weapons. Each weekend there were reports of massacres of innocent people. Forty-four

[19]Seven of the 26 organisations participating in the multi-party negotiations were opposed to naming the election date at that stage.The PAC was opposed but remained in the negotiations.
[20]The Independent Board of Inquiry, which reported on the level and causes of violence in South Africa in 1993, warned against making a simplistic assumption that all of the violence was caused by rivalry between the ANC and IFP. It stated that in Katlehong and Tokoza, the two townships that accounted for more than half of the deaths in the PWV region, an ongoing taxi dispute caused much violence. There was also, it stated, enforced colonialisation of areas of the townships by either the ANC or the IFP that resulted in residents either paying special taxes or fleeing their homes. (Independent Board of Inquiry, *Report for October and November 1993*, pp14-17.)

people were killed during four days of fighting in Katlehong and Tokoza in the first week of July. The SAP reported that between 3 July and 26 August, 544 people died violently in the east Rand townships.[21] The manner in which many of these people died was illustrated on 7 November when gunmen shot dead 10 teenagers and injured eight others in a homestead in Nqutu, in the middle of Inkatha-dominated territory in KwaZulu-Natal. The chief of the homestead was a member of the ANC, as were nine of the victims.[22] The NUM had its own fears in this respect. It became particularly apprehensive towards the end of 1993 about the public pronouncements of members of the white Mine Workers' Union, threatening to use force in their attempt to stop the negotiating process. In May, the MWU had aligned itself with the Afrikaner Volksfront that had recently been formed by 21 parties and organisations to demand the creation of an Afrikaner state. The threats were directed at the achievement of that demand. The NUM treated them seriously because of the ease with which white miners who supported the creation of a Volksfront could obtain commercial explosives.[23]

The NUM secretariat, moreover, feared that the violence would spill over into the mine hostels. The danger spots were those mines where Zulus were employed and, therefore, where UWUSA was likely to be present. Mine managers had been quick to respond to requests for negotiating rights from UWUSA even though it never represented more than six per cent of the black labour force in any mine. The NUM felt that an already volatile situation could easily get out of control. There had been trouble earlier in the year between Inkatha supporters and NUM members at the Beatrix gold mine in the Orange Free State. The situation had been defused when the Inkatha supporters temporarily left the mine but it flared up when they returned in April and two mineworkers were killed in the fighting.[24]

In the light of the NUM's experience, Motlatsi decided to approach the mining houses about the matter. He sent a memorandum to each of them explaining the reason for his fear and proposals to deal with its causes. He wrote:

[21]South African Institute of Race Relations, 1993/1994, p666.

[22]Independent Board of Inquiry, October and November 1993, pp13-14. Phillip van Niekerk, writing in *The Observer* of 6 March 1994, claimed that the massacre had been planned in Inkatha's offices in Johannesburg and from there the actual perpetrators had been chosen and armed.

[23]*NUM NEWS*, December 1993-January 1994, p7.

[24]*New Nation*, 16-22 April 1993.

It is difficult to avoid the conclusion that there is a correlation between the democratisation process and the indiscriminate slaying of ordinary people. There are groups in our society who are prepared to commit any crime to destabilise the process by making it impossible to conduct the first democratic election on 27 April 1994. The nearer we get to that date, the more desperate and callous their crimes will become. And if the election is held as planned, then they will endeavour to invalidate its results by a campaign of destruction. It would be unwise not to prepare for the most pessimistic scenario. The current spate of unprovoked attacks on unarmed and innocent people began in Natal and then moved to the townships of the PWV. It is unlikely that they will stop there as the desperation of those wishing to preserve white rule increases. No doubt attempts will be made to extend the havoc into industry. The mining industry, with its workers incarcerated in compounds and housed on a tribal basis in hostels, is a ready-made target for those wishing to make South Africa both uncontrollable and ungovernable. It is the special responsibility of the Chamber of Mines and the National Union of Mineworkers to prevent this happening.[25]

Motlatsi then explained how they should fulfil that responsibility. He stated:

The Chamber of Mines and the National Union of Mineworkers should act immediately to establish Peace Committees at every hostel, mine and mining house which would be responsible for identifying trouble flash points before they were ignited and taking immediate action to render them harmless. Their task would be to anticipate trouble and not to wait until it had broken out. To do this the committees would have to be intimately acquainted with events in the hostels, have the ability to act swiftly and possess the authority to take preventative action. Their decisions would have to have the full backing of management. Indeed, it should be written into their constitutions that mine managers should be obliged to implement all unanimous decisions of the Peace Committees.

He went on to describe the size and composition of each of the committees and detailed their powers to collect evidence, summon witnesses, reach balanced judgements and impose penalties. Motlatsi suggested, finally, that the system should be backed by a Code of Conduct that (a) prohibited all black and white employees from bringing weapons of any kind into the mine compounds; (b) prohibited them from taking instruments from the mine that could be used as weapons into the hostels; (c) made racial provocation, whether between white and black or black and black, a punishable offence; and (d) made any other act punishable that was likely to cause factional disturbances. There must, he concluded, "be no concessions in the fight against mine violence".[26] The proposals had hardly been formulated when an occasion arose for their implementation. In July 1993, a faction fight broke out between Mozambican mineworkers and others at the Hartbeestfontein gold mine in Klerksdorp. Nineteen mineworkers were killed and more than 100

[25] James Motlatsi, NUM, *Memorandum: The Prevention of Violence in the Mines*, 1993.
[26] *Ibid.*

were injured. As a consequence of the conflict, 3 300 Mozambicans left the mine to return home. The mine, owned by Anglovaal, employed more than 12 000 black mineworkers who were accommodated in hostels that were divided along tribal lines and administered by *indunas*. The labour organisation of the mine had remained unchanged since the 1970s and contained all the ingredients for tribal conflict. This particular conflict, however, seriously disrupted production and persuaded the manager to work more closely with the NUM in making dramatic changes in its hostel organisation. He introduced a hostel system based on working sections rather than ethnicity and replaced the *indunas* with democratically elected hostel administrators. Moreover, he responded to Motlatsi's memorandum by forming a mine Peace Committee consisting of five members from the NUM and five from the management and with corresponding sub-committees at shaft level. The Peace Committee's initial task was to ensure that the Mozambicans returned to the mine by the end of July as well as to maintain peaceful relations between the various tribal groups. In general, then, Motlatsi's exhortation to set up Peace Committees had some effect for when he addressed the union's National Congress in February 1994 he was able to say: "We have already taken steps to deal with the question of violence in the compounds by setting up Peace Committees in many mines in collaboration with some of the mining houses."

However, despite Motlatsi's concern about the possible spread of violence to the mine compounds and the possibility of it obstructing the negotiating process, the NUM continued to act in the hope that it would not. It participated fully in the election campaigning initiated by COSATU, carried on with its voter education scheme and endeavoured to ensure that as many black mineworkers as possible would be eligible to vote. As a part of its election campaign, COSATU held a special congress in September 1993 at which it announced the names of 20 trade union leaders who would be nominated to be included in the top 51 per cent of the ANC's own National Electoral list. Marcel Golding, the NUM's assistant general secretary, and Godfrey Oliphant, a full-time NUM steward at the Finsch diamond mine in the Kimberley region who had also been the second vice-president of COSATU since 1991, were on that list. The NUM's own vice-president, Elijah Barayi, had stood down as president of COSATU in 1991, but he entered the ANC's final top 100 Parliamentary list independently because of his popularity as an inveterate and long-standing ANC advocate.[27]

[27]Barayi had been the president of COSATU from its formation. John Gomomo from

Those on COSATU's list who held office resigned their positions so the congress proceeded to replace them. Jay Naidoo, the general secretary, was on the list and Sam Shilowa, the assistant general secretary, replaced him. Zwelinzima Vavi, the national organising secretary who had previously been a NUM official, took Shilowa's place. There were also replacements lower down the line.

These acts by the COSATU congress seemed to indicate that it believed that the procession to the democratic elections was irreversible, but that was not the case. The Negotiating Forum had produced a draft of a constitution for the transitional period to which COSATU had two serious objections. Firstly, it objected to the clause guaranteeing the tenure of civil servants. As all civil servants were white and the vast majority were Afrikaners, the question as to what would happen to them after the election was, in Slovo's words, one "of fundamental importance to the very delicate post-election period". He explained that "an intelligent handling of this area needs to take seriously the imperative of limiting dangers of counter-revolution and chaos that could ensue from just dumping existing civil servants in accordance with some mechanical policy".[28] COSATU, however, was concerned about the absence of any mention of affirmative action. The second issue of importance was the equality given to the workers' right to strike and the employers' right to lock out in the transitional constitution. The feeling in COSATU about these issues was so intense that it threatened to call a general strike on 15 November unless the Negotiating Forum changed the clauses. The ANC and SACP representatives on the Negotiating Forum were placed in an ambivalent position by COSATU's demand because they felt that they would have to support such a strike in order to maintain the solidarity of the alliance. After lengthy negotiations, COSATU compromised, firstly by agreeing to a clause about the civil service that allowed for affirmative action whilst addressing the job security fears of the present employees and, secondly, by accepting that whereas the right to strike was a fundamental right within the constitution, the right to lock out was not fundamental but would be permitted subject to the terms of the prevailing labour legislation. The South African Employers' Consultative Committee on Labour Affairs endorsed this compromise.[29] The strike threat was withdrawn, whereupon Joe Slovo, who was a

NUMSA had replaced him.

[28]Slovo, Fourth Quarter 1993, p11.

[29]South African Institute of Race Relations, 1993/1994, p75.

member of the Negotiating Forum, said that he believed that COSATU's intervention had "had a positive outcome on the negotiating process" and that the "capacity for this kind of pressure will remain critical in the coming weeks, months and years".[30]

[30]Slovo, Fourth Quarter 1993, p7.

CHAPTER EIGHTEEN

THE UNION AND THE ELECTION 1993-1994

Ongoing Issues

Whilst the profound question of the future of apartheid was being answered in dramatic though uncertain terms, black mineworkers still had to contend with their own industrial relations issues by confronting the mine owners as they had done each year since 1983. Nothing that was happening through the political negotiations impinged in any way on that relationship. So in May 1993, the ritual of wage negotiations began, as in previous years, and, on this occasion, ended in September. The union had submitted a demand for a 30 per cent increase in wages for all black mineworkers in both gold and coal mines because, it declared, its members would not tolerate another year of wage rises that amounted to only about one-third of the rate of inflation. The Chamber of Mines was not impressed and responded with miniscule offers of four per cent and six per cent increases for coal and gold mineworkers respectively. By August, the NUM had reduced its pay demand by between 35 and 65 per cent but that still did not satisfy the Chamber of Mines and it seemed that a strike ballot would be held. At the last minute, however, the two parties agreed to refer the case to a conciliation board through which a settlement of about nine per cent, approximately the rate of inflation, was reached. Another facet of the agreement involved the reversal of the Chamber of Mines' opposition to the agency shop when it agreed to establish one for mineworkers in the first eight job categories. Under the agreement those who refused to join the union had to pay into a joint union/management fund for charities and social welfare organisations. The union members had their best deal for three years.

Following the settlement with the Chamber of Mines, the union took on the long-standing and difficult task of gaining full union recognition at eight mines owned by Gold Fields.[1] Hitherto, the union had attempted to gain recognition by conforming to the agreement with the Chamber of Mines and organising at least 50 per cent plus one of the labour forces in each of Gold Fields' mines but, though it had

[1]The mines were: New Clydesdale, O'Kiep, Deelkraal, Driefontein, Venterspost, Libanon, Leeudoorn and Kloof.

achieved that level of membership in some of the mines, it had not achieved recognition. Gold Fields, it stated at the end of 1993, had denied it access to its mines, refused to process stop-order forms, falsified union membership figures, intimidated members and supporters and used its 'private army' to assault them.[2] It had also continued to house mineworkers along tribal lines and to use the discredited system of *indunas* to control them.

On this occasion, NUM decided to campaign from within each of the mines with various forms of mass action at local, regional and national level. It started with a march of 9 000 mineworkers from the Deelkraal gold mine in Carletonville, led by Motlatsi who presented a list of demands to the manager. On the list were the abolition of the *induna* system, ending the trespass system that constrained the movement of black mineworkers but not that of white miners, stopping having different change rooms for black and white workers, allowing mass meetings to be held in the hostels, extending visiting rights for wives so that their transport difficulties could be taken into account and providing the union with mine facilities so that it could do its job more effectively. Motlatsi insisted that the demands should be met by the end of January 1994. The union, however, was still protesting beyond that time limit. On 4 February, with the support of COSATU, it organised a mass picket outside the company's head office in Johannesburg to protest not only against its attitude to the union but also against its negligent approach to safety standards. NUM representatives had never been included in investigations into the causes of mine accidents. Indeed, as Archie Palane, the NUM's regional organiser, explained, the union often had to rely on press reports for information about them.[3] The new campaign, however, was no more successful than the old one. Gold Fields remained inflexibly hostile to black trade unionism and the NUM appeared to be unable to do anything about it.

The State of the Union

Mining had the highest industry-wide trade union density in South Africa but, by 1994, it appeared to have reached its limits. The target of 400 000 members that the union had set itself in 1989 continued to be elusive for whenever it made gains, for instance from the successful

[2] *NUM News*, December 1993-January 1994.
[3] *Business Day*, 1 February 1994.

Campaigning for recognition by Gold Fields, February 1994.

Shaming Gold Fields, February 1994.

recruitment campaigns in the platinum mines and then in the electricity supply industry following the NUM's amalgamation with the Electricity Workers' Union, they were offset by retrenchments. When the union's Eighth National Congress was held in February 1994, the membership was 281 000, only about 12 000 more than it had been at the 1991 congress. It had organised only just over one-half of the black work force in the mining and energy industries. There were, according to Motlanthe, 358 000 'free riders' who benefited from the NUM's negotiations without contributing to its maintenance. Nonetheless, he was optimistic. He asked the congress to map out a way that would take the union to the elusive 400 000 by 1997.[4] As a result, a detailed plan was worked out to reach a 75 per cent rate of organisation for the industry as a whole. The achievement of their prime target was brought forward to the end of 1995. Each branch was provided with the statistics relating to its own mine labour force and branch membership and was given a target, showing exactly how many it had to recruit in order to reach it. The plan, however, did not explain why previous targets had been unfulfilled or how the new targets were to be reached.

The NUM had the lowest rate of full-time organisers to members of any union in the country. In part, this had occurred because, once access had been granted, the high concentrations of mineworkers in the hostels of the gold mines had seemingly made recruitment relatively easy. There did not seem to be a pressing need to employ full-time organisers. A number of the gold mines were enormous concentrations of people. The largest mine in 1993 was Vaal Reefs with 37 000 black mineworkers. Free State Geduld came second with 22 000 while Western Deep Levels had just over 20 000. Hartbeestfontein and Harmony, each of which employed more than 12 000 mineworkers, followed them. All in all, looking at 46 gold, platinum, copper and lead mines at the end of 1993, 12 had more than 10 000 mineworkers each and 17 employed between 5 000 and 10 000. The coal mines were very much smaller. Out of 15 collieries, 11 ranged in size from 500 to more than 1 000.[5] Size, however, was a double-edged thing for whilst high concentrations of men in single-sex hostels were relatively easy to organise, they were also susceptible to management manipulation and propaganda. The union had extended its organisation in one respect by increasing the total number of full-time shaft stewards from 12 in 1991 to 117 in 1994. The increase, however, occurred

[4]NUM, *Eighth National Congress Report*, February 1994, p20.
[5]The Employment Bureau of Africa Limited (TEBA), *Report and Financial Statements for the Year Ending 31 December 1993*, pp20-22.

mainly in Rustenburg's platinum mines for they had more than half of all full-time stewards. Impala Platinum alone had 50 of them. There were more in the small regions with low concentrations of mineworkers than in the large ones with the big mines. The Orange Free State region, for instance, with the highest concentration of large mines, had only one mine, Harmony, with a full-time shaft steward. The enormous Vaal Reefs complex had only two such stewards. There were no others in the Klerksdorp region, though it was the third-largest in the union. Yet each of 10 collieries in Witbank had one full-time steward. This disparity had occurred partly because of variation in local militancy and partly because organising had not been given a sufficiently high priority by the secretariat in the allocation of the union's resources.

Gwede Mantashe, who was the co-ordinator of the national organisation department, had a staff of 42 to deploy amongst a black mining work force of about 545 000 that was distributed unevenly throughout the country, sometimes in remote rural areas. He relied on a small core of experienced officials whom he moved around the regions in a struggle to create effective organisational networks. Recruitment necessarily took place on a mine-by-mine basis, thus giving much scope for mine managers to express their dislike of the NUM, despite the policies of their respective mining houses. This was done in different ways during the early 1990s. One way was the continued employment discrimination against unionised South Africans in favour of non-union Mozambicans. Though the size of the mine labour force declined by almost 30 000 between 1991 and 1994, equalling five per cent of the total, the number of Mozambicans increased significantly in absolute terms and as a percentage of the total labour force. In 1994, they comprised 17,8 per cent of the total labour force compared with 13,4 per cent in 1991.[6] On the other hand, the number of mine recruits from South Africa fell by almost 30 000. A factor that impeded recruitment at the margin was that after the removal of the 'colour bar' black mineworkers moved increasingly into semi-skilled and skilled job categories that were represented exclusively by the white unions. The NUM had tried to stop this seepage by demanding the right to negotiate on behalf of all black workers in the mines beyond wage categories one to eight, but the Chamber of Mines had refused to give way.

However, it was not simply mine managers, on their own initiative, who resisted the expansion of the NUM. The mining houses did it through the application of particular policies over which the mine

[6] *Ibid*, 1992, p19 and 1995, p19.

managers had little or no influence. One such policy was outsourcing or subcontracting. An ILO report described the practice in the following way: "Mines have always contracted out some work. Traditionally, this was confined to construction, shaft-sinking and access development, as well as specialised jobs that could not be handled by the regular work force. However, in the 1980s and 1990s mines have turned to subcontracting out other 'non-core functions' such as catering, grounds maintenance and office cleaning. In the 1990s, mines have increased their use of contractors to handle stoping, vamping and sweeping, haulage construction, wire meshing and lacing. The trend towards more contract labour, and towards using contractors for mining labour, has been linked to the mines' drive for higher productivity, flexibility and cost-cutting. Among the reasons for the trends are that contract workers are typically not unionised and do not receive wages negotiated between unions and mining companies. Contractors have tended to discourage their workers from joining unions. Employees of contractors are generally not covered by death and funeral benefit schemes or retirement saving schemes negotiated by unions with mining companies."[7] In essence, subcontracting was a complex form of casual labour with all the disadvantages for workers that implied. Wherever possible the casual workers were prevented from mixing with regular mineworkers so that there would be no exchange of ideas about trade unionism. A gold mine in Carletonville, for instance, fenced off a hostel that was used exclusively for contract workers.[8] In other mines the casual workers were transported to and from the mines each shift so that there would be no mixing with ordinary mineworkers. As the influence of the NUM increased amongst regular mineworkers, outside agencies were given more responsibility for mining activities. In some instances, whole shafts were 'outsourced' to them. In consequence, the number of subcontracted workers in gold mining rose from three per cent of the labour force in 1987 to 10 per cent in 1994. In coal mining the change was from five per cent to 16 per cent. Underground subcontracted labour in 1994 amounted to nearly 30 000 workers.[9] Members of the NUM began to regard contract workers as a

[7]Guy Standing, John Sender and John Weeks, *Restructuring the Labour Markets: The South African Challenge*, ILO, Geneva, 1996, p302. Also see Bridget Kenny and Andries Bezuidenhout, *Contracting, Complexity and Control: An Overview of the Changing Nature of Subcontracting in the South African Mining Industry*, Sociology of Work Unit (SWOP), University of the Witwatersrand, 1998.
[8]Kenny and Bezuidenhout, 1998.
[9]Standing, Sender and Weeks, 1996, p302.

threat to their jobs as well as to their conditions of work and there was increasing hostility between them. For the NUM organisers, recruiting the subcontracted workers was not an easy option.

The union's ability to develop its organisation and administration had always been constrained by its lack of resources though, in the case of education, health and safety and legal affairs, the constraints had been removed by generous financial donations that had flowed without a break from Sweden, Norway, Finland and the Netherlands since 1983. Once democratic elections were in sight, however, the unions in those countries informed the NUM that their financial support would end in 1996. This decision focused the mind of the NEC on how to be self-sufficient. It asked itself how those donor unions, with total memberships that were smaller than some of the NUM's branches in the gold and platinum sectors, could accumulate surplus resources to the extent that they could assist other unions. It decided to try and copy their financial practices. One result was a decision to establish an investment fund.[10] This decision was made easier by the fact that by 1994 the union was benefiting from the increase in subscriptions from R2 to one per cent of wages each month as a result of a 1991 congress resolution. It was well on the way to financial self-sufficiency. In 1993, it saved and invested an unprecedented 32 per cent of its income. The value of its net assets rose sharply between 1991 and 1994 from just over R4 million to R30 million. Moreover, it opened its own residential training college in the Yeoville suburb of Johannesburg on 4 December 1993, the 11th anniversary of the founding of the union, and appointed a director to run it.[11] In addition, early in 1994, it purchased a new building in the centre of Johannesburg for its head office and had it internally reconstructed to meet its own requirements. The NUM was beginning to display its new opulence.

The various head office departments did not immediately benefit from the union's newly-discovered affluence but they were given security and scope to reformulate their functions. Education was made

[10]The NEC established an investment company called the Mineworkers' Investment Company (MIC) in July 1995.

[11]Cyril Ramaphosa opened the college. It was subsequently named the 'Elijah Barayi Memorial Training Centre' after the deceased vice-president of the union. It had four lecture rooms that were able, in total, to provide space for 180 students. Its kitchen and dining room were able to feed 60 students at a time while 16 could be provided with residential accommodation. There were small hotels in the vicinity of the college where other students could be accommodated. Its primary function at first was providing rented conference facilities to other organisations. Only later did it play an integral part in the union's educational programme.

a union priority when it committed itself first to improving the literacy and numeracy of its members and then to raising their general level of education to the equivalence of the school standard seven to eight. This commitment was seen as a necessary stage in the democratisation process. The mandate of the Project and Development Unit was also altered and expanded. Hitherto, its function had been to provide income opportunities for retrenched mineworkers in the rural areas by establishing production co-operatives. By 1992 it had assisted in the formation of 60 such co-operatives and the creation of 500 jobs. This scale of operation, however, was sufficient in itself because the levels of income it generated were low. It was also miniscule compared with the number and frequency of retrenchments. After a process of evaluation and consultation with other organisations, the NEC agreed to extend the mandate of the unit to include the integration of returned mineworkers into their communities and local structures and to support them in co-operative self-employment strategies without, as formerly, requiring them to organise in formal collective production units. Under the new guidelines, as many individuals were assisted in six months during 1993 as had been serviced in the whole period between then and the formation of the unit in 1988.[12] The unit was able to establish Development Centres in different regions with finance from retrenchment funds administered jointly by the NUM and mining houses. Its extension in size and scope encouraged the NEC to contemplate allowing it to operate independently of the union.[13]

The other main service departments were for legal affairs and health and safety at work issues. The volume and character of the work of the Legal Department were largely determined by the extent to which the union could mount legal challenges to the decisions of the mine owners in relation to their impact on NUM members. It did not take up every individual case because, given the large numbers involved in, say, unfair dismissal cases or wage disputes, that would have entailed prohibitive costs. What it did, therefore, was to pursue what it hoped would be precedent-setting cases through which judgments would be broadly applied. Most of the cases initiated by the Legal Department went to the Industrial Court or private arbitration. The latter course became fairly common practice after the NUM and Anglo-American had concluded an agreement that provided for the automatic referral of

[12]NUM, *Eighth National Congress Report*, 1994, pp35-37.

[13]The unit was registered as the Mineworkers' Development Agency in 1995 and from then operated independently of the NUM.

individual dismissal cases to private arbitration or private conciliation boards. In the period between August 1991 and 1994, 573 cases, involving just over 13 000 mineworkers, were referred to an already overburdened Industrial Court. The union's legal officers handled 272 of the cases while 301 were passed on to private attorneys. Seventy-two per cent of the cases were eventually settled out of court. Some of those that remained within the legal system sometimes lasted for years. A legal dispute involving ERGO, a branch of Anglo-American in the union's Wits region, which arose during the 1987 national strike, was still being fought in 1994. Another case being pursued had started with mass dismissals at the Doornfontein gold mine in Carletonville during 1991.[14] The length of the disputes, as well as the cost of private litigation, made such cases expensive. The payment of legal fees was consistently by far the biggest single item of expenditure in the union's accounts and there were no signs of it declining. Between January and April in 1994, there were 15 legal cases pending that had arisen from what the union described as "mine repression".[15]

Whereas the work of the Legal Department was a response to the infringement of the conditions of employment of black mineworkers, that of the Compensation, Safety and Health Department (COSH) had grown out of the violation of their conditions of work that had resulted in diseases, injuries and deaths. Those violations, as is illustrated in Part Three, had continued unabated during the life of the union. The responsibility for dealing with that situation, in 1994, lay with the union's National Health and Safety Sub-committee that supervised the work of COSH, in the head office as well as the regions. It had six full-time staff members operating from Johannesburg. In addition, Carletonville, Klerksdorp and the Orange Free State each had a full-time safety officer while the PWV had two such officers. Motlanthe complained to the Eighth National Congress that there was a very uneven emphasis on health and safety issues throughout the union, with some regions ignoring them completely. This was illustrated by the regional responses to the annual celebration of Kinross Day on 16 September for each year only five or six of the regions had ever held rallies. Twelve of the regions did not have any full-time health and safety officers. The only branches where health and safety agreements had actually been implemented were in the regions serviced by a full-time health and safety officer.

[14]NUM, *Eighth National Congress Report*, pp28-33.
[15]*NUM News*, December 1993-January 1994.

An important achievement for COSH was the settlement of the struggle for compensation for the victims of the Kinross disaster in 1986. Gengold, the owners of the Kinross gold mine, had agreed in August 1992 to make a payment of R2,6 million. Of this sum, R1,4 million was to be paid into a trust established for the benefit of the families of dependents who had already instituted legal claims. R1 million was to be held for the benefit of dependents who had not instituted claims but who might wish to do so. The remaining R200 000 was to cover the union's legal costs. Nevertheless, the union regarded the settlement as an inadequate means of compensating for the suffering of the families of the victims during the six years of litigation. Another of COSH's achievements was its role in persuading the government to establish, after nine years of agitation, a Commission of Inquiry Into Safety and Health in the Mining Industry. It was hoped that the recommendations of an inquiry would transform the future of health and safety in the mines by giving black mineworkers legal rights, such as the right to refuse to work in dangerous places, and by giving the NUM the legal entitlement to participate with the Chamber of Mines and individual mining houses, in determining the industry's health and safety policies. The advent of HIV/AIDS, however, gave the future a troubled, unpredictable appearance. As a part of the 1993 wages agreement with the Chamber of Mines, NUM members who were HIV-positive were given new rights and protection against victimisation. The union, however, was left with the responsibility of advising its members about the disease, counselling those who were infected, ensuring that they were aware of their rights and with demanding not only the latest treatment but also the infrastructural provisions to make the treatment effective. Even if the new inquiry met all of its demands, the future work of COSH, though different, would be no less vital than in the past.

The co-ordination of the work of the various union departments in the head office and the supervision of their effectiveness were the responsibility of the secretariat.[16] The cohesion that had existed within the secretariat when Ramaphosa was the general secretary had disintegrated on his departure and did not reappear. The elevation of Golding to take Ramaphosa's place and then Golding's defeat in the subsequent election about eight months later had soured relationships in it. After Golding had been included in COSATU's electoral list in September,

[16]The secretariat consisted of the president, general secretary and assistant general secretary who were all full-time officials, plus the vice-president and national treasurer who were part-time.

he began to devote less of his time to union work and more to electioneering and preparing himself for a new job in Parliament. Then, on 15 January 1994, Elijah Barayi died after a short illness, thus leaving Motlatsi as the only survivor of the original secretariat. Barayi, however, had played a diminishing role in the union for some time previously, partly because of ill-health and partly because he was preparing for retirement. Throughout his time as vice-president he had always been employed as a personnel assistant at the Blyvooruitzicht gold mine in Carletonville.

The Eighth National Congress in February 1994 confirmed the election of Motlanthe as general secretary and filled the positions that had been held by Barayi and Golding. Senzeni Zokwana, who had been first the secretary and then the chairperson of the Orange Free State region, was elected to take Barayi's place. Gwede Mantashe, the national industrial organiser, defeated Archie Palane, the regional co-ordinator in Carletonville, to become the new assistant general secretary. The positions of president and national treasurer were not contested so that Motlatsi and Nkuna continued in the posts. The new secretariat, that was to lead the union into the democratic elections, had a distinctly different political flavour compared with any previous one for the general secretary, assistant general secretary and vice-president were all members of the SACP. This tended to generate suspicions of communist intrigues in the union and, therefore, criticisms of those particular union officials. The suspicions and criticisms were reciprocated so that there was an underlying current of tension in the head office that spread to the regions.

The changes in the composition of the secretariat influenced the manner in which the administration in the head office was organised. Ramaphosa had intervened in the details of its day-to-day operations and, through those interventions, had exercised control over it. A relatively stable and committed staff had, moreover, served him. His departure to the ANC created an administrative vacuum that was not filled. There was no-one in the new secretariat with his status, skills and inclination to intervene in administrative matters. At the same time, the composition of the administrative staff was changing. A number of staff members left the union to take up positions in the ANC and the SACP. Some were transferred to other organisations. A report on the state of the head office administration written in 1991 had stated that it had "suffered a big loss and the comrades were never replaced. Those who were left were given more tasks than they could handle."[17] Its members

[17]NUM, Administration and Co-ordination, *Strategic Planning for 1991.*

Gwede Mantashe, the newly elected assistant general secretary, addressing a meeting in Witbank during the 1987 strike.

Senzeni Zokwana, the newly elected vice-president, addressing a strike meeting at Buffelsfontein in Klerksdorp.

complained that they had ceased being treated as a part of the union organisation and were regarded simply as typists, clerks, receptionists and the like. This demeaning of their roles led them to fear accepting responsibility and inhibited them from taking initiatives. The administration was described as being in a state of collapse. Between then and 1994, when there was a head office staff of 11, computers were introduced and the quality of work improved. There was, however, an acute problem of supervision in the department and different attempts were made to resolve it, including the use of outside consultants.

The control of finance was a central feature of the work of the head office. It was undertaken by a staff of five under the supervision of a finance sub-committee, formed by the NEC in 1991. The finance department came under constant criticism from other departments for being slow in making payments mainly through its insistence on receiving corroborative documents, such as receipts, before handing out money. It was often regarded as being a hindrance to their work. It suffered, however, from the absence of any form of central control over expenditure. There was a multiplicity of ways in which money could be spent at regional and national level and frequently the finance department knew nothing about them until it received invoices or claim forms. Then came the inevitable delays as the union's budgetary requirements were examined and attempts were made to rectify imbalances. The advent of relative affluence did not ease the work of this department, as it had done in other cases, but brought a need for greater budgetary controls and enforcement.

Finance, however, was a source of contention for reasons other than the need to balance the books. Its allocation was used as a means to generate accusations of favouritism and, thereby, dissension. Late in 1993, an anonymous 11-page document was circulated around the head office that criticised the secretariat for neglecting to restructure the union, for encouraging favourite cliques and regions in the allocation of resources and for being profligate in expenditure on itself. The document claimed, though without evidence, that it represented the "thoughts and views of almost all staff and those workers who were open about their feelings on the subject".[18] It alleged that the secretariat was not interested in restructuring the union but was satisfied with operating on a crude 'grace and favour' basis by, for instance, rewarding its friends with overseas trips and favouring the regions, mainly large ones, where its friends were located. The document claimed that the

[18]Anon, *Whither NUM?* preface, nd.

secretariat paid no attention to the grievances of the staff in the belief that all it needed to do was to increase salaries at regular intervals.

An intention of the document was to contrast the poverty of the ordinary union members with the luxurious lifestyles of the secretariat by focusing on the benefits they received, such as housing subsidies, housing allowances and the provision of cars that eventually became their own personal possessions. It singled out Motlatsi for special denigration. He was accused of using union money to purchase a new house and of equipping it with furniture from the union's head office. The document, in effect, was an attack on the probity of the union president and resembled earlier anonymous attempts to discredit him, along with Ramaphosa, through stories about their personal behaviour on overseas delegations. Moreover, like the earlier attempts, this one was without any credible supporting evidence. It was a smear and was ignored by the members. Nothing, in consequence, resulted from the document. It was not a propitious time for those who intended to create divisions in the union for whatever reasons. The main concern amongst the members leading up to the elections, as in the case of most black people in general at that time, was to consolidate their unity in the struggle against apartheid.

Preparing for the Democratic Elections

The disaffection amongst a section of the NUM's staff was a symptom of the stress in the union's administration but, in effect, it was a minor distraction from the actual work of the union that was taking place in the branches and regions and, at that particular time, at the Eighth National Congress, where delegates were determining policies that could impact on the whole of South African society. The Klerksdorp region had submitted a resolution that demanded the common ownership of the mines. This had always been the implicit policy of the NUM since it had adopted the *Freedom Charter* but on this occasion the Klerksdorp region wanted it to be explicit and present in the election manifesto of the ANC and its alliance partners. The congress's response was: "During the stage of government of national unity the ANC should appoint its own commission, including representatives of the NUM, to draw up a detailed blueprint for the operation of a commonly owned mining industry." Nelson Mandela, in his address to the congress on its last day, did not refer to common ownership but he endorsed a section in the ANC's RDP policy statement which proposed that the state should own mineral rights and lease them out to the mining companies.[19] Even before his speech, this proposal had evoked widespread criticisms from the corporate sector of the mining industry. The Chamber of Mines had responded with an article by Nic Segal, a member of its executive committee, in which he condemned the policy by explaining the historical complexities of the ownership of mineral rights in South Africa and, in doing so, emphasising the risks to the stability of the industry that would accrue from disturbing them.[20] Mining house officials had spoken of their anger at the proposal and had described the ANC's mining advisers as "romantic idealists".[21] The pressure from the mine owners had its desired effect. There was no reference to either the state ownership of the mines or mineral rights in the ANC's revised RDP documents.

On the first day of the congress, the delegates voiced their concern that all black mineworkers, whether or not they were South African citizens, should be allowed to vote in the election if they so wished. It was an issue the union had raised the previous year. They were opposed to the suggestion that they should be required to produce what they defined as "unnecessary documents", such as birth certificates or police clearance reports, and asked that the matter should be referred to the TEC as a matter of urgency. The TEC, which had been established on 26 November 1993, had held its first meeting in Cape Town on 8 December. Motlatsi moved the motion and it was carried unanimously. However, Mandela abruptly dismissed their demand in his address to the congress a couple of days later. He told the delegates that the laws would not be changed to allow foreign migrant workers to vote. He said that migrants from neighbouring states who did not qualify for permanent residence status or citizenship, including the union president, James Motlatsi who was from Lesotho, could not take part in the election. "You may not like it," he added, "but you have to accept it because they must comply with the existing laws of the country."[22] He added, however, that the identity documents issued by the homelands governments, such as Bophuthatswana and the Ciskei, would be recognised and that the residents of those areas should simply register to vote in the same manner as South African citizens were required to do. Many mineworkers were covered by this latter provision. However, during December 1993, a few weeks before Mandela spoke, the task of

[19] *The Star*, 10 February 1994.

[20] *Business Day*, 3 February 1994. Only in South Africa and the USA, amongst the major mining countries, were mineral rights not owned by the state.

[21] *Sunday Times*, 30 January 1994.

[22] *Business Day*, 10 February 1994.

answering questions concerning voting rights and of overseeing the conduct of the elections was given to an Independent Electoral Commission comprising prominent black and white South Africans and including five foreign observers. It was possible, therefore, that Mandela's word was not the last to be spoken on the question of voting rights for foreign black mineworkers.

For the rest of their time during the congress, the delegates explained, often forcefully, the grievances they considered should be redressed during the tenure of the government of national unity that it had been agreed was to be established following the democratic elections for the following five years. They stated, for example, that "the ANC should press the government of national unity (i) to adopt measures that will end the migrant labour system; (ii) make urgent provisions for the education and training of blacks at all levels of skills and competence; and (iii) provide a framework for the full participation of the union in the running of the mining industry during the interim period". The congress made a number of proposals about the spread and quality of educational opportunities and the need for industrial development and reconstruction. It wanted the government to establish a Reconstruction Fund "to lend money to government departments such as education and health, to democratic local authorities for infrastructural development and housing; to new rural development agencies; and to similar bodies, including non-governmental organisations, committed to development". In order to ensure that its aims were met it declared that the "union should jealously guard the tripartite alliance which will ensure that the ANC-led government retains a working class base". It wanted to maintain the integrity of the alliance in order "to make progress towards socialism".[23] At the conclusion of the congress, Motlatsi pledged financial support from the NUM for the ANC's election campaign even though that was illegal under the existing labour legislation. In the time that remained between the congress and the planned election, the union aimed to devote its time and resources to campaigning for the ANC.

By the time the NUM delegates had returned home the future political shape of the country was already enshrined in law. Sufficient unity had been achieved on the Negotiating Council for a Bill, after four drafts, providing for a transitional constitution to be ratified by Parliament. The result was a compromise between the ANC's desire for a unitary

[23]All references to resolutions in this and the previous two paragraphs are taken from NUM, *Resolutions Adopted at the Eighth National Congress of the National Union of Mineworkers Held From 7-9 February 1994.*

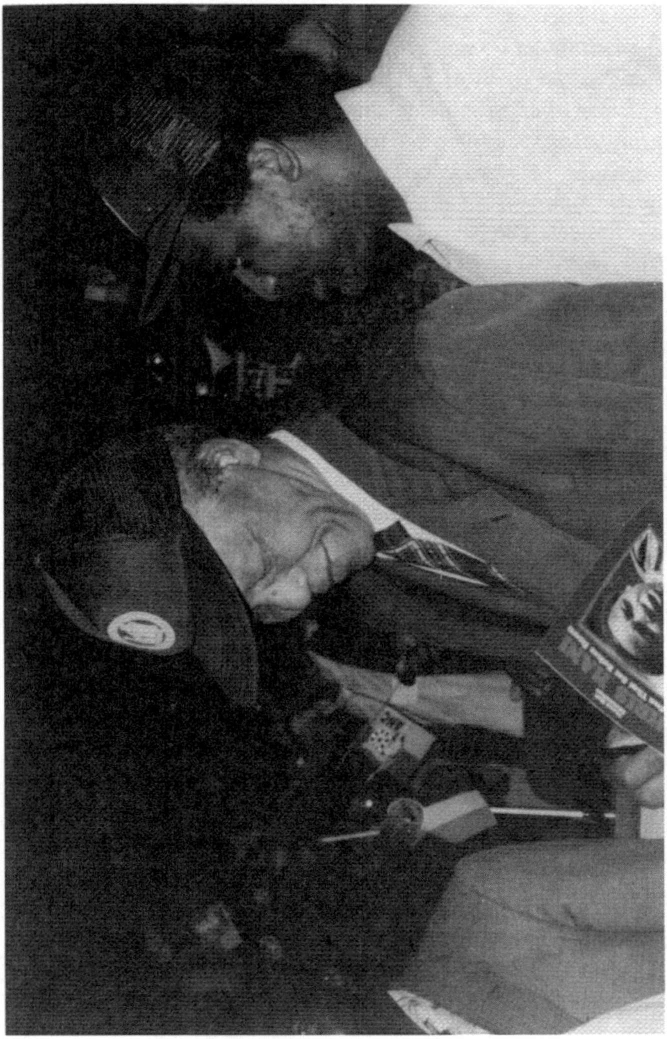

Nelson Mandela and James Motlatsi at the 1994 NUM National Congress.

state and the NP's insistence on a form of federalism and protection for the interests of the white population. The character of Parliament, with a National Assembly and Senate, had been defined. Nine provinces, incorporating the former 'homelands', had been delineated and 11 languages had been acknowledged as the official languages. The provincial governments were to have extensive powers but within a system where, in the event of a conflict of interest, national legislation took precedence over provincial acts. There was to be compulsory power-sharing during the interim period. For this purpose, a multi-party cabinet, consisting of representatives of all parties that had more than 20 seats in the National Assembly of 400 members, was to form the interim government. The National Assembly was to elect the president and any party with 80 or more seats could select an executive deputy president. These constitutional provisions were accompanied by a list of unalterable fundamental rights for protection against unfair discrimination on the grounds of "race, gender, sex, ethnic or social origin, colour, sexual orientation, age, disability, religion, conscience, belief, culture or language". Whereas there was provision to amend the constitution provided two-thirds of the National Assembly and the Senate in a joint sitting agreed, there was no formal provision to amend any of the fundamental rights. The task of drawing up a permanent constitution was given to a Constitutional Assembly made up of a joint sitting of the National Assembly and the Senate. A Bill had been passed in December 1993 that restored South African citizenship to the residents of the 'independent homelands' as from 1 January 1994 and provided for the reincorporation of the territories into South Africa on the day of the election. Steps were already being taken to merge the South African Defence Force with Umkhonto we Sizwe.

The main opposition to the battery of legislative provisions recommended by the Negotiating Council came from the Freedom Alliance, formed in October 1993 by the joint efforts of the Afrikaner Volksfront, the IFP and the governments of Bophuthatswana, the Ciskei and KwaZulu-Natal. It demanded exclusive powers for the provinces, a clear delineation and protection for their borders and the right not only to levy their own taxes but also to write their own constitutions and compile their own respective bills of rights. In short, it wanted an unadulterated federal system. This was fairly clear-cut for the IFP because its demands related to an existing territory but that was not the case with the Afrikaner Volksfront for it insisted on the creation of a *volkstaat* that had no clearly delineated geographical identity. Each of them held bilateral talks with the government and the ANC in the hope that the interim constitution could be amended to make self-determination a possibility. Mandela, however, told the NUM congress in February that the demand for a *volkstaat* could never be granted. He said he was prepared to meet with the Afrikaner Volksfront but only to allay their fears. There was little, therefore, that the Freedom Alliance could hope to achieve from negotiations without compromising its own demands. Its remaining option was to prevent the election from going ahead and that is what it aimed for. In order to do that, it could either try to discredit the election by refusing to take part in it or it could actually destroy the election by creating so much havoc through violence that it would be unsafe for people to cast their votes. In the remaining weeks up to the election both courses were tried.

With the approach of the deadline on 12 February 1994 for political parties to register for the election, both the NP and the ANC put increased pressure on the Freedom Alliance to register, but with no effect.[24] The election, of course, could legally have gone ahead without it but, in order to gain legitimacy, the ANC and the NP felt that the election had to be inclusive. They wanted Inkatha, in particular, to participate because of its strong regional base. The main consequence of the campaign of persuasion, however, was a hardening of attitudes. There was much talk about the need to resort to violence. Terreblanche, the leader of the AWB, told a meeting of Afrikaners in January that they would "seize freedom through violent means if necessary". At the same meeting, the leader of the Conservative Party, Ferdi Hartzenberg, declared that after the failure of negotiations the "Afrikaner nation would now use other methods to gain freedom within its own state".[25] The chairman of the Freedom Alliance, General Constand Viljoen, added his voice by saying that if Afrikaners wanted a *volkstaat* before the election they would have to seize it violently. A spokesperson for the Freedom Alliance thought they were sliding into a civil war.[26] There had been a demand in the apartheid Parliament in December 1993 for an inquiry into an "alleged right-wing war plan that included the provision of mass graves for blacks and the mobilisation of right-wing forces loyal to the AWB".[27]

There was much activity behind the rhetoric. The AWB claimed that it had access to 60 000 troops and was prepared to deploy them to Bophuthatswana if required. The IFP had its 'self-protection units'

[24]Nineteen political parties did register in time.
[25]South African Institute of Race Relations, *Race Relations Survey*, 1994/95, Political Developments, Key Points.
[26]*The Star*, 9 February 1994.
[27]South African Institute of Race Relations, 1994/95, Security, Key Points.

and, at one point, had 1 200 people, including 200 hostel-dwellers, being trained in the use of firearms in the PWV. Third force groups in collusion with IFP forces continued to engage in random killings. Indeed, in March, the government directed the Commission of Inquiry Regarding the Prevention of Public Violence and Intimidation (the Goldstone Commission) to investigate allegations that high-ranking officers in the SAP were involved with Inkatha officials in gun-running and 'third force' activities. An organisation called the Boer Republican Army was implicated in bomb blasts on railway lines, power lines and petrol stations in mining areas in the western Transvaal and the Orange Free State. The flood of AK-47 machine guns into South Africa from Mozambique made the situation more dangerous. The number of casualties reached unprecedented levels. In March, 552 people were killed through political violence throughout South Africa, of whom 82 per cent died in KwaZulu-Natal and the PWV region.

Various attempts were made to stem the violence. Following on from the signing of the National Peace Accord in 1991, a National Peace Secretariat had been formed in October 1993 which was responsible for establishing a special peace radio station called 'Peace 2000 FM' and for forming 11 regional and 200 local peace committees. These committees, however, were not effective in the areas of high tension where they were most needed. The most ambitious gesture was the creation by the TEC in January 1994 of a National Peacekeeping Force to "assist in the maintenance of peace and public order in the period leading up to and during the general election". It was comprised of equal numbers of the military forces and policing units under the control of the TEC.[28] Almost immediately, however, there were reports of ill-discipline, drunkenness and disobedience among its members.[29] Nonetheless, one of its units was deployed in the townships of Katlehong, Tokoza and Vosloorus but, after a week, its members were confined to barracks because they were unable to maintain order and the SADF was returned to patrol the streets.[30]

In February, the TEC introduced a scheme to counter the violence in the east Rand that included replacing the Inkatha self-protection units there with members of the SADF. Though Buthelezi had been consulted beforehand, the IFP rejected the scheme and the self-protection units stayed in place. The ANC had its own self-defence

[28]These were the SADF, the SAP, MK and the defence forces of the Transkei and Venda.
[29]*The Star*, 9 February 1994.
[30]The National Peacekeeping Force was disbanded by the government in June 1994.

units patrolling the townships. In effect, the units from both organisations were a part of the problem rather than its solution because wherever one of them was present it provoked violent retaliatory measures from the other. The ANC had already recognised that this was the case early in January and had tried to disband its own units, but without success, for some of them had begun to operate autonomously and were out of control.[31] Nothing seemed to go according to plan for either the TEC or the government. The government issued call-up papers to SADF reservists and ex-soldiers but 57 per cent of them did not respond. A last desperate measure was the government's declaration of a state of emergency in KwaZulu-Natal on 31 March when, despite the opposition of the IFP, the government deployed a large number of troops there. It was a matter of conjecture as to whether any of these measures to reduce the incidence of violence had any effect for it is not known what it would have been if they had not been tried. The TEC was moved by this uncertainty to increase the number of police allocated for election duties from 60 000 to 100 000.

The political tensions that arose from the rejection of the transitional constitution by the Freedom Alliance spilled over into the mine compounds wherever Zulus worked and, in consequence, UWUSA was present. UWUSA had always used threats and intimidation to extend its influence in the compounds but it was not always responsible for initiating violent clashes. Wherever it operated it created an unstable situation because by its very presence it challenged the hegemony of the NUM. The members of the NUM had always been intolerant of the presence of a hostile union in the compounds but the situation of virtual open warfare outside between the ANC and the IFP had extended their intolerance to its limits. Thus UWUSA's use of violence as a tactic combined with the NUM's complete intolerance of its presence in the early months of 1994 was a ready-made recipe for violent compound clashes. The Commission of Inquiry into fighting between UWUSA and NUM members at the New Denmark colliery in Secunda early in the year described this shared responsibility. In its report it stated that part of the blame lay with the hostility of NUM members towards Zulus. Nearer the election, there was a similar situation when NUM members prevented Zulus belonging to UWUSA from entering the Cinderella hostel at the ERPM gold mine near Boksburg. The Zulus retaliated and, in the fracas, a number of mineworkers were killed. The mine manager's solution was to dismiss the Zulu-speakers but that did

[31] *Business Day*, 7 January 1994.

not end the dispute because they applied to be reinstated through the Industrial Court, thus converting a political dispute into a labour one.[32] Resorting to a legal remedy was an unusual tactic for UWUSA but it did not signify its conversion to legitimate methods. It continued with its policy of intimidation and provocation.

A further consequence of the political overspill was the proliferation of armed right-wing groups, particularly in the Orange Free State and the west Rand, to which white miners belonged. The union's Klerksdorp office was virtually destroyed by a bomb at the beginning of February 1994. Motlatsi blamed the white Mine Workers' Union though he was not sure that commercial explosives, to which the white miners had access, had been used in that case.[33] The union, he said, would debate its reaction to the bombing and its plans to guard its property and members at its forthcoming Eighth National Congress when the question of security would be added to the agenda. "The NUM," he said, "is not going to be a sitting duck for the right-wing miners and their cohorts." He explained his feelings more fully at the congress. "The situation," he said, "is made more unstable and unpredictable by the fact that the police, whose loyalty is questionable, have said that they may not have the forces to prevent violence or even to enforce the ban on carrying weapons at political meetings. I do not think that armed whites can win a war or even a battle but they can create a lot of damage before they are defeated. We are facing a dangerous situation and we must take immediate action to deal with it ... I propose, therefore, that the congress should debate an emergency resolution that will enable us to protect the property of our union and the lives of our members and officials ... we are not without resources in the matter of security. Our task is to mobilise them ..."[34] The delegates, however, did not really need that advice for they had always been willing to defend themselves against violence from whatever quarter it came.

The Freedom Alliance began to fragment during March and that seriously altered the political scene. The first sign was internal dissension in Bophuthatswana, the government of which was a key member of the alliance. This had consequences for the NUM because it had been struggling against the opposition of the government to gain a foothold

[32]The New Denmark and ERPM cases are described in *The Report of the Commission of Enquiry Into the Recent Violence and Occurrences at the East Driefontein, Leeudoorn and Northam Mines*, the Myburgh Commission, October 1996.
[33]*Business Day*, 3 February 1994.
[34]NUM, Eighth National Congress, February 1994.

in the platinum mines there. The union's fragile position had been threatened by the passing of a new Industrial Relations Act in November 1993 that outlawed South African trade unions and provided for criminal sanctions against employers who recognised them. The immediate effect was to make the managers at Impala Platinum more bellicose in their attitude towards the local NUM representatives. The NUM, at that time, had about 29 000 members at Impala Platinum. When it was rumoured that there were to be 10 000 retrenchments during 1994, the local officials organised a demonstration during which, the management claimed, mine property was damaged. The mine security force had retaliated with rubber bullets and tear gas and the two leading shaft stewards were dismissed. The union interpreted the dismissals as victimisation and called for a strike in all of the shafts on 5 January 1994. On that day, however, the mineworkers failed to respond to the strike call and the local officials, in a seriously weakened position, were compelled to take their grievances directly to the mine manager.[35] The mineworkers' solidarity had been fractured by the fear of mass dismissals by the manager as well as the consequences of defying Bophuthatswana's anti-trade union law. The union, therefore, had a direct interest in the events that began to unfold in the Bophuthatswana 'homeland' during March 1994.

Lucas Mangope, the president of Bophuthatswana, had copied and utilised the repressive methods of the apartheid state that had created him and his government. Throughout his rule he had denied the Tswana people any democratic rights and had given them no legitimate say in whether or not Bophuthatswana should be reintegrated into South Africa. He was intent, it appeared, on perpetuating his personal rule and for that he needed to retain Bophuthatswana's nominal independence. This had led him to join the Freedom Alliance in opposition to the democratic elections. When, on 9 March, Judge Kriegler, the head of the Independent Electoral Commission, asked him three questions, namely whether he would accord political parties the right to campaign, agree to undertake voter education and allow polling booths to be set up during the election, he answered "no" to each one. He also banned the ANC from campaigning. However, he was fearful of the immediate future and, according to General Viljoen, had asked for military assistance from the Afrikaner Volksfront but with the proviso that AWB troops should be excluded. Viljoen responded positively, he said, because he wanted to rescue Mangope from an alleged ANC/SACP plot to destroy

[35]*Business Day*, 6 January 1994.

Bophuthatswana and its people.[36]

It was becoming clear, however, that on the question of reintegration into South Africa, Mangope did not reflect the views of the majority of the people of Bophuthatswana. On the same day as he rejected Kriegler's requests, virtually the whole of his civil service went on strike in favour of reintegration. All but three of the country's hospitals had been on strike for three weeks over the same issue. Students were demonstrating in support of the strikers in the streets of the capital, Mmabatho. By the next day, Thursday, 10 March, the situation had degenerated into anarchy as people rampaged through the urban areas, smashed shops and looted their contents. The demonstrators of all kinds were shot at and tear gassed by the police and Mangope's security force who, by that time, were backed by a motley army of white racists who had congregated at Mmabatho's airport in response to Viljoen's summons to them.

It was estimated that about 5 000 AWB members had arrived in Bophuthatswana to join the Afrikaner Volksfront troops, despite Mangope's request to Viljoen to keep them out. They had responded to broadcasts by Terreblanche on Radio Pretoria, the AWB's clandestine radio station, and had travelled to Mmabatho's airport in their farm trucks, armed with hunting rifles, shotguns and pistols.[37] A number of them, however, wearing their Nazi-like insignias, broke away from the main contingent at the airport and drove through the streets of Mmabatho in a convoy of cars, waving their weapons in the air, "yelling racial abuse at the locals and taking pot-shots at groups of people, killing and wounding several".[38] This behaviour, described by the Afrikaans newspaper, *Beeld*, the following day, as a "*kaffirskiet-piekniek*" (nigger-shooting picnic), so outraged the local police and defence force that they mutinied and changed sides.[39] The police marched to the South African embassy to demand reintegration and demonstrated alongside the people they had been shooting at the day before. With the police and the army opposed to them and the local youths stoning them, the AWB contingent was forced to leave in a humiliating manner. Some of them put on an act of bravado and fired back at the youths from their convoy but this provoked the police and security force to return their fire. One car, a Mercedes with three occupants, was hit. One of its occupants was killed on the spot. The

[36]Phillip van Niekerk, *The Observer*, 13 March 1994.

[37]See Sparks, 1997, chapter 14 for a detailed description of these events.

[38]*Ibid*, p211.

[39]*Ibid*.

other two, lying injured, were killed deliberately at point-blank range by a black police officer in a much-publicised incident.[40] This marked the conclusion of an episode during which at least 60 people were killed and 300 were injured. The white right-wing forces returned to their homes humiliated and demoralised. That was not the end of white violence but it was the beginning of the end. It was also the termination of Mangope's rule and of the nominal independence of Bophuthatswana, for, thereafter, it became subject to the authority of the TEC and the laws of South Africa. The black mining labour force at Impala Platinum suddenly had the same legal rights of association as black mineworkers in South Africa.[41]

The fragmentation of the opposition to the elections was completed over the next two weeks. Firstly, while the events in Bophuthatswana were actually unfolding, General Viljoen relinquished his leadership of the Afrikaner Volksfront and broke away to form the Freedom Front party. He took with him a number of members of the Afrikaner Volksfront, moderate Conservative Party Members of Parliament and influential figures from the Transvaal and Orange Free State farming communities. The new party was immediately registered for the election, just in time to meet the revised deadline. Then, 10 days later, a strike of civil servants and threats of mutiny in the police and security force in the Ciskei prompted the military ruler, Brigadier Oupa Gqozo, to request the TEC to take over the administration of the territory. He resigned at the end of the day and plans began to be laid to hold the elections there.

Thus, from 23 March, the Conservative Party and the IFP were the only political parties that had still not registered for the election. With the formation of the Freedom Front, however, the Conservative Party was no longer a unified political force and it ceased to be of any significance to either the ANC or the NP. The IFP was, as always, a different matter. It was strongly entrenched in KwaZulu-Natal but it had to compete with two other equally entrenched institutions for the political allegiance of its people. The province had a monarchy, supported by Zulu traditionalists, with a monarch who was not always in agreement with Buthelezi. In the most troublesome area, it also had an ANC organisation that had been built and led by Harry Gwala, a trade unionist and highly popular stalwart of the SACP whose involvement in the ANC stretched back to the 1950s. Gwala, at the time, was the

[40] See *ibid* for a first-hand account of the incident and the events leading up to it.

[41] Bophuthatswana was included in the North West Province for the election. The ANC received 83,3 per cent of all the votes cast there.

chairperson of the Natal Midlands region of the ANC. He adopted a strategy, when necessary, of confronting violence with force and of even using force as a pre-emptive weapon. Through this strategy, the ANC provided many villages, where its supporters and their families lived, with military-type protection. Without Gwala's determination, tenacity and ruthlessness it is possible that the ANC would not have survived as a provincial organisation during those turbulent times.[42] As it was, there was sufficient support for the ANC to warrant it challenging the IFP for control of the province. Some even thought it could win if the IFP's tactics of intimidation could be controlled. The violence, then, that reached its highest-ever intensity towards the end of March 1994, was primarily between Zulus themselves.

Though the IFP was only a partial reflection of Zulu aspirations, it was, nonetheless, a sufficiently vital component in the process leading to the election to convince the ANC-led alliance and the government to join forces in a final attempt to persuade Buthelezi to take part in it. They made a number of concessions to him. The TEC again extended the deadline for registration. De Klerk recalled the old Parliament in order to sanction amendments to the interim constitution that would strengthen the authority of the provincial governments, as an inducement to him. On the other hand, Mandela made overtures to the Zulu King Goodwill Zwelithini in an endeavour to cause divisions between him and Buthelezi and thus weaken Buthelezi's position. Mandela, however, failed to convert Zwelithini in time to influence Buthelezi's decision. Mandela and De Klerk then agreed to Buthelezi's request for international mediation over his demand for independence but the mediators left within 48 hours of their arrival after failing to agree on their terms of reference.[43]

The point was reached where the ANC could see no alternative but to hold the election without Inkatha. The question, then, was how free elections could be held in hostile territory. This, according to Ramaphosa, made it into a military issue. "We had," he said, "to look at how to rein in the IFP, particularly the warlords and the hit squads, and we started identifying the camps where they were training people."[44]

[42]Gwala's membership of the Communist Party was suspended in July 1994 for his alleged involvement in 'hit squads' in the province.

[43]The mediators were Henry Kissinger, the former US Secretary of State, and Lord Carrington, the former British Foreign Secretary. Buthelezi wanted to include the date of the election on their agenda but the ANC would, in no circumstances, allow this to be changed.

[44]Sparks, 1997, p223.

Subsequently, a task force appointed by the TEC managed to disrupt the IFP's training operations by causing it to abandon its main camp at Mlaba where approximately 5 000 Zulus were being trained to become 'soldiers for the Inkatha Freedom Party'. Without adequate military backing, Buthelezi was left with only one option, which his friend, the Kenyan professor Washington Okumu, advised he should take. That was to take the IFP into the election and fight for change from within the system. He accepted that advice on 19 April, eight days before the election was due to be held. He was most probably encouraged to take this step by a secret agreement he had reached with De Klerk, with the assent of Mandela, to transfer three million acres to the Zulu kingdom to prevent them from falling under the jurisdiction of a new national government.[45] An immediate consequence of Buthelezi's decision was a significant fall in the level of violence.

A special session of Parliament was held on 25 April, the day before the voting was due to start, to pass legislation enabling a late registration to be made. The IFP's name was not on the voting paper so stickers were printed containing its name and attached to the bottom of 80 million ballot papers. Apart from the fact that this task caused delays in the delivery of voting papers, the election went ahead with relatively few upsets, though not without fears of white violence.[46] Archbishop Desmond Tutu, commenting on his feelings when he queued to cast his vote at a black township near Cape Town, said that his fear that white groups would disrupt the election process persisted to the election itself. There were few police around, he added, so that a couple of white men with AK-47s could have created havoc at his polling station.[47]

The implications of this first truly democratic election affected everyone in South Africa. There were not, nor could there have been,

[45]Leonard Thompson, *A History of South Africa*, 1995, p253.

[46]The result of the ballot in KwaZulu-Natal was that the ANC received only 32,2 per cent of the votes compared with the IFP which received 50,3 per cent. There was much controversy about the ballot in KwaZulu-Natal due to the presence there of illegal voting stations in predominantly Inkatha-controlled areas. Hundreds of ballot boxes, some of which were stuffed with fraudulent ballot papers, turned up at the count and caused the Independent Electoral Commission to delay publication of the result (*The Guardian*, 4 May 1994). Ramaphosa listed five points in his explanation of the ANC's failure to win the province. The result, he wrote, was due to: 1: the high incidence of voting irregularities that favoured the IFP; 2: the fact that the undecided vote contained more IFP support than the opinion polls had suggested; 3: the belief amongst voters that the violence would escalate if the IFP did not do well; 4: the restrictions on the ANC's capacity to campaign caused by the violence; and 5: defects in the ANC's campaign (*Secretary-General's Report*, African National Congress, December 1994).

[47]Desmond Tutu, *Taking a Stand*, Radio 4, BBC, 19 February 2002.

any disinterested bystanders. In effect, what had happened up to 27 April was everyone's history in South Africa. In that sense it was an integral part of the history of black mineworkers. Some people, of course, were involved in the process more than others and, consequently, may have suffered more. Many had interests in the outcome that were additional to its generalised effects for individual South Africans. Members of the NUM were amongst that group. They were particularly concerned about the future of Bophuthatswana with its platinum mines, the Ciskei, from which many of them came and had their homes, and KwaZulu-Natal, where there were coal mines and a union regional office - but few members, and the west Rand and the Orange Free State where white miners, belonging to right-wing groups, were fighting for the continuation of white supremacy. In addition, they had a special interest in the election process itself for a number of them did not possess South African identity documents and were not, therefore, entitled to vote under the law. This omission, however, was rectified in time for the election. In spite of Mandela's uncompromising statement at the NUM's National Congress in February, the Independent Electoral Commission ruled that all black mineworkers who had been employed in the mines since before 1986 would be allowed to vote whether or not they possessed South African identity documents. James Motlatsi, and a number of Lesotho nationals who had assisted in establishing the union, benefited from this new dispensation.

Black mineworkers who went to the polls were conscious that the election, though crucial, was the first of a number of staging posts on the road to emancipation. They had itemised their expectations along that road and had outlined the necessary conditions for their fulfilment. They still believed in the goals of the *Freedom Charter* and were strong supporters of the RDP. They knew there would be difficult struggles ahead but expected their own, new democratic government to ease a way through them in some instances and to resolve them in others. However, their perception of the source of their oppression and exploitation had always been blurred and confused by their treatment under the apartheid system and, in consequence, they had no clear notion of the workings of the capitalist system. Understandably, they believed that without apartheid their lives in the mines would be transformed for the better. They were yet to learn the reality of the class struggle. It was, therefore, with some optimism that James Motlatsi, the president of NUM from its beginning, and Cyril Ramaphosa, its founding general secretary, joined to cast their votes at the polling station in the Kloof gold mine in Westonaria in an act of symbolism to remind themselves and members of their union that it had been at the same Kloof mine, in July 1982, that black mineworkers had first demanded a union that became the National Union of Mineworkers.[48]

[48]The results of the election in terms of the percentage of the votes cast were as follows: African National Congress – 62,7 per cent; National Party – 20,4 per cent; Inkatha Freedom Party – 10,5 per cent; Freedom Front – 2,2 per cent; Democratic Party – 1,7 per cent and the Pan-Africanist Congress – 1,2 per cent. Ramaphosa, who was at the top of the ANC's national electoral list, became a Member of Parliament in addition to his job as the Secretary General of the ANC. The two other NUM members on that list, namely Marcel Golding and Godfrey Oliphant, were also elected.

Part Three

TAKING STOCK OF UNRESOLVED LEGACIES

CHAPTER NINETEEN

THE CASUALTIES OF EXPLOITATION 1994

The NUM, by 1994, had a relatively high profile in its relations with the mining houses and the government. Though it was not formally involved in decision-making where health and safety issues were concerned, it was able to make its views about them known to the mining houses and was positioned in some mines to intervene with managements. Motlatsi, speaking on behalf of the union, was being provided with an increasing number of public platforms from which to elaborate the union's safety policies and to criticise those of the mining houses. The NUM, moreover, was able to exert political pressure on the government through its status as an influential constituent within the Tripartite Alliance that was backing the ANC in its power-sharing arrangement with the National Party. This arrangement, expressed through the Transitional Executive Council, took the ANC close to the source of government power. Within a time-span of four years, the potential for political action in support of the NUM had been transformed. It was no longer excluded from direct consultation with the government and did not even have to wait in line behind the Chamber of Mines when lobbying it.

Yet despite such politically propitious circumstances and its stature as the representative of more than half of all black mineworkers, the NUM's influence was not evident in those areas it had consistently deemed to be vital for the well-being of its members, namely the operation of the migrant labour system as a transmission belt for contagious diseases and the organisation of underground work that led to brutally high casualty rates. Modifications had occurred in both areas due to the transforming qualities of internal contradictions under the impact of the liberation struggle on the apartheid state, changes in the international demand for minerals and the advent of new technology, but none of them had lessened the risks to the well-being of the black mineworkers. The migrant labour system had altered its substance without losing its lethal role as a transmitter of diseases while the labour process continued to operate as if black labour power was an innate, disposable factor of production that could readily be replaced. Taking stock, therefore, of what had and had not been achieved since the early years of the union raised important questions for the NUM.

The Migrant Labour Issue

By 1994, much of the juridical basis of apartheid had been dismantled while, independently, the substance, as against the formal structure, of the migrant labour system had been altered. Once the turnover of labour had been reduced by the mine owners in the late 1970s in order to alleviate their short-term labour supply problems, longer-run consequences became apparent, almost as if the mine owners, and later on the NUM, were bystanders. Black mine labour was stabilised as mineworkers consistently renewed their contracts and treated the end of a contract not as the termination of employment but as the beginning of a finite holiday interlude to be renewed each year.

This process had been consolidated by an increase in the skill content of mine labour that intensified the emerging self-perception of mineworkers as 'career' workers rather than 'casual migrants'. Enhanced skills were required in order to operate the machines that were introduced to cope with the problems of mining at increasing depths and to reduce costs as lower-grade ores were encountered. There were pressures on managements to improve shaft-sinking and development techniques, counter the heat problem, provide stronger roof supports, speed up the underground transport system and increase the speed of rock-cutting. Trackless vehicles were introduced which provided greater versatility in the gullies and stopes. The stopes were widened for manoeuvrability and the number of access tunnels to them was reduced. Pneumatic drills were replaced by hand-held high pressure hydraulic ones that used water or a mix of oil and water. In the larger stopes, 'jumbo' drills, with moveable arms that could drill more than one hole at a time and which ran on tyres, were introduced. Blasted rock was moved from the face by high-pressure water jets, continuous scrapers or trackless load-haul-dump trucks, depending on the gradient and the width of the stopes and gullies. Where these methods were introduced, production depended less on the number and physical efforts of mineworkers than on their skill in handling machines. Fewer mineworkers were required but more attention was paid to their qualifications. For instance, at Vaal Reefs only mineworkers who had reached at least standard eight at school, and who had passed two aptitude tests, were considered suitable for training as drivers of trackless vehicles.

The degree of labour stability that this situation created was indicated by the fact that, by 1990, as many as 90 per cent of black mineworkers from South Africa and 70 per cent of those from foreign neighbouring countries were employed by all of the mining houses on standard annual renewable contracts with fixed leave provisions.[1] The number of novices had declined from about 35 per cent of the mine labour force in the mid-1970s to less than 10 per cent in 1990.[2] In 1976, only 14 per cent of those who worked in the gold mines had possessed more than 10 years of mining experience, whereas by 1990 the comparable figure was 37 per cent. During roughly the same period, their average age had risen from 26 years to 37 years.[3] All the indices pointed to the creation of a permanent black mine labour force that was changed only by the loss of labour through injuries, deaths and retirements.

All the elements in the work situation encouraged black mineworkers to stay in their jobs for as long as possible. With a tightening labour market caused by a much-reduced labour turnover, dissatisfied mineworkers were no longer able to leave their jobs with a reasonable prospect of getting more congenial ones in other mines. Nor, if they were dismissed, could they readily get comparable jobs in other industries because their skills were not transferable. In any case, in the country as a whole there was an economic depression with an unemployment rate amongst black workers averaging about 41 per cent in 1994.[4] The loss of a mining job could, therefore, mean the loss of any form of wage employment and, thereby, lead to economic disaster for a mineworker and his family in the rural areas where traditional subsistence farming had been virtually destroyed and where there was almost no alternative wage work. A survey of former mineworkers in Lesotho in the early 1990s found that none could survive through farming alone. Half of the sample had no cash income at all and only one was able to utilise the skill he had acquired through mining. The plight of those who were forced to subsist from the land was illustrated by the fact that whereas agriculture occupied 70 per cent of the work force in Lesotho in 1994, it contributed only eight per cent of its gross national product.[5]

In such circumstances, a secure job in mining was a highly attractive proposition, despite its relatively low pay and dangerous working conditions, as was illustrated by the desperation shown by first-time young job-seekers in the rural areas. Many thousands of young black men converged on the TEBA recruitment offices in search of work. There were tales of queues waiting at the offices, camping near them

[1]Wilmot G James, 1992, pp28-29.
[2]Crush, Jeeves and Yudelman, 1991, p119.
[3]Steinberg with Seidman, September 1995, p7.
[4]Standing, Sender and Weeks, 1996, p116.
[5]Steinberg and Seidman, 1995, pp36-37.

for days on end and using any means that might secure them a job. Stories circulated about corrupt dealings as TEBA officials exploited the situation by selling employment certificates to the highest bidders. Sometimes, those without money attacked the officials in desperation, with the result that security fences were constructed around the offices so that they began to resemble some of the mine compounds to which the applicants for jobs hoped to be sent.

All of those changes had taken place with very few alterations, however, in the formal structure of the migrant labour system. Mineworkers continued to arrive at the mines without their families and to live in single-sex hostels, on contracts lasting from 12-18 months. The proportion of the total black labour force employed by the Chamber of Mines who were migrants had fallen only marginally from 88 per cent in 1988 to 83 per cent in 1990 but then it stabilised up to 1994 at the 1990 figure.[6] Only the senior black personnel, who were formally recognised as permanent monthly-paid workers and who constituted most of those not classified as migrants, were given married quarters. For the large number of mineworkers who still lived in the hostels many things remained unchanged. They continued to have the familiar grievances about overcrowded sleeping accommodation, spoiled food, inadequate leisure facilities, too few liquor outlets, insufficient communications with their families and a lack of accommodation for visiting wives. But with no foreseeable possibility of achieving the union's policy for the abolition of the migrant labour system or the replacement of the hostels by community housing provisions for married and single men, the members of the NUM began to adapt to their situation.

Those who had left their homes with every intention of sustaining their families and accumulating cash in order to build up their rural homesteads were drawn into a life that was much more diverse and stimulating than they had ever experienced at home. The freedom they had to move in and out of the compounds enabled them to visit and establish relationships in the black townships. Some entered into semi-permanent relationships with township women and fathered children. Their interest in the economies of their villages, therefore, that had been instrumental in driving them to the mines in the first place, tended to decline soon after they had written the first letter or sent a remittance home. In some instances they had gone to the mines immediately following their marriages and left behind wives who had

[6]*Ibid*, p29.

to settle for a meagre existence for most of their remaining procreative lives, perhaps helped by taking on a lover or two who gave them money or presents. An increasing number of rural marital relationships, in consequence, were wrecked. In desperation, some discarded wives travelled to the mines and stood at the waiting places to search for their husbands to remind them of their family responsibilities. Those wives who decided to stay near the mines usually moved into the rapidly growing shantytowns where they lived alongside mineworkers who had established themselves in shacks with township women. Thus in one way or another the mineworkers had begun to move the focus of their lives away from their rural homes. In that way, they hastened the deterioration of the rural areas. The result was that their dissatisfaction with hostel life no longer always added up to demands for its complete abolition. They appeared to be ambivalent about the system they had hitherto condemned as evil and destructive for it gave them new freedoms and a relatively attractive alternative lifestyle.

The NUM, of course, had to face up to realities. Despite recovering from the effects of the 1987 strike and overcoming management repression, increasing its size and enhancing its status through engaging in industry-wide discussions about the future of mining and, perhaps most important of all, being a power-broker behind the ANC and the SACP in the movement towards a negotiated settlement with the government over the future shape of South Africa, the NUM had no prospect of enforcing its long-time policy for the abolition of the migrant labour and hostel system by establishing community housing and allowing individual mineworkers to decide whether or not they wanted to migrate or live with their families near their place of work. Such a policy implied that there had to be a sufficient supply of married quarters and properly constructed single quarters to satisfy demand at whatever level it reached. However, neither the government nor the mine owners made any concessions. The apartheid government, almost to its last breath, regarded the circular flow of black labour as essential for its policy of separate development while the mine owners still saw it as the means of ensuring a continuous supply of cheap labour. The mine owners, moreover, declined to finance the construction of new mine townships as a step towards dismantling the migrant labour system. The overall effect was to leave black mineworkers and their families to cope both with the continuation of problems associated with the migrant labour system and with the emergence of new ones. Nothing, therefore, of real material significance to black mineworkers had changed during the previous decade. The hostel system remained as a source of faction violence. Mineworkers, cut off from their families for long periods, continued to engage in promiscuous sexual relations which led to the spread of VD and HIV/AIDS. Infectious diseases continued their rotation between mine compounds and villages and mine compounds.

The Housing Question

Whenever the mine owners had been questioned about ending the migrant labour system they had pointed out that they could not bear the high cost of building new townships or improving and extending existing ones. The government, moreover, was unwilling to ease that burden by providing social infrastructures for them. But, in the early 1990s, however, the NUM was not arguing for permanent family houses for all mineworkers. Its argument was that the mines should cease employing contract migrant labour and should leave mineworkers to decide for themselves where and how they wanted to live. This would mean abolishing single-sex hostels and providing a broad range of flexible housing options that would enable some mineworkers to continue commuting to the rural areas and others to settle in mine townships. It would also entail reconstructing hostels to accommodate single men, with facilities for wives to visit them, in order to meet the needs of the 40 per cent of mineworkers who came from foreign countries and for those from within South Africa who wanted to retain their land rights in the rural areas. The extent of family housing provisions, therefore, would most probably change over time, depending on the proportion of the labour force that opted for it.

The NUM did, however, want rented houses on the same subsidised terms as those given to white miners. But by then, the mining houses also had the option of legally building houses for sale for black mineworkers. It was an option that had been available only since 1983 when the government, recognising that it could no longer even pretend to prevent African urbanisation, enacted a 99-year leasehold system of tenure for black people. This possibility would considerably ease the burden of cost. The government's regulation that restricted family housing in mine compounds to three per cent of the labour force was rescinded. Thus, from the late 1980s, there had been no legal restriction to prevent the mining houses from providing permanent homes to all of their black employees.[7] This meant that, for the first time, South

[7]Catherine Laburn-Peart, "Housing as a Locus of Power", in Jonathan Crush and Wilmot James, eds, *Crossing Boundaries*, South Africa, 1995, pp37-38.

African black mineworkers could legally live with their families in mine townships.

In the light of these changes in the law, Anglo-American had, in 1986, conducted a survey to assess the housing needs of its black employees.[8] It found that 40 per cent of the South African and 29 per cent of the foreign mineworkers wanted to settle in family accommodation at or near the mines within the next five years. But only 47 per cent of the married South Africans, for whom the houses were intended, wanted to have their families with them and only 63 per cent of those said they would move permanently to married quarters.[9] When, however, the NUM conducted its own industry-wide survey late in 1990, it found that 62 per cent of black mineworkers wanted to live with their families near to the mines where they worked. This was a significant number considering that many mineworkers in the Orange Free State were able to visit their homes at weekends on a monthly, or even weekly, basis and were not keen on changing the pattern of their lives.

The housing problem in the mining industry was part of the wider one of providing affordable accommodation to the black population at large. It had been estimated in the late 1980s that 84 per cent could not afford housing without a subsidy. Indeed, almost half could not afford to make any financial contribution at all. Various estimates were made about the number of houses that needed to be built but the scale of the problem can be gauged from the fact that more than half of the black population lived in squatter settlements. The building of houses was only part of the problem, for with every township or township extension, water needed to be supplied, sanitation laid on, roads built and electricity provided. The provision of this social infrastructure was the responsibility of the state but it had eased itself out of any responsibility for house-building when it made it possible for black people to own houses in 1983. From that year it had switched its emphasis from rental to private ownership and stopped building houses for black people. Thereafter black housing became the concern of private developers aided by housing associations and mortgage companies. Because of insuperable financial obstacles for the majority of the

[8]Dillo Mashale, then head of the NUM's Housing Unit, questioned the competence of employers' organisations to conduct valid surveys on this issue. He believed that the majority of mineworkers preferred the abolition of the hostel system and the establishment of mining towns but, knowing that the mining houses were opposed to it, were reluctant to say so in case they jeopardised their jobs. It all depended, he said, on who was asking the questions. (Conversation between the author and Mashale at the 10th National Congress of the NUM, April 2000.)

[9]Crush, Jeeves and Yudelman, 1991, p174.

black population, houses became accessible to only about 10 per cent of them. The average size of their loans was R30 000. Building societies required substantial deposits, would not loan if the repayments exceeded 25 per cent of a borrower's income and refused to loan for non-conventional houses. Home loans were extended from October 1989 through a scheme devised by the Urban Foundation which enabled the private insurance industry to provide cover for home loans between R12 000 and R35 000 but these would still be beyond the means of almost 60 per cent of the black population. Even the cost of a tiny unimproved site with water and sewage facilities in the Transvaal was twice the amount the poorest 20 per cent of the black population could afford. The emphasis on private house ownership and the financial obstacles faced by individual black people ensured that little was done to alleviate the housing shortage for them in general. About two million houses were needed to accommodate the homeless but in 1988 only 45 000 were built. Almost half of the money invested in housing from public and private sources went on houses costing more than R40 000. This meant that they were far too expensive for the majority of black people.[10]

The main features of the national situation were replicated in the mines. The mine owners evaded the full cost of providing houses by proposing to build them for sale and offering subsidised mortgages to those mineworkers who wished to settle near the mines with their families. But, as in the case of black people in general, even the prices of subsidised loans were too high for most black mineworkers. Shortly after the end of the 1987 strike both Anglo-American and Rand Mines had made announcements about the provision of houses. Anglo-American declared that it intended to build 24 000 houses over a three-year period at four locations and would provide subsidised loans at a five per cent interest rate. Rand Mines, on the other hand, stated that it intended to incorporate a housing subsidy into the basic wage scales so that any black mineworker would have an opportunity to build his own house. Its scheme was confined to the coal division and intended for 4 500 mineworkers in the first instance. Neither the NUM nor the mineworkers themselves were consulted in either case. By the end of 1988, Gencor and JCI had also announced home ownership schemes. Only Gold Fields displayed no interest.[11] Anglo-American reached

[10]For details about the housing problem, see South African Institute of Race Relations, 1987, 1988, 1989 and 1990.

[11]Crush, Jeeves and Yudelman, 1991, pp174-175.

agreements with private building contractors to build houses with prices ranging from R20 000 to R80 000. It required deposits of only two-and-a-half per cent, instead of the usual five per cent, and fixed interest rates at five per cent with payments to be made over 20 years. The purchasers were to be responsible for the down payment, the mortgage repayments, property taxes and maintenance costs. But, the NUM reported in 1990, even the basic, lowest-cost R20 000 house was too expensive for 86 per cent of all black mineworkers. By the end of 1989, only 2 000 mineworkers, representing fractionally more than one per cent of its black labour force, had taken up Anglo-American's home-ownership offer. The proportion had risen to almost three per cent by 1992 and to six per cent by the following year. Anglo-American, therefore, was a long way from fulfilling its aim of accommodating 10 per cent of its labour force in family houses. A similar scheme, offered by JCI, was taken up by nine per cent of its black work force by 1993. It was estimated that no mineworker on a wage grade between one and eight would be able to afford to buy a conventional house under an Anglovaal housing scheme that had three options. The cheapest option was a serviced site on which the owner was entitled to build his own dwelling. It was calculated that 42 per cent would not even be able to afford that option and that those who could would most probably end up with a shack on a serviced site.[12] Apart from the financial cost, a major obstacle to home ownership was the fact that given the contraction in the mining industry, and gold in particular, no mineworker could be sure of retaining his job even in the near future. Nor, where the mines that were closed were in isolated areas, did they want to be left with their own homes in abandoned townships.

As an alternative to the housing subsidy schemes, the mine owners decided to pay living-out allowances to those mineworkers who chose not to live in the hostels. When the idea was first raised in 1989, both Ramaphosa and Motlatsi were opposed to it because they believed that it would enable the mine owners to pass the responsibility for providing houses on to individual mineworkers and would lead to an untidy growth of mine shanty towns.[13] It was implemented nonetheless but in a piecemeal fashion without any discussion with the union and often in a discriminatory manner. The allowances that were paid to mineworkers who chose to live out varied between mines. The Chamber of Mines calculated at the time that each mineworker received food and

[12]See Steinberg and Seidman, 1995, pp28-33 for brief details of these housing schemes.
[13]Discussion between Ramaphosa, Motlatsi and the author in 1989.

accommodation to the value of R210 every month. Its actual cost to them, however, was estimated to be 60 per cent of that amount. The mines, however, gave either a flat rate of R50 a month to all mineworkers who chose to live out or a rate that varied between R64 and R167. In the early 1990s, the mine owners began to lose interest in both housing schemes and living-out allowances, and, instead, offered wage supplements to those who did not want to live in the hostels. By 1993, Anglo-American was offering mineworkers a wage supplement of between R200 and R250 a month, depending on their respective wage grades. Anglovaal, on the other hand, was giving a choice of two options. A mineworker could either receive a minimum wage allowance of R120 or a 20 per cent wage increase. Genmin, JCI and Rand Mines offered similar schemes.

None of these payments was sufficient to enable a mineworker to purchase a house so many who received them chose the least costly housing option which, as Ramaphosa anticipated, was a shack in a squatter camp near to their respective mines. Squatter camps began to grow around mines. Women moved there but not usually the wives of mineworkers. Foreign mineworkers who had been dismissed and lost their South African residency rights, but who did not want to return to their homes, also moved in. The inhabitants were a diverse collection of people, some of whom worked in the mines, some who did not, some who did not work at all and others who lived off crime. Money circulated but little of it came from wage work. One of these camps, called Freedom Park, sprang up near to the Wildebeestfontein South platinum mine hostel north of Rustenburg. It had no plan, mud tracks instead of roads and no fresh water, sanitation or electricity. It became a sprawling mass of tin and wood shelters, which gave no protection from either the rain or the cold nights and hot days. It had residents but no medical facilities and children but no schools. It represented the least desirable solution to the problem of accommodation. But most of the mineworkers who wished to escape from the hostels had no other option.

The NUM made little impact on any of these schemes, though it debated and reached decisions over a range of related issues at its Seventh Congress in 1991. It emphasised the need to stimulate economic development in the rural areas, on the one hand, and for the mines to invest in local industries with effective community infrastructures, on the other. These were regarded as the preconditions for dismantling the migrant labour system. It wanted to end piecemeal living-out allowances and, instead, for all black mineworkers to be given an equal housing allowance that was incorporated into the basic wage and was sufficient to cover family living costs, including that of

Freedom Park Squatter Camp.
One of the housing options for black mineworkers in Rustenburg Platinum mines.

Another view of Freedom Park's living accommodation.

acquiring accommodation. On the assumption that these steps would be taken, the delegates insisted that all single-sex hostels should be upgraded for family and bachelor use by the end of 1993. Following the congress, the NUM had discussions with a number of mines about alternative provisions to the hostel system but little progress was made.

By the time the Eighth Congress was convened in February 1994, the union was resigned to the continuation of migrant labour and the use of single-sex hostels for at least the next quarter of a century. It continued its opposition but focused its attention on democratising and upgrading hostel accommodation. By then, however, a new development had occurred that led to slow but systematic progress. In 1992, the NUM formed a Housing Unit with the assistance of Rooftops, a Canadian non-governmental agency that specialised in providing housing facilities for the poor. Its purpose was to develop a housing policy that would assist in implementing the union's resolution to eliminate the hostel system. A Rooftops worker, Sue Moorhead, was seconded to become the head of the Housing Unit and arranged for the training of housing officials. By 1994, the unit had negotiated hostel constitutions allowing for hostel resident participation with the Impala Platinum Mines, the President Steyn gold mine and Kriel colliery and was involved in negotiation with seven other companies, including Eskom, the parastatal electricity-provider.[14] With the full participation of the union, Hostel No 7 at Vaal Reefs had been converted into family units while Hostel No 1 had been upgraded to provide accommodation for two or three people in each room. Both the President Steyn and Winkelhaak gold mines not only had hostel constitutions but also were operating Hostel Residence Committees.[15] The unit acted as a COSATU delegate to the Working Group on Hostels which had established a National Hostels Co-ordinating Committee to process applications for government funds to upgrade public sector hostels. It was pressing to have private sector hostels, for example those in mine compounds, included in the government's public sector redevelopment programme.[16]

The Eighth National Congress also returned to the other housing themes that had been debated three years earlier. On this occasion, however, it laid down a 'Framework of Principles on Housing Forums and Hostels' that was intended to ensure that there would be uniform practice throughout the industry and that all mineworkers would be

[14]NUM Eighth National Congress Report, 1994, pp38-39.

[15]For achievements between 1991 and 1994, see the *Housing Unit Report to the NUM National Congress*, 1997.

[16]NUM Eighth National Congress Report, p39.

treated equally. It stated that all mineworkers should have the right to affordable and secure housing, in the sense that they would not lose their homes if they lost their jobs. It added that every mineworker should be able to choose whether to buy a house, rent accommodation or participate in a social housing scheme. Social housing was a new concept in South Africa and, therefore, to black mineworkers. It was available in two non-profit-making forms. The first was where the land and houses were entirely owned by the community, co-operatives or housing associations. The second allowed for limited individual ownership within that same collective context. Both allowed for democratic decision-making. In both forms, the responsibility for applying for loans or mortgages from banks and other financial institutions and for negotiating with the mines over access to land or the transfer of hostels and other mine property was taken over by the housing organisations. Social housing was a means of providing affordable accommodation to the lowest-paid mineworkers in a manner that was consistent with the Framework of Principles. The NUM, with the Chamber of Mines and the government, began working on possible social housing pilot projects. The Eighth National Congress, in the light of these discussions and impatient for results, decided that all single-sex hostels had to be dismantled by the end of 1994.

The Casualties

It was shown in Part One how congested, unhealthy mine hostels, on the one hand and poverty-stricken rural areas on the other hand, had been the breeding grounds for infectious diseases that had been spread in a continuous circle by the migrant labour system acting as a transmission belt. Dust diseases to which mineworkers were subjected increased the severity of the infections but also, because they had long gestation periods, created legacies in that their symptoms usually appeared after the mineworkers had left the mines. The diseases, however, did not simply appear as legacies in the 1990s because, as the previous section illustrates, there had been no lapse in the conditions that generated and perpetuated the spread of infectious diseases. The circle of infection had continued to take its devastating toll on the lives of black mineworkers and their families.

Various factors had interacted to consolidate and intensify the hold that contagious diseases exercised over the lives of the mine hostel residents. Their primary occupational disease problem resulted from the spread of silicosis due to their increased exposure to dust. They worked for much longer periods than hitherto in dust-exposed jobs. There was, at the same time, a greater volume of circulating dust due to the increased use of machines underground. The dust hazard, moreover, was made worse by the ventilation difficulties encountered as the mines became deeper. The deterioration in living conditions in the rural areas had accelerated the spread of pulmonary tuberculosis that, in the case of former mineworkers, frequently combined and interacted with silicosis. Dust diseases by themselves were seriously disabling and could result in death but, in combination with tuberculosis, they were incurably lethal.

The number of former mineworkers facing this frightening prospect was high and was increasing. A study of a random sample of black mineworkers from the Libode district of the eastern Cape who had been employed in the mines between 1969 and 1980 showed that more than two-thirds had silicosis, pulmonary tuberculosis or a combination of the two.[17] A second study of a random sample of mineworkers from Thamaga in Botswana, another labour resource area, who had worked in the mines between 1960 and 1980 showed that 56 per cent complained of a variety of chest problems associated with mine work, about 30 per cent had silicosis while almost eight per cent had progressive massive fibrosis for which there was no cure. The report of this study concluded that there was a potential for "a silicosis epidemic on a grand scale".[18] There may, of course, have been such an epidemic because, as the Libode study showed, almost 57 per cent of those selected for the sample had already died of diseases, which, as in previous years, had been unspecified.

The Occupational Diseases in Mines and Works Act of 1973 had made provision for chest diseases in living mineworkers to be reported and for the cardio-respiratory organs of dead mineworkers to be examined so that the complete incidence of lung diseases amongst mineworkers could be compiled. From those official statistics, it was estimated that about 128 500 mineworkers had been certified as having acquired an occupational disease between the mid-1970s and the early 1990s. Between 1980 and 1989, from 1 500 to 3 500 cases had been reported annually. The NUM, however, claimed that these figures were gross underestimates. Normally, black mineworkers were reluctant to visit

[17]*South African Medical Journal*, Volume 86, No 5, May 1996. The preliminary results of this study showed that 36 per cent had pulmonary tuberculosis (with or without silicosis), 55 per cent had silicosis (with or without pulmonary tuberculosis) and only 32 per cent showed no sign of these diseases at all.

[18]*Occupational and Environmental Medicine*, 1997, No 54, pp19-26.

clinics because, like those who suffered from other ailments, they were suspicious of management motives in providing such facilities. In any case, the first signs of tuberculosis were not alarming. It was regarded as normal for underground workers to cough. The NUM's contention was supported by a study of active black gold mineworkers that had been carried out in 1987. It was based on routine mass radiography and revealed that the prevalence of silicosis was four times higher than those figures suggested. But even that was likely to have been an underestimation because the miniature x-rays lacked sensitivity, particularly where both silicosis and tuberculosis were present, and were, in any event, often poorly and inaccurately interpreted by the medical staff.[19]

Where there were incorrect diagnoses it was not just a statistical matter, a case of under-recording, because every time a mineworker who was suffering from an occupational disease was recorded as being fit he lost his compensation rights and failed to get proper medical surveillance and treatment.[20] But even those who were diagnosed as suffering from an occupational disease often failed to get compensation. In a review of a number of individual cases of former asbestos mineworkers in Lebowa whose radiographs demonstrated extensive disease, it was shown that compensation was hardly ever awarded. Indeed, in the mines in the Pietersburg area where the asbestos dust levels were high no steps had been taken to ensure that the mineworkers had benefit examinations.[21] Most of those in the Botswana sample mentioned above had not previously been recognised as having an occupational lung disease and very few had received compensation. This, it seemed, was a reflection of the general situation. Most doctors and hospitals in the recruitment areas appeared to have been unaware or unconcerned about their obligations to black mineworkers under the Act. When a member of the National Centre for Occupational Health in Johannesburg had visited most of the hospitals in the eastern Transvaal lowveld in 1991 he found that almost all of them were unaware of their obligations. The same situation was found when two members of

[19]*Ibid.*

[20]Lump sum compensation for a black mineworker under the Act amounted to less than one year's wages. By 1989, compensation awards were calculated according to a specific formula which differentiated between temporary and permanent injuries and were paid as lump sums or monthly pensions. However, no matter how they were calculated, the compensation payments were too small to maintain a family because the wages of which they were a proportion were themselves inadequate.

[21]C A Davies, "Personal View", National Centre for Occupational Health, *South African Medical Journal*, Volume 84, March 1994.

the Epidemiology Research Unit in the Medical Bureau for Occupational Diseases in Johannesburg visited several hospitals in and around Umtata in the Transkei. They concluded that "there seemed to be little awareness regarding the provisions of the Act on the part of doctors and health administrators in a range of institutions".[22]

The primary responsibility for detecting signs of occupational lung ailments that might lead to compensation awards lay with the mine doctors. It was often at this point that their management responsibilities stood in the way of their clinical judgments. They tended to be reluctant to increase the demands on the Compensation Fund, to which the mining houses contributed, by enabling all those with occupational diseases to make legitimate compensation claims. The result was that less than 10 per cent of those who suffered from such a disease received compensation and adequate treatment and few, therefore, were cured of it.[23] Inadequate radiology and medical surveillance in the mines meant that, because the period of maturation of the diseases lasted from six to 20 years, many mineworkers left the mines before they showed any symptoms and also, therefore, without any diagnosis. The maturation period for progressive massive fibrosis was at least 15 years so it, too, was hardly ever diagnosed in the mines. In the absence of medical facilities in the rural areas very few of these cases were subsequently diagnosed and the diseases were left to take their deadly toll. In, perhaps, a misguided consideration for mineworkers, the mine doctors tended to under-report cases of silicosis because of the legal requirement that those suffering from it had to be repatriated, thereby losing their sole source of income. They reasoned that it was preferable for a mineworker to struggle on working than have to face certain unemployment with only a small lump sum payment with which to maintain his family. As it was estimated that one in every five mineworkers with silicosis would develop progressive massive fibrosis, the toll from that disease alone was likely to be heavy.[24]

But it was not only the defects in diagnosis that caused so many former mineworkers to suffer from extreme forms of lung diseases. High exposures to toxic quartzite respirable dust was the root cause of silicosis and that, too, was seriously underestimated by the mine authorities, thus giving the impression that the mines had the dust problem under control. This was not, however, the conclusion reached through

[22]*South African Medical Journal*, Volume 86, No 5, May 1996.

[23]R L Cowie, M E Langton and M E Becklake, "Pulmonary Tuberculosis in South African Gold Miners", *American Review of Respiratory Diseases*, 139, 1989, ppl 086-9.

[24]*Occupational and Environmental Medicine*, 1997.

independent studies. When the Commission of Inquiry Into Safety and Health in the Mining Industry investigated the question of dust levels in 1994 it found that there was very little public information about them. It reported that "the results of the many thousands of dust measurements which have been made in recent years have not been analysed and published in a form which makes it possible for experts in the field to describe the trends accurately, nor is it possible without access to the raw data to determine whether current sampling strategies and methods used to measure dust levels succeed in identifying the highest exposures accurately". It concluded, taking into account that the industry relied on "old-fashioned methods of dust sampling" and on the basis of available evidence, that "dust levels have remained roughly the same over a period of about 50 years".[25] A similar conclusion was reached in the study of the prevalence of occupational lung disease in former mineworkers from Botswana. "It seems," the Botswana study stated, "that there has been no significant change in dust exposure underground in gold mines for at least 50 years."[26] These statements amounted to a dramatic condemnation of the safety culture in the mines.

The social costs of such managerial neglect had always been borne by the families in the rural areas. This had been ensured by the repatriation of sick mineworkers with or without compensation payments. The practice of repatriation had been eased a little after 1985 when some casualties were retained by the mines and treated with chemotherapy. But for many, it remained the same. The absence of diagnoses before and after leaving the mining industry ensured that sick mineworkers would return to their homes as if there had been no change in policy. Many were unable to get work and, therefore, received no incomes. In the Botswana study almost 47 per cent were unemployed after an average spell of 15 years in the mines. The burden that mine diseases placed on the rural communities was incalculable.

The compounds in the early 1990s were, then, generally no safer from contagious and occupational diseases than they had been for the previous 40-50 years. It was estimated, for example, that death rates "from all causes or from disease had not changed substantially between 1940 and 1980" and that in the early 1990s there was "no evidence to indicate a decline in the prevalence or severity of any occupational disease in the mining industry during the past 20

[25]Commission of Inquiry Into Safety and Health in the Mining Industry (Leon Commission), 1995, Report, Volume 1, paras 4.6.4-4.6.7.
[26]*Occupational and Environmental Medicine*, 1997.

years".[27] Indeed, in some respects the position had worsened due to the greater use of machinery underground. It was not until 1989 that hearing loss was recognised as a problem by the government and regulations were promulgated to establish the infrastructure to screen it. A study of hearing loss taken three years earlier, that had highlighted the need for new regulations, showed that two out of every three rock drillers would have significant hearing loss after 10 years of service. Within three years of the promulgation of the regulations the number of known cases of hearing loss had trebled.[28]

Little had changed in other ways. The new recruits from the rural areas were still generally physically weak from malnutrition and indigenous debilitating diseases, such as bilharzia, when they entered the mines. Just like most of those who had preceded them, initially they found it exhausting to perform unremittingly hard work for long periods without breaks, and that increased their proneness to infection. Some of them may have gained an element of immunity from those infectious diseases that were endemic in the rural areas but there were likely to be some recruits arriving in the early 1990s who had gained no immunities or who might have latent or early non-diagnosed symptoms of tuberculosis that made them susceptible to influenza or colds. In any event, in the early 1990s it was estimated that about 4 000 black mineworkers developed tuberculosis each year and that the longer they worked in the mines, the more likely they were to acquire it.[29]

Many hostel residents, therefore, remained vulnerable to infection of one kind or another. It was necessary to make only one slip, allowing, say, a disease-carrier into a hostel, for an epidemic to develop. It was difficult to control all the sources of possible infection because, as a number of services, such as the provision of water, sanitation and food, were centralised, it was relatively easy for a solitary infection to circulate around the hostel blocks in a compound.[30] Moreover, once an

[27]Leon Commission, *Report*, 1995, Volume 1, paras 4.5.12 and 4.5.18.

[28]*Ibid*, para 4.5.17.

[29]*Ibid*, para 2.3.17.

[30]This had been illustrated by an outbreak of cholera in two mines early in 1974. Cholera was endemic in Malawi, Mozambique and Angola so the Chamber of Mines introduced a large-scale cholera surveillance programme. Nonetheless, cholera entered 10 mines but as a localised epidemic in two of them. Faecally contaminated water was the source in both cases. In one mine, 'healthy' carriers of the disease passed the infection to others through their sweat in the acclimatisation chamber while in the other mine, mineworkers underground drank water from a hose, the nozzle of which trailed in a contaminated waste water ditch. In each case, the infection spread quickly through the hostels. There were, however, no deaths. (*South African Medical Journal*, 14 December 1974, pp2 557-2 560.)

infection had been identified it was difficult to prevent it from spreading because there were no facilities in the crowded hostels to enable the carriers to be isolated. It was clear, therefore, that once an infectious disease arrived in a mine, then the hostel remained the main transmission point. The industry, therefore, retained its historical function as a generator and facilitator of disease in the circle of infection.

The HIV/AIDS Alarm

The feature of the migrant labour system that epitomised for black mineworkers its most unpalatable and undesirable qualities was their incarceration in single-sex hostels, apart from their families and kinfolk, for long periods of their contracts. As a result of the separation from their wives and village female friends, and particularly since the late 1970s when they were given greater freedom to move in and out of the compounds, they formed relationships with prostitutes and other women in the township, through which many of them contracted sexually transmitted diseases, which, until the mid-1980s had been gonorrhoea and syphilis.[31] Both of these diseases could be cured by penicillin but this treatment was rarely available in the rural areas. From 1986, however, the list of sexually transmitted diseases included a newly discovered one for which there was no treatment anywhere and which was ultimately invariably fatal.

The new disease, the Acquired Immune Deficiency Syndrome (AIDS), had been first identified in South Africa when two homosexual men died of it in December 1982. By the end of 1984, the number of cases had increased to 16, of whom five had died. Thirty-four cases were known by August 1986 but there was no information about the number of deaths.[32] AIDS was not at that time a notifiable disease. It was caused by the Human Immunodeficiency Virus (HIV) that destroyed the body's immune system and made it vulnerable to a number of chronic, life-threatening diseases. Most people diagnosed as HIV positive could expect to develop full-blown AIDS within a period of between two and 10 years. In the 1990s, in the developed countries the average time between the final diagnosis of AIDS and death from it was one to two years but within developing countries it was half of that

[31]The manner in which this came about and the extent of the spread of venereal diseases are described in chapter 3.
[32]South African Institute of Race Relations, 1986, pp798-799.

because their inhabitants were exposed to a wider range of infections that were stimulated by AIDS. The most vulnerable people in general were those who were marginalised, suffered from discrimination, were most deprived of resources and who lived in the kind of conditions that facilitated the spread of other sexually transmitted diseases.[33]

The virus could be transmitted through contaminated blood and from infected mothers to their infants but the most common form of transmission was unprotected sexual intercourse with an HIV-carrier. Unlike the situation in the USA and Britain where HIV infections were spread initially through homosexual contacts, the primary source among the black population of South Africa was heterosexual intercourse. This was particularly so in the mines where homosexual activity, though reportedly common in the hostels, did not pose such health risks for it tended to be monogamous and involved thigh contact rather than anal penetration.[34] The most vulnerable men were those who, for whatever reason, participated in promiscuous, unprotected sex. They included the migrant black mineworkers who often had multiple sexual partners. Women figured equally with men in this equation but were exposed to greater risks because, for biological reasons, they were up to 10 times more likely to get infected with every sexual encounter than were men.[35] The women most at risk were the village wives, abandoned wives, divorcees, girlfriends, illegal township women and prostitutes. Indeed, they included all women who possessed insufficient or no financial support for themselves and their families. The women who took on sexual partners out of economic necessity were in a position of weakness that left them with no control over how their sexual activity should be conducted.

Black mineworkers were particularly vulnerable to the spread of HIV because they lived in an environment where venereal diseases (VD) and tuberculosis, both of which interacted with it, often lethally, were not only firmly established but were spreading. The combination of those diseases with HIV constituted the greatest threat to their health. Those suffering from the conventional strains of venereal diseases were most likely to contract the HIV infection. The incidence of VD in all of the mining sectors had risen significantly throughout the 1980s and was continuing to rise in the 1990s, despite a fall in the size of the black mining labour force. The rate, moreover, was high. It was claimed that

[33]Mary Crewe, "Learning AIDS", *Work in Progress*, December 1993, p27.
[34]Jochelson, Mothibeli and Leger, 1988, p20.
[35]Crewe, December 1993, p28.

in the early 1990s more cases of gonorrhoea were reported at the Leslie Williams Memorial Hospital in Carletonville in a week than were seen in the whole of Norway in a year.[36] The extent to which VD and HIV were combined was indicated by evidence from the same hospital where about 30 per cent of those attending its clinic had both diseases.[37]

The emergence of HIV had a devastating impact on the spread of tuberculosis amongst black mineworkers because it greatly increased its progression from a latent to an active state and transformed the already high incidence into the dimension of an epidemic. The correlation between the two diseases was such that an effective tuberculosis programme might have extended the lives of HIV-infected mineworkers while an AIDS preventative programme could have slowed down the rate of progression of tuberculosis. A failure to do either left the diseases to work rampantly on each other. In harness, they made mineworkers feel so ill that their normal reluctance to go voluntarily to hospital clinics disappeared. At the Ernest Oppenheimer Hospital in Welkom in the early 1990s, for instance, as many as 60 per cent of all cases of tuberculosis with HIV had attended voluntarily.[38]

The Chamber of Mines had been alerted to the possible immediate threat AIDS posed to its labour force when it received information from the South African Institute of Medical Research that a number of blood samples it had received from Malawi in the early part of 1986 had indicated an extremely high HIV prevalence there. The institute, aware that the mines employed many Malawians, had advised the Chamber of Mines to conduct research amongst its migrant workers to obtain a picture of the extent and distribution of HIV amongst them. It then established an AIDS Centre through the sponsorship of the mining industry.[39] In response to the advice it had received, the Chamber of Mines began a screening process of its labour force in June 1986. Over 300 000 blood samples were collected from black and white employees on its gold and platinum mines. A sample of almost 30 000 specimens was selected according to the territorial origin of the workers and their race group. The general conclusion of the study was that "the prevalence of HIV infection in the mining population is comparable

[36]Brian Williams and Catherine Campbell (eds), *HIV/AIDS Management in South Africa: Priorities for the Mining Industry*, Epidemiology Research Unit, Johannesburg, 1996, p112.
[37]*Ibid*, p113.
[38]*Ibid*, pp117-119.
[39]Chamber of Mines, "The AIDS Crisis: A Mining Industry Perspective", *Mining Survey*, Number 2, 1991.

with the norm for any low-risk population".[40] No proven cases of AIDS were found in the industry. The detailed results had shown that the incidence of HIV was low amongst migrant workers from Lesotho, Mozambique, Swaziland and South Africa but relatively high amongst workers from Botswana and Malawi. The majority of carriers had come from Malawi. At about the same time, blood specimens had been taken from 1 200 black women, of whom 94 were believed to be prostitutes, who all lived in areas near to the mines, but the results had all been negative.

Through the survey, it was found that 130 Malawian mineworkers were HIV positive. The government immediately adopted the attitude that foreign migrants were the prime source of the infection and demanded that all of the infected Malawians should be repatriated even though 53 of them were found to be fit enough to continue working. It politicised the issue by labelling AIDS primarily as a black disease. AIDS, it asserted, was a 'black peril' that was brought into South Africa from the north by foreign migrants who constituted "a reservoir of infection capable of contaminating the entire country".[41] This belief moulded its subsequent practical responses for they were concentrated on foreign migrant workers. It did not test foreign workers who were already in the country but insisted on testing all black workers who entered it. It introduced regulations for the isolation and compulsory treatment of sufferers and carriers of AIDS and then, on 30 August 1987, gazetted further regulations that described AIDS and HIV infection as a disease "which will render a person a prohibited person".[42] AIDS, moreover, was characterised as being a disease of marginal groups who engaged in immoral promiscuous behaviour, gay sex and prostitution and who were generally licentious. This exonerated the mainstream white communities from any responsibility for tackling the problem and allowed them to perceive it as being primarily a black issue.

The initial responses of the Chamber of Mines and the NUM to the news about the Malawians differed from that of the government. They jointly insisted that all actual employees who were carriers of the virus should be clinically assessed and counselled and, if fit, they should be allowed both to continue working in the mines and to return to the mines after the end of their contracts. They believed that only those who were clinically unfit should be discharged and repatriated.

[40]See Jochelson, Mothibeli and Leger, 1988, for these and other details of the Chamber of Mines' screening process.
[41]Jochelson, Mothibeli and Leger, 1988, p45.
[42]South African Institute of Race Relations, 1987-88, p316.

However, both parties contended that no recruits who were found to be HIV positive should be allowed to enter the mines and that all new recruits from 'AIDS-prevalent areas' should be screened before being allowed to sign on for work.[43] In the case of Malawi, the Chamber of Mines wanted to test mine recruits for HIV before they left the country but Dr Banda, the Malawian president, had refused to give his permission for that to be done. Instead, it introduced blanket testing for all recruits from Malawi and other high-risk countries once they were at the recruitment centres in South Africa. Through a series of measures, the Chamber of Mines achieved the same effect for Malawian mineworkers as would have resulted from repatriation. It refused to take on novices from Malawi and re-employed only those with experience. In some cases, Malawians who had gone on home leave were blocked from returning to the mines. By 1990, their number recruited by TEBA had fallen to 13. The following year it had dropped to zero and, henceforth, Malawi ceased to be listed by TEBA as a recruitment area. This policy resulted in a sharp fall in the number of Malawians in the mines from 20 989 in 1987 to 2 703 in 1989 and then to a mere 37 in 1991. There were reductions in the employment of mineworkers from other countries that had relatively high HIV infection rates. For instance, the number of mine recruits from Botswana, where seven per cent of the total population were infected, fell by almost 40 per cent between 1986 and 1992.[44] The Botswana town of Francistown, which was the staging post for the tropical mine recruits from the north, contained 34 per cent of all HIV positive people in Botswana. Lesotho, on the other hand, although heavily reliant on migrant labour, had the lowest reported HIV rate in southern Africa at one per cent.[45] A partial explanation for this might have been that Sotho mineworkers, because of the proximity of the gold mines in the Orange Free State to Lesotho, were often able to make relatively regular visits to their homes.

The approach of the Chamber of Mines diverged from that of the NUM because of the NUM's insistence that the migrant labour system facilitated the spread of AIDS. Marcel Golding, speaking for the NUM in 1986, had called for the abolition of single-sex hostels because, he said, they promoted the spread of AIDS.[46] The Chamber of Mines,

[43] *The Citizen*, 29 August 1986.

[44] These figures do not reflect the precise incidence. In Botswana, HIV infection was concentrated in the 20-40 age group. (*Africa South and East*, ASE Publications, Yeoville, Johannesburg, Dec 93-Jan 94, p10.)

[45] *Ibid*, p8 and p10.

[46] South African Institute of Race Relations, 1986, p799.

however, had refused to acknowledge that there was any particular link between the spread of HIV and conditions in the mining industry and was reluctant, therefore, to engage in special programmes directed at mining which might link AIDS and mining in the public mind. On different occasions in 1987 and 1988 it had asserted that the migrant labour system did not contribute to the spread of AIDS.[47] It gave various reasons for its attitude. One was that, as the prevalence of HIV in the mining industry, except amongst Malawians, was no greater than in the country at large, the hostel system could not be a contributory factor. It also asserted, at a time when AIDS was believed to be transmitted through homosexual activity, that single-sex hostels had not contributed to HIV "because, contrary to popular belief, homosexuality is not seen to be a widespread problem in the compounds".[48]

Apart from its screening process, the Chamber of Mines launched public awareness and education campaigns in the belief that the issue of AIDS was a personal matter in terms of its cause and cure. It distributed pamphlets, leaflets, posters and videos and broadcast exhortations over the radio. In addition to trying to generate an awareness about AIDS to adults and children, it and the mining houses separately distributed free condoms to mineworkers and, through Anglo-American funding, gave financial assistance to the American-based Population Services International which sold and distributed condoms throughout South Africa. Individual mines introduced their own 'AIDS Awareness' campaigns involving leaflets, pamphlets, radio programmes in the indigenous languages, videos, puppet shows, billboards near taxi ranks, bus stops and hostels, trained peer educators and lectures by social workers and medical staff. The campaigns were accompanied, in some mines, by making free condoms available on request or, in others, by placing them at compound entrances so that mineworkers could freely take as many as they wished. The mine managers, however, ended their activities, and their responsibilities, at the mine gates. The victims of AIDS had to leave the mines and return to their rural homes, as did the victims of other compound-generated and occupational diseases, and from that point on the mine authorities had nothing further to do with them. In the case of AIDS, however, they were not able to offload all of their healthcare costs on to the rural communities because the mines were obliged to treat HIV positive mineworkers who suddenly

[47]See Catherine Campbell and Brian Williams, *HIV/AIDS and the Productivity Sectors in the SADC Region: A Case Study of the Mining Industry*, Epidemiology Research Unit, Johannesburg, November 1996.
[48]*Financial Times*, 22 April 1988.

developed serious AIDS-related diseases whilst they were actually working in the mines. As the mines provided free healthcare services and AIDS treatment was expensive, the cost to the mining houses could be high in such cases. This factor reinforced their determination to introduce pre-employment testing while, at the same time, suggesting that the state should bear the cost of treating AIDS-related diseases.[49]

There was a scarcity of reliable data about the incidence of HIV amongst black mineworkers. After the Chamber of Mines had conducted its industry-wide survey in 1986, there was no other systematic collection and analysis on the prevalence of HIV on the South African mines. The only significant figures came from the numbers tested for HIV at the Sexually Transmitted Disease clinics that were present on all of the mines affiliated to the Chamber of Mines and which offered free and confidential testing, counselling and treatment to all patients suffering from sexually transmitted diseases. There were between 2 000 and 3 000 such tests each month, amounting to a total of 120 000 between 1986 and 1991.[50] But the tests were voluntary and were made only on people obviously at risk. Gold Fields supplemented the data from the clinics with figures from its own tests at its Leslie Williams Memorial Hospital in Carletonville, but it was the only mining house to do so. Anglo-American had phased out its annual medical reports in 1983; Rand Mines had done likewise in 1986 while Gencor had kept its medical reports confidential. The figures, in any event, would have been incomplete because, as was the case with venereal diseases, about one-third of all mineworkers who felt they needed to be tested for HIV went to private practitioners because they feared that they would be dismissed if they were found by management to be HIV positive. But it was not simply a distrust of management that deterred mineworkers from seeking treatment at the STD clinics. They also feared discrimination from other mineworkers because of the stigma attached to being HIV positive as information about it was circulated. However, the most significant explanation as to why black mineworkers avoided the STD clinics was that they associated them with the apartheid system.

All of the AIDS campaigns, especially those of the government, were seriously tainted by racism. The intervention of the government suggesting that AIDS was a black phenomenon, the prescriptive and authoritarian character of the publicity messages, the absence of black people from any involvement in the formulation of the awareness

[49]Campbell and Williams, November 1996, p10.
[50]Chamber of Mines, *Mining Survey*, No 2, 1991, pp24-26.

campaigns and the lack of any cultural or even linguistic sensitivity in their presentation, combined to create a prejudice amongst black people against the health education campaigns. Advocacy of the use of condoms was associated in the minds of black people with long-standing attempts by apartheid governments to control black population growth and was, therefore, resisted. Mineworkers, moreover, had an implicit distrust of anything advocated by management. They refused to believe that managers were serious about combating AIDS for if they were, as one worker stated, "they would give us at least four days every month to be with our wives".[51]

There were many ways in which the interests of the black mineworkers had been neglected. They had never been asked for their opinions. Indeed, all of the policies had been formulated and enforced from the top down. In many instances, they had been tested for HIV without their informed consent and had not been told afterwards whether or not they were infected. No-one had consulted them about the formulation of HIV educational programmes with the result that they tended to be Eurocentric and failed to be understood by the majority of mineworkers who were illiterate in English. Insufficient account was taken of their different languages and cultures in explaining the ways in which to prevent or cope with AIDS. Then there were practical questions, relating to the welfare of those suffering from full-blown AIDS, for which mineworkers and their families needed immediate answers. Should those suffering from AIDS be kept in hospital or should they be sent home to die? If they were to be sent home, then what kind of care should they receive and who should be responsible for administering it? Without answers to these and related questions, the pain and disruption from having AIDS was compounded by the kind of official contempt for black casualties that had always characterised their lives.[52]

In general, avoiding HIV did not figure high on the list of priorities of black mineworkers. Every working day they were exposed to dangers which could kill them so that another risk, which might not affect them for five to 10 years, did not seem to be so important. It was, in any case, simply a matter of adding another disease to the list they

[51]Jochelson, Mothibeli and Leger, 1988, p40.

[52]All of these questions were raised by Fleur Plimmer in her contribution to the workshop on HIV/AIDS in the mining industry on 6-13 October 1995 held by the Epidemiology Research Unit at the Medical Bureau for Occupational Disease, Braamfontein, Johannesburg. (See Williams and Campbell, 1996, pp58-60.) Plimmer was the Health and Safety Co-ordinator for the NUM at the time.

already had to contend with. At that time, no mineworkers had actually seen anyone suffer from AIDS, let alone die of it, and few knew anything about its characteristics.[53] Not all of them were convinced that it could not be cured. Some even believed that traditional healers could cure it. In the main, however, those who had been informed about AIDS tended to believe they could do little to protect themselves from infection. Experience within the apartheid system had conditioned them to be fatalistic. Although the NUM asserted that the cause of the spread of HIV infection lay in the migrant labour system, they had little evidence to indicate that their union was able to do anything about it. Moreover, they had personal reasons for ignoring the advice to take precautions by wearing condoms when having sex. Mine work and compound life generated so much stress that they sought every possible means of relief. Alcohol was one such means and there was a strong association between drinking and practising unsafe sex. But the physical, flesh-to-flesh intimacy of unprotected sex was the most satisfying source of relief from tension.[54] Information about AIDS was, in that situation, largely irrelevant as an influence on sexual practices. In a survey of four gold mines in 1989, it was found that although there was a broad understanding of preventative measures, due to publicity through the industry's news media, medical staff and friends, as many as 67 per cent said that they never used condoms.[55] It was not surprising, therefore, that the infection rates were roughly doubling every 12-15 months. From 1991, on the basis of the tests at the STD clinics, it was estimated that by 1996, there were would be about 45 000 HIV positive mineworkers, amounting to about 15 per cent of the mine labour force. It was also estimated that approximately 4 000 of those would fall ill each year over the next 10 years.[56]

Given the NUM's inability to alter the single-sex hostel system, there was little it could do apart from engage in the publicity campaigns for safe sex launched by the Chamber of Mines. It concentrated on the human rights issues raised by the mines' screening and testing measures, such as seeking protection against unfair dismissal and discrimination.

[53]Fleur Plimmer, "Employee Perspectives on HIV in the Mining Industry", Williams and Campbell, 1996, pp58-60.

[54]See Catherine Campbell, "The Psychosocial Context of HIV Transmission on the Gold Mines: Implications for HIV Education Programmes", in Williams and Campbell, 1996, pp76-91, for a discussion of mineworkers' perceptions of AIDS.

[55]Chamber of Mines, *Mining Survey*, No 2, 1991, p26.

[56]Susan Foster, "Economic and Financial Implications of HIV", in Williams and Campbell, 1996, p108.

It also attempted to ameliorate the screening process by dealing with the question of confidentiality and installing grievance procedures. In doing so, it moderated the suspicions black mineworkers had of the white-initiated AIDS campaigns. It conducted its own publicity. The union's health and safety department compiled a simple explanatory statement about HIV/AIDS at the end of 1991 for distribution to union members. It was a stark description that concluded with the statement: "The union is sitting on an explosion."

The NUM concluded an industry-wide agreement with the Chamber of Mines, in August 1993, to provide guidelines that aimed to minimise the effects and to contain the spread of AIDS whilst, at the same time, protecting HIV positive mineworkers from discrimination, victimisation, harassment and any adverse consequences, such as dismissal or the denial of alternative work. The agreement resulted from an undertaking in the 1991 review of wages and conditions which stated that there should be an industry-wide policy on AIDS. It was based on the recognition that HIV infection and AIDS should be approached on the same basis as any other serious ailment in that those who suffered should receive comparable treatment and benefits. The agreement also stated that the results of the many epidemiological studies that were being conducted at the time should not be used as a basis for discriminating against any group of mineworkers, such as, for instance, Malawians. Fundamental to the agreement was the provision that the Chamber of Mines would take no action without consulting the NUM, which, in turn, would consult with its members.[57] The spread of AIDS, however, continued unremittingly. When Motlanthe presented his secretarial report to the Eighth National Congress in February 1994, he described its progression as a matter that presented the NUM with new challenges. It was not only a subject, he stated, "which is concerned with sexual issues. It has altered employment practices with respect to recruitment, access to medical aid and insurance policies, and job security. AIDS has also put issues such as living conditions and housing firmly on the organising agenda. We have to rise to the task of educating all members about AIDS and to beat off the attack on working conditions which the AIDS epidemic has precipitated."[58]

[57]For details of the agreement, see Chamber of Mines, *Newsletter*, Issue Number 5/6: 1993, p4.
[58]NUM, Eighth National Congress Report, February 1994.

The Daily Carnage

The NUM had been pressing the government to appoint a commission to examine health and safety in the mines since the Hlobane disaster in 1983. However, it was not union pressure but another major disaster in which 53 mineworkers were killed in a methane gas explosion at Middelbult colliery in Secunda on 13 May 1993 that propelled the government to establish the Commission of Inquiry Into Safety and Health. A month after the disaster, the Minister of Mineral and Energy Affairs announced the appointment of the commission to be chaired by a retired judge, Mr Justice R N Leon, assisted by three academic assessors. The terms of reference for the Leon Commission were to recommend improvements in the legislative framework that was based on the 1911 Mines and Works Act and to propose how such changes should be implemented in the light of the existing conditions in the mining industry and the level of international standards.

Little had changed in the union's safety priorities between its formation and the appointment of the commission. When the health and safety department of the NUM submitted its proposals to the new Safety and Health Commission, its list of grievances was broadly the same as that which Ramaphosa had drawn up and publicised immediately after the Hlobane disaster.[59] Black mineworkers still had no legal right to participate in the determination of health and safety policies in the mines, to have access to information about them and to refuse to work in dangerous conditions. There were major hazards, such as rock bursts and rock falls, where the law was silent. And where it was not silent, as in cases of negligent killing or infringements of mining regulations, and where the consequential deaths could rise into hundreds, its sanctions were still the equivalent of those imposed for breaches of road traffic regulations.[60]

The clearest sign of progress was where the union had succeeded in persuading some mine managers to accept that safety was a negotiable issue and to agree to the formation of mine health and safety committees on which the NUM was represented. To a large extent, this was a rebuilding exercise following the destruction of similar committees during the 1987 strike. Its progress, however, was slow. By 1994, the union had signed 25 branch safety agreements with a further 28 in the process of

[59]See chapter 5 for Ramaphosa's list.

[60]See chapters 5 and 9 respectively for details of the penalties imposed after the 1983 Hlobane and 1986 Kinross disasters.

negotiation. Only eight of the signed agreements were in the goldmines. No branches had been able to sign an agreement in any coal mines or with any of Gold Fields' gold mines. All in all, the signed and still-to-be-signed agreements comprised only a fraction of the 258 union branches but they covered a majority of the union members. Their significance at the time lay in their potential rather than in their actual achievements.

There was a distinction in the general accident picture since the Hlobane disaster between the trends in the rates of injuries and deaths. Largely because of changes in mining technology, there had been a considerable fall in the number of serious injuries in all mines from 15 528 in 1984 to 8 532 in 1993, representing a 29 per cent drop in the injury rate. In the gold mines alone, the injury rate had fallen by one-quarter.[61] The death rate, however, had remained relatively constant. In the gold mines there had been 588 deaths in 1984, representing a fatality rate of 1,15 whereas in 1993 there were still 418 deaths and, because the labour force was smaller, a fatality rate only fractionally below at 1,13.[62] In the coal mines, during the same period, the number of deaths increased from 73 to 90, causing the fatality rate to rise from 0,63 to 1,71. The fatality rate for all mines actually rose from 1,01 in 1984 to 1,03 in 1993. The yearly average number of black fatalities was 680,which meant, on average, that at least 10 mineworkers were killed every week.[63]

These statistical distinctions, however, made little impression on the black mineworkers for, no matter how they were viewed, the overall situation caused considerable anxiety. Many of them believed when they entered the shaft at the beginning of a shift that they might not emerge from it alive. It was of little consolation to them that there was less likelihood of being injured so long as their chances of being killed remained unchanged. In any event, from their point of view, the number of injuries at the lower rate in 1994 was still excessive. They were both depressed and angered by the fact that during the previous 10 years, 1 492 mineworkers had been killed and 30 649 had been seriously injured.[64] During August 1993, about 10 000 NUM members in the Orange Free State marched on the offices of the Department of Mineral and Energy Affairs in a demonstration for stricter

[61]Leon Commission, 1995, p19.

[62]*Ibid.*

[63]The ILO compared the mine fatality rates for 19 countries on an annual basis between 1989 and 1991. Out of the 19 only Turkey, Yugoslavia, Papua New Guinea and Pakistan had worse rates than South Africa. (ILO Report, *Safety and Health in Mines*, 1994, No 51, p13.)

[64]South African Institute of Race Relations, 1993/1994, pp593-594.

underground safety measures. They handed in a memorandum listing those figures and demanding the appointment of more mining inspectors, the provision of free boots with toe caps, the reduction of underground noise levels and full pensions for those who had been totally disabled. Kgalema Motlanthe, the acting general secretary, endorsed the reason for the demonstration when he reported to the Eighth National Congress in April 1994.[65] "The rate," he stated, "at which accidents continue to occur in the industry remains unacceptable … despite the fact that in the 1990s the number of workers employed on the mines was substantially lower than in the late 1980s there has been no change in the number of people killed … in the industry."

Disasters continued to occur as in the past. That was likely to be the case, so long as the number of small-scale accidents remained high for there was always the possibility that periodically something in the environment, such as undetected methane gas, an inflammatory substance or some undisturbed coal dust, would create an explosion or cause a fire that would raise the death rate beyond the level of five and convert an accident into a disaster. The persistent trend towards deeper mining, moreover, ensured that seismic movements would increasingly cause unpredictable rock bursts that were likely to result in many deaths. The Hlobane disaster in 1983 had marked the beginning of the disaster period for the NUM. From the end of that year until April 1994 there had been 60 accidents in which six or more mineworkers were killed and five in which more than 30 had died. Altogether, 820 mineworkers had been killed in those disasters. Almost 80 per cent of the disasters had occurred in the gold mines. Nine of them had happened in 1990 alone, including one at Vaal Reefs in September when 21 mineworkers were killed. Twelve were killed the following month at Western Deep Levels. In 1991, four disasters, in which a total of 27 people were killed, were caused by rock bursts. This was by far the most common cause and accounted for 29 disasters and one-third of the fatalities in the 10 years prior to 1994. The government mining specialists were pessimistic about their ability to stop this carnage. The Government Mining Engineer had stated in 1992: "Rock falls and rock bursts continued with no improvement despite extensive safety campaigns … being conducted throughout the year."[66] This was pessimistic news for it meant that in 1994, as in every other year before that, black mineworkers would face the risk of injury or death whenever they went underground. In effect, as always, they were subsidising the production of minerals with their own health and lives.

[65]NUM *Triennial Report*, *1991-1994*, p24.

[66]Department of Minerals and Energy Affairs, *Annual Report*, 1992, p42.

TABLE 3

ACCIDENT DATA FOR THE SOUTH AFRICAN MINES 1984-1993[67]

Type of mine	Year	Fatalities	Fatality rate	Injuries	Injury rate
ALL MINES	1984	740	1,01	15 528	21,20
	1985	706	0,94	14 820	19,60
	1986	800	1,02	12 709	16,30
	1987	753	0,97	11 473	14,71
	1988	674	0,91	11 357	15,38
	1989	735	0,99	10 097	13,63
	1990	675	1,05	9 858	15,38
	1991	604	0,98	9 103	14,74
	1992	552	1,02	8 801	16,19
	1993	578	1,03	8 532	15,21
GOLD MINES	1984	588	1,15	13 736	26,90
	1985	539	1,03	13 168	25,10
	1986	702	1,27	11 624	21,10
	1987	547	0,98	9 969	17,87
	1988	510	0,97	9 576	18,14
	1989	549	1,06	8 561	16,46
	1990	526	1,25	8 195	19,43
	1991	461	1,19	7 571	19,51
	1992	407	1,12	7 585	20,94
	1993	418	1,13	7 375	20,00
COAL MINES	1984	73	0,63	840	7,20
	1985	93	0,78	806	6,80
	1986	66	0,55	709	5,90
	1987	121	1,05	550	4,86
	1988	53	0,50	435	4,07
	1989	54	0,52	361	3,50
	1990	50	0,52	404	4,21
	1991	42	0,56	361	4,80
	1992	46	0,66	339	5,90
	1993	90	1,71	279	5,32

[67]Leon Commission, 1995, p19.

TABLE 4

MINE ACCIDENTS IN WHICH FOUR OR MORE PEOPLE WERE KILLED 1984-1994[68]

Name of mine	Persons killed	Persons injured	Type of accident	Date of accident
Blyvooruitzicht	7	-	Rock burst	1984
ERPM	6	-	Rock burst	1984
ERPM	8	-	Rock burst	1984
Vaal Reefs	6	-	Inundation	1984
Buffelsfontein	8	-	Fire	1984
Ermelo Mine	6	-	Explosion	1984
Middelbult Colliery	33	10	Explosion	1985
Western Deep Levels	9	-	Rock burst	1985
ERPM	6		Rock burst	1985
Vaal Reefs	7	-	Explosion	1985
ERPM	6	-	Rock burst	1985
ERPM	17	20	Rock burst	1985
Deelkraal	10	4	Explosives	1986
Kinross	177	230	Fire	1986
Randfontein Estates	7	-	Winding plant	1986
Randfontein Estates	7	3	Fall of ground	1987
Buffelsfontein	8	4	Rock burst	1987
Western Deep Levels	9	5	Rock burst	1987
St Helena	62	5	Explosion	1987
ERPM	6	13	Rock burst	1987
Ermelo Mine	34	10	Explosion	1987
Coalbrook Colliery	12	4	Inundation	1987
Vaal Reefs	8	7	Fall of ground	1988/01/05
Western Deep Levels	4	-	Rock burst	1988/02/18
Doornfontein	4	18	Rock burst	1988/03/28
Harmony	10	5	Conveyance	1988/05/14
Western Deep Levels	7	-	Rockburst	1988/06/03

[68]Source: NUM Health and Safety Department, July 2000.

Rustenburg Platinum	4	-	Conveyance	1988/06/16
ERPM	4	1	Rock burst	1988/06/18
Iscor Works	6	-	Burning	1988/06/29
West Driefontein	4	3	Fall of ground	1988/08/16
Western Deep Levels	6	-	Rock burst	1988/09
Western Deep Levels	7	9	Fire	1988/10
Rustenburg Platinum	6	-	Conveyance	1988/10
Sasol	12	1	Fire	1989/01
Kloof	4	2	Rock burst	1989/01
Vaal Reefs	10	13	Fire	1989/03
Kloof	4	6	Rock burst	1989/04
Blyvooruitzicht	5	21	Rock burst	1989/04
Kloof	6	-	Excavations	1989/05
Kloof	10	-	Drowning	1989/07
ERPM	4	1	Strainburst	1989/08
West Driefontein	14	4	Rock burst	1989/08
Western Deep Levels	5	2	Rock burst	1989/11
Sub Nigel	9	8	Conveyance	1989/12
Libanon	5	-	Excavations	1989/12
Deelkraal	5	1	Conveyance	1990/03
Elandsrand	6	2	Rock burst	1990/06
Doornfontein	5	12	Rock burst	1990/06
Jackson Anthracite	5	-	Fall of ground	1990/06
President Steyn	10	12	Explosives	1990/08
Vaal Reefs	21	17	Explosives	1990/09
East Driefontein	5	-	Methane explosion	1990/10
Western Deep Levels	12	5	Rock burst	1990/10
Western Deep Levels	4	1	Rock burst	1990/10
Western Deep Levels	9	-	Rock burst	1990/11
Doornfontein	7	-	Rock burst	1990/11
Kloof	4	-	Rock burst	1990/12
Western Holdings	4	-	Fall of ground	1991/04
Free State Saaisplaas	7	5	Rock burst	1991/05
Buffelsfontein	7	16	Fire	1991/06
President Brand	4	7	Rock burst	1991/07
Western Deep Levels	6	-	Rock burst	1991/09
Blyvooruitzicht	7	5	Rock burst	1991/10
Freddies	6	-	Fall of ground	1991/11

Leeudoorn	4	1	Fall of ground	1991/12
Hartebeestfontein	7	68	Conveyance	1991/12
Deelkraal	5	2	Rock burst	1992/01
Western Deep Levels	7	12	Rock burst	1992/01
Western Deep Levels	10	1	Rock burst	1992/02
Western Deep Levels	4	-	Rock burst	1992/02
Venterspost	5	4	Conveyance	1992/02
Western Deep Levels	4	1	Rock burst	1992/04
West Driefontein	4	-	Rock burst	1992/05
Vaal Reefs	7	3	Rock burst	1992/06
Western Deep Levels	4	2	Rock burst	1992/07
Doornfontein	9	5	Rock burst	1992/10
Ermelo Mine	6	2	Explosion	1992/11
Vaal Reefs	5	-	Inundation	1992/11
Okiep Copper	4	-	Inundation	1992/11
Libanon	6	-	Rock burst	1992/12
Western Holdings	7	4	Rock burst	1993/01
Western Deep Levels	5	2	Rock burst	1993/01
Middelbult Colliery	53	-	Explosion	1993/05
Vaal Reefs	5	3	Fall of ground	1993/05
Buffelsfontein	6	6	Rock burst	1993/05
Leeudoorn	4	1	Rock burst	1993/08
Western Deep Levels	4	1	Rock burst	1993/09
Western Deep Levels	6	-	Rock burst	1993/09
Leeudoorn	5	2	Rock burst	1993/11
Vaal Reefs	4	7	Rock burst	1994/01
Leeudoorn	11	1	Rock burst	1994/01
Harmony	9	-	Inundation	1994/02
Gloria Colliery	16	1	Fire	1994/02
Deelkraal	4	3	Rock burst	1994/05
West Driefontein	5	1	Conveyance	1994/05
Western Deep Levels	4	2	Rock burst	1994/06
First Wesgold	4	-	Inundation (gas)	1994/08
Eersteling	6	-	Fall in shafts	1994/08
Doornfontein	5	-	Explosion	1994/10
Western Deep Levels	6	6	Rock burst	1994/11/03

CHAPTER TWENTY

THE KILLER DUST

Asbestos Fibres

When the suffering from accidents and diseases is assessed in the overall picture of mining in South Africa, then those who have worked in the asbestos mines and mills are in the worst-case scenario. They have been exposed to varying extents to similar hazards as those met in the mining of other minerals and, because all of the asbestos deposits were located in the rural areas, they also experienced the intensity of rural poverty throughout their working lives. The uniqueness of their suffering, however, has been caused by the deadly nature of the dust they have inhaled in the course of mining and producing asbestos. Asbestos dust contains fibres that have made it the most pernicious, pervasive and destructive of all mineral dusts.

South Africa possessed extensive and varied deposits of asbestos that had been mined from the early years of the 20th century. The commercial value of asbestos had depended on the qualities of its fibres that differed in thickness, length and chemical composition. In these terms, there were six distinct asbestos minerals, but only three have been exploited commercially.[1] The differences between the three types of fibres have enabled them to be put to as many as 3 000 different industrial uses.[2] The most frequently used fibre was chrysotile, known as white asbestos. It had the shortest fibres with the most fire-resistant and flexible qualities that made it easy to spin and weave in textiles. It was found in the Barberton area of the eastern Transvaal. Then there was crocidolite, or blue asbestos, with fibres of varying lengths and a high tensile strength that could withstand acid and sea water better than the others. It was mined in the asbestos mountains of the northern Cape and along a 90-kilometre stretch of the Drakensberg Escarpment in the northern Transvaal. The third type, amosite, a brown asbestos, lay in the same ore body as crocidolite in the northern Transvaal though, as the deposits stretched out to the east along the banks of the Olifants

[1]The average lengths of fibres ranged between 10 and 100 mm. In exceptional cases, fibres were found that were 300 mm in length. See H P Hart, "Asbestos in South Africa", *Journal of the South African Institute of Mining and Metallurgy*, Volume 88, January-December 1988, pp185-186.

[2]C J van der Merwe in *South Africa's Mineral Industry*, Minerals Bureau, Johannesburg, 1988, p141.

River to Penge, they became pure amosite.[3] Amosite possessed the longest fibres and had qualities that differed slightly from those of crocidolite.[4]

Different Asbestos-related Diseases

The length, thickness and chemical substance of asbestos fibres, however, accounted for very much more than the different industrial uses of asbestos. They also gave asbestos dust its distinctive health-threatening qualities. The routes by which the fibres in the dust entered and damaged the body were by inhaling or ingesting them. Inhalation, which was the most important route, depended on the diameter of the fibres but their length determined their ability to generate disease. The longer the fibres, the more dangerous to health they were. Some were too wide to inhale at all while those with the smallest diameters penetrated deepest into the lungs and often caused permanent damage to them. The most dangerous ones were those that got lodged in the body and could not be secreted as mucus.

For many decades after asbestos mining in South Africa had begun, neither the government nor mine owners had acknowledged that the dust it generated carried any exceptional health risks. They only admitted that where the ore bearing the asbestos fibres contained quartz then there would be the risk of contracting silicosis. Yet, outside of South Africa there were early diagnostic developments which revealed that asbestos dust had distinctive health-damaging qualities. In 1906, it had been noted at a UK inquiry into compensation for industrial diseases that asbestos caused its own brand of disease. And in 1919, the Prudential Insurance Company in the USA had refused to issue asbestos workers with life insurance policies.[5] There followed sporadic and unco-ordinated inquiries into pulmonary diseases amongst asbestos mineworkers until, in 1927, the term 'asbestosis' was coined to describe the distinctive physical consequences of inhaling asbestos dust. The following year, a South African pathologist had reported on two cases that related pulmonary impairment specifically to asbestos. He commented that it "has been known for some time that workers exposed to the dusty

[3]Amosite got its name from the village of Amosa which, in turn, was an acronym for the Asbestos Mines of South Africa, a subsidiary of the Anglo-American Corporation.
[4]For a description of asbestos mining in South Africa, see Allen, 2003, Volume II, Chapter Two.
[5]Irving J Selikoff and Douglas H K Lee, *Asbestos and Disease*, 1978, p21.

atmosphere arising from some processes involved in the preparation of asbestos materials suffer from pulmonary disability".[6] This view had been repeated at an international conference in Johannesburg in 1930 when it was stated that South African physicians had been aware of the existence of asbestosis for some time but had not published their findings.[7] G F Slade, a medical officer at a Transvaal chrysotile mine, who had investigated the incidence of respiratory impairment in workers employed in the mining and milling of chysotile between 1926 and 1930, had found a high proportion of workers with advanced physical signs of asbestosis.[8]

At that time asbestosis was seen as a variant of pneumoconiosis but, in the 1930s, medical authorities in Britain and the USA were gradually acknowledging that asbestos dust was not a version of coal or silica dust but had its own pernicious qualities. The Medical Inspector of Factories in Britain had presented a report to Parliament in 1930 establishing the fact that "the inhalation of asbestos dust over a period of years results in the development of a serious type of fibrosis of the lungs". The American government was also alerted to the dangers of asbestos mining. The asbestos mine owners in South Africa, with extensive international links, must have known about these revelations. Indeed, some of their representatives had discussed the matter in public at the Seventh Commonwealth Mining and Metallurgical Congress in Johannesburg in 1961.[9] Then, in 1971, at the annual shareholders' meeting of Charter Consolidated, the British company that mined asbestos in South Africa, a shareholder asserted that asbestosis incapacitated five black workers every week.[10] So, in the period when

[6]F W Simpson, "Pulmonary Asbestosis in South Africa", *British Medical Journal*, 1, 885, 1928.

[7]Selikoff and Lee, 1978, p23.

[8]The evidence was published in a doctoral thesis for the University of Witwatersrand in 1931. For a summary of its conclusions see M A Felix, Jean-Patrick Leger and Rodney I Ehrlich, "Three Minerals, Three Diseases: Asbestos Mining and Disease in South Africa", in M Mehlman and A Upton, eds, *Advances in Modern Environ Toxicology: The Identification and Control of Environmental and Occupational Diseases: Asbestos and Cancer*, Princeton, USA, 1994, Volume XXII, pp269-270.

[9]F E Keep, "Amphibole Asbestos in the Union of South Africa", *Transactions of the Seventh Commonwealth Mining and Metallurgical Congress*, 1961, Volume I, p107 and p113. One such representative spoke of the need "to reduce the amount of dangerous exposure of personnel to asbestos-dust-laden air..." while another reminded the Congress that "asbestos dust can be dangerous".

[10]South African Institute of Race Relations, 1971, p229. The South African Department of Mines denied the allegation and stated that the total number of certifications for asbestosis and asbestosis with tuberculosis in 1970 was less than half of that figure.

South Africa was becoming a major asbestos producer, the relationship between asbestos dust and serious pulmonary impairment had been firmly established in the scientific literature.[11]

Asbestosis involved a slow thickening and scarring of the lining of the lungs. If this was all that happened it was called pleural asbestosis. There were no visible symptoms of this disease, such as breathlessness, and only x-rays could detect it. Pleural asbestosis, however, usually developed into parenchymal asbestosis over time and when that happened the scarring affected the lung tissue itself, damaged its operation and caused breathlessness, a persistent cough and made exertion difficult. If the exposure to asbestos dust was prolonged or intense it could eventually result in death through heart failure, the collapse of the lungs or lung cancer but it was usually fatal only in conjunction with other diseases such as tuberculosis and pneumonia. If these had, however, been the only consequences, then asbestosis would have continued to be regarded as a version of pneumoconiosis or silicosis and treated similarly by the mines. Mine managers would have believed that dust suppression measures, through the use of water and ventilation, would have been sufficient to prevent it occurring.

Autopsies on people who had died after being exposed to asbestos dust, however, made it clear that it could lead to potentially wider and more devastating health consequences than asbestosis. They revealed that asbestos dust could have a diffused effect around the lungs, causing the pleura to thicken, become seriously impaired and lead to lung cancer.[12] It had been recognised in the USA in 1935, and confirmed in the UK 10 years later, that there was a link between asbestosis and lung cancer.[13] But it was not until the 1950s that scientific studies proved that lung cancer was a complication of asbestosis. Then it was shown that between 50 and 60 per cent of all asbestosis sufferers eventually died of lung cancer.

There was a further ominous revelation in the mid-1950s when Dr Chris Sleggs, the superintendent of the tuberculosis hospital in Kimberley, noticed that some of his patients who lived in the blue

[11]See Maria Anne Felix, *Environmental Asbestos and Respiratory Disease in South Africa*, a thesis submitted to the Faculty of Medicine, University of the Witwatersrand, for the degree of Doctor of Philosophy, Johannesburg, 1997, for a detailed analysis of the effects of asbestos dust in Mafefe in the northern Transvaal.

[12]The pleura is the thin transparent membrane enveloping the lungs.

[13]Jonathan Myers, *Asbestos and Asbestos-related Disease in South Africa*, SALDRU Working Paper No 28, Cape Town, June 1980, p22. This is a detailed and highly informative booklet about the subject.

asbestos region of the northern Cape, but who had not worked in the asbestos mines, were suffering from mesothelioma. This was a particularly virulent and fatal form of cancer of the lungs or abdomen that was not an aggravated version of asbestosis. It took from 20-40 years to develop, as did lung cancer, but in the interim, unlike lung cancer, it showed no symptoms. People could contract mesothelioma by inhaling a tiny amount of dust on a single occasion and show no clinical or radiological signs of its presence during the long incubation period. Once it reached a diagnosable stage, however, death came within about 14 months.

Dr Sleggs had reported his findings to the Pneumoconiosis Conference in Johannesburg in 1959 and had then collaborated with J C Wagner, a pathologist, and P Marchand, a thoracic surgeon, in an investigation of the relationship between mesothelioma and asbestos. They jointly and separately presented papers confirming and elaborating the findings by Sleggs. Their conclusions generated considerable international interest. Hitherto it had been thought that mesothelioma was a rare disease but the prompt international responses from specialists in Britain, the USA, Germany, Holland, Canada, Australia and Italy revealed that this was not the case. The information that was collected not only confirmed that there was a connection between the disease and asbestos but showed that there was a direct correlation between the spread of mesothelioma and the growth of the asbestos industry. Asbestos, it was discovered, was the only known cause of mesothelioma in about 80 per cent of the cases. This firmly and finally established asbestos as a killing mineral.

The Mine Owners' Responses

The revelations made at the 1959 Pneumoconiosis Conference about the prevalence of mesothelioma prompted it to initiate further studies. As a result, the government-sponsored Council for Scientific and Industrial Research established the Pneumoconiosis Research Unit. Its first project was a field survey into the incidence of mesothelioma in the northern Cape and the amosite asbestos range in the northern Transvaal. The project was financed jointly by the asbestos mining companies and the South African Cancer Association. The results encouraged the director of the unit to extend the survey to the Transvaal chrysotile mines. Until the 1960s, all the cases of mesothelioma had been concentrated amongst people in the northern Cape crocidolite blue asbestos region. This led to the presumption that white asbestos and, possibly, amosite were safe. In support of this, there were no known cases of mesothelioma amongst the ordinary black population of Johannesburg or in the north-eastern Transvaal where deposits of chrysotile were being worked.[14] The mine owners wanted that presumption to prevail. The adverse international publicity relating to blue asbestos had caused a sharp fall in the demand for it and they did not want the same fate to befall the markets for white and brown asbestos by research that showed a correlation between them and mesothelioma. They were alarmed and reacted to defuse and contain the situation.

First, they, and the Cancer Association, decided not to continue financing the Pneumoconiosis Research Unit's project. This meant the ending of all of its fieldwork. The mine owners refused to fund further research except that which conformed to their wishes. Then they embarked on a campaign to denigrate the authors of the research findings by accusing them of "trying to destroy a valuable export industry for self-aggrandisement". They were "told that they were meant to be looking into silicosis, not cancer, and it was made abundantly clear to them that they were no longer welcome on the company properties they had to visit if they were to conduct any further research".[15] Then they tried to restrict the publication of the first year's survey after it had been completed in April 1962. The report, as well as confirming the link between mesothelioma and asbestos dust, had reached the unpalatable conclusion that dust levels in the regions were so high that "people who live or have lived in proximity to asbestos mines or mills are in danger of contracting asbestosis even though they have had no industrial exposure to asbestos dust inhalation".[16] However, after having, at first, said "no" to publication, the asbestos employers and the Chamber of Mines proposed a compromise. They would have no objection to publication, they said, so long as "the cancer hazard was passed off as tuberculosis".[17] They wanted, at all costs, to avoid having the stigma of cancer attached to asbestos mining.

And for some years they succeeded. The Pneumoconiosis Research Unit's work on mesothelioma in Prieska in the northern Cape was resumed but under the title of "An Investigation Into Possible Air Pollution by Asbestos Dust", with no reference to mesothelioma. After two years, the director of the unit explained the fate of the report. In a letter dated 23 June 1964, he wrote: "… it was subsequently proposed

[14]Selikoff and Lee, 1978, pp29-30.

[15]Laurie Flynn, "South Africa Blacks Out Blue Asbestos Risk", *New Scientist*, 22 April 1982.

[16]Felix, Leger and Ehrlich, 1994, pp271-272.

[17]Flynn, 22 April 1982.

by the directorate of the unit that the survey as such be 'finalised' by the submission of the findings that had been made and a statistical analysis of these. Such a 'report' would not be published or made available outside the unit, other than to sponsors and the various members of the working committees that had been concerned with the conduct of the survey. This proposal has now been carried out."[18] The intrusion of the mine owners into scientific research matters did not end there. Thereafter, they played a critical role, not simply in determining whether or not particular research findings should be made public, but in determining the nature and scope of research in general in the industry. As one research group commented: "Apart from vetting research proposals and offering 'editorial advice', mine management representatives were privy to research results prior to publication ... Studies that were performed relied heavily upon management co-operation..."[19]

The mine owners, still fearing the effects on the demand for asbestos from its association with cancer, received some solace in 1973 from Dr Ian Webster of the National Research Institute for Occupational Diseases. He had asserted that there were unsolved problems in the relationship between asbestos and malignancy and that, possibly, asbestos alone did not cause it.[20] At that time there was still a wide disparity between the number of people suffering from mesothelioma in the Cape blue asbestos region compared with those in the Transvaal blue and amosite regions which, Webster maintained, could possibly be explained by the contamination of asbestos by another mineral that was present in the Cape but not in the Transvaal. He suggested that it could be either manganese or iron. His hypothesis, if proved, would have exonerated asbestos as the sole cause of mesothelioma but it was never substantiated.

The attempts by the mine owners to suppress the publication of critical papers continued, nonetheless. In 1978, Dr Leslie Irwig, from the National Research Institute for Occupational Diseases, proposed to read two papers at a conference organised by the New York Academy of Sciences. The research on which the papers were based was limited because it did not include any black victims but it still upset the mine owners. The first paper, emphasising that amosite was just as likely as

[18] *Ibid.*

[19] Felix, Leger and Ehrlich, 1994, pp271-272.

[20] Ian Webster, "Asbestos and Malignancy", *South African Medical Journal*, 3 February 1973, pp165-171. Dr Webster had been involved in the research into mesothelioma in the early 1960s and had been given the uncomfortable task of telling the asbestos mine owners of the research findings. (Flynn, 22 April 1982.)

crocidolite to cause asbestosis, presented no real problems to the mine owners. The second paper, however, was a different matter because it raised the spectre of cancer by concluding that people living in the vicinity of asbestos mines but not necessarily working in them "died more often than usual from cancer".[21] The South African Medical Research Council, under pressure from the asbestos industry, instructed Irwig and his co-author, Dr Hannes Botha, not to read the second paper. It was neither read nor ever subsequently published.[22]

The Spread of the Dust

However, despite the endeavours of the asbestos mine owners, it had already become apparent by 1963 that asbestos dust was a danger to health wherever and whenever it was inhaled. More than half of the reported cases concerned people who had never worked in the mines or mills at all but had simply lived in their vicinity. For example, in the first 33 recorded cases of mesothelioma in South Africa, seven were white women who had never worked with asbestos and had only been exposed to its dust in the environment.[23] Thus, depending on how dust was spread, hundreds of thousands of people who lived in asbestos-producing areas were at risk. Its most frightening quality, however, was that whereas asbestosis and lung cancer developed after a prolonged exposure to asbestos dust, mesothelioma could be contracted through only a very brief acquaintance with it.[24] This meant that children playing occasionally on dumps of asbestos dust, visitors walking along roads constructed from asbestos tailings or people living in houses where asbestos was used as a building material were all possible victims. Any exposure was a risk. Some of the hundred cases of

[21]Flynn, 22 April 1982.
[22]*Ibid.*
[23]J C Wagner, C A Sleggs and Paul Marchand, "Diffuse Pleural Mesothelioma and Asbestos Exposure in the North-Western Cape Province", *British Journal of Industrial Medicine*, Volume 17, 1960, p270.
[24]There are a number of case histories relating to people who have died of mesothelioma after a relatively brief contact with blue asbestos dust. One such case concerned Hilary Mitchell, a medical sociologist from South Africa who had moved to England with her husband, Professor Clyde Mitchell, and whilst there died of mesothelioma. Her only contact with asbestos dust had been when, at the age of 11 years, she had visited her father's small asbestos mine in southern Botswana for a brief period during her school holidays. Forty-five years later she had constant bouts of pleurisy and the first symptoms of mesothelioma were diagnosed. She died two years later in 1976 at the age of 58 years. (Conversation between the author and Professor Mitchell.)

mesothelioma known by 1964 had been associated with the asbestos area of the northern Cape for as little as a few days or, at the most, weeks.[25] There were also instances of asbestosis in factories that did not use asbestos but which were situated close to those that did.[26] Whereas it was possible for black mineworkers to reduce the risks of contracting other dust-related diseases by having relatively long periodic breaks in their work cycle, they could not, in the same way, protect themselves from mesothelioma. As it was not possible to stop the disease progressing by moving away from the presence of dust, regulatory or suppression measures alone, whether in the mines, mills or environment, were not sufficient safeguards for none could guarantee 100 per cent protection. Neither black nor white people were safe. In the asbestos mines, there was no safe dust exposure level.

The difficulty was that wherever asbestos was mined, there was dust. The asbestos was found in banded ironstone formations from which it had to be separated and cleaned of grit and dust impurities until it was as pure as possible. The quality of the finished mineral depended on its purity. Dust was inevitably and invariably generated at every stage of the extraction and purifying process. The extent and intensity of the exposure of mineworkers to it depended, in the first instance, on the means used to separate the ore from the waste and they, in turn, depended on whether the rock was difficult to drill and blast; whether the fibre-bearing ore could be sorted from the waste underground or on the surface; whether the production process was labour- or machine-intensive and over what distances and in what quantities the ore had to be moved in each of the separation and extraction processes.

During the drilling and blasting some asbestos was invariably shattered and dispersed in the air. Where it was possible, the initial sorting was done underground in order to save on hoisting costs. The waste was then used to fill in as support for the stopes. As this was done within confined working places with low roofs the mineworkers were exposed to high levels of dust. Where underground sorting was not possible, then it was done on the surface and the waste was returned underground to fill in. Dust was released into the air every time the waste was moved. The asbestos-bearing rock which was sent to the surface, or which was already there after the waste had been separated, had to go through another process of separation in order to isolate the fibres. This began with a difficulty in the case of crocidolite and

[25]Ian Webster, "Asbestosis", *South African Medical Journal*, October 1964, p872.
[26]*Ibid*, p870.

amosite asbestos because the fibres and the rock to which they were attached had a similar chemical composition so that, except for the fibrous nature of the asbestos, it was difficult to distinguish between them.

All of the underground black mineworkers were men but on the surface and in the mills, women and children were employed on such jobs as sorting, crushing and packaging the fibres. Groups of black women had traditionally dislodged the fibres from the ore with 'five-pound' hammers called 'cobbers', but as the mines expanded they used machines to do the 'cobbing'. Whichever method was used, some of the fibres were chipped away and entered the dust. After machines were introduced the women were employed primarily on dusty mill tasks. The women, moreover, frequently carried their babies to work on their backs where they, too, were exposed to asbestos dust. Though it was illegal to employ children in the mills they were the preferred choice for crushing the fibres because of their size and agility.[27] After the fibres had been separated from the ore they had to be sorted and that was best done by sight and hand. Then they were dried as part of the process of purification. At first they were sun-dried on a thin bed on a concrete floor and were turned over by hand. It was impossible to control the dust in that situation and to prevent workers from being contaminated by it. After 1956, however, the process was improved in the large mills through the introduction of mechanical dryers using hot air.

The need to keep the fibres dry varied between the mines in the north-eastern Transvaal where the climate was wet and those in the northern Cape where, with its arid land and hot, dry climate, the dust was easily airborne and water had to be used underground as a dust suppression measure. However, wherever water had to be used to suppress dust, it had to be used sparingly so as not to wet the fibres unduly. It was not, therefore, an effective dust-suppression measure. It was important in all of the mines to prevent water from collecting in dams or pools through which the ore would have to be scraped. If any of it did get wet, then it had to be collected to be specially dried. The subsequent crushing and grinding not only created asbestos dust but attached more dirt and grit to the fibres which then had to be removed by processes that put more dust and fibres into the air. One of the ways used to remove the last vestiges of grit and dust was to pass the finely ground ore and fibres over a shaking screen which caused the fine grit

[27]Felix, Leger and Ehrlich, 1994. See also Flynn, 1992, pp192-194.

and dust to pass through, leaving the fibres behind. Every attempt to achieve quality control created dust. It had been estimated in 1961 that 6,5 per cent of the tonnage of ore that went through the mills ended up as dust and that between 20 and 31 per cent of the actual fibres were lost as waste and dust.[28]

Once the ore reached the surface, the dust that was generated by the various processes of crushing, drying, separating, re-crushing, air separation, transporting and manufacturing entered the atmosphere and settled in the environment. These emissions were critically significant for the health of those working on the various processes and living in their immediate vicinity. However, the principal source of dust in the environment was from the dumps made up of waste from the mills and the uses to which that waste was put. When the mills were built and had begun to accumulate their waste dumps in the northern Cape towns of Koegas, Prieska, Kuruman and Griquatown after the First World War no consideration was given to their location in relation to the residential areas. The large mill at Kuruman, for instance, was less than 300 metres from the main street of the town. That situation remained largely unchanged.

The black populations in the vicinity of the mines and mills lived in mine compounds and townships built on and with asbestos tailings.[29] In the period of expansion after the Second World War there had been as many as 31 mines near to Kuruman, linked by roads partly made from asbestos waste and served by people who lived in houses built with asbestos materials. Pathologists, writing about asbestos-related diseases in 1960, stated that "in the early days the manager and labourers lived within a few yards of their places of work, and even today the non-white prefers to live as close as possible, and the children play on the dumps from the mine and mill".[30] For some time after mining had stopped at the White Bank blue asbestos mine, near Kuruman, "there were mountains of asbestos tailings stretching as far as the eye could see".[31] The dust, however, was not confined to the vicinity of the mills and dumps. In the 1960s there was such "a huge plume of dust" over the White Bank mine that local aeroplane pilots were able to navigate by it.[32] The fibres became trapped in the trees and the grass; they clung to food, clothes and buildings and contaminated the rivers and lakes until the water

[28]Keep, 1961, p112.

[29]Flynn, 1992, Chapter 9, describes the spread of asbestos dust in the northern Cape.

[30]Wagner, Sleggs and Marchand, 1960, p262.

[31]Flynn, 1992, p169.

[32]*Ibid.*

turned blue. They were conveyed by the winds to the white suburbs and entered the lungs of their unsuspecting white residents.

It was a similar situation in the Pietersburg asbestos fields that stretched from the town of Pietersburg eastward into Swaziland, where the villages were clustered around the mines and mills and in the vicinity of about 43 of the tailings dumps. The mills, which were sufficiently open to allow the wind to blow through and carry the dust into the open, were situated alongside streams in the valleys and were served by tracks that eventually became the main roads. As the mine owners made no attempts to house and feed them, the black mineworkers constructed their own settlements close to the mills. Over time the settlements became established villages but without piped water or electricity and with very few other service amenities. In the Mafefe district, for instance, 30 villages served 12 mills and eight mines.[33] Twenty-seven of the villages were polluted by asbestos dust from 20 tailings dumps that bordered the villages, the main roads and the riverbanks.[34] Of these, only four had communal water taps, two had pumps with which to draw borehole water, while the remaining 21 had to obtain their water from one of two rivers that were contaminated with asbestos dust by effluent from the mines and mills.

Asbestos dust was everywhere in Mafefe, as it was in the northern Cape. And, because it was indestructible, it remained with its potent qualities wherever it settled. Asbestos waste was used instead of river sand for cement brick-making in three local brickyards. Many people actually made their own bricks and road surfaces with asbestos tailings. Women mixed asbestos tailings with cow dung to plaster the walls of their homes or the floors of their '*lapas*' (courtyards) to give them an aesthetically pleasing smoky blue appearance. Water that already contained asbestos dust ran down asbestos cement roofs into tanks and then through asbestos cement pipes. There was no escape from the dust. No-one in the villages was aware that it was dangerous to inhale asbestos dust and no-one, therefore, took any precautions. The tailings dumps were never fenced off to prevent people from disturbing them. Altogether, seven of the 12 schools, 18 of the 34 public buildings and one in three of the homes in Mafefe were said to have been contaminated with asbestos dust.[35]

[33]Felix, 1997, pp61-62.
[34]Maria Anne Felix, "Risking Their Lives in Ignorance: The Story of an Asbestos-polluted Community", in Jacklyn Cock and Eddie Koch (eds), *Going Green*, Cape Town, 1991, pp35-36. See also Felix, 1997, pp29-30 and chapter 6.
[35]Felix, 1997, p146 and p153.

The Suppression of Dust Through Regulation

Despite the evidence about the effects of asbestos dust on health, there was no consensus in the South African medical profession about it. This lack of agreement gave the mine owners a legitimate reason to procrastinate about introducing measures to regulate it. The procrastination, however, began with the government that throughout was concerned more with encouraging the extension of asbestos mining than with protecting mineworkers and their families from the dust it generated. There had been some underground dust sampling in the asbestos mines since 1940 but it was not until 1947 that surface dust in the mills was sampled. The next step came in 1953 when the Government Mining Engineer required asbestos mines to be registered so that medical assessments could be enforced and workers would be entitled to compensation. The first specific dust standard, however, was not imposed until after the 1956 Mines and Works Act was passed. Then the maximum allowable concentration of dust was 45 fibres per millilitre of air but this standard had still not been applied 13 years later, even though it was excessively lax by international norms.[36]

The South African standard was a very rough approximation that gave workers in the dust-exposed areas of the mines and mills relatively little protection. The fibre content in the air was measured by an official walking around a designated area carrying a sampling device which was either a konimeter or perhaps, after 1962, a thermal precipitator. Both of the instruments gave underestimated readings. Their measures were approximations that did not monitor actual personal exposures to dust. They were also average ones. Thus high dust values such as those experienced by drillers, sweepers and lashers were lost in the averaging process. The significance of the readings depended, therefore, on the choice of the area to be sampled and the time in the shift when they were taken. The standard had been improved when the tolerable limit was reduced, first to 12 and then to 10 fibres per millilitre of air, between 1973 and 1976. This was still unacceptable by international standards. In the USA, the limit was two fibres per millimetre of air. This became the international standard.

Very little notice, however, was taken of any of the dust regulations designed for the South African asbestos mines at that time. There was virtually no pressure to conform to them by the Government Mining

[36]Felix, Leger and Ehrlich, 1994, p281. "A fibre needle was defined as a rod-like particle with a width of less than 5 microns and a length between 5 and 100 microns" (*ibid)*. A micron is one-millionth of a metre.

Engineer. The Mines Inspectors had been infrequent and irregular visitors to the mills and mines. In the Mafefe area, for example, only 35 inspections had been made during the 24 years between 1949 and 1973.[37] That did not even amount to a rudimentary surveillance of compliance with the dust regulations. The government's attitude to atmospheric pollution by asbestos dust had been even more neglectful. The first regulation in this sphere of mining was the Atmospheric Pollution Prevention Act of 1965 but its terms covered only gold mining. It was not until a media campaign in 1984 had publicised the pollution of residential areas by asbestos tailings dumps that the Department of Mineral and Energy Affairs initiated an abatement programme on waste dumps.[38]

By the mid-1980s, increasing international concerns about the future of asbestos mining had persuaded the government to increase its scrutiny of the asbestos mining companies to ensure that they conformed to its dust control regulations. In general, the companies had responded by making changes at all stages of the production process. Underground mining techniques were altered to prevent dust from rising; an ore-washing system was introduced; conveyors and screens were enclosed and subjected to negative pressures to prevent dust from rising; dust-suppressers were attached to machines that generated dust; the dust that was extracted was passed through air-ducts to bag filters to be disposed of dry or mixed with water so that it could be piped away as sludge. Safeguards were introduced in the processing of the final product for instead of using women and children to compact the fibres with their hands and feet it was now pressure-packed in dust-proof bags, shrink-wrapped and either palletised or transported in containers.

A monitoring system was introduced to check the effectiveness of their dust suppression measures. Air quality readings had to be taken each month and sent in quarterly reports to the Government Mining Engineer. All new recruits had to take an induction course that explained how to handle asbestos and check for dust hazards. Once the recruits were working they were given periodic medical checks, including x-rays. The Griqualand Exploration and Finance Company (GEFCO) had established shared primary and occupational healthcare facilities for its White Bank and England blue asbestos mines near Kuruman. The company had based a clinic at its White Bank mine that

[37]Felix, 1997, p38.

[38]*Ibid*. The programme was carried out by the Institute for Reclamation Ecology at the University of Potchefstroom.

was equipped with consulting rooms, an x-ray facility, eye-testing equipment, an audiometer and sound booth, a dispensary and a five-bed ward. It employed a general practitioner for 25 hours a week and a full-time medical staff comprising a nursing sister, three nurses, three clinic assistants and a radiography assistant. Routine chest x-rays were taken and, during 1991, lung function testing was introduced. It appeared that everything was in place to keep a detailed and up-to-date case history for each employee. As a result of these changes, managers and their scientific advisers had begun to believe that the occupational risk from asbestos was minimal, that the risk to the general public was virtually non-existent and that "asbestos-related diseases [would] in due course disappear altogether..."[39]

Changes in the Ownership of Asbestos Mines

All of the owners of asbestos mines over the length of their history bore responsibility for asbestos pollution. However, because of the long incubation period of some asbestos-related diseases, the owners since the 1950s had become responsible for their impact in the 1980s and 1990s. During that time the mines had changed hands a number of times. Multinational conglomerates were regularly selling or buying mines to maximise their profits. After 1948, the dominant company was the Anglo-American Corporation through its subsidiaries, Cape Asbestos in Prieska, Asbestos Mines of South Africa in Penge and Cape Asbestos Fibres that was responsible for marketing the output from the mines. The other principal firm was the Dominion Blue Asbestos Company in Kuruman, owned by the Swiss and Belgian company, Everite-Eternit. By the mid-1970s, the ownership of the mines had changed and was dominated by three major conglomerates, two of which were foreign-owned.[40] In 1976, the sole domestic company was the General Mining and Finance Corporation that controlled 42 per cent of South Africa's production of all types of asbestos. It had acquired the Griqualand Exploration and Finance Company (GEFCO) that mined blue asbestos. It also owned Msauli Asbes Beperk that dominated the production of chrysotile. The second conglomerate was the British-based firm, Charter Consolidated, which owned Cape Industries. The third was Everite-Eternit, still mining blue asbestos but with worldwide asbestos

[39]Hart, January-December, 1988, p197.
[40]Myers, June 1980, pp6-10.

interests. Lonrho had a marginal interest in asbestos mining and there were a number of small individually owned mines in the Pietersburg fields, north-west of Penge.

During the 1980s the distribution of ownership was directly influenced by the sharp decline in world demand for asbestos due to the international pressure on governments from trade unions and consumer groups to ban its use in manufacturing. A number of Western governments had responded by imposing complete or partial bans on its use. As about 95 per cent of South Africa's production had been exported, it was directly and dramatically affected by the bans. The demand for its asbestos fell by almost 60 per cent between the peak production year of 1977 and 1992. In 1977, there had been 34 mines, employing 21 665 mineworkers. By 1992, there were only six mines with a total labour force of 3 176. Penge, the only mine producing amosite, was closed down in June 1992 when the Japanese government forbade its use in the manufacture of calcium silicate insulation boards. Following its closure there remained only two mines in the northern Cape, producing crocidolite, and three chrysotile mines in the Barberton area. By 1994, the total labour force for the five mines had fallen to 2 230.

The foreign-based multi-national companies operating in South Africa experienced most of the pressures from the international campaigners and they tried to get out of asbestos production as quickly and quietly as possible. They closed mines, sold them or "even in some cases [tried] to give them away to nominally independent black homelands".[41] But there were South African-based companies willing to take over. GEFCO became part of Gencor when it was formed through a merger between the General Mining and Finance Corporation and Union Corporation in 1980. Whilst in Gencor, GEFCO acquired the large Penge amosite mine. However, it quickly broke away to become an independent company and, in 1981, acquired a monopoly of blue asbestos production when it took over the asbestos mining interests of the Swiss/Belgian Dominion Blue Asbestos Company. In August 1988, GEFCO and Msauli Asbes Beperk, still the main producer of chrysotile, were bought by a newly-formed consortium called Hanover Mining Holdings, which then controlled virtually the whole of the minuscule asbestos industry in South Africa.

[41]Flynn, 1992, p202.

Estimating the Extent of the Problem

There were difficulties in estimating the prevalence of asbestos-related diseases. When the mineworkers became sick they were usually returned to their rural homes where the diseases developed without treatment because there was no medical infrastructure with diagnostic or treatment facilities. Secondly, where the gestation period was long, as in the case of lung cancer and mesothelioma, black mineworkers, who might have contracted one or other of those diseases, frequently died in their homes of other causes before showing any obvious symptoms of either one of them. But even if that had not been the case and they had shown the symptoms, they would have died without diagnosis, treatment or recognition. Although South African legislation, at the time, required that deceased miners and mineworkers should receive post-mortem examination of their cardiovascular organs, subject to permission from their next of kin, in practice, the only black mineworkers to receive them were those who died whilst in employment because there was no follow-up system in the rural areas to ensure compliance with the legislation.[42]

The lack of data about the medical histories of black mineworkers was a severe handicap to those scientists who wanted to study the extent and characteristics of asbestos-related diseases. Even their service records, kept by the companies, were rudimentary, as the Medical Bureau for Occupational Diseases in Johannesburg had discovered in 1981 when it set up a project to examine the mortality of workers in the crocidolite and amosite mines between 1946 and 1948. There was so little information about the morbidity of black mineworkers that they had to be excluded entirely from the project even though some of them "were known to have been exposed to exceptionally high fibre concentrations".[43] Reliable explanations about the incidence of the diseases amongst black mineworkers therefore depended on field studies by independent medical research groups. One such group had gone to the Havelock chrysotile mine and mill in November 1978 in order to conduct an epidemiological study. It was one of the first occasions when working black mineworkers were included in the sample to be studied.

The Havelock mine was owned by the British firm, Turner and

[42]Felix, Leger and Ehrlich, 1994, p275.

[43]G K Sluis-Cremer, F D K Liddell, W P D Logan and B N Bezuidenhout, "The Mortality of Amphibole Miners in South Africa, 1946-80", *British Journal of Industrial Medicine*, 1992, 49, p567.

Newall. It was situated in Swaziland, about one kilometre from the South African border. The fibres, however, had always been transported for treatment, by an aerial ropeway over the border to Barberton, the nearest South African town, and from there they had been despatched for manufacturing purposes to Turners Asbestos Products at Jacobs in Natal. The mine and mill had been visited in 1976 by Dr Lewinsohn, a medical consultant for Turner and Newall, in order to "assess the position with regard to asbestos and health".[44] Havelock was a segregated company town in which the mine provided housing, schools, a small cottage hospital and a general medical practice run by two medical officers. It also supplied the residents with a free supply of wood fuel for fires on which they cooked their own food. The fires were lit outside the houses, except when it was cold. In the cold weather there was so much smoke pollution that people died of carbon monoxide poisoning. There was also a high incidence of chronic bronchitis, tuberculosis, bilharzia and venereal diseases.

The mine, Lewinsohn discovered, had instituted a 'dusting and spillage' programme in 1951 but it had never formally put it into operation in order to review health hazards or institute dust control measures. There had never been any dust sampling. The manager had tentatively recognised the presence of dust by recently appointing a dust sampler but while Lewinsohn was there he had not yet started work. Masks had been issued for anyone working in dusty conditions but, according to Lewinsohn, they were unsuitable for asbestos dust. The two medical officers showed little interest in the health hazards of asbestos mining and did not make regular visits to the mine or mill. The asbestos dust levels underground were moderated by the damp conditions but silica dust was released by the ore blasting and, in consequence, there was a high prevalence of pneumoconiosis amongst the mineworkers.

The conditions in the mill were so chaotic that Lewinsohn was reluctant to enter. He wrote: "Going through the mill to the storage shed … climbing over piles of fibres was frightening. The operator sitting in a glass box … was covered in fibre. I was astonished to see a man

[44]The details of his visit were recorded in documents belonging to Turner and Newall which the Chase Manhattan Bank in New York had obtained by court order in the USA in its litigation with Turner and Newall over the use of asbestos in a Chase Manhattan Bank. Chase Manhattan had distributed the documents to interested parties throughout the world. They dated from 12 April 1976 to 28 March 1985. Copies of the documents were sent to Leigh, Day and Co, a legal firm in London, that was beginning legal proceedings against Turner and Newall for compensation for some of its former employees in South Africa. The solicitor, Martyn Day, passed copies of the documents relating to the Havelock mine to the author.

poking at a blocked screw conveyor and creating clouds of dust all about him ..."[45] A medical officer had been appointed in 1975 but he was so busy with other commitments that he was able to spare only five hours a week to asbestos health hazards. Conditions in the mill at Havelock, Lewinsohn concluded, "were dusty and needed improving". Yet despite this, the manager claimed that there was no serious asbestos disease at Havelock. As there had never been an epidemiological study there, and only one case of asbestosis had ever been reported, it was difficult to contradict him. Nonetheless, he had taken some steps to control the emission of dust in the mine. A dust monitoring system had been introduced. Little progress, however, had been made in the mill. Dust was still being cleared away with brushes and shovels rather than with vacuum techniques because, the manager claimed, the Swazis found it difficult to learn how to handle a vacuum cleaning system.

Dr Lewinsohn wrote a highly critical report of the conditions in the mine and mill and, as a result, members of the Pneumoconiosis Unit of the British Medical Research Council visited Havelock in November 1978 in order to conduct an epidemiological study. In its final report it had revealed an entirely different picture from that presented by the management.[46] Two groups had been examined. The first comprised 271 men who had worked at Havelock for 10 or more years. The study showed that 29 per cent had simple pneumoconiosis, just over 23 per cent suffered from asbestosis and 17 per cent had chronic bronchitis. Nineteen per cent of the sample who had worked in the mine and 54 per cent of those who had worked in the mill had asbestosis. In the second group, consisting of 273, more than 14 per cent had asbestosis. Overall, 102 out of the 544 who were examined suffered from asbestosis. There were no cases of mesothelioma or lung cancer in the sample. That study was followed up during 1979 by a medical examination of all the remaining employees in the mine and mill until, by the end of the year, the whole work force of 2 200 had been seen. There were 183 cases of asbestosis and 154 cases of possible asbestosis. As most of the possible cases would fairly soon become actual ones it meant that more than 15 per cent of the work force had contracted asbestosis. There were still no cases of mesothelioma but there was one instance of lung cancer.

During the investigation it was discovered that two wives were also suffering from asbestosis, so clinical examinations and x-rays were

[45]Memorandum to Mr H S D Hardie, 15 April 1976.

[46]The final report was published in the *British Journal of Industrial Medicine*, 1982, volume 39, pp338-343.

offered to the wives of deceased workers who had been living at Havelock for 15 years or more. Of the 175 widows who were examined, almost 13 per cent had symptoms of asbestosis. This rate was not much less than that for the workers themselves. They had absorbed asbestos dust both through shaking dusty overalls before washing them and from its presence in the atmosphere. It was decided after that to offer clinical examinations to the wives of all the workers. The mill manager, who had worked at Havelock for 28 years and who was in part responsible for the negligent approach to the asbestos dust problem, was diagnosed in September 1984 as suffering from lung cancer as a complication of asbestosis.

This episode of change in the history of the treatment of the dust hazard at Havelock ended with the introduction of new dust filter masks, the installation of a system of dust monitoring and an employee surveillance programme in which every worker had a medical examination, including an x-ray, every two years except for asbestotics who were examined every year. A new means of recording medical and occupational data for each employee was introduced so that a case history was available for every worker diagnosed as having an asbestos-related disease. A medical officer was engaged with special responsibility for those diseases. Finally, management agreed to provide alternative, less dusty jobs for those who had definite or possible asbestosis. These changes were symptomatic of the anxiety of the managers of asbestos mines throughout the country to distance asbestos from the stigma of being labelled a killing mineral but they were too late for the thousands still living who had worked in the mines and mills and whose cases the NUM was beginning to follow up.

The Involvement of the NUM

Until the late 1980s, the NUM had no history of involvement in the asbestos mines and had given no indication that it was aware of the risks black mineworkers there were taking. In 1988, it had only three small asbestos branches. Two of them were in the Nelspruit region of the eastern Transvaal. At Gencor's African Chrysotile Asbestos Mine it had organised 84 per cent of the 1 600 members of the black labour force and was recognised by Gencor. The other branch in Nelspruit was at the Kaapsehoop Asbestos Mine. It had no recorded membership and no recognition agreement. The third branch was at a GEFCO mine near Kuruman. It, too, had no recorded membership but it had an access agreement to allow it to enter the mine. The NUM had shown no signs of being bothered by its peripheral role in the asbestos mines and had been willing to allow the Black Allied Mining and Construction Workers' Union a free hand to organise them.[47] Concerns about asbestos mineworkers had never been raised at any of its congresses.

This situation began to change when Manne Dipico, a NUM official since 1985, was made responsible for the Kimberley and the northern Cape region in 1992.[48] He began a campaign to organise the asbestos mineworkers in his area but by then the industry had shrunk to a fraction of its peak size in the mid-1970s. Dipico, however, recognised that the union's primary task was not, in fact, recruiting the unorganised but catering for the needs of the many mineworkers who had been retrenched after being exposed to asbestos dust. He was soon given an opportunity to get started.

In 1991, only three of the 31 asbestos mines that had existed in the northern Cape were still operating. They were White Bank, Merencor and England, located near Kuruman. In November and December of that year, the owners, GEFCO, began to close the White Bank operation and retrenched about one half of its labour force of just over 1 000. The remaining 547 were retrenched the following March and April. All of the retrenched workers were given exit medical checks in order to assess whether or not they were entitled to compensation payments. A number of the second batch, however, felt that the screening process had been faulty and that, in consequence, they had been denied their right to obtain compensation. They complained to Dipico who took up their case on behalf of the NUM. At his request, the Industrial Health Research Group from the University of Cape Town agreed to check GEFCO's medical screening system by examining the entire second batch of retrenched mineworkers. He organised the project with the help of three shaft stewards and with the full co-operation of GEFCO.

Doctors M F Abdullah and S J Manjra went to Kuruman on behalf of the Industrial Health Research Group and examined the case histories of the retrenched mineworkers. They found that the clinic at White Bank placed a special emphasis on asbestos-related diseases and was

[47]The Black Allied Mining and Construction Workers' Union had attempted to organise asbestos mineworkers during the 1980s and had succeeded at the Penge mine in the north-eastern Transvaal. It had campaigned over the issue of health and safety but had seen the solution in the closure of the mines. This union had become an important affiliate of the black consciousness Azanian Congress of Trade Unions with a signed-up membership of 20 000 by the mid-1980s. (Kraak, 1993, p193.)

[48]On his release from prison in 1990, Dipico had returned to the NUM as its National Education Co-ordinator until being made area organiser in the northern Cape.

properly equipped to conduct an effective screening process.[49] Until nine months previously the clinic had used mass miniature radiography (MMR) for the statutory annual chest x-rays and the exit screening process but the x-rays had been neither sufficiently clear nor detailed to enable lung diseases to be effectively diagnosed. For that purpose it was necessary to have full-size chest x-rays.[50] The company, therefore, had purchased a full-size x-ray machine along with the means to undertake lung function tests.

The team examined 1 550 x-rays at White Bank and found that 20 per cent were unreadable, 60 per cent were of poor quality but just readable and only 20 per cent were of good quality. It discovered that the assistants who had taken the x-rays had not had any formal training. They had learned 'on the job' and were unaware of what constituted good quality in x-rays. Even the doctor responsible for reading the x-rays was not qualified for that task. The team found that 106 out of the 547 retrenched mineworkers had radiological evidence of asbestosis so applications for compensation were made on their behalf. However, in order to assess for lung impairment it was necessary for them to be given lung function tests. The situation with regard to these was also unsatisfactory.[51]

The lung function machine installed by GEFCO nine months beforehand was not calibrated so that the results were inaccurate. The technician responsible for operating the machine had received no training and was unable to complete an acceptable test. Neither he nor anyone else in the mine could read an acceptable test. The team found that 95 per cent of the lung function tests that were carried out were unusable. The story of incompetence did not end there. The company had bought a personal computer containing medical software but no-one in the clinic was trained to use it. The filing system was chaotic. The chest x-rays were filed haphazardly while the medical records were so inadequate and poorly kept that it appeared that no-one in the clinic really understood the nature of asbestos-related diseases.

This litany of neglect was reinforced by the team's assessment of working conditions at the Merencor mine and mill. They found that there was no culture of safety either amongst management or workers.

[49]M F Abdullah and S J Manjra, *The Kuruman Asbestos Report*, Industrial Health Research Group, University of Cape Town, 1992.

[50]At Merencor they had the facility to perform full-sized x-rays, but the machine was old and all of the x-rays were of such poor quality that it was almost impossible to make reliable diagnoses.

[51]Abdullah and Manjra, 1992, p16.

There was very little respiratory protection and the principal means of controlling the dust was through the lavish use of water. Water, of course, was a crucial safety measure but it had to be accompanied by good 'housekeeping'. The asbestos fibres that fell on to the floor were not cleared away and were converted into sludge that ran through defects in the flooring from one level to another. The asbestos droplets, which contaminated the clothing of the workers, dried quickly and released the fibres into the air, thus negating the purpose of the water in the first instance. The situation had many contradictions. For instance, the company had an environmental pollution laboratory at its offices in Kuruman with a full-time technician who was responsible for monitoring airborne dust. Dust samples were sent regularly to France for quality control. Yet little was done outside to prevent environmental contamination. For instance, the asbestos tailings dumps at White Bank and Merencor were not fully covered to prevent the fibres blowing away.

When the Penge mine was also closed, in June 1992, the NUM went through the same screening exercise as at Kuruman in order to ensure that the 591 mineworkers who were dismissed were properly medically examined and given compensation if it could be shown they were entitled to it. The union again approached the Industrial Health Research Group which sent Dr M F Abdullah and Dr Mohamed Jeebhay to Penge on 20 October 1992.[52] Although the Penge mine was also owned by GEFCO it had a much more efficient screening standard than the clinic at the Kuruman mines. Only two out of 568 tests were unreadable so that it was possible to make a fairly accurate assessment of the condition of the workers. It found that 259, or 45 per cent of the total, had confirmed asbestos-related diseases and that a further 19 per cent were borderline cases. GEFCO, moreover, had given the workers lung function tests every two years and had tested each retrenched worker before he left the mine premises. But this part of the process had only a veneer of efficiency. The tests were made on the correct type of machine and the results were recorded on a computer. But the machine was not properly calibrated and its time signals were inaccurate. The test results, the Industrial Health Research team discovered, were unusable.

The evidence that was collected for the NUM about the levels and prevalence of asbestos-related diseases at Kuruman and Penge was

[52]See M F Abdullah and M Jeebhay, *Medical Screening at the Penge Amosite Mine*, prepared for the National Union of Mineworkers, Industrial Health Research Group, University of Cape Town, 22 October 1992.

substantiated by the findings of a simultaneous study by Dr Maria Anne Felix from the National Centre for Occupational Health in Johannesburg. In 1987, Dr Felix had begun a study of the environmental impact of asbestos pollution in the rural area of Mafefe.[53] It was the first study of its kind. Indeed, until 1992, there was no data available about people exposed to the Transvaal crocidolite asbestos dust. When the project started, all of the asbestos mines had been closed but the 30 villages that had housed the workers and their families were still there. So was the dust.

Dr Felix estimated that 52 per cent of the occupationally exposed adults in Mafefe and 34 per cent of those with only environmental exposure suffered from asbestos-related pleural disease. Ten per cent of all adults in Mafefe were likely to suffer from respiratory impairment.[54] All the signs were that large proportions of the populations in asbestos mining areas were likely victims of asbestos pollution. This led Dr Felix to conclude: "The mining and milling of asbestos in South Africa [had] led to three local epidemics in which tens of thousands of persons suffered premature death from asbestosis, pleural thickening, lung cancer and mesothelioma, following occupational and environmental exposure." Moreover, she added, there were thousands of former mineworkers waiting to be assessed for compensation.[55] This alarming prognosis endorsed the view of the Industrial Health Research Group from the University of Cape Town that had forecast an epidemic of asbestos-related diseases within 10 or so years.[56]

The Question of Compensation

The predominant question now facing the inhabitants of all asbestos mining areas was not how to protect themselves from the diseases associated with asbestos, for that was beyond their control. The question was how to get treatment and subsistence when one of the diseases made its appearance. In the main, there was little possibility of them getting treatment because of the paucity and low quality of medical services in the rural areas. Thus they were left with the struggle for subsistence. This depended, by and large, on their ability to obtain

[53]It was this study that was developed into the thesis presented by Dr Felix for the degree of Doctor of Philosophy at the University of the Witwatersrand in 1997. See Felix, 1997.
[54]*Ibid*, p211.
[55]*Ibid*, p31 and p207.
[56]Abdullah and Manjra, 1992, p5.

compensation. There were no provisions for compensation for those who suffered from environmental exposure to asbestos dust. For former and existing mineworkers, however, it was different provided they had radiological evidence to support their claims. They had the legal right to submit claims for occupational compensation to the Medical Bureau of Occupational Diseases in Johannesburg. A panel from the bureau then assessed each radiograph and decided whether or not compensation should be paid. Former mineworkers, however, were unable to get over the first hurdle. That is, they could not obtain radiological evidence of their diseases. As in the case of other mining occupational diseases, the absence of diagnosis facilities in the rural areas left the victims to suffer physically and, with their families, to live in poverty. Few of them had sufficient information to enable them either to apply for compensation or even to know how to go about making an application.

The asbestos companies had submitted claims for compensation on behalf of their employees in the past but they had done so inefficiently, to the disadvantage of the claimants. For instance, in the case of GEFCO in Kuruman, the records had been so poorly organised that the medical team from the Industrial Health Research Group at the University of Cape Town could not discover how many cases had been submitted and how many had been accepted and rejected. There had been no systematic process for appealing against rejection. This muddle and incompetence were symptomatic of much of the administrative work of the mine owners. The medical team eventually presented claims for 105 mineworkers and made recommendations about how they should be processed. There was also incompetence at the Penge mine. There the company misread the x-rays for the 259 cases prepared by the team and had submitted compensation claims for less than half of them. Even amongst those cases it had submitted, there were 18 who were not medically entitled to receive compensation.

As a result of the participatory nature of the Mafefe project, community health committees were formed which submitted compensation claims to the clinic established at the Dr Mphahlele Hospital by the National Centre for Occupational Health. The hospital was an hour's drive from Mafefe and the clinic was held once a month. Before the clinic was opened in November 1991 not a single claim for compensation had been made from Mafefe. Indeed, not a single case of occupational lung disease had been reported by the hospital. Moreover, during the first year of the clinic's existence, no doctor at the hospital had referred a case for compensation. On the other hand, the clinic itself had submitted 204 claims to the Medical Bureau of Occupational

Disease.[57] It seemed that, until the National Centre for Occupational Health had intervened, hospitals in general and doctors in particular had been either unaware of or uninterested in legislative provisions for occupational compensation.[58] In consequence, thousands of former mineworkers who were ill from one or other of the asbestos-related diseases were waiting to be assessed for compensation.

Meanwhile, an interest was being generated in Britain about the complicity of British companies in South Africa in spreading asbestos dust and then ignoring the consequent spread of diseases. Attention was being focused on Cape Industries Limited that had possessed asbestos mining subsidiaries in South Africa. The initial issue was whether a British firm could be pursued for compensation in a British court on behalf of its South African employees. If that were possible, then the claimants would be freed from the constraints of South African law that not only limited compensation to relatively small fixed sums but also applied it in a racist manner in that the amounts awarded varied considerably between black and white workers.[59] Under British law the amounts of compensation were determined by the courts, could be excessively high and were not racially discriminatory. However, because the legal costs of taking a compensation claim to court in Britain were also high, black South African claimants could proceed only if they received legal aid from the British state.

Nonetheless, a British law firm, Leigh, Day and Co, specialising in legal aid and compensation cases, decided to pursue the matter and try to persuade the English courts to accept jurisdiction. Once that was decided it aimed to seek legal aid for the South African claimants. At that point, the NUM became involved, though at first in a small way. James Motlatsi, the NUM's president, met with the solicitor, Martyn Day, during May 1994 to hear about the possibility of obtaining compensation for the South African mineworkers. Day wanted details about six former mineworkers who had been employed by Cape Industries Limited so that he could begin a test case. He asked Motlatsi to assist with identifying possible claimants and persuading them to make detailed statements about their experiences and suffering. Thus began a long, slow haul through the English legal system.

[57]Felix, 1997, p31.

[58]*South African Medical Journal*, May 1996.

[59]In 1980, a black worker would have received R1 000 as compensation for asbestosis whereas a white claimant would have got R12 000. A black worker with TB only would have been given R600 but the payment to a white worker would have been R5 000. (Myers, 1980, pp58-59.) In 1992, compensation for a black worker with asbestosis was slightly over R3 000.

TABLE 5

RSA ASBESTOS FIBRE PRODUCTION[60]
(Kilotons)

1970-1994

YEAR	TYPE			TOTAL
	CHRYSOTILE	AMOSITE	CROCIDOLITE	
1970	52,80	97,30	137,23	287,41
1975	99,75	88,41	164,75	352,89
1977	111,58	66,98	299,97	379,53
1980	106,94	51,65	118,15	276,74
1981	76,77	56,83	102,34	235,94
1982	81,13	43,45	87,23	211,84
1983	93,02	40,66	87,44	221,12
1984	75,41	33,24	58,74	167,39
1985	92,32	37,86	34,07	164,25
1986	91,00	36,00	11,85	138,85
1987	101,72	26,03	7,32	135,07
1988	113,47	20,33	11,89	145,69
1989	115,50	26,10	23,70	165,30
1990	103,40	26,10	31,50	161,00
1991	101,60	27,40	31,60	157,60
1992	103,70	5,10	24,50	133,30
1993	92,40	-	11,60	104,00
1994	85,90	-	6,30	92,10

[60]Department of Mineral and Energy Affairs, *South Africa's Mineral Industry*, 1988 and 1994/5, Republic of South Africa.

BIBLIOGRAPHY

A selection of references from Volumes I to III inclusive of *The History of Black Mineworkers in South Africa*

GENERAL REFERENCES

African Mine Workers' Union, *Statement to the Witwatersrand Mine Native's Wages Commission*, June 1943

Agency for Industrial Mission (AIM), *Another Blanket: Report on an Investigation Into the Migrant Situation*, June 1976

Agency for Industrial Mission, *With Dirty Shoes: Draft Report of AIM Migrant Labour Seminar*, June/July 1984

Alberts, B C and Ortlepps, J A L, "Iron Ore Mining in South Africa", *Transactions of the Seventh Commonwealth Mining and Metallurgical Congress*, Volume II, Johannesburg, 1961

Andor, L E, *Aptitudes and Abilities of the Black Man in Sub-Saharan Africa, 1784-1963: An Annotated Bibliography*, South African Council for Scientific and Industrial Research, Johannesburg, 1966

Allen, F J, "Remarks on the Physique of Colliery Natives, the Nature of Their Work and its Effects on Health, With Special Reference to the Suggested Lowering of Physical Standards", *Proceedings of the Transvaal Mine Medical Officers' Association*, July 1935

Allen, F J, "Morbidity and Mortality on Coal and Gold Mines for 1941", *Proceedings of the Transvaal Mine Medical Officers' Association*, October 1943

Allen, V L, *Trade Union Leadership*, London, 1957

Allen, V L, "Management and Labour in Africa", *The Listener*, London, 15 August 1963

Allen, V L, "The Study of Trade Unionism in Tropical Africa", *Journal of Modern African Studies*, Volume 7, No 2, 1969

Allen, V L, *Future Industrial Relations: English-Speaking Africa*, ILO, Geneva, 1972

Allen, V L, "The Meaning of the Working Class in Africa", *Journal of Modern African Studies*, Volume 10, No 2, 1972

Allen, V L, *Social Analysis: A Marxist Critique and Alternative,* London, 1975, Yorkshire, 1982

Anglo-American Corporation, *The Patterson Job Evaluation Handbook,* Personnel Systems Department, October 1984

Anglo-American Corporation/National Union of Mineworkers,

Reaping the Whirlwind?, Johannesburg, May 1986

Associated Scientific and Technical Societies of South Africa, *Uranium in South Africa, 1946-1956*, Two Volumes, Joint Symposium Papers, Johannesburg, 1957

Association of Mine Managers of the Transvaal, "Some Aspects of Deep-Level Mining ... With Special Reference to Rockbursts", *Papers and Discussions*, 1933

Atmore, A, Chirenje, J M and Mudenge, S I, "Firearms in Central Africa", *Journal of African History*, XII, 4, 1971

Atmore, A and Sanders, P, "Sotho Arms and Ammunition in the Nineteenth Century", *Journal of African History*, XII, 4, 1971

Axsel, K, "Diamonds", in *South Africa's Mineral Industry*, Minerals Bureau of South Africa, 1987

Badenhorst, G P and Bakker, D, "Responsibility for Safety and Health in the Production Environment", *Mine Safety and Health Congress Proceedings,* Chamber of Mines, 1988

Barcza, M and Lambrechts, J de V, "Ventilation and Air-Conditioning Practice in South African Gold Mines", *Transactions of the Seventh Commonwealth Mining and Metallurgical* Congress, Volume I, Johannesburg, 1961

Baskin, J, *Striking Back: A History of COSATU*, Johannesburg, 1991

Batson, E, (ed), *The Social Survey of Cape Town 1941-1944,* University of Cape Town

Batson, E, *The Poverty Datum Line in Salisbury*, University of Cape Town, 1944

Beinart, W, *The Political Economy of Pondoland, 1860-1930*, Cambridge, 1982

Benjamin, P, "Trade Unions and the Industrial Court", *South African Review* 4, 1987

Bettison, D G, "The Poverty Datum Line in Central Africa", *The Rhodes Livingstone Journal*, No 27, Manchester University, June 1960

Biesheuvel, S, "Personnel Selection Tests for Africans", *South African Journal of Science*, Volume 49, No 1, August 1952

Blainey, G, "Lost Causes of the Jameson Raid", *The Economic History Review*, Volume XVIII, 1965

Bonner, P, "The Transvaal Native Congress, 1917-1920", in Marks, S and Rathbone, R, *Industrialisation and Social Change in South Africa,* London, 1982

Bonner, P, "The 1920 Black Mineworkers' Strike: A Preliminary Account", in Bozzoli, B, (ed), *Labour, Townships and Protest: Studies in the Social History of the Witwatersrand,*

Johannesburg, 1979

Botha, J B C, *Report: Autopsies: Kinross Mine Disaster*, 30 September 1986

Bozzoli, B, (ed), *Labour, Townships and Protest: Studies in the Social History of the Witwatersrand*, Johannesburg, 1979

Brooks, A, *From Class Struggle to National Liberation: The Communist Party of South Africa, 1940-1950*, MA thesis, University of Sussex, 1967

Brown, David, "The Basements of Babylon: English Literacy and the Division of Labour on the South African Gold Mines", *History Workshop*, University of the Witwatersrand, nd

Bryson, R H, Clausen, H T and Moyers, W H, "Brackish Waters of the Mines of the OFS Goldfields", *Transactions of the Seventh Commonwealth Mining and Metallurgical Congress*, Volume III, Johannesburg, 1961

Buell, R L, *The Native Problem in Africa*, USA, 1928

Bundy, C, *The Rise and Fall of the South African Peasantry*, London, 1979

Bunting, Brian, *Moses Kotane, South African Revolutionary*, London, 1975

Burke, G and Richardson, P, "The Migration of Miners' Phthisis Between Cornwall and the Transvaal 1876-1918", in Bozzoli, B, *Labour, Townships and Protest: Studies in the Social History of the Witwatersrand*, Johannesburg, 1979

Cairncross, B and Dixon, R, *Minerals of South Africa*, Geological Society of South Africa, Johannesburg, 1995

Campbell, C and Williams, B, *HIV/AIDS and the Productivity Sectors in the SADC Region: A Case Study of the Mining Industry,* Epidemiology Research Unit, Johannesburg, November 1996

Campbell Pitt, L T, "Winding Plant of the Gold Mines of South Africa", *Transactions of the Seventh Commonwealth Mining and Metallurgical Congress*, Volume III, Johannesburg, 1961

Cartwright, A P, *Doctors of the Mines*, Cape Town, 1971

Chaskalson, M, Jochelson, K, Seekings, J, "Rent Boycotts and the Urban Political Economy", *South African Review* 4, 1987

Chamber of Mines of the Transvaal (later of South Africa), *Annual Reports*, 1897-1994

Chamber of Mines, *Mining Survey*, various, 1966-1991

Chamber of Mines, *Newsletter*, No 5/6, 1993

Chamber of Mines, *Statistical Tables*, 1991, 1992

Chamber of Mines, *Review*, 1987-88

Chemical Workers' Industrial Union, *Struggle for Workers' Rights: A*

History of the Chemical Workers' Industrial Union, Durban, 1994
Chimutengwende, C C, *South Africa: The Press and Politics of Liberation*, 1987
Christian Action, *Prisoners of Apartheid*, London, 1965
Christie, P, *The Right to Learn: The Struggle for Education in South Africa*, Johannesburg, 1991
Clarke, D G, *Labour From Rhodesia to the South African Mines: A Study in the International Division of a Labour Reserve*, SALDRU Working Paper, No 6, October 1976, Cape Town
Cobbett, W and Cohen R, (eds), *Popular Struggles in South Africa,* London, 1988
Cole, D T, "Fanagalo and the Bantu Languages in South Africa", *African Studies*, Volume 12, No 1, March 1953
Cole, M, *South Africa*, London, 1961
Congress of South African Trade Unions (COSATU), *Bi-Annual Report*, 3rd National Congress, Johannesburg, 1989
Cook, A, *Akin to Slavery: Prison Labour in South Africa*, International Defence and Aid Fund, London, 1982
Coovadia, H M and Benatar, S R, (eds), *A Century of Tuberculosis: South African Perspectives*, Cape Town, 1991
Cooper, R A, "Mineral Constituents of Rand Concentrates", *Journal of the Chemical, Metallurgical and Mining Society of South Africa*, January 1924
Cowan, E, "Is Job Evaluation Scientific?", *South African Labour Bulletin*, Volume 10, No 4, January-February 1985
Crewe, M, "Learning Aids", *Work in Progress*, Johannesburg, December 1993
Crowley, M L and Du Plessis, M, *Black Employee Mobilisation in the South African Gold Mining Industry, 1987-1988*, Human Resources Laboratory, Chamber of Mines, April 1990
Crush, J and James, W, (eds), *Crossing Boundaries. Mine Migrancy in a Democratic South Africa*, Cape Town, 1995
Crush, J, Jeeves, A and Yudelman, D, *South Africa's Labour Empire: A History of Black Migrancy to the Gold Mines*, USA and South Africa, 1991
Davies, R H, *Capital, State and White Labour in South Africa, 1900-1960*, Brighton, England, 1979
Davies, R H, "The 1922 Strike and the Political Economy of South Africa", in Bozzoli, B, *Labour, Townships and Protest: Studies in the Social History of the Witwatersrand*, Johannesburg, 1979

Davies, R, O'Meara, D and Dlamini, S, *The Struggle for South Africa: A Reference to Movements, Organisations and Institutions,* Volumes 1 and 2, London, 1985

De Kiewiet, C W, *A History of South Africa,* London, 1941

Delius, P, "Migrant Labour and the Pedi, 1840-1880", in Marks, S and Atmore, A, (eds), *Economy and Society in Pre-Industrial South Africa*, London, 1980

Delius, P, "Sebatakgomo: Migrant Organisation, the ANC and the Sekhukhuneland Revolt", *Journal of Southern African Studies*, Volume 15, No 4, October 1989

Dempster, E L, "Microclimate Acclimatisation Trial at Vaal Reefs Exploration and Mining Company", *Papers and Discussions, 1982-1983*, Association of Mine Managers of South Africa

Denoon, D J N, "The Transvaal Labour Crisis, 1901-1906", *Journal of African History*, Volume VII, 3, 1967

De Ridder, J C, *The Personality of the Urban African in South Africa,* London, 1961

De Vries, P and Radise, J, *The Effects of Mine Procedures and Practices on the Quality of Life of Black Mineworkers on Four Mines,* Human Resources Laboratory, Chamber of Mines, Research Report 5/80, February 1980

Dreyer, C J B and Robinson, H A, "Occurrence and Exploitation of Amphibole Asbestos in South Africa", in Riordan, P H, (ed), *Geology of Asbestos Deposits*, New York, 1981

Dropkin, G and Clark, D, *Past Exposure: Revealing Health and Environmental Risks of Rössing Uranium*, Namibia Support Committee, London, 1992

Du Toit, M A, *South African Trade Unions: History, Legislation, Policy,* Johannesburg, 1976

Edwards, A M and Silk, M H, *Platinum in South Africa,* Special Publication, No 12, Council for Mineral Technology, 1987

Edwards, E P, "Native Labour Control and Payment of Wages and Bonuses", *Papers and Discussions*, Association of Mine Managers of the Transvaal, 1942-45

Edwards, I E, *Towards Emancipation: A Study in South African Slavery*, Cardiff, 1942

Eisner, H S and Leger, J P, "The International Safety Rating System in South African Mining", *Journal of Occupational Accidents*, 10, 1988

Eisner, H S and Leger, J P, "Safety in South African Mines: An Analysis of Accident Statistics", *Journal of the South African Institute of Mining and Metallurgy,* Volume 88, No 1, January 1988

Eisner, H S and Leger, J P, "Safety in South African Mines II: Aspects of Supervision and Inspection", *Mine Safety and Health Congress Proceedings*, Chamber of Mines, 1988

Ellis, S and Sechaba, T, *Comrades Against Apartheid: The ANC and the South African Communist Party in Exile*, London, 1992

Elphick, R and Giliomee, H, (eds), *The Shaping of South African Society, 1652-1820*, Cape Town, 1979

Erasmus, J S and Baucom, K L, *Fanakalo: Through the Medium of English*, Language Laboratory Course, Anglo-American Corporation, Johannesburg, 1976

Erasmus, J S and Penny, E R, "Fanakalo Teaching With Tape Recorders …", *Papers and Discussions*, Association of Mine Managers of South Africa, 1966-1967

Farley, J, *Bilharzia: A History of Imperial Tropical Medicine,* Cambridge, 1991

Felix, M A, "Risking Their Lives in Ignorance: The Story of an Asbestos Polluted Community" in Cock, J and Koch, E, (eds), *Going Green: People, Politics and the Environment in South Africa*, Cape Town, 1991

Felix, M A, *Environmental Asbestos and Respiratory Disease in South Africa*, PhD Thesis, Faculty of Medicine, University of the Witwatersrand, 1997

Felix, M A, Leger, J P and Ehrlich, R I, "Three Minerals, Three Diseases – Asbestos Mining and Disease in South Africa", in Mehlman, M and Upton, A, (eds), *Advances in Modern Environ Toxicology: The Identification and Control of Environmental and Occupational Diseases: Asbestos and Cancer*, Princeton, USA, 1994, Volume XXII

Fine, B and Rustomjee, Z, *The Political Economy of South Africa,* 1996

First, R, *The Farm Labour Scandal*, A New Age Pamphlet, Johannesburg, 1959

First, R, "The Gold of Migrant Labour", *Africa South in Exile*, Volume 5, No 3, April-June 1961

First, R, *Black Gold: The Mozambican Miner, Proletarian and Peasant,* New York, 1983

First, R, *117 Days*, London, 1965

First, R, Steele, J and Gurney, C, *The South African Connection: Western Investment in Apartheid*, London, 1972

Flynn, L, "South Africa Blacks Out Blue Asbestos Risk", *New Scientist,* London, 22 April 1982

Flynn, L, *Death is Part of the Process*, Granada TV, UK, 13 October 1986

Flynn, L, *Studded With Diamonds and Paved with Gold: Miners, Mining Companies and Human Rights in Southern Africa,* London, 1992

Flynn, L, *Dust to Dust*, Granada TV, UK, March 1994

Forman, L and Sachs, E S, *The South African Treason Trial*, London, 1957

Fox, F W. "The Food and Feeding of Mine Native Labourers", *Proceedings*, Transvaal Mine Medical Officers' Association, September 1941

Frankel, H S, *Capital Investment in South Africa: Its Course and Effects*, UK, 1938

Fraser, A E, "Coal Mining in South Africa", *Transactions of the Seventh Commonwealth Mining and Metallurgical Congress*, Volume II, Johannesburg, 1961

Friedman, S, "Chamber of Mines' Policy and the Emerging Miners' Unions", *South African Labour Bulletin*, Volume 8, No 5, April 1983

Friedman, S, *Building Tomorrow Today: African Workers in Trade Unions*, 1970-1984, Johannesburg, 1987

Gay, J, "Wage Employment of Rural Basuto Women: A Case Study", *South African Labour Bulletin*, Volume 6, No 4, November 1980

Gemmill, J A, "Native Labour on the Gold Mines", *Transactions of the Seventh Commonwealth Mining and Metallurgical Congress,* Volume I, Johannesburg, 1961

Giles, G S, "Diamond Mining Practice in South Africa", *Transactions of the Seventh Commonwealth Mining and Metallurgical Congress*, Volume II, Johannesburg, 1961

Gregory, T, *Ernest Oppenheimer and the Economic Development of Southern Africa*, London, 1962

Grout, R W and Lechmore-Oertel, R L, "A Brief History of the Progress of Efficiency in South African Collieries in the Last Seventy Years", *Journal of the South African Institute of Mining and Metallurgy*, May 1958

Hailey, Lord, *An African Survey*, London, 1938

Halpern, J, *South Africa's Hostages*, London, 1965

Harries, P, "Slavery, Social Incorporation and Surplus Extraction: The Nature of Free and Unfree Labour in South-East Africa", *Journal of African History,* Volume 22, 1981

Hart, H P, "Asbestos in South Africa", *Journal of the South African Institute of Mining and Metallurgy,* Volume 88, 1988

Hellmann, E, "The Application of the Concept of Separate

Development to Urban Areas" in Kirkwood, K, (ed), *St Antony's Papers*, Volume X, African Affairs, 1959

Helliker, K, Roux, A and White, R, "*Asithengi*: Recent Consumer Boycotts", *South African Review* 4, 1987

Hetherington, N A, "Labour Supply and the Genesis of South African Confederation in the 1870s", *Journal of African History,* Volume 20, 1979

Hill, F G and Denkhouse, H G, "Rock Mechanics Research in South Africa, With Special Reference to Rockbursts and Strata Movement in Deep-Level Gold Mines", *Transactions of the Seventh Commonwealth Mining and Metallurgical Congress,* Volume II, Johannesburg, 1961

Hindson, D, *Pass Controls and the Urban African Proletariat,* Johannesburg, 1987

Hirson, B, *Year of Fire, Year of Ash: The Soweto Revolt: Roots of Revolution?* London, 1979

Hirson, B, *Yours for the Union: Class and Community Struggles in South Africa*, 1930-1947, London, 1989

Hirson, B, *Revolutions in my Life*, Johannesburg, 1995

Hocking, A, *Randfontein Estates: The First Hundred Years*, South Africa, 1986

Hodgson, H F and Sewell, E J B, "Mining Practice at Premier (Transvaal) Diamond Mining Company", *Papers and Discussions*, Association of Mine Managers of South Africa, 1960-61

Horner, D and Kooy, A, *Conflict on South African Mines*, *1972-1976*, SALDRU Working Paper, No 5, Cape Town, August 1976

Horner, D and Kooy, A, *Conflict on South African Mines*, *1972-1979*, paper presented to the 5th World Congress of the International Industrial Relations Association, Paris, 1979

Horrell, M, *South African Trade Unionism*, South African Institute of Race Relations, Johannesburg, 1961

Humphries, S V, "A Consideration of the Advisability of Treating Syphilis in Mine Natives", *Proceedings of the Transvaal Mine Medical Officers' Association*, February 1937

Hunt, G F and Daniel, R, "Water Problems at Wesselton and Dutoitspan Mines", *Papers and Discussions*, Association of Mine Managers in the Transvaal, 1952-53

Iliffe, J, *The African Poor*, Cambridge, 1987

Industrial Health Research Group, *The Kuruman Asbestos Report,* University of Cape Town, 1992

Industrial Health Research Group, *Medical Screening at the Penge*

Amosite Mine, University of Cape Town, 1992

Innes, D, *Anglo: Anglo-American and the Rise of Modern South Africa,* Johannesburg, 1984

International Labour Organisation, *Safety and Health in Mines,* Report No 51, Geneva, 1994

James, W G, *Our Precious Metal: African Labour in South Africa's Gold Industry*, 1970-1990, Cape Town, 1992

Jamieson, D M, Pearce, M P and Plumstead, E R A, "The Evolution of Shaft Design and Sinking Technique in South Africa", *Transactions of the Seventh Commonwealth Mining and Metallurgical Congress*, Volume II, Johannesburg, 1961

Jeeves, A H, *Migrant Labour in South Africa's Mining Economy: The Struggle for the Gold Mines' Labour Supply, 1890-1920,* Canada and South Africa, 1985

Jochelson, K, Mothibeli, M and Leger, J P, *Heading for Crisis? The Implications of the Migrant Labour System for the HIV Epidemic*, Sociology of Work Programme, University of the Witwatersrand, 1988

Johnstone, F A, *Class, Race and Gold: A Study of Class Relations and Racial Discrimination in South Africa*, London, 1976

Johnstone, F A, "The IWA on the Rand: Socialist Organising Among Black Workers on the Rand, 1917-18", in Bozzoli, B, (ed), *Labour, Townships and Protest: Studies in the Social History of the Witwatersrand*, Johannesburg, 1979

Karis, T and Carter, G M, (eds), *From Protest to Challenge: A Documentary History of African Politics in South Africa, 1882-1964*, Volumes 1-4, California, USA, 1972-1977

Kark, S L, "The Social Pathology of Syphilis in Africans", *South African Medical Journal*, 23(5), January 1949

Kasrils, R, *Armed and Dangerous: My Undercover Struggle Against Apartheid*, London, 1993

Katz, E, "White Workers' Grievances and the Industrial Colour Bar, 1902-1913", *The South African Journal of Economics*, Volume 42, 1974

Katz, E, *The White Death*: *Silicosis on the Witwatersrand Gold Mines*, 1886-1910, Johannesburg, 1994

Keep, F E, "Amphibole Asbestos in the Union of South Africa", *Transactions of the Seventh Commonwealth Mining and Metallurgical Congress*, Volume I, Johannesburg, 1961

Kenny, B and Bezuidenhout, A, *Contracting, Complexity and Control: An Overview of the Changing Nature of Subcontracting in the South African Mining Industry*, Sociology of Work Unit,

University of the Witwatersrand, 1998

Kibble, S, *The External Role of the South African State: The Case of Labour Migration from Malawi*, PhD thesis, University of Leeds, 1989

Kimble, J, "Labour Migration in Basutoland, c 1870-1885", in Marks, S and Rathbone, R, (eds), *Industrialisation and Social Change in South Africa*, London, 1982

Knight, M, "The Mines Death Benefit Scheme", *South African Labour Bulletin* Volume 9, No 4, February 1984

Korf, C W, "Stick-and-Pillar Support on Union Section, Rustenburg Platinum Mines", *Papers and Discussions*, 1978-1979, Association of Mine Managers of South Africa

Kraak, G, *Breaking the Chains*, London, 1993

Kubicek, R V, *Economic Imperialism in Theory and Practice*, USA, 1979

Kuper, L, "African Nationalism in South Africa, 1910-1964", in Wilson, M and Thompson, L (eds), *Oxford History of South Africa,* Volume 2, 1870-1966, Oxford, 1971

Labour History Group, *Durban Strikes*, Salt River, South Africa, July 1987

Laburn, C M and McNamara, J K, *Weekend Visits Home by Migrant Workers*, Research Report, No 7/81, Chamber of Mines, 1981

La Hausse, P, *Brewers, Beerhalls and Boycotts: A History of Liquor in South Africa*, Johannesburg, 1988

Lambert, R V, *Political Unionism in South Africa: The South African Congress of Trade Unions*, 1955-1965, PhD thesis, University of the Witwatersrand, 1988

Lang, J, *Bullion Johannesburg: Men, Mines and the Challenge of Conflict*, Johannesburg, 1986

Leger, J P, "Safety and the Organisation of Work in South African Gold Mines: A Crisis of Control", *International Labour Review,* Volume 125, No 5, September-October, 1986

Leger, J P, "From Hlobane to Kinross: Disasters and the Struggle for Health and Safety on the Mines", *South African Review* 4, 1987

Leger, J P, "Key Issues in Safety and Health in South African Mines", *South African Sociological Review*, Volume 2, No 2

Leger, J P, *From Fatalism to Mass Action*, Sociology of Work Programme, University of the Witwatersrand, July 1988

Leger, J P, "Trends and Causes of Fatalities in South African Mines", *Safety Science*, 14, 1991

Leger, J P, "*Talking Rocks": An Investigation of the Pit Sense of*

Rockfall Accidents Amongst Underground Gold Miners, PhD thesis, University of the Witwatersrand, 1992

Leger, J P and Mothibeli, M, "South African Miners' Perceptions of Safety 1984-1987", *Mine Safety and Health Congress Proceedings*, Chamber of Mines, Johannesburg, 1988

Lerumo, A, *Fifty Fighting Years: The South African Communist Party,* 1921-1971, London, 1971

Levy, N, *The Foundations of the South African Cheap Labour System,* London, 1982

Lewis, J, *Industrialisation and Trade Union Organisation in South Africa, 1924-55: The Rise and Fall of the South African Trades and Labour Council*, Cambridge, 1984

Lipton, M, "Men of Two Worlds", *Optima*, Volume 29, No 2/3, Anglo-American Corporation, Johannesburg, November 1980

Lipton, M, *Capitalism and Apartheid, South Africa, 1910-1986,* England, 1986

Luckhardt, K and Wall, B, *Organise... or Starve! The History of the South African Congress of Trade Unions,* London, 1980

McAteer, J D, *Miners' Manual*, Washington, USA, 1985

Mandela, N R, *No Easy Walk to Freedom*, Oxford, 1965

Mandela, N R, *Long Walk to Freedom*, London, 1994

Manganyi, N C and Du Toit, A, (eds) *Political Violence and the Struggle in South Africa*, London, 1990

Maré, G and Hamilton, G, *An Appetite for Power: Buthelezi's Inkatha and South Africa*, Johannesburg, 1987

Maree, J, *An Analysis of the Independent Trade Unions in South Africa in the 1970s*, PhD thesis, University of Cape Town, 1986

Marks, S and Atmore, A, "Firearms in Southern Africa: A Survey", *Journal of African History*, Volume XII, No 4, 1971

Marks, S and Atmore, A, (eds), *Economy and Society in Pre-Industrial South Africa*, London, 1980

Marks, S and Rathbone, R, (eds), *Industrialisation and Social Change in South Africa*, London, 1982

Marshall, J, "The Present-day Treatment of Syphilis", *South African Medical Journal*, 29 January 1949

Martinson, M J, "Occupational Health and Safety in Underground Mines", (in two parts), *Journal of the Mine Ventilation Society of South Africa*,Volume 29, Nos 1 & 2, 1976

Martinson, M J, "Heat Stress in Witwatersrand Gold Mines", *Journal of Occupational Accidents*, 1976-1977, Amsterdam

Martinson, M J, "A Catastrophic Methane Explosion in a Main Vertical Shaft Serving an Orange Free State Gold Mine", *Proceedings of the Fourth US Mine Ventilation Symposium*, University of California, USA, 1989

Masilela, A M, "The Reaction of African Mineworkers to Aptitude Testing", *Bulletin*, Volume 2, No 2, National Institute for Personnel Research, 1950

Mayer, P, *Townsmen or Tribesmen*, Cape Town, 1961

Mbeki, G, *South Africa: The Peasants Revolt*, London, 1964

Mbeki, T, *Keynote Address by President Thabo Mbeki*, NUM, 10th National Congress, Pretoria, April 2000

McNamara, J K, "The Development of a Labour Infrastructure: Migration Routes to the Witwatersrand Gold Mines and Compound Accommodation, 1889-1912", *South African Labour Bulletin*, Volume 4, No 3, May 1978

McNamara, J K, *Black Worker Conflict on South African Gold Mines,* PhD thesis, University of the Witwatersrand, 1985

McNamara, J K, "Inter-Group Violence Among Black Employees on South African Gold Mines: 1974-1986", *South African Sociological Review*, Volume 1, No 1

McNamara, J K, *The Underlying Causes and Resolution of Labour Conflict at Matla Colliery*, Trans-Natal Coal Corporation Ltd – NUM, November 1988

Meli, F, *A History of the ANC: South Africa Belongs to Us*, London, 1989

Mendelsohn, R, "Blainey and the Jameson Raid: The Debate Renewed", *Journal of Southern African Studies*, Volume 6, No 1, October 1979

Metcalf, P E, "Regulation of Radiological Protection in the South African Mining Industry: Past, Present and Future", in Kvasnicka, J, (ed), *International Workshop on Radiological Protection in Mining*, Volume A, April, 1988, Darwin, Australia

Metcalf, P E, "Licensing of Activities Involving Nuclear Hazard Material Within the Mining Industry", *Mine Safety and Health Congress Proceedings*, Chamber of Mines, Johannesburg, October/November 1991

Mine Medical Officers' Association of the Transvaal (later South Africa), *Proceedings*, 1921-1982

Mine Medical Officers' Association of South Africa, 1921-1971, *Proceedings of the Jubilee Congress of the Association*, March 1971

Momberg, A and McNamara, J K, *The Tribal Representative System on Gold Mines*, Research Report, No 10/77, Chamber of Mines,

1977

Moodie, T D with Ndatshe, V, *Going for Gold*, Johannesburg, 1994

Moroney, S, "The Development of the Compound as a Mechanism of Worker Control 1900-1912", *South African Labour Bulletin,* Volume 4, No 3, May 1978

Murray, C, "The Effects of Migrant Labour: A Review of the Evidence From Lesotho", *South African Labour Bulletin*, Volume 6, No 4, November 1980

Murray, C, *Families Divided: The Impact of Migrant Labour in Lesotho*, Cambridge, 1981

Myburgh, J F, "Patterns of Workplace Behaviour by Officials and Members of the National Union of Mineworkers", *Memorandum,* Anglo-American Corporation, nd

Myers, J, *Asbestos and Asbestos-Related Diseases in South Africa*, SALDRU Working Paper, 28, Cape Town, June 1980

National Union of Mineworkers, *National Congress Reports*, 1982-1994

National Union of Mineworkers,. *Five Brave Days in August – The African Mine Workers' Union and the Miners' Strike of 1946*, Johannesburg, 1986

Nattrass, J, *The South African Economy: Its Growth and Change*, Cape Town, 1988

Newbury, C, *The Diamond Ring: Business, Politics and Precious Stones in South Africa*, 1867-1947, Oxford, 1989

Newman, S C, "A Review of Tunnelling Techniques in the Gold Mining Industry of South Africa", *Transactions of the Seventh Commonwealth Mining and Metallurgical Congress,* Volume II, Johannesburg, 1961

O'Meara, D, "The African Mine Workers' Strike and the Political Economy of South Africa", *The Journal of Commonwealth and Comparative Politics*, Volume XIII, 1975

Omond, R, *The Apartheid Handbook*, London, 1986

Orenstein, A J and Webster, I, "Pneumoconiosis – A Review", *Transactions of the Seventh Commonwealth Mining and Metallurgical Congress*, Volume I, Johannesburg, 1961

Oxley, J, *Down Where No Lion Walked: The Story of Western Deep Levels*, Johannesburg, 1989

Packard, R M, *White Plague, Black Labour,* Natal, 1990

Pampallis, J, *Foundations of the New South Africa*, London, 1991

Parsons, J A, *Wages in Gold Mining*, 1970-1978, Human Resources Laboratory, Chamber of Mines, Research Report 29/79, June 1979

Pearson, P, *The Social Structure of a South African Gold Mine Hostel,* BA dissertation, University of the Witwatersrand, 1975

Peires, J B, *The House of Phalo: A History of the Xhosa People in the Days of their Independence*, Johannesburg, 1981

Peires, J B, *The Dead Will Arise*, Johannesburg, 1989

Perold, J, "The Historical and Contemporary Use of Job Evaluation in South Africa", *South African Labour Bulletin*, Volume 10, No 4, Jan/Feb 1985

Phelan, J M, *Apartheid Media: Disinformation and Dissent in South Africa*, USA, 1987

Pinkey, E T and Westwood, R J, "Uranium in South Africa", *Transactions of the Seventh Commonwealth Mining and Metallurgical Congress,* Volume III, Johannesburg, 1961

Plaatje, S, *Native Life in South Africa*, London, 1916

Pretorius, P G D, "Underground Training and Acclimatisation at City Deep", *Papers and Discussions,* 1966-67, Association of Mine Managers of South Africa, 1968

Price, R M, *The Apartheid State in Crisis*, Oxford, 1991

Rafel, R, "Job Reservation on the Mines", *South African Review* 4, 1989

Ramdin, R, *The Making of the Black Working Class in Britain,* London, 1987

Richardson, P, *Chinese Mine Labour in the Transvaa*l, London, 1982

Richardson, P, "The Natal Sugar Industry in the Nineteenth Century", in Beinart W, Delius, P and Trapido, S, (eds), *Putting a Plough to the Ground*, Johannesburg, 1986

Richardson, P and Van Helton, J-J, "The Gold Mining Industry in the Transvaal, 1886-99", in Warwick, P, (ed), *The South African War: The Anglo-Boer War, 1899-1902*, UK, 1980

Riley, E, *Major Political Events in South Africa, 1948-1990*, Oxford, 1991

Rogers, C C, "A Talk on the Heat Tolerance Test and Microclimate Acclimatisation", *Papers and Discussions, 1978-1979,* Association of Mine Managers of South Africa, 1981

Rose, H J M, "The Introduction of Self-Contained Self-Rescuers Into the South African Mining Industry – An Update", *Mine Safety and Health Congress Proceedings*, Chamber of Mines, 1988

Roux, E R, *Time Longer Than Rope*, London, 1948

Rushton, M W, "Native Labour Control and Conservation", *Papers and Discussions, 1942-45*, Association of Mine Managers of the Transvaal

Russell, C, "The AIDS Crisis: A Mining Industry Perspective", *Mining*

Survey, No 2, Chamber of Mines, 1991
Rycroft, A and Jordaan, B, *A Guide to South African Labour Law,* South Africa, 1993
SALDRU, *South Africans Rich and Poor: Baseline Household Statistics*, Project on Living Standards and Development, Cape Town, August 1994
Sampson, A, *Mandela: The Authorised Biography*, Great Britain, 1999
Sandbrook, R and Cohen, R, (eds), *The Development of an African Working Class*, London, 1975
Schapera, I, *Government and Politics in Tribal Societies*, London, 1956
Scott, R, *The Development of Trade Unionism in Uganda*, Kampala, 1966
Seekings, J, *The UDF: A History of the United Democratic Front in South Africa 1983-1991*, UK, 2000
Segal, R and First, R, (eds), *South West Africa: Travesty of Trust,* London, 1967
Seidman, G, "If Harmony Closes, Will the Last One to Leave Turn Out the Lights? The Social Impact of Downscaling on the Goldfields Area", *Preliminary Report*, Sociology of Work Unit, University of the Witwatersrand, April 1993,
Seidman, A and Makgetla, N S, *Outposts of Monopoly Capitalism,* London, 1980
Selikoff, I J and Lee, D H K, *Asbestos and Disease*, New York, 1978
Sewell, E J B, "Native Labour and Payment on Premier Diamond Mine", *Papers and Discussions*, Association of Mine Managers of South Africa, 1960-61
Shillington, K, "The Impact of the Diamond Discoveries on the Kimberley Hinterland" in Marks, S and Rathbone, R, (eds), *Industrialisation and Social Change in South Africa*", London, 1982
Simons, H J and R E, *Class and Colour in South Africa*, 1850-1950, England, 1969
Simpson, F W, "Pulmonary Asbestosis in South Africa", *British Medical Journal*, 1, 885, 1928
Sinclair, W E, "Mining Methods on the Kimberley Diamond Fields", *Papers and Discussions*, Association of Mine Managers in the Transvaal, 1948-49
Slovo, J, *The Unfinished Autobiography*, Johannesburg, 1995
Sluis-Cremer, G K, Liddell, F D K, Logan W P D and Bezuidenhout, B N, "The Mortality of Amphibole Miners in South Africa, 1946-80", *British Journal of Industrial Medicine*, 49, 1992
Smalberger, J M, "The IDB and the Mining Compound System in the

1880s", *The South African Journal of Economics*, Volume 42, 1974

Smalberger, J M, "The Role of the Diamond Mining Industry in the Development of the Pass Law System in South Africa", *The International Journal of African Historical Studies*, Volume 9, No 1, 1976

South African Communist Party, (SACP), *A Distant Clap of Thunder: Fortieth Anniversary of the 1946 Mine Strike*, London, 1986

South African Communist Party, *The African Communist*, 1961-1993

South African Institute of Race Relations, *A Survey of Race Relations in South Africa*, 1950-1994, (published annually), Johannesburg

Somerville, K, *Southern Africa and the Soviet Union*, London, 1993

Sparks, A, *Tomorrow is Another Country: The Inside Story of South Africa's Negotiated Revolution*, UK, 1995

Sparks, A, *The Mind of South Africa: The Story of the Rise and Fall of Apartheid*, UK, 1990

Standing, G, Sender, J and Weeks, J, *Restructuring the Labour Markets: The South African Challenge,* ILO, Geneva, 1996

Steinberg, J with Seidman, G, *Gold Mining's Labour Markets: Legacies of the Past, Challenges of the Present*, Labour Studies Research Report 6, Sociology of Work Unit, University of the Witwatersrand, Johannesburg, 1995

Stephenson, A, "Scientific Management With Particular Application to a Witwatersrand Gold Mine", 1930. Reprinted in *Paper and Discussions*, 1942-45, Volume 2, Association of Mine Managers of the Transvaal

Stocken, C G, "The Diamond Deposits of the Sperrgebeit, South West Africa", *Oranjemund*, January 1963

South African Communists Speak: Documents From the History of the South African Communist Party, 1915-1980, (various contributors), London, 1981

Strom, G W, *Migration and Development: Dependence on South Africa: A Study of Lesotho*, Sweden, 1986

Strydom, N B, "Age as a Causal Factor in Heat Stroke", *Journal of the South African Institute of Mining and Metallurgy*, November 1971

Strydom, N B, Kotze, H F, Van der Walt, W H and Rogers, C C, "Effect of Ascorbic Acid on Rate of Heat Acclimatisation", *Journal of Applied Physiology*, Volume 41, No 2, August 1976

Taylor, J, *The Pattern and Consequences of Declining Mine Labour Recruitment in Botswana*, ILO, Geneva, 1986

The Employment Bureau of Africa (TEBA), *Reports and Financial Statements*, 1977-1995

Thompson, L, *A History of South Africa*, USA, 1995

Trades Union Congress, *Annual Report*, UK, 1975

Transformation Resource Centre, *Work for Justice*, a Quarterly Journal, Maseru, Lesotho, December 1985, June 1987 and an undated post-1987 strike issue

Truth and Reconciliation Commission of South Africa, *Reports*, Volumes 1-5, 1998

Turrell, R V, "The 1875 Black Flag Revolt on the Kimberley Diamond Fields", *Journal of Southern African Studies*, Volume VII, No 2, 1981

Turrell, R V, *Capital and Labour on the Kimberley Diamond Fields*, 1871-1890, Cambridge, 1987

Van Coller, D L, "Job Evaluation and the Changing Wage Pattern", *Journal of the South African Institute of Mining and Metallurgy*, August 1974

Van der Horst, S, *Native Labour in South Africa*, London, 1942

Van der Merwe, C J, "Asbestos", in *South Africa's Mineral Industry*, 1988, Minerals Bureau, Johannesburg, 1989

Van Helten, J-J, "German Capital, the Netherlands Railway Company and the Political Economy of the Transvaal, 1886-1990", *Journal of African History*, Volume XIX, 3, 1978

Van Niekerk, P, "NUM: First Legal Mine Strike", *South African Labour Bulletin*, Volume 10, No 2, October/November 1984

Van Onselen, C, "Reactions to Rinderpest in Southern Africa, 1896-97", *Journal of African History*, Volume XIII, 3, 1972

Van Onselen, C, *Chibaro: African Mine Labour in Southern Rhodesia*, 1900-1933, London, 1976

Van Onselen, C, *Studies in the Social and Economic History of the Witwatersrand* 1886-1914, Volumes 1 and 2, Johannesburg, 1982

Van Rensburg, J C J, "Dust Suppression on Continuous Miners", *Mine Safety and Health Congress Proceedings*, Chamber of Mines, 1988

Visagie, M E, Du Plessis, J P, Groothoe, G, Alberts, A and Laubscher, N F, "Changes in Vitamin A and C Levels in Black Mineworkers", *South African Medical Journal*, 11 December 1974

Von Lieres, B, *The 1987 NUM Strike*, BA dissertation, University of the Witwatersrand, March 1988

Wagner, J C, Sleggs, C A and Marchand, P, "Diffuse Plural

Mesothelioma and Asbestos Exposure in the North Western Cape Province", *British Journal of Industrial Medicine*, Volume 17, 1960
Walker, I L and Weinbren, B, *2 000 Casualties: A History of the South African Labour Movement*, Johannesburg, 1961
Warwick, P, (ed), *The South African War: The Anglo-Boer War, 1899-1902*, UK, 1980
Warwick, P, *Black People and the South African War, 1899-1902,* Johannesburg, 1983
Webster, D, "Repression and the State of Emergency", *South African Review 4*, 1987
Webster, E, (ed), *Essays in Southern African Labour History,* Johannesburg, 1978
Webster, I, "Asbestosis", S*outh African Medical Journal*, October 1964
Webster, I, "Asbestos and Malignancy", *South African Medical Journal*, February 1973
Wessels, B, "Fanakalo: The *Lingua Franca* of the Mining Community", *Mining Survey*, Chamber of Mines, No 1/ 4, 1986
Westcott, G and Wilson, F, (eds), *Economics of Health in South Africa,* Volumes 1 and 2, Johannesburg, 1979
Williams, B and Campbell, C, (eds), *HIV/AIDS Management in South Africa: Priorities for the Mining Industry*, Epidemiology Research Unit, Johannesburg, 1996
Williams, E, *Capitalism and Slavery,* USA, 1944
Williams, L S, "The Food and Feeding of Mine Native Labourers", *Proceedings of the Transvaal Mine Medical Officers' Association*, July and August 1941
Wilson, F, *Labour in the South African Gold Mines, 1911-1969*, Cambridge, 1972
Wilson, F, *Migrant Labour in South Africa*, Johannesburg, 1972
Wilson, F and Ramphele, M, *Uprooting Poverty*, South Africa, 1989
Wilson, M and Thompson, L, (eds), *The Oxford History of South Africa*, Volumes 1 and 2, Oxford, 1969 and 1971
Witwatersrand Native Labour Association (WNLA), *Annual Reports,* 1948-1977
Woddis, J, Africa: *The Way Ahead*, London, 1963
Woods, D, *Biko*, London, 1987
Worden, N, *Slavery in Dutch South Africa*, Cambridge, 1985
Worger, W H, *South Africa's City of Diamonds: Mine Workers and Monopoly Capitalism in Kimberley*, 1867-1895, USA, 1987
Wyndham, C H, "A Survey of the Causal Factors in Heat Stroke and Their Prevalence in the Gold Mining Industry", *Journal of the*

South African Institute of Mining and Metallurgy, November 1965
Wyndham, C H, Gonin, R and Reid, R D W, "Seasonal Variations in Acute Respiratory Diseases and Meningitis in Black Miners Living in Hostels", *South African Medical Journal*, 26 August 1978
Young, S, "Fertility and Famine: Women's Agricultural History in Southern Mozambique", in Palmer, R and Parsons, N, *The Roots of Rural Poverty in Central and Southern Africa,* London, 1972

OFFICIAL AND GOVERNMENT SOURCES

African National Congress, *Secretary-General's Report*, 49th National Conference, 17-21 December 1994
African National Congress, *The Reconstruction and Development Programme* (RDP), March 1994
Commission Appointed to Enquire Into.Acts of Violence Committed by Natives at Krugersdorp, Newlands, Randfontein and Newclare, *Report*, UG47/1950
Commission for the Socio-Economic Development of the Bantu Areas Within the Union of South Africa, *Summary of the Report,* UG61/1955
Commission of Enquiry Into Events at Bofokeng North Mine During March 1992, (Pretorius Commission), *Report*, December 1992
Commission of Enquiry Into Riots in Durban, *Report*, UG36-49
Commission of Enquiry Into the Disturbances in the Witsieshoek Native Reserve, *Report*, UG26-51
Commission of Enquiry into the Recent Violence and Occurrences at the East Driefontein, Leeudoorn and Northam Mines, (Myburgh Commission), *Report*, October 1996
Commission of Enquiry Into Violence at Vaal Reefs Mine During 1986 (Bregman Commission), *Report*, Johannesburg, 12 May 1989
Commission of Enquiry on Occupational Health, *Report*, R/P55/1976
Commission of Enquiry Regarding Safety in Mines (Marais Commission), *First Interim Report, Supervision of Mines for Safety*, RP.21/1963
Commission of Inquiry Into Labour Legislation (Wiehahn Commission), *Report*, Part 1, 1979 and Part 6, 1981
Commission on Native Education, *Report*, 1949-51, UG53/51
Committee of Inquiry Into the Position of Marginal-Profit Mines

(Marais Committee), *Report*, 1990
Committee of Inquiry Into the Violence at the President Steyn Gold Mine in Welkom During November 1991 (Goldstone Commission), *Report,* April 1992
Cost of Living Commission, *Report on the Investigation of Accounts of Mine Storekeepers on the Witwatersrand and as to the Fixing of Prices of the Chief Articles of Native Trade in the Witwatersrand Area, 1918*, UG4/19, Cape Town, 1919
Department of Native Affairs, *Report*, 1948-49, UG51-1950
Department of Mineral and Energy Affairs, *Minerals-South Africa. Statistical Tables 1980-1999*, Bulletin B1/2000
Department of Mineral and Energy Affairs, *Annual Report*, 1992
Department of Mineral and Energy Affairs, *Report of the Commission of Inquiry Into Safety and Health in the Mining Industry* (Leon Commission), Volume 1, 1994
Department of Mineral and Energy Affairs, *South Africa's Mineral Industry*, 1988, 1994/5
Department of Mines, *Annual Reports*: 1947 to 1973
Government Mining Engineer, *Annual Reports*, 1948-1965
House of Commons Select Committee on Expenditure (Trade and Industry Sub-Committee) *Wages and Conditions of African Workers Employed by British Firms in South Africa*, Minutes of Evidence, HMSO, London, 1973-74
Inter-Departmental Committee of Enquiry, *Report Into Riots on Mines in the Republic of South Africa*, Pretoria, 1975
Joint Inquest and Inquiry Into the Hlobane Disaster, February 1984, *Report*, 1984
Joint Inquiry and Inquest Into the Coalbrook Disaster, *Transcript of the Evidence*, January 1960
Mines and Works Act, 1911, *Amendment Act*, 1926
Native Economic Commission, 1930-1932, *Report*, UG22/1932
Native Grievances Inquiry, 1913-14, *Report*, UG37/1914
Native Labour Regulation Bill, *Report of the Select Committee*, Cape Town, 1911
Native Labour Regulation Act, 1911
Native Laws Commission, *Report,* 1946-48, UG-1948
Penal and Prison Reform Commission, *Report* 1947, UG47-1947
Regional Court of the Northern Transvaal, *Official Transcript: The State* v *Kinross Mines and Others,* Witbank, 21 September 1987
Sedition Trial, *Rex* v *Andrews and Others: Transcript*, March/May 1948
Special Commissioner Appointed to Enquire Into the Boycotting of the

Rand Storekeepers by Natives in 1918, *Report,* UG4/1918, Cape Town, 1918

Suppression of Communism Amendment Bill, *Memorandum*, 1951

Social and Economic Planning Council, *The Native Reserves and Their Place in the Economy of the Union of South Africa*, Report, 9, UG32/1946

Social and Economic Planning Council, *Economic Aspects of the Gold Mining Industry*, Report, 11, 1947

Social and Economic Planning Council, *The Economic and Social Conditions of the Racial Groups in South Africa*, Report, 13, UG53/1948

Thorion Commission, *Eighth Interim Report of the Commission of Inquiry Into Alleged Irregularities and Mis-Application of Property in Representative Authorities and the Central Authority of South West Africa*, September 1985

Transvaal Labour Commission, *Minutes of Proceedings and Evidence,* Cd 1896, 1904

Witwatersrand Disturbance Commission, *Report*, Cd 7112, December 1913

Witwatersrand Mine Native Wages Commission, *Report*, UG21-1944

SUBJECT INDEX

A

B

C

D

E

F

G

H

I

J

K

L

M

N

O

P

R

S

T

U

V

W

X

Z

NAME INDEX

A

B

C

D

E

F

G

H

I

J

K

L

M

N

O

P

Q

R

S

T

U

V

W

X

Y

Z